Advanced Micro & Nanosystems
Volume 7
LIGA and its Applications

Further Reading

Other AMN Volumes

Brand, O., Fedder, G. K., Hierold, C., Korvink, J. G., Tabata, O., T. Tsuchiya (eds.)

Reliability of MEMS

2008
ISBN 978-3-527-31494-2

Kockmann, N., Brand, O., Fedder, G. K., Hierold, C., Korvink, J. G., Tabata, O. (eds.)

Micro Process Engineering

Fundamentals, Devices, Fabrication and Applications

2006
ISBN 978-3-527-31246-7

Advanced Micro & Nanosystems
Volume 7

LIGA and its Applications

*Edited by
Volker Saile, Ulrike Wallrabe, Osamu Tabata,
and Jan G. Korvink*

WILEY-VCH Verlag GmbH & Co. KGaA

Volume Editors

Prof. Dr. Volker Saile
Institut für Mikrostrukturtechnik
Forschungszentrum Karlsruhe
Hermann-von-Helmholtz-Platz 1
76344 Eggenstein-Leopoldshafen
Germany

Prof. Dr. Ulrike Wallrabe
University of Freiburg–IMTEK
Georges-Köhler-Allee 102
79110 Freiburg
Germany

Prof. Dr. Osamu Tabata
Department of Mechanical Engineering
Kyoto University
Yoshida Honmachi Sakyo-ku
606-8501 Kyoto
Japan

Prof. Dr. Jan G. Korvink
Laboratory for Design of Microsystems
University of Freiburg–IMTEK
Georges-Köhler-Allee 102
79110 Freiburg
Germany

Series Editors

Oliver Brand
School of Electrical and Computer Engineering
Georgia Institute of Technology
777 Atlantic Drive
Atlanta, GA 30332-0250
USA

Prof. Dr. Gary K. Fedder
ECE Department & Robotics Institute
Carnegie Mellon University
Pittsburgh, PA 15213-3890
USA

Prof. Dr. Christopher Hierold
Chair of Micro- and Nanosystems
ETH Zürich
ETH Zentrum, CLA H9
Tannenstrasse 3
8092 Zürich
Switzerland

Prof. Dr. Jan G. Korvink
Laboratory for Design of Microsystems
University of Freiburg–IMTEK
Georges-Kühler-Allee 102
79110 Freiburg
Germany

Prof. Dr. Osamu Tabata
Department of Mechanical Engineering
Kyoto University
Yoshida Honmachi Sakyo-ku
606-8501 Kyoto
Japan

■ All books published by Wiley-VCH are carefully produced. Nevertheless, authors, editors, and publisher do not warrant the information contained in these books, including this book, to be free of errors. Readers are advised to keep in mind that statements, data, illustrations, procedural details or other items may inadvertently be inaccurate.

Library of Congress Card No.: applied for

British Library Cataloguing-in-Publication Data
A catalogue record for this book is available from the British Library.

Bibliographic information published by Die Deutsche Bibliothek
Die Deutsche Bibliothek lists this publication in the Deutsche Nationalbibliografie; detailed bibliographic data are available on the Internet at http://dnb.ddb.de

© 2009 WILEY-VCH Verlag GmbH & Co. KGaA, Weinheim, Germany

All rights reserved (including those of translation into other languages). No part of this book may be reproduced in any form – by photoprinting, microfilm, or any other means – nor transmitted or translated into machine language without written permission from the publishers. Registered names, trademarks, etc. used in this book, even when not specifically marked as such, are not to be considered unprotected by law.

Typesetting SNP Best-set Typesetter Ltd., Hong Kong
Printing betz-druck GmbH, Darmstadt
Binding Litges & Dopf Buchbinderei GmbH, Heppenheim
Cover Grafik-Design Schulz, Fußgönheim

Printed on acid-free paper
Printed in the Federal Republic of Germany

ISBN: 978-3-527-31698-4

Contents

List of Contributors   *VII*

1   **Introduction: LIGA and Its Applications**   *1*
    Volker Saile

2   **X-ray Masks for LIGA Microfabrication**   *11*
    Yohannes Desta and Jost Goettert

3   **Innovative Exposure Techniques for 3D Microfabrication**   *51*
    Naoki Matsuzuka and Osamu Tabata

4   **Hot Embossing of LIGA Microstructures**   *69*
    Mathias Heckele and Matthias Worgull

5   **Exposure and Development Simulation for Deep X-ray LIGA**   *103*
    Jan G. Korvink, Sadik Hafizovic, Yoshikazu Hirai, and Pascal Meyer

6   **Design for LIGA and Safe Manufacturing**   *143*
    Ulrich Gengenbach, Ingo Sieber, and Ulrike Wallrabe

7   **Commercialization of LIGA**   *189*
    Ron A. Lawes

8   **Polymer Optics and Optical MEMS**   *205*
    Jürgen Mohr

9   **Refractive X-ray Lenses Produced by X-ray Lithography**   *233*
    Arndt Last

10 **RF Applications**   *243*
    Sven Achenbach and David M. Klymyshyn

11 **Evolution of the Microspectrometer**   *281*
    Reiner Wechsung, Sven Schönfelder, and Andreas Decker

Advanced Micro & Nanosystems Vol. 7. LIGA and Its Applications.
Edited by Volker Saile, Ulrike Wallrabe, Osamu Tabata and Jan G. Korvink
Copyright © 2009 WILEY-VCH Verlag GmbH & Co. KGaA, Weinheim
ISBN: 978-3-527-31698-4

12 **Actuator Manufacture with LIGA Processes** *297*
Todd Christenson

13 **Development of Microfluidic Devices Created via the LIGA Process** *323*
Masaya Kurokawa

14 **Application of Inspection Devices** *337*
Yoshihiro Hirata

15 **The Micro Harmonic Drive Gear** *351*
Reinhard Degen and Rolf Slatter

16 **Microinjection Molding Machines** *395*
Christian Gornik

17 **Filled Resist Systems** *415*
Thomas Hanemann, Claas Müller, and Michael Schulz

18 **Dramatic Downsizing of Soft X-ray Synchrotron Light Source from Compact to Tabletop** *443*
Hironari Yamada, Norio Toyosugi, Dorian Minkov, and Yoshiko Okazaki

19 **PTFE Photo-fabrication by Synchrotron Radiation** *453*
Takanori Katoh and Yanping Zhang

Index *469*

List of Contributors

Sven Achenbach
University of Saskatchewan
Department of Electrical and
  Computer Engineering
57 Campus Drive
Saskatoon
Saskatchewan, S7N 5A9
Canada

Todd Christenson
HT Microanalytical, Inc.
3817 Academy Parkway S, NE
Albuquerque
NM 87109
USA

Andreas Decker
Boehringer Ingelheim
  microParts GmbH
Hauert 7
44227 Dortmund
Germany

Reinhard Degen
Micromotion GmbH
An der Fahrt 13
55124 Mainz
Germany

Yohannes Desta

Ulrich Gengenbach
Forschungszentrum Karlsruhe
Institut für Angewandte Informatik
76021 Karlsruhe
Germany

Jost Goettert
Louisiana State University
Baton Rouge, LA 70803
USA

Christian Gornik
Battenfeld Kunststoffmaschinen
  GmbH
Wiener Neustädter Str. 81
2542 Kottingbrunn
Austria

Sadik Hafizovic
ETH Zürich
Zürich
Switzerland

Thomas Hanemann
University of Freiburg
Department of Microsystems
  Engineering (IMTEK)
Georges-Koehler-Allee 102
79110 Freiburg
Germany
and
Forschungszentum Karlsruhe
Institut für Materialforschung III
76021 Karlsruhe
Germany

Mathias Heckele
Forschungszentrum Karlsruhe
Institut für Mikrostrukturtechnik
76021 Karlsruhe
Germany

Yoshikazu Hirai
Kyoto University
Department of Micro
  Engineering
Japan

Yoshihiro Hirata
Sumitomo Electric Industries, Ltd.
Electronics Business Unit
Japan

Takanori Katoh
Sumitomo Heavy Industries, Ltd
2-1-1 Yatocho
Nishitokyo
Tokyo188-8585
Japan

David M. Klymyshyn
University of Saskatchewan
Department of Electrical and
  Computer Engineering
57 Campus Drive
Saskatoon
Saskatchewan, S7N 5A9
Canada

Jan G. Korvink
Institute for Microsystem Technology
  (IMTEK)
Albert-Ludwigs-Universität Freiburg
Georges-Köhler-Allee 103
79110 Freiburg
Germany

Masaya Kurokawa
Starlite Co., Ltd.
Micro Device Development
  Department
Miniature Precision Products Devision
2222, Kamitoyama, Ritto City
Shiga 520-3004
Japan

Masaya Kurokawa
Starlite Co. Ltd.
Kusatsu 520-3004
Japan

Arndt Last
Forschungszentrum Karlsruhe
Institut für Mikrostrukturtechnik
76021 Karlsruhe
Germany

Ron A. Lawes
Micro-Nanotechnology Engineer
12 Leamington Drive
Faringdon
Oxfordshire SN7 7JZ
UK

Naoki Matsuzuka
Kyoto University
Graduate School of Engineering
Department of Micro Engineering
Yoshida-Honmachi
Sakyo-ku
Kyoto 606-8501
Japan

Pascal Meyer
Universität Karlsruhe
Institut für
 Mikrostrukturtechnik
Postfach 3640
76021 Karlsruhe
Germany

Dorian Minkov
Ritsumeikan University
Synchrotron Light Life Science
 Center
1-1-1 Nojihigashi
Kusatsu City
Shiga 525-8577
Japan

Jürgen Mohr
Forschungszentrum Karlsruhe
Hermann-von-Helmholtz
Platz 1
76344 Eggenstein-Leopoldshafen
Germany

Claas Müller
University of Freiburg
Department of Microsystems
 Engineering (IMTEK)
Laboratory for Process
 Technology
Georges-Koehler-Allee 103
79110 Freiburg
Germany

Yoshiko Okazaki
Ritsumeikan University
Synchrotron Light Life Science
 Center
1-1-1 Nojihigashi
Kusatsu City
Shiga 525-8577
Japan

Volker Saile
Universität Karlsruhe and
 Forschungszentrum Karlsruhe
Institut für Mikrostrukturtechnik
76021 Karlsruhe
Germany

Michael Schulz
University of Freiburg
Department of Microsystems
 Engineering (IMTEK)
Georges-Koehler-Allee 102
79110 Freiburg
Germany
and
Forschungszentrum Karlsruhe
Institut für Materialforschung III
76021 Karlsruhe
Germany

Sven Schönfelder
Boehringer Ingelheim
 microParts GmbH
Hauert 7
44227 Dortmund
Germany

Ingo Sieber
Forschungszentrum Karlsruhe
Institut für Augewandte Informatik
76021 Karlsruhe
Germany

Rolf Slatter
Micromotion GmbH
An der Fahrt 13
55124 Mainz
Germany

List of Contributors

Osamu Tabata
Kyoto University
Graduate School of Engineering
Department of Micro
 Engineering
Yoshida-Honmachi
Sakyo-ku
Kyoto 606-8501
Japan

Norio Toyosugi
Photon Production Laboratory Ltd
4-2-1 (808) Takagai Cho Minami
Omihachiman City
Shiga 523-0898
Japan

Ulrike Wallrabe
Universität Freiburg
Iustitut für Mikrosystemtechnik
Arbeitsgruppe Mikroaktorik
Georges-Köhler-Allee 102
79110 Freiburg
Germany

Reiner Wechsung
Boehringer Ingelheim
 microParts GmbH
Hauert 7
44227 Dortmund
Germany

Matthias Worgull
Forschungszentum Karlsruhe
Institut für Mikrostrukturtechnik
76021 Karlsruhe
Germany

Hironari Yamada
Ritsumeikan University
Synchrotron Light Life Science Center
1-1-1 Nojihigashi
Kusatsu City
Shiga 525-8577
and
Photon Production Laboratory Ltd
4-2-1 (808) Takagai Cho Minami
Omihachiman City
Shiga 523-0898
Japan

Yanping Zhang
Sumitomo Heavy Industries, Ltd
2-1-1 Yatocho
Nishitokyo
Tokyo188-8585
Japan

1
Introduction: LIGA and Its Applications

Volker Saile

1.1 LIGA Background *1*
1.2 The Current Status of LIGA *4*
1.3 Challenges for the Future *8*
   References *9*

1.1
LIGA Background

The first publication on LIGA technology – LIGA is a German acronym for lithography, electroplating and polymer replication – appeared in 1982, more than 25 years ago [1]. Since then, the LIGA technique has been successfully used in many research projects. Also components for industrial customers have been produced, but still on only a rather limited scale. A compilation of many relevant references for LIGA may be found in review papers, for example, Refs [2, 3]; the LIGA basics and technical limits are well documented and explained in textbooks, such as Ref. [4].

The core process in LIGA is deep X-ray lithography at a synchrotron radiation source. X-ray lithography (XRL) as such was proposed in the 1970s for semiconductor patterning and first demonstrated by IBM using synchrotron radiation at the German National Laboratory DESY in 1975 [5]. The results generated enormous interest among semiconductor manufacturers and also funding agencies. This enthusiasm was fueled by the hope of having found a technology for replacing optical lithography that was assumed to be at the end of its resolution capabilities in the 1970s or 1980s. Complete technology platforms for making chips with X-rays were developed in the United States, Japan and Europe [6] in the 1980s and large companies prepared for production in the early 1990s [7]. However, after spending a huge amount of money, far in excess of $1 billion, XRL was abandoned by the semiconductor industrial community in the mid-1990s. 'X-ray' became a synonym for wasting gigantic resources and extreme ultraviolet (EUV) lithography emerged as the new candidate for next generation lithography (NGL). Today, the

Advanced Micro & Nanosystems Vol. 7. LIGA and Its Applications.
Edited by Volker Saile, Ulrike Wallrabe, Osamu Tabata and Jan G. Korvink
Copyright © 2009 WILEY-VCH Verlag GmbH & Co. KGaA, Weinheim
ISBN: 978-3-527-31698-4

future of EUV lithography is also open. The bottom line of the history of XRL is that an established technology, namely optical lithography, stayed ahead for over 20 years by continuous improvements to performance levels that nobody could imagine in the 1970s. However, it was not so much the performance that kept optical lithography ahead of XRL but more the confidence and trust of the production people in well-established technology rather than using disruptive processes. The rapid rise of XRL in the 1980s and 1990s originally also boosted LIGA, but later with the fall of XRL the viability of LIGA was also questioned.

The origins of LIGA are in nuclear technologies: the Nuclear Research Center KfK in Karlsruhe, Germany, had been developing new methods for separating uranium isotopes since the 1970s. Their specific approach was based on nozzles that exploit the mass-dependent centrifugal forces for a spatial separation of the two relevant isotopes. For an efficient separation process, nozzles with critical dimensions of the order of a few micrometers and a very high aspect ratio were required. After evaluating various manufacturing options for mass producing such devices, but, with limited success, the Karlsruhe team, headed by Becker and Ehrfeld, contacted the promoters of XRL in Germany, headed by Heuberger and Betz. One of the key capabilities of XRL that was actually not used in semiconductor applications is the large depth of penetration of such radiation. This property can be exploited for exposing very thick resist up to several millimeters and the actual penetration depth can be tailored to specific resist heights by varying the photon energy of the X-rays. The joint team, with a background in electrical engineering and lithography on one side and nuclear and mechanical engineering on the other, understood the enormous potential of using XRL at short wavelengths for fabricating devices with extreme resolution and very high aspect ratio. In the first publication in 1982 [1], all relevant features and properties of LIGA were described already (Figures 1.1 and 1.2).

LIGA was originally a technology for a single product only, namely the uranium separation nozzles. After a dedicated synchrotron radiation source for mass production of such nozzles was proposed in 1985 by the Karlsruhe laboratory, the customer for the nozzles, a consortium for constructing a uranium separation facility in Brazil, terminated the project. In addition, nuclear R&D in Germany was dramatically downsized. Among the survivors of the golden days of nuclear engineering was the LIGA technology, now often being oversold as the greatest technology for fabricating high-resolution, high aspect ratio devices. Research in developing and in using novel structures, devices and entire systems made with LIGA flourished in the 1990s, several synchrotron radiation facilities all over the world added LIGA beamlines and laboratories and the Forschungszentrum Karlsruhe (FZK) finally received permission and funding in 1995 to build their own synchrotron ring for supporting the LIGA activities. In the United States, Henry Guckel of the University of Wisconsin, Madison, brought LIGA to the new world. His laboratory contributed many new ideas and novel concepts in the 1990s. The DARPA-funded, so-called 'High-MEMS Alliance' involving the Center for Advanced Microstructures, CAMD, in Baton Rouge, LA, the University of

"X-ray lithography using synchrotron radiation has been applied in a multi-step process for the production of plastic molds to be used in the fabrication of separation nozzles by electrodeposition. For characteristic dimensions of a few microns a total height of the nozzle structure of about 400 µm has been achieved. Structural details of about 0.1 µm are being reproduced across the total thickness of the polymer layer. The surface finish of the metallic separation nozzles produced by electrodeposition was equivalent to the high quality of the polymer surface. The separation-nozzle systems fabricated by the described method allow an increase by a factor three of the gas pressure in separation-nozzle plants as compared to the present standard. This results in considerably savings in the enrichment of ^{235}U for nuclear power production."

Figure 1.1 Abstract of the 1982 LIGA publication, Ref. [1].

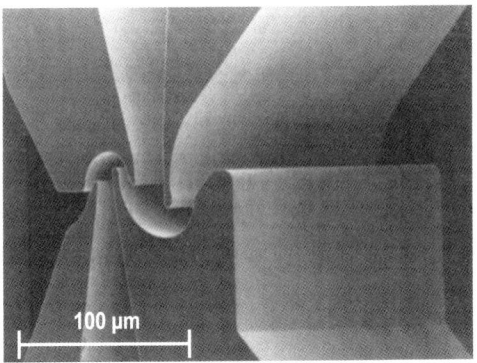

Figure 1.2 Uranium separation nozzle fabricated with LIGA; see Ref. [1].

Wisconsin, MCNC of North Carolina, IBM and others established a sound basis and a network for LIGA in the United States. Nevertheless, the expectations raised by the early promoters of the commercialization potential of LIGA were far too optimistic and they have not yet been realized. Therefore, after more than 25 years of LIGA, we should find answers to critical questions such as: will LIGA follow the path of XRL, that is, become obsolete for commercial implementation after more than 20 years of R&D? And if not, why not?

This long history of LIGA and the valid questions on the future of the technology led the Editors of this book to invite distinguished experts in LIGA to present reviews on the current status of the technology, on applications, on equipment and also on new ideas.

1.2
The Current Status of LIGA

Over the years, the science and technology community followed numerous attempts to establish LIGA activities all over the world. Fascinating LIGA pictures and results were frequently used in glossy brochures to convince agencies, politicians and the public to fund proposals on new synchrotron facilities. The postulate that high-tech research facilities lead directly to economic development is in fact intriguing but often too simple minded in reality. Professional LIGA work requires a significant infrastructure with cleanrooms and equipment, supporting laboratories, mask-making capabilities and, probably most important, highly qualified staff for operation of these facilities. Furthermore, reproducibility, manufacturability and acceptable cost turned out to be major challenges – issues that fall more in the area of competence of production engineers rather than scientists. As a consequence of insufficient financial or human resources or of a lack of patience of industry or government customers, we could follow rise and fall of several LIGA activities, for example, at LURE in France, at LNLS in Brazil, at SRS in the United Kingdom and, most recently, at Sandia National Laboratories, Livermore, CA in the United States. Synchrotron facilities with currently active LIGA programs are listed in Figure 1.3, where some activities are rather small whereas others are significant. Also included in Figure 1.3 are companies and organizations that have direct access to synchrotron facilities for manufacturing commercial LIGA devices for their customers.

When we discuss LIGA, we must distinguish between three rather different approaches that are usually all summarized under the name LIGA (see Figure 1.4):

Canada:	CLS – SyLMAND
China:	NSRL
Germany:	ANKA, BESSY-II
Italy:	Elettra
Japan:	NewSUBARU, Rits SR
Korea:	PLS
Russia:	VEPP-III
Singapore:	SSLS
Taiwan:	SRRC
USA:	ALS, CAMD

Facilities with Activities in Deep X-ray Lithography

(a)

ANKA COS, Karlsruhe, Germany
Axsun Technologies Inc., Livermore, CA, USA
BESSY Anwendezentrum, Berlin, Germany
MEZZO Corp., Baton Rouge, LA, USA
ht micro, Albuquerque, NM, USA

LIGA Manufacturers and Service Providers

(b)

Figure 1.3 (a) Synchrotron radiation facilities with active deep X-ray lithography activities. (b) Companies and organizations offering LIGA manufacturing capabilities.

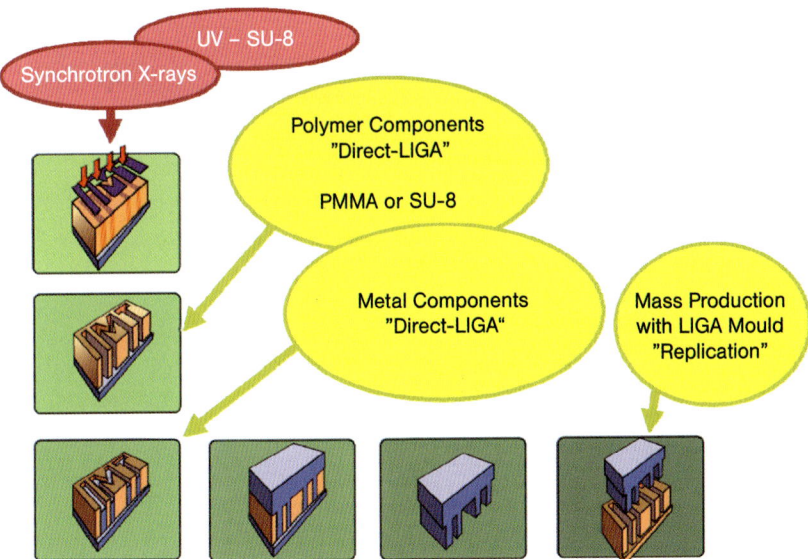

Figure 1.4 LIGA process sequence and variations including UV-LIGA and Direct-LIGA concepts. A thick resist layer (brown) is exposed with X-rays or UV radiation through a mask. After development polymer parts are available. Electroplating (blue) yields metal components. After over-plating, a mold can be separated from the substrate and employed in mass production schemes such as injection molding or hot embossing.

- The classical LIGA process sequence where deep X-ray lithography (DXRL) is used for fabricating a mold to be used in injection molding or hot embossing mass production schemes (see Figure 1.5a).
- Deep X-ray lithography where each individual component is produced lithographically just as devices in semiconductor manufacturing ('Direct-LIGA' [8]); typical resists are poly(methyl methacrylate) (PMMA) and also EPON SU-8. For an example, see Figure 1.5b and c.
- Replacing the X-ray source, a synchrotron, by UV radiation and exposing a specific resist system, EPON SU-8. The applicability of standard quartz masks is the major advantage of UV-LIGA over X-ray LIGA, but at the expense of a decrease in quality and the restriction to SU-8 resist.

Examples of the capabilities of DXRL for research applications, for the classical LIGA process sequence and for components fabricated by Direct-LIGA are displayed in Figure 1.5b, a and c, respectively.

Without any question, LIGA offers technical features – *precision, spatial resolution, aspect ratio and sidewall roughness* – *superior to any other microfabrication technology*. The results presented in the chapters of this book clearly demonstrate the outstanding capabilities and potential of LIGA. Nevertheless, it is still a niche technology with a rather limited spectrum of commercial applications.

1 Introduction: LIGA and Its Applications

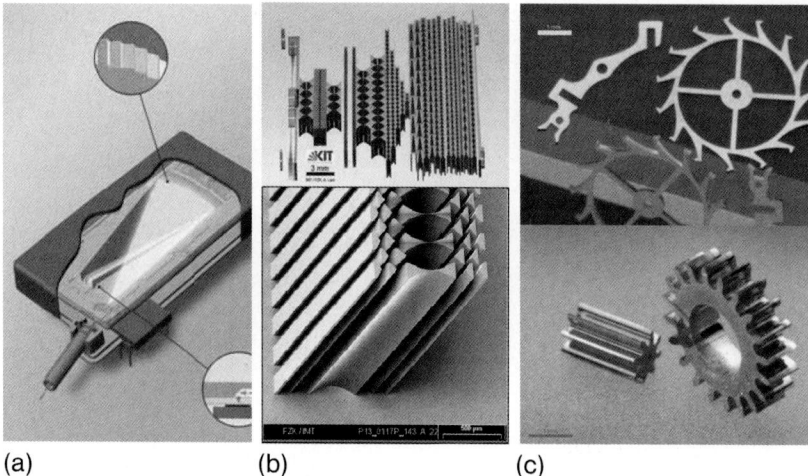

Figure 1.5 (a) LIGA: commercially available microspectrometer manufactured by injection molding with a LIGA mold. For details, see Chapter 11. (b) Research with DXRL: X-ray lenses fabricated with LIGA. Such lenses are employed in novel X-ray microscopes. For details, see Chapter 9. (c) Direct-LIGA: gold gears for wrist watches and nickel gears for gear trains for micro- and mini-motors. For a discussion on the commercialization aspects for such devices, see Chapter 7.

Why is this so? There is a variety of reasons – some have a technical background, some are cost related and some are strategic or even psychological. Among the most prominent objections to LIGA are the following:

- *LIGA is too expensive:* The cost for LIGA is often overestimated; see also Chapter 7. X-rays at a synchrotron source are available at prices of the order of €200 per hour; such prices are based on the operational cost of a synchrotron facility without including depreciation as is usual for government investments. When compared with UV lithography, the main difference in cost is in the mask, which for a high-quality X-ray mask is in the range €10 000–20 000.

- *LIGA is too slow:* This statement is probably confirmed by the experience of many LIGA customers, but it is not so much technology related, but rather related the research-dominated environment of many LIGA laboratories.

- *LIGA is unreliable:* As with all high technologies, the transition from research to production is painful, requires time and demands significant resources for establishing high quality and high yield in production.

- *LIGA did not deliver; its promoters oversold its capabilities:* This is true and was due to the understandable enthusiasm of the early LIGA researchers.

- *LIGA lacks granularity – it is linked to a very few synchrotron sources:* LIGA shares this deficiency with XRL, where granularity had been declared as a highly important issue in the 1990s. Companies prefer wafer-steppers at their own facility rather than to expose wafers elsewhere. The only other option is to purchase an entire high-price, high-cost storage ring with many slots for steppers. The granularity issue is a convincing reason for a production scheme with UV-LIGA.

- *LIGA offers a limited range of materials:* Compared with other technologies, LIGA offers, in fact, a broad range of materials from polymers to metals, alloys and ceramics. The stability of the associated processes, however, often deserves additional R&D efforts.
- *Other technologies such as DRIE and other advanced silicon processes became competitive:* This is true, but as far as specific technical features are concerned, such as aspect ratio and sidewall roughness, LIGA is technically superior.
- *UV-based LIGA is less expensive:* For low-volume production this is true, but it is at the expense of quality.
- *Integration of LIGA components with CMOS is not straightforward:* This is a disadvantage of LIGA. Silicon microcomponents can be monolithically integrated into CMOS circuits for classical MEMS devices.
- *LIGA lacks prototyping capabilities:* This is in fact a LIGA deficiency. For fabricating any serious LIGA prototype, at least an X-ray mask must be provided; such masks are expensive and their production can be time consuming, especially when high-resolution e-beam writing is required and a low absorber height mask must be copied with X-rays to a production mask with sufficient absorber thickness.

The criticism of LIGA, as summarized above, is partly justified, but some key issues deserve a more detailed discussion. The presumably high costs have been emphasized again and again, but the exposure costs per hour at more or less all synchrotrons worldwide are actually low. Furthermore, LIGA is still mostly conducted in research laboratories, where cost reduction is not the primary focus. Significant cost reduction potential can be identified for several processes of the LIGA sequence, such as

- Automated inspection and quality control; this is a straightforward cost-saving measure.
- Cost reduction in mask fabrication. This objective is less important for mold fabrication but essential for the production of a small or medium number of substrates with components by Direct-LIGA.
- Higher throughput requires either shorter exposure times and/or larger areas to be exposed simultaneously. The development and qualification of resists with higher sensitivity than PMMA are under way worldwide, and SU-8 has demonstrated its capabilities. However, serious problems with the quality and, in particular, the reproducibility of its chemical composition have limited the application of this resist system in production so far.

Major efforts are under way at FZK and other LIGA facilities to improve mask making, to increase the usable area and to reduce costs. Exposure of large areas is also being pursued by developing new scanners and beamlines for synchrotrons. A first such exposure facility with the project name FELIG [9], a partly automated fabrication line for Direct-LIGA parts, will be available for production at the FZK synchrotron ANKA in 2008. Furthermore, a major R&D project has been launched in Germany under the name INNOLIGA, involving the LIGA laboratories FZK and BESSY, and also a resist manufacturer and a polymer research institute. This joint effort aims at major improvements of SU-8, in particular, to make this resist compatible with the requirements for industrial production.

From a technology point of view, LIGA is today competitive with, if not superior to, most other high aspect ratio manufacturing technologies. Also, cost is not necessarily the limiting factor for many applications requiring LIGA-type features and quality; such cost can be reduced significantly as indicated above. From a principle point of view, cost is not necessarily the bottleneck, provided that R&D to this end is properly supported. More important are process stability and control with associated turn-around times, throughput and yield. For better control of the individual processes and the whole process sequence, IMT at FZK has implemented a rigorous quality management system in compliance with ISO 9001 and has been certified since 1999. Nevertheless, the LIGA process sequence is long, maybe too long, and the financial resources spent over the past 25 years on optimizing each process step for improving the overall yield has been very insufficient. Further, process development as needed here is mostly incompatible with the mission of universities and public research institutes that are evaluated on the basis of their accomplishments in science. How much it really costs to establish a process sequence with associated processes and equipment has been best demonstrated by the semiconductor industry.

1.3
Challenges for the Future

LIGA has delivered excellent research results not obtainable with other techniques and, as such, LIGA is valuable and important. As far as commercial applications and manufacturing are concerned, LIGA has not yet fully delivered what its early promoters promised and it was oversold in the early days. As a consequence, LIGA has an image problem: the perception in industry is that LIGA is expensive, slow and unreliable. Even if this is mostly not true, we have to accept that 'perception is reality'. The major difference from XRL for chip making is that LIGA does not serve a single product market and also there is no 'killer' application (yet). However, LIGA has the potential to penetrate at least many niche markets and has done so in several cases, as demonstrated in this book. The basic problem of LIGA is the lack of resources for establishing professional manufacturing. There are still significant efforts required for cutting costs and time and improving reliability. This will require additional time and money. In any case, LIGA will remain a valuable technology for science and for niche applications, always with the potential for large-volume mass manufacturing in future.

This book on 'LIGA and its Applications' should serve as a comprehensive status report on LIGA today, more than 25 years after the milestone publication of Ref. [1]. The Editors of this book were able to obtain contributions from authors representing the worldwide LIGA community. These experts convincingly demonstrate significant progress in technology development and in applications. Especially impressive and encouraging is the large number of authors from industry; this would not have been possible only a few years ago. The message is that LIGA products are on the market, not yet in large volume but with fairly optimistic perspectives. That LIGA is very much alive and productive can also be followed in the two key conferences in this field: the bi-annual HARMST (High Aspect Ratio Micro-Structure Technology) workshop, that was organized in Besancon, France, in 2007 and will be held next in Saskatoon, Canada, in 2009 and COMS (Commercialization of Micro and Nano Systems Conference), in Puerto Vallarta, Mexico, in 2008 and Copenhagen in 2009), an annual conference with a special session on high aspect ratio technology and commercialization.

The Editors are aware that there are unavoidable gaps in the coverage of LIGA, for example, electroplating technologies for microstructures is such a missing topic [10]. One could also include some new developments that are not directly LIGA but are interesting in the context of micro- and nano-patterning. An example is interference lithography with coherent soft X-rays, as pursued at the Swiss Light Source in Switzerland [11]. However, we are also convinced that the book presented here will not be the last book on LIGA and that an updated version will be published after the next couple of years of flourishing LIGA research and applications.

Finally, the Editors are deeply grateful to authors of the individual chapters for their articles and patience. With their excellent contributions they have provided a great service to the micro- and nanotechnology community.

References

1. Becker, E.W., Ehrfeld, W., Münchmeyer, D., Betz, H., Heuberger, A., Pongratz, S., Glashauser, W., Michel, H.J. and von Siemens, R. (1982) Production of separation-nozzle systems for uranium enrichment by a combination of X-ray lithography and galvanoplastics. *Naturwissenschaften*, **69**, 520–3.
2. Khan Malek, C. and Saile, V. (2004) Applications of LIGA technology to precision manufacturing of high-aspect-ratio micro-components and -systems: a review. *Microelectronics Journal*, **35**, 131–43.
3. Wallrabe, U. and Saile, V. (2006) LIGA technology for R&D and industrial applications, in *MEMS: a Practical Guide to Design, Analysis and Applications* (ed. J.G. Korvink), William Andrew Publications, Norwich, NY, pp. 853–99.
4. (a) Menz, W., Mohr, J. and Paul, O. (2000) *Microsystem Technology*, 1st edn, Wiley-VCH Verlag GmbH, Weinheim.
   (b) Menz, W., Mohr, J. and Paul, O. (2005) *Mikrosystemtechnik für Ingenieure*, 3rd edn, Wiley-VCH Verlag GmbH, Weinheim.
5. Spiller, E., Eastman, D., Feder, R., Grobman, W.D., Gudat, W. and Topalian, J.

(1976) The application of synchrotron radiation to X-ray lithography. *Journal of Applied Physics*, **47**, 5450–49.

6 Heuberger, A. (1985) X-ray lithography. *Microelectronic Engineering*, **3**, 535–56.

7 Hauge, P.S. (ed.) (1993) *IBM Journal of Research and Development*, **37**, 287–474.

8 Saile, V. (1998) Srategies for LIGA Implementation, 6th International Conference on Micro Electro, Opto, Mechanical Systems and Components, Potsdam, December 1–3, 1998, in *Micro System Technologies 98*, Reichl, H., Obermeier, E. (eds.), VDE-Verlag GmbH, Berlin, pp. 25–30.

9 Meyer, P., Klein, O., Arendt, M., Saile, V. and Schulz, J. (2006) Launching into a golden age (3)–gears for micromotors made by LIG(A) process. 11th Annual International Conference of Micro and Nano Systems (COMS 2006), St Petersburg, FL, 27–31 August 2006, Tuesday Session, pp. 42–3. See also CD 'COMS2006 Proceedings' provided by MANCEF Foundation.

10 Some aspects on micro-electroplating can be found in Guttmann, M., Schulz, J. and Saile, V. (2005) Lithographic fabrication of mold inserts, in *Microengineering of Metals and Ceramics, Part I: Design, Tooling and Injection Molding (Advanced Micro and Nanosystems)*, Vol. **3**, Wiley-VCH Verlag GmbH, Weinheim, pp. 187–219.

11 Solak, H.H., David, C., Gobrecht, J., Wang, L. and Cerrina, F. (2002) Multiple-beam interference lithography with electron beam written gratings. *Journal of Vacuum Science and Technology B*, **20**, 2844–8.

Solak, H.H. (2006) Nanolithography with coherent extreme ultraviolet light, *Journal of Physics D: Applied Physics*, **39**, R171–R188.

2
X-ray Masks for LIGA Microfabrication

Yohannes Desta and Jost Goettert

2.1 Introduction *11*
2.2 **Mask Substrate and Absorber Material Selection** *14*
2.2.1 X-ray Transmission Characteristics of Substrate Materials *14*
2.2.2 X-ray Attenuation Characteristics of Absorber Materials *16*
2.2.3 Thermo-elastic Properties of Mask Substrates *20*
2.2.4 Surface Quality of Mask Substrate Materials *22*
2.2.5 Optical Transmission *23*
2.3 **Mask Architecture** *24*
2.3.1 Mask Substrate Thickness *24*
2.3.2 Absorber Thickness *26*
2.4 **X-ray Mask Fabrication Methods** *28*
2.4.1 Titanium Membrane Mask Fabrication *29*
2.4.2 Silicon Nitride Mask Fabrication *31*
2.4.3 Graphite Mask Fabrication *33*
2.4.4 Beryllium Mask Fabrication *35*
2.4.5 Glass Mask Fabrication *37*
2.4.6 Alternative Approaches for X-ray Masks *38*
2.4.7 Gold Electroplating for X-ray Masks *39*
2.5 **Summary and Conclusion** *44*
References *45*

2.1
Introduction

In the first step of the LIGA process [1], the pattern on an X-ray mask is transferred to a resist material by use of X-ray lithography. The X-ray mask is essentially composed of microstructures made of an X-ray absorbing material on a highly X-ray transparent substrate. In a typical X-ray mask, a variety of high-resolution patterns with critical dimensions ranging from sub-micrometer [2–4] to several millimeters [5–7] can be present.

As the starting point for any LIGA process, an X-ray mask is perhaps the most critical piece of the LIGA process as it allows users to convert their design into a real structure. It is, however, also the 'process bottleneck' as there is not a standard solution for the mask architecture as is the case in optical lithography where a transparent substrate (typically quartz glass) carries the design information in the form of a patterned film of chromium [8].

Employing X-ray photons for patterning microstructures is in principle similar to the use of UV light. However, a suitable light source such as an electron storage ring is a much more complicated instrument than a UV discharge lamp, significantly larger and more expensive to operate [9]. Inside a vacuum chamber, charged particles are forced on to a circular orbit traveling nearly at the speed of light, thereby generating a continuous radiation spectrum, so-called synchrotron radiation [10]. Synchrotron radiation forms a homogeneous narrow fan in the direction of propagation with a Gaussian-shaped, energy-dependent profile perpendicular to the orbit plane. The mask–substrate assembly is periodically scanned across this radiation fan, shadow-casting the mask pattern into the radiation sensitive X-ray resist as illustrated in Figure 2.1.

In addition to X-ray lithography, synchrotron radiation is also a powerful tool for many fields in basic and applied science, including physics, chemistry and biology, and a number of dedicated synchrotron light sources have been built and are now operated, serving the growing demands of a diverse scientific community [9].

Already in the 1970s, the use of synchrotron radiation as 'short-wavelength' light source was intensively discussed for lithographic patterning of small microelectronic circuits [11, 12]. X-ray lithography was a serious contender for the next generation lithography (NGL) and dedicated compact synchrotron light sources such as HELIOS 1 [13] (designed and built by Oxford Instruments, Oxford, UK) were operated successfully by researchers at IBM [11]. In the early 1980s, a second property of X-rays, their large penetration depth into matter, was then considered

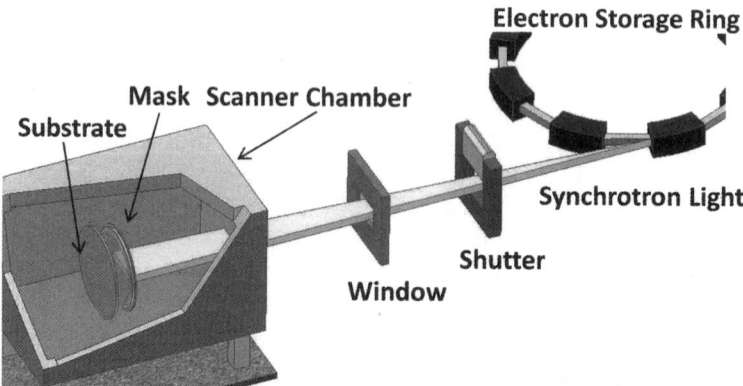

Figure 2.1 Schematic of an X-ray beamline attached to an electron storage ring. At the end of the beamline, an exposure station or X-ray scanner houses a linear motion stage carrying the mask and substrate assembly perpendicular or at a desired angle across the synchrotron beam.

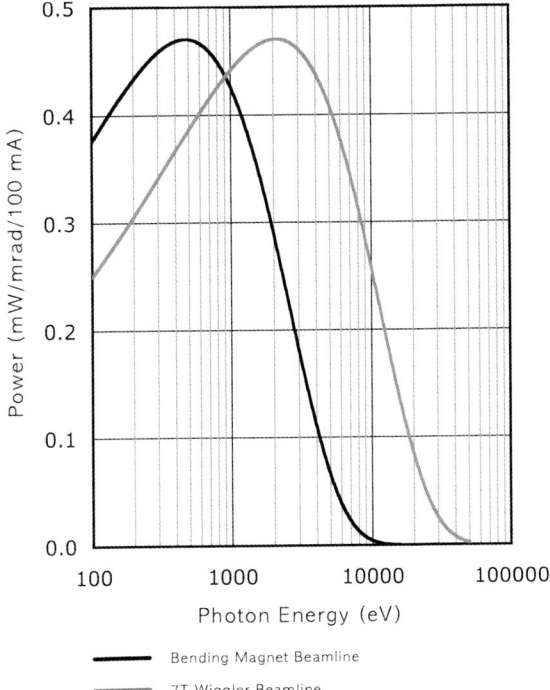

Figure 2.2 Power output for an average current of 100 mA as a function of photon energy for different CAMD source points. The low-energy part of the 'white spectra' is typically absorbed by a vacuum window made from beryllium resulting in photon energies of the exposure spectrum ranging from 2 to 10 keV for bending magnet sources and up to 40 keV for the wiggler source [20].

for patterning of tall microstructures for optical, mechanical and fluidic applications using deep X-ray lithography (DXRL) shadow printing [14] – a strong effort today practiced at different synchrotron light sources worldwide by users in academia and industry alike [15–19].

The Center for Advanced Microstructures and Devices (CAMD) at Louisiana State University (LSU) in Baton Rouge, LA, USA, is one of these dedicated synchrotron radiation facilities and CAMD beamlines will be used as example to discuss the mask material requirements in this chapter [20]. CAMD provides two different sources used for X-ray lithography – a bending magnet source with relative low photon energies up to 10 keV and an insertion device (wiggler) source providing X-ray photons of up to 40 keV energy. The power spectra for CAMD's bending magnet and wiggler sources are shown in Figure 2.2 and will be used in this chapter as a reference for X-ray mask properties. The bending magnet spectrum can also be adjusted for soft X-ray lithography by inserting two grazing incidence mirrors (silicon coated with 120 nm chromium) cutting off the high-energy portion of the spectrum (a grazing angle of 1° results in a high energy cut-off energy of ~4.5 keV).

This chapter discusses various aspects of X-ray mask fabrication suitable for the LIGA process. It will not address specific solutions and issues associated with the fabrication of microelectronic devices utilizing soft X-ray lithography techniques [12, 21]. The chapter will address mask substrate material choices and possible fabrication processes and briefly discuss lithographic performance including resolution and sidewall quality. In general, the design and fabrication process of an X-ray mask must take into account the spectrum of the X-ray source and characteristic requirements of the microstructures for which the mask is intended.

2.2
Mask Substrate and Absorber Material Selection

The primary role of an X-ray mask substrate is to provide structural support for the absorber pattern needed to form an image during X-ray lithography, while the absorber material should efficiently block the incoming X-ray radiation [12]. The mask substrate and absorber materials also play major roles in dissipating the heat generated by the absorption of radiation by both the mask substrate and absorber material. Since X-ray masks are subjected to high levels of radiation over long periods of time, the construction materials must have high radiation resistance and show little change in X-ray and optical transmission as well as mechanical properties. The following sections discuss possible mask substrate and absorber material choices considering X-ray and optical transmission as well as thermomechanical material properties.

2.2.1
X-ray Transmission Characteristics of Substrate Materials

The X-ray attenuation properties of any material is described by the Beer-Lambert law as

$$P(x) = P_0 e^{-x/L} \tag{1}$$

where P_0 is the incident X-radiation power per unit area, x is the material thickness and L is the attenuation length [22–24]. The attenuation length is simply the inverse of the mass attenuation coefficient, μ, and depends on the photon energy and on the type of material. Equation 1 can be expanded further and presented as a function of the wavelength, λ, such that

$$P(x) = \int_0^\infty \frac{\partial P_0(\lambda)}{\partial \lambda} e^{-\mu(\lambda)x} d\lambda. \tag{2}$$

At a thickness equal to the attenuation length, $x = L$, the radiation retains $1/e$ of the original intensity while the remainder is absorbed by the material. The attenuation lengths of candidate mask substrate materials were computed using a web-based tool hosted by the Center for X-ray Optics (CXRO) (Berkeley, CA, USA) [25] and are shown in Figure 2.3. The attenuation length generally increases with increasing

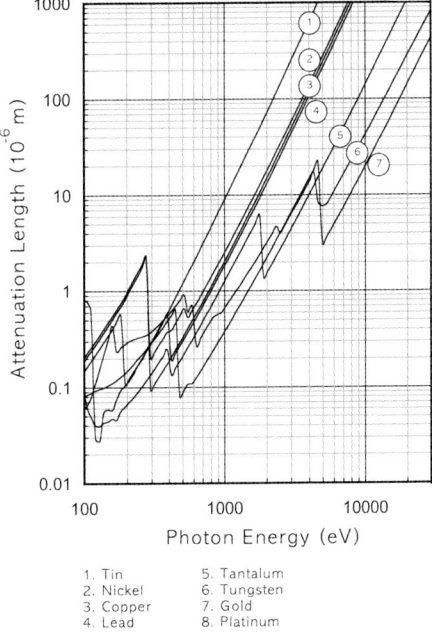

1. Tin   5. Tantalum
2. Nickel   6. Tungsten
3. Copper   7. Gold
4. Lead   8. Platinum

Figure 2.3 X-ray attenuation lengths of suitable mask substrate materials.

photon energy, indicating that materials become 'more transparent' for higher photon energies. However, jump-like changes in the material absorption indicating higher absorption with increasing photon energy are observed and are caused by photon energies sufficient to remove inner shell electrons leaving an electron hole [22]. Radiation emitted when these holes are filled with outer shell electrons is called fluorescence radiation [26]. This radiation is characteristic for each material, influencing the patterning accuracy of the X-ray lithography process and thus becoming an additional selection criterion for the mask substrate material [27–29].

Synchrotron light sources produce a continuous spectrum of radiation ranging from infrared to hard X-rays often referred to as *white light* which needs to be considered for the material transmission. An effective transparency, T_{eff}, can be defined for a set of exposure conditions as the ratio of the dose deposited at the bottom of a resist with a mask substrate in place to the dose deposited without a mask substrate and written as

$$T_{eff} = \frac{\int G\left(\frac{\lambda_c}{\lambda}\right) \mu_R e^{-\mu_R x} e^{-\mu_M d_M} d\lambda}{\int G\left(\frac{\lambda_c}{\lambda}\right) \mu_R e^{-\mu_R x} d\lambda} \tag{3}$$

where G is the spectral distribution function of the radiation incident upon the X-ray mask, λ_c is the critical wavelength of the synchrotron source, μ_R and μ_M are the attenuation coefficients of the resist and mask, respectively, and x and d_M are

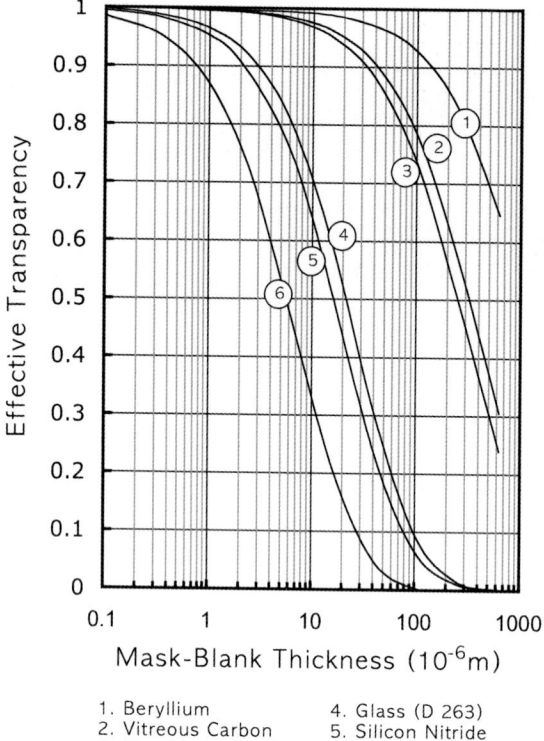

1. Beryllium
2. Vitreous Carbon
3. Graphite (DFP-3)
4. Glass (D 263)
5. Silicon Nitride
6. Titanium

Figure 2.4 Effective transparency of selected mask substrate materials at the CAMD bending magnet beamlines.

the thickness of the resist and mask substrate, respectively [30]. Attenuation lengths and effective transparencies of commonly used mask substrate materials are shown in Figures 2.3 and 2.4, respectively for the photon energy range 100 eV–30 000 keV.

In order to achieve high transmission, the attenuation length should be large for the energy range relevant for X-ray lithography. As can be concluded from Figure 2.4, this will result in different acceptable substrate thicknesses for the various materials ranging from a few micrometer thin membranes (e.g. titanium and silicon) to several hundred micrometer thick substrates in the case of beryllium when considering 75% effective transmission as an appropriate performance.

2.2.2
X-ray Attenuation Characteristics of Absorber Materials

An effective absorption, A_{eff}, can be defined for an X-ray absorber material such that $A_{\text{eff}} = 1 - T_{\text{eff}}$, where T_{eff} is the effective transmission given by Equation 3.

The spectral distribution, G, used in determining A_{eff} is the radiation that is incident upon the absorber material after passing through the front-end window,

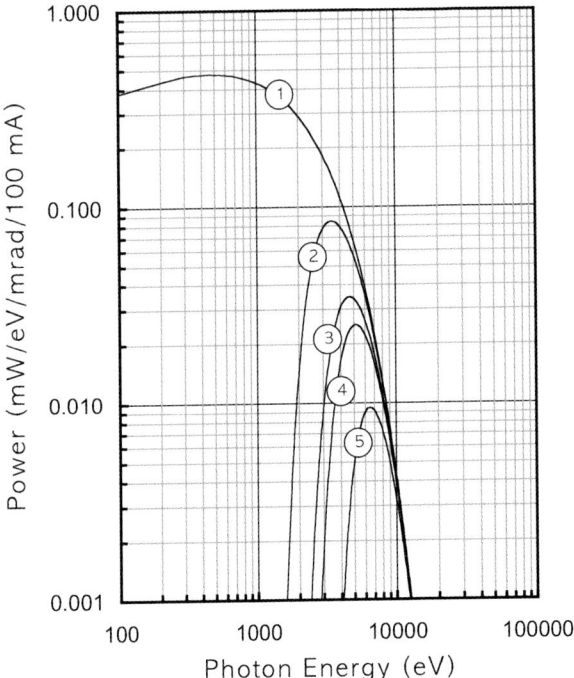

1. Raw spectrum
2. Front end window (220 μm Be)
3. Filter (10 μm Al)
4. Mask blank (400 μm Be)
5. Resist (500 μm PMMA)

Figure 2.5 Spectral distribution of photons generated at the XRLM3 bending magnet beamlines at CAMD: 1.3 GeV ring energy; 2.928 m bending magnet radius; 10.35 m distance from the source point; 1.66 keV critical energy.

filter and mask substrate materials. The effect on the exposure spectrum from CAMD's bending magnet source is shown in Figure 2.5 considering typical materials required for exposing a 500 μm thick poly(methyl methacrylate) (PMMA) resist. The insertion of several materials into the radiation spectrum effectively shifts the exposure spectrum to harder X-ray photons requiring a thicker absorber to block the radiation efficiently [31].

Eight metals ranging in density from 7.31 to 21.45 g cm^{-3} are considered for use as absorber materials (Table 2.1).

Tin, lead, tantalum, tungsten and gold have been used previously as absorbers either as alloys or in elemental form. Nickel and copper are considered because they are commonly used as structural materials in LIGA, while platinum has high density and can be electroplated. The absorption characteristics and effective absorption curves of the candidate metals are shown in Figures 2.6 and 2.7, respectively.

In addition to the absorption characteristics, the ease of formation and the typical processing conditions for the absorber material have to be taken into

2 X-ray Masks for LIGA Microfabrication

Table 2.1 Densities and absorption characteristics of candidate X-ray absorber materials.

Material	Density (g cm^{-3})	Thickness for 80% absorption	Patterning method
Nickel	8.908	12.7	Electroplated
Copper	8.920	12.5	Electroplated
Tin	7.310	5.5	Electroplated
Lead	11.34	3.7	Electroplated
Tantalum	16.65	3.1	Dry-etched
Tungsten	19.30	2.8	Dry-etched
Gold	19.32	2.2	Electroplated
Platinum	21.45	2.1	Electroplated

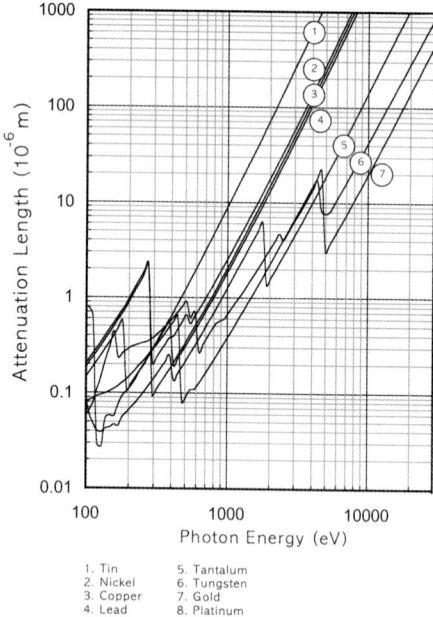

1. Tin
2. Nickel
3. Copper
4. Lead
5. Tantalum
6. Tungsten
7. Gold
8. Platinum

Figure 2.6 X-ray attenuation lengths of candidate X-ray absorber materials.

account when deciding on a material. Electroplating and reactive ion etching (RIE) are two methods that are commonly used to form absorber layers for X-ray masks. All of the candidate absorber materials listed in Table 2.1 with the exception of tantalum and tungsten can be electroplated with varying levels of difficulty from aqueous solutions with plating rates ranging from 1 to 300 µm h^{-1} at bath temperatures ranging from 20 to 100 °C.

A lead–tin alloy (90:10) formed by electroplating was reported by Fischer *et al.* as an inexpensive absorber material [32]. However, they reported that for an average photon spectrum of 20 keV, a 40 µm thick layer of lead–tin alloy provides a contrast ratio of only 20 for a 1 mm thick layer of X-ray resist, which is insuffi-

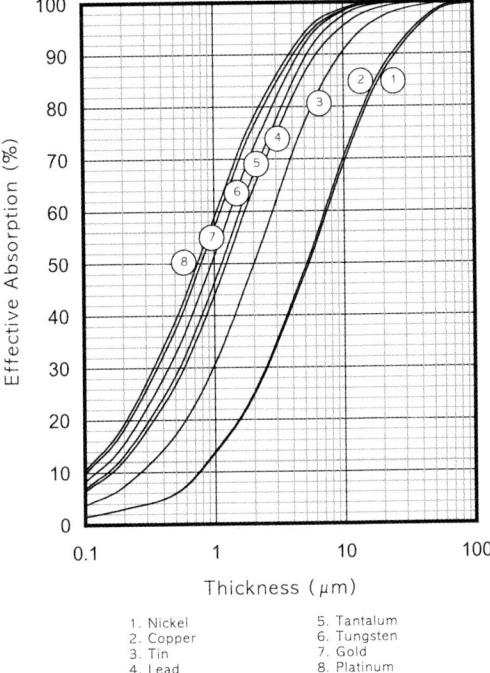

Figure 2.7 Effective absorption of various X-ray absorber materials for the XRLM1 beamline at CAMD (1.3 GeV).

1. Nickel
2. Copper
3. Tin
4. Lead
5. Tantalum
6. Tungsten
7. Gold
8. Platinum

cient for both PMMA and SU-8 lithography. Copper, nickel and tin can also be easily electroplated but, similarly to the lead–tin alloy reported by Fischer et al., they all have relatively long attenuation lengths, as shown in Figure 2.6. The use of materials with long attenuation lengths increases the necessary thickness of the photoresist mold patterned lithographically on the mask substrate, impacting resolution and the smallest feature sizes (see Section 2.4).

Tantalum and tungsten are both refractory metals with relatively high densities but, with the exception of chromium, refractory metals can only be electroplated from molten-salt baths at temperatures as high as 850 °C [33], which makes them incompatible with mask substrate and photoresist materials. However, tantalum [34], tungsten [35, 36] and lead [37] can be deposited with physical vapor deposition (PVD) and patterned with RIE to form absorber patterns for X-ray masks. High-aspect ratio absorber patterns with sub-micrometer critical dimensions that are usually susceptible to structural collapse can be produced by using all dry processes such as PVD and RIE; however, both of these processes are very slow, making it difficult to form absorber layers greater than 3 µm in thickness.

Gold and platinum have the highest effective absorption among the candidate absorber materials. Therefore, a relatively thin layer of one of these materials can be used as an absorber; for example, only 2.2 µm thick gold or platinum is needed to provide an effective absorption of 80%, as opposed to 12.6 µm of nickel or 3.7 µm lead for the CAMD bending magnet source. Gold can be electroformed easily from cyanide- or sulfite-based baths with temperature ranging from 25 to 80 °C [38–42] into patterned photoresist layers ranging in thickness from sub-micrometer [38, 43] to over 100 μm [44]. Platinum can be electroplated from aqueous solutions based on ammine or acid chloride from 30 to 90 °C [33]. However, since gold electroplating is extensively used in jewelry, a myriad of well-established processes are commercially available, making gold electroplating the method of choice for forming X-ray mask absorber structures (for more details, see Section 2.4.6).

2.2.3
Thermo-elastic Properties of Mask Substrates

The absorption of photons in the X-ray mask substrate and absorber during lithography leads to an overall increase in the temperature of the mask, which may be a significant source of pattern distortion [27, 45]. Mask substrates are also exposed to temperatures as high as 150 °C during the soft bake and post-exposure bake cycles of the mask fabrication process and must be able to withstand these large temperature variations without damage or, in the case of membranes, loss of the tensile stress.

Vladimirsky et al. used an X-ray mask with *built-in* thermal resistors to measure the temperature rise during exposure at a soft X-ray beamline at the Synchrotron Radiation Center (University of Wisconsin-Madison [46]) and showed that for highly transparent masks with low absorber coverage, the maximum change in temperature, ΔT_{max}, could be limited to less than 10 °C [47]. Zetter reported ΔT_{max} as high as 70 °C in titanium membrane masks [48]. The temperature rise in X-ray masks was also studied at CAMD using thermocouples mounted directly on the mask surface shielded by an absorber material. Figure 2.8 shows the temperature history during exposure of a beryllium mask with an approximately 75% absorber coverage at the XRLM1 and XRLM4 beamlines, respectively. The lowest ΔT_{max} of 2 °C was measured at the XRLM1 beamline at 120 mA ring current and the largest ΔT_{max} of 30 °C was measured at the XRLM4 beamline at 137 mA ring current. XRLM1 is a bending magnet beamline with a power flux at the mask of $2.98 \times 10^{-3}\,W\,cm^{-2}\,mA^{-1}$, whereas XRLM4 is a wiggler beamline with a power flux of $2.70 \times 10^{-2}\,W\,cm^{-2}\,mA^{-1}$.

Especially critical is the situation with thin membrane masks. As the temperature of the mask membrane increases during exposure, it expands and decreases the membrane tension. The expansion of the mask frame can be neglected because its holder is water cooled and kept at room temperature during exposure. To minimize pattern distortion, the expansion of the mask per unit length due to ΔT_{max} must be less than the pre-existing strain in the mask membrane; this condition sets a lower limit for the pre-stress of a membrane, $\sigma_{membrane}$. On the other hand, the yield strength of the membrane, σ_{yield}, must be greater than the pre-stress in the membrane to avoid plastic deformation. These two conditions can be summarized as

$$\frac{\Delta T_{max}\alpha E}{1-v} < \sigma_{membrane} \leq \frac{\sigma_{yield}}{n} \tag{4}$$

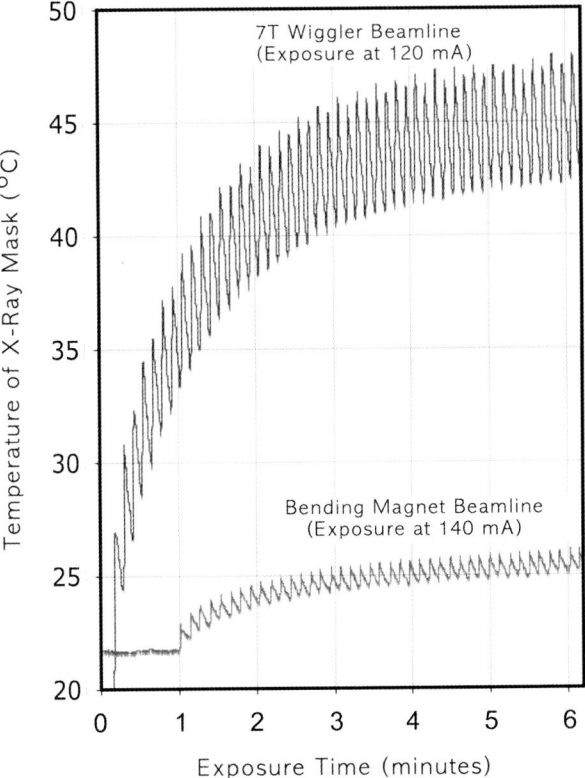

Figure 2.8 Temperature rise in an X-ray mask with 35 μm thick gold absorber on a 500 μm thick beryllium substrate with 75% absorber coverage.

where n is a factor of safety and α, E and ν are the coefficient of thermal expansion, modulus of elasticity and Poisson's ratio of the membrane material, respectively. In practice, absorber patterns contain structures of various asymmetric shapes and sizes and since the thickness of the absorber layer is usually much larger than the thickness of the membrane, the thermoelastic behavior of the absorber material dominates.

Unlike membranes, thick substrates do not need to be pre-stressed for stability. Thick substrates expand and contract in response to the temperature changes during use – only constrained by the strength of the bond to the mask frame. Therefore, minimizing mask distortion by selecting a material with high thermal conductivity in addition to a low coefficient of thermal expansion is imperative. The maximum temperature rise also depends on exposure conditions such as scan speed, acceleration, dwell time, scan length and beam current; all of these parameters must be optimized to minimize ΔT_{max}.

2.2.4
Surface Quality of Mask Substrate Materials

Absorber patterns are typically produced by using a combination of lithography and electroplating. Low surface roughness, low porosity and good adhesion to metals and photoresists are some of the critical and important surface characteristics for mask substrates [49]. An electrically conductive surface that is also suitable for electroplating can simplify the fabrication process by eliminating the need to deposit an electroplating base.

Beryllium can be formed into a membrane for masks [4, 50] or obtained as a thick substrate (>250 μm) from Brush Wellman (San Jose, CA, USA), the largest beryllium product manufacturer in the world, with various surface finishes ranging from a 'super best buff' to a polished 'mirror finish' with mean average roughness of 1.4 and 0.1 μm, respectively. The high roughness 'super best buff' finish is suitable for masks fabricated with X-ray lithography as the roughness enhances the adhesion to PMMA resist [51], whereas the 'mirror finish' beryllium is best for photolithography. Although beryllium has low resistivity, the native oxide prevents successful electroplating, requiring the beryllium surface to be activated by either a chemical pickle with concentrated nitric acid or an anodic pickle in phosphoric acid [33], which are both incompatible with photoresist materials. An electroplating base, such as gold, needs to be deposited by physical vapor deposition (PVD) to make the surface suitable for electroplating.

Graphite is one of the allotropes of carbon and has a hexagonal sheet-like structure. Graphite is found in various forms, including pyrolytic graphite and rigid graphite. The sheet-like structure is maintained in pyrolytic graphite with a very weak bond between the sheets. As a result, pyrolytic graphite cleaves very easily along the weak bonds parallel to the hexagonal plane; this property, which is sometimes referred to as 'flaky', makes pyrolytic graphite unsuitable for use as an X-ray mask substrate. On the other hand, rigid graphite is composed of graphite particles with typical grain sizes ranging from 1 to 20 μm, compressed to form a bulk material. Rigid graphite possesses low resistivity and can be readily electroplated with good adhesion. Rigid graphite is also highly porous and exhibits high surface roughness [52]. Regardless of the undesirable high surface roughness (for photolithography) and high porosity, X-ray masks have been successfully produced on graphite [7, 53–55].

Vitreous carbon, also called glassy carbon, is an amorphous form of carbon produced by the thermal decomposition of crosslinked polymers. Vitreous carbon has low density (1.4 g cm^{-3}), no porosity and possesses excellent X-ray transmission characteristics [56, 57]. A major challenge with vitreous carbon substrates is the poor adhesion to metal layers deposited by either PVD or electroplating, which is still research in progress despite first promising results [58].

Silicon nitride and borosilicate glass have low surface roughness and show no porosity and are most suitable for photolithography. These materials are not electrically conductive, requiring the deposition of a thin metallic plating base, typically a Cr–Au or Ti–Au combination, which does not change the mirror-like surface quality [59].

Titanium membranes are formed by sputter deposition and to improve the adhesion of photoresists and electroplated metals to titanium; the surfaces are treated with an alkaline peroxide (AP) solution resulting in an approximately 500 nm thick layer of titanium dioxide (TiO_2). Although the roughness of the titanium increases as a result of the AP treatment, high-quality sub-micrometer resolution patterns have been produced with electron beam, X-ray and photolithography methods [4].

2.2.5
Optical Transmission

Employing X-ray lithography in an aligned exposure process in order to build more complex, multi-level microstructures, for example anchored seismic masses for acceleration sensors, mask substrates with good transmission in the visible spectral range are required [60, 61]. Depending on their thickness, mask substrates made from glass, silicon nitride, silicon carbide and polyimide provide acceptable optical transmission, whereas substrates made from beryllium, graphite, vitreous carbon and titanium are opaque and require additional efforts. As an example, Figure 2.9 shows a section of a 1 µm thick silicon nitride membrane mask carrying cross-hair alignment structures. There are three fields – one with a broken membrane (a), one where the Cr/Au seed layer is etched away after plating (b) and one where the seed layer is still covering the nitride membrane (c). The images underneath each of the fields show the alignment pattern (cross-hair) projected on to a line pattern on a substrate illustrating the acceptable transparency of the silicon nitride membrane. In the case where the seed layer is still remaining on the membrane, only the defocused, reflected image of the cross-hair marker is seen

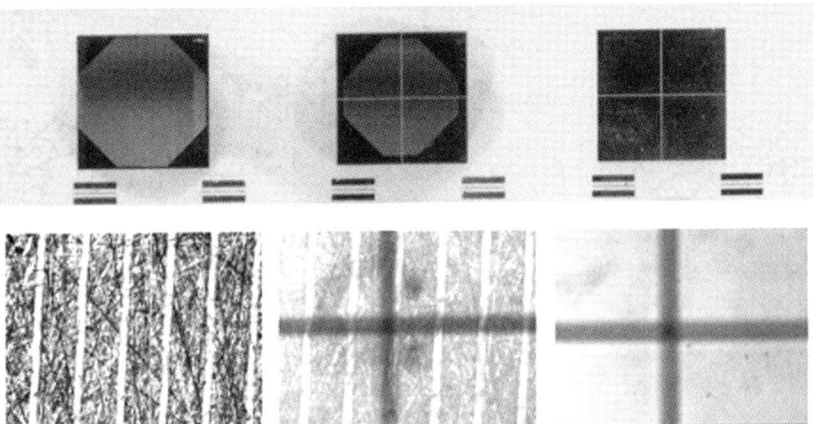

Figure 2.9 Three individual alignment markers (cross-hair) on a silicon nitride membrane (top) and the corresponding microscope images when aligning the mask relative to an underlying substrate carrying a line pattern (bottom) (see text for further details).

in the microscope and the underlying substrate is not visible. Although the image captured through membrane (b) lacks some contrast compared with the image with no membrane, it is sufficient to align substrate and mask properly, typically with 1–2 µm overlay accuracy.

Methods have been devised by various researchers to create optically transparent alignment markers even on opaque mask substrates. In one approach developed for titanium and beryllium membrane masks at the IMT (Karlsruhe, Germany), the membrane material under the alignment markers is selectively etched to create stencil alignment markers [4]. An alternative method was developed at CAMD for opaque mask substrates by mounting the substrate on an optically transparent glass ring and patterning the alignment markers on the glass ring at the same time as the high aspect ratio microstructures are formed on the mask substrate [62].

2.3
Mask Architecture

A critical factor in determining the thickness of a mask substrate material is the effective transparency of the material, T_{eff}, indicating how well X-ray photons are transmitted through the mask substrate and are absorbed in the underlying resist. Another important property is the absorber thickness determining the absorption contrast, C_{absorp}, provided by an absorber of a given thickness. Dose definitions used in this chapter are illustrated in Figure 2.10. The X-ray photons pass through mask substrate areas not covered with an absorber material with minimum losses resulting in a high exposure dose. However, as we typically employ X-ray lithography for patterning of tall, up to several millimeters thick structures, absorption in the resist material itself needs to be considered. This is done by defining two different exposure dose values, D_{TR} and D_{BR}, indicating the absorbed dose at the top of the resist facing the mask and at its bottom at the substrate–resist interface, respectively. In areas, where absorber structures are blocking X-ray photons, less dose is deposited in the same time, resulting in an exposure dose D_{TA}. It should be noted that there will always be some radiation 'leaking' through the absorber, resulting in an exposure in the underlying resist. This 'leaking' is strongly dependent on the light source and its characteristic exposure spectrum and may require an X-ray mask with different absorber thickness for different sources.

2.3.1
Mask Substrate Thickness

As discussed in Section 2.2, X-ray mask substrates can be classified into two categories: thick substrates and membranes. Thick substrates have sufficient structural stiffness to be handled and processed without additional support and range in thickness from 100 to 1000 µm. Conversely, membrane-type substrates lack structural stiffness and must be mounted on a frame or formed *in situ* with built-in tensile stress during processing and use. The X-ray transmission of mask sub-

2.3 Mask Architecture

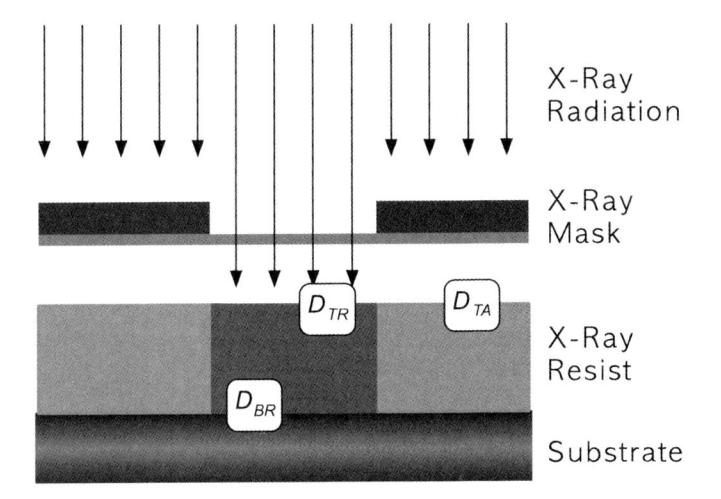

D_{TR} = Dose deposited at the top of the resist

D_{BR} = Dose deposited at the bottom of the resist

D_{TA} = Dose deposited at the top of the resist under the absorber

Figure 2.10 Definition of doses used in the calculation of the absorption contrast of a mask.

Table 2.2 Effective transparencies at the XRLM1 beamline with various beamline elements for a 500 μm thick layer of PMMA.

Beamline and exposure elements	Effective transparency (T_{eff}) (%)	Dose ratio of 500 μm PMMA (R)
None (white spectrum)	N/A	261
220 μm Be (front-end window)	100	23
400 μm Be mask substrate	75	9
10 μm Al filter	51	5

strate materials can be characterized with an effective transmission, T_{eff}, as defined by Equation 3. An acceptable transmission for a given synchrotron exposure spectrum can be used to determine the maximum thickness.

The choice of an acceptable T_{eff} is based on a few practical considerations, which can best be illustrated by use of an example. If only the *white* exposure spectrum from a bending magnet beamline at CAMD is used to expose a 500 μm thick PMMA resist without considering absorption from vacuum window or X-ray mask substrate, the top-to-bottom dose ratio, R, will be equal to 261 with a T_{eff} of 100%. As shown in Table 2.2, R decreases from the initial value of 261 to 9 on inserting

Table 2.3 Properties and thickness of suitable mask substrate materials.

Material	Density (g cm^{-3})	Thickness for 75% T_{eff} (µm)	Availability (µm)	Type of substrate	Commercial source	Cost ($)
Beryllium	1.8	400	>250	Self-supportive	Brush & Wellman Corp. (San Jose, CA, USA)	500–2000
Vitreous carbon	1.4	105	>100	Self-supportive	Goodfellow Corp. (UK)	~120
Graphite	1.7–1.8	100	>150	Self-supportive	POCO, Inc. (Decateur, TX, USA)	~15
Glass	2.3	10	30, 80	Membrane	Schott GmbH (Germany)	~100
Silicon nitride	2.4	6	1	Membrane	DIMES (The Netherlands)	~200
Titanium	4.5	2	2.8	Membrane	IMT (Karlsruhe, Germany)	N/A

a 220 µm Be vacuum window and a 400 µm Be mask substrate yielding a T_{eff} of only 75%. In order to achieve a dose ratio of 5 or less, typically used for X-ray lithography exposures, an additional 10 µm thick aluminum filter needs to be inserted, yielding a T_{eff} of only 51%. The same level of transparency of 51% can also be achieved by using a 700 µm thick Be substrate and no Al filters instead of the 400 µm thick Be mask combined with a 10 µm thick Al filter. However, inserting an Al filter in front of the mask results in effective absorption of low-energy X-ray photons in the filter and reduces the thermal load on the mask.

In Table 2.3, the thicknesses for different materials used as mask substrates are listed for an acceptable transparency of 75% for the CAMD bending magnet exposure spectrum. This can vary slightly for other sources and therefore is only used as an approximate guideline. Other factors, such as the availability of a material in a given thickness and cost, are also included in the table. All of the materials except titanium membranes are commercially available from different vendors. Although X-ray masks made on titanium membranes can be procured commercially, the mask substrate alone is not available. Beryllium is the most expensive material of the six, and graphite has the lowest cost.

2.3.2
Absorber Thickness

Of the number of possible absorber materials illustrated in Figures 2.6 and 2.7, gold is the preferred choice due to its short absorption length and ease of formation and will be used as example to discuss the minimum thickness requirements for use as an absorber on an X-ray mask.

The relative efficiency of an X-ray mask in transmission and the absorption of radiation during lithography are measured by a dimensionless parameter called contrast. Two types of contrasts can be defined for X-ray masks: the transmission contrast, C_{trans}, is the ratio of the total power transmitted through the mask substrate (P_M) to the total power transmitted through both the absorber and the mask substrate (P_A), whereas the absorption contrast, C_{absorp}, is defined as the dose absorbed at the bottom of the resist in the exposed area (D_{BR}) to the dose absorbed at the top of the resist in the shielded region (D_{TA}) (for details, see Figure 2.10) [63].

The transmission contrast depends on the properties of the mask and the spectrum of radiation used for lithography, but the absorption contrast also depends on the absorption length and thickness of the resist. Hence absorption contrast is a better indicator of a mask's suitability for a particular lithographic task and will be simply referred to as the 'contrast' in this text and denoted by C, such that

$$C = \frac{D_{\text{BR}}}{D_{\text{TA}}} \tag{5}$$

The white, continuous exposure spectrum from a synchrotron light source is a function of the electron energy and the bending magnet radius [64]. An absorption contrast curve, which is independent of the type of mask substrate used, can be generated for a given beamline, resist thickness and dose ratio, R. The absorption contrast curve relates the thickness of the absorber to the contrast at a constant dose ratio. These curves have been calculated for different resist thickness ranging from 250 to 1000 µm and a dose ratio of 5 for CAMD's bending magnet beamlines and are shown in Figure 2.11.

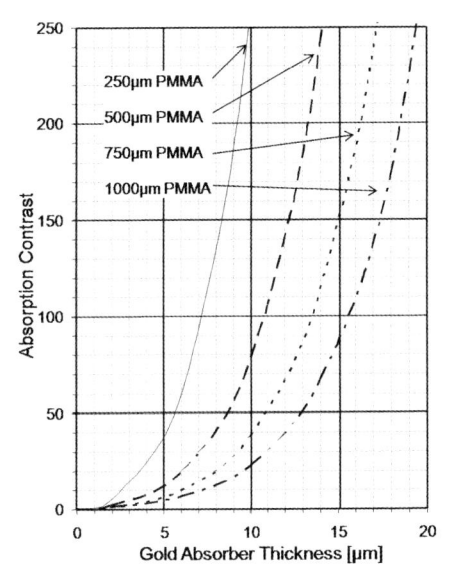

Figure 2.11 Contrast characteristic curves for the bending magnet beamlines at CAMD at a top-to-bottom dose ratio of 5 for different thickness of PMMA resist.

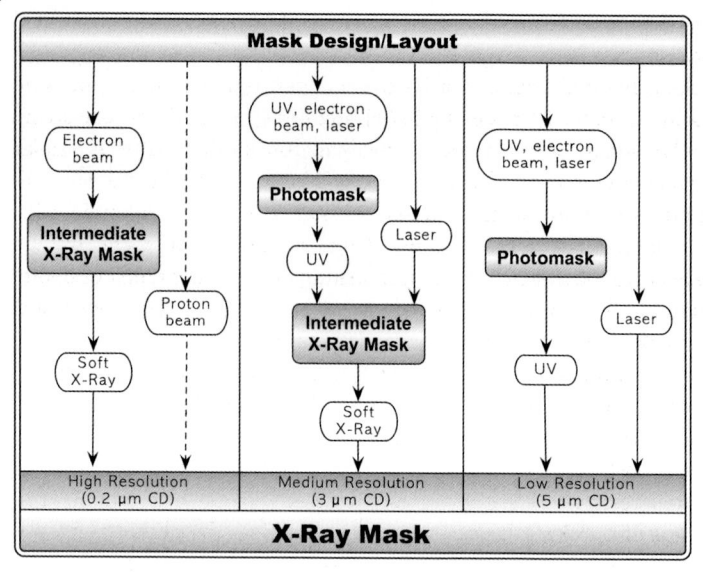

Figure 2.12 Overview of X-ray mask fabrication methods.

These curves serve as a useful reference to determine the minimum thickness of the absorber for an X-ray mask to estimate the contrast provided by a mask for a particular exposure. For example a 13 μm thick layer of gold is sufficient to provide a contrast of 50 for resist thicknesses of up to 1000 μm with a top-to-bottom resist dose ratio of 5. If a higher contrast is required, the gold thickness needs to be thicker, for example a contrast of 250 requires a nearly 20 μm thick gold absorber for a 1000 μm thick PMMA resist.

2.4
X-ray Mask Fabrication Methods

X-ray masks for LIGA can be roughly classified into three categories based on the smallest lateral dimension of the absorber structures present, commonly referred to as the critical dimension (CD), as high-, medium- or low-resolution. The choice of fabrication process depends on the type of mask to be fabricated and is summarized in Figure 2.12.

High-resolution X-ray masks for high aspect ratio microstructures possess CDs as small as 0.2 μm and can only be fabricated using a two-step process. In the first step, an *intermediate mask* is made by using a combination of electron-beam (e-beam) lithography and gold electroplating. Soft X-ray lithography is then used to transfer the pattern from the intermediate mask into a thicker layer of PMMA followed by gold electroplating forming the final X-ray mask, commonly referred to as a *working mask*, which is now suitable for deep X-ray lithography. Typically, the intermediate mask has an absorber thickness between 2 and 3 μm, whereas the working mask is made from 15–25 µm thick resist templates, resulting in gold absorber thickness up to 20 µm. A good example of a high-resolution working mask is used to make a hot-embossing tool for a microspectrometer with a focusing diffraction grating of 0.2×2 µm steps [65, 66]. Recently, van Kan and coworkers at the National University of Singapore produced high-resolution patterns similar to those found on working masks by proton-beam lithography raising the possibility of making high-resolution working masks with a single-step process [67–69].

Medium-resolution X-ray masks are fabricated using a two-step process but the *intermediate masks* are made with either laser-writing or UV lithography of photoresist layers ranging from 3 to 10 µm in thickness *in lieu* of e-beam lithography. The working masks are made by a process similar to high-resolution working masks, that is, soft X-ray lithography followed by gold electroplating. Due to the limitations imposed by the laser-writing and UV lithography, the CD of medium-resolution X-ray masks is limited to approximately 3 µm.

Low-resolution masks are made by UV lithography of thick photoresists such as SU-8 negative tone resist [70, 71] and gold electroplating. Patterns as small as 5 µm CD with thickness of up 50 µm can be patterned using SU-8 photolithography and are routinely used to fabricate X-ray working masks in a single-step process.

This section discusses some commonly used fabrication methods in more detail in order to provide information about the complexity and also the limits of the different methods. This also confirms the initial statement that there is no standard X-ray mask fabrication method but that there are a number of possible solutions and that the type of X-ray mask used for a particular applications is mainly determined by the specifications, especially CD and sidewall roughness.

2.4.1
Titanium Membrane Mask Fabrication

The fabrication process of titanium membrane X-ray masks has been well established for over two decades [4]. These masks are typically fabricated by using the two-step, high-resolution mask fabrication method illustrated in Figure 2.13.

For the intermediate mask, a ~2.5 µm thick titanium film is sputter deposited on a silicon wafer used as a membrane carrier and already coated with a 100 nm thin carbon layer used as an anti-adhesion film [step (a)]. During the sputtering process, the deposition conditions are closely monitored and controlled such that the resulting titanium film possesses tensile stress. Next, the titanium surface is wet oxidized by use of an AP treatment [72], followed immediately by spin-coating of a ~2.5 µm thick e-beam lithography resist layer, typically PMMA, and baking. Subsequently, the design pattern is written into the resist using a high-energy (up to 100 keV) e-beam writer [43] [step (b)]. After development, the pattern is filled with electrodeposited gold up to 2 µm thick. The remaining PMMA resist is then stripped and the titanium membrane is lifted off the silicon carrier by gluing it to an Invar frame [step (c)]. This completes the fabrication process of the intermediate mask, which is next transferred into a thicker resist layer using soft X-ray lithography. The SEM pictures in Figure 2.14 illustrate typically gold structure quality on the intermediate mask.

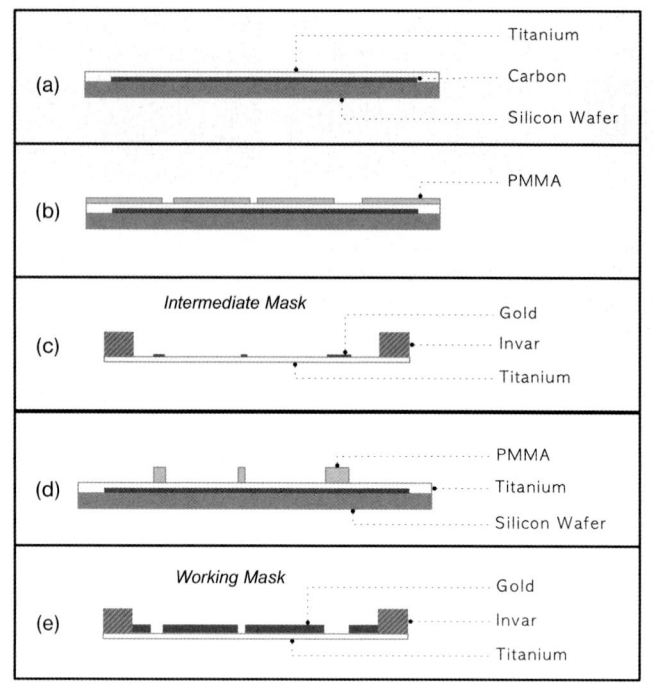

Figure 2.13 Fabrication process of titanium membrane masks using a two-step fabrication process (for details see text).

Figure 2.14 High-resolution intermediate mask patterns produced by use of a 100 keV e-beam writer and electroplating with a gold thickness of 1.8 μm. (a) 700 nm wide lines and spaces; (b) 1 μm diameter posts; (c) 20 μm diameter gear with teeth length of 1.25 μm.

The fabrication of the working mask starts out in a similar way by forming another 2.8 μm thick wet-oxidized titanium layer similarly on a silicon wafer carrier. Instead of spin-coating, a fairly thick resist layer (typically 30 μm) of PMMA resist is formed by polymerizing a mixture of high molecular weight PMMA powder dissolved in methyl methacrylate (MMA) with the aid of benzoyl peroxide (BPO), N,N-dimethylaniline (DMA) and γ-methacryloxypropyltrimethoxysilane

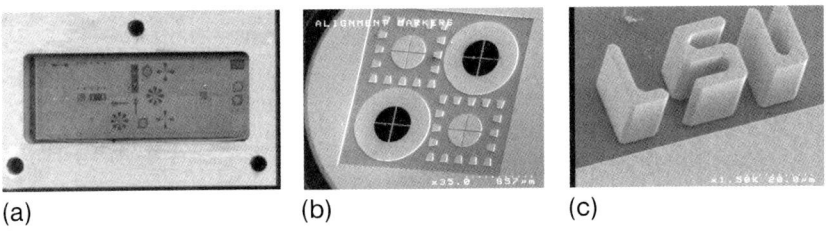

Figure 2.15 Ti membrane X-ray working mask with 17 μm thick gold absorber. (a) Overall mask mounted on Invar frame; (b) stencil-type alignment marker; (c) example of gold microstructure on mask.

(MEMO) [72]. BPO and DMA serve as initiator and hardener, respectively. The addition of MEMO causes crosslinking of the PMMA, which slightly lowers the sensitivity of the PMMA but significantly increases resist contrast. MEMO also helps in the formation of chemical bonds to the TiO_2 layer formed by the AP treatment [73]. The PMMA layer is annealed at 115 °C with slow ramp-up and cool-down cycles to relieve the internal stress that is generated during the polymerization process [74]. The PMMA resist is lithographically patterned using soft X-ray lithography with a bottom dose of $4\,kJ\,cm^{-3}$ and developed with the GG developer [75] [step (d)]. Electroplating using a sulfite-based gold electroplating solution forms the absorber pattern with a thickness up to 20 μm. Similarly to the intermediate mask, the PMMA resist is stripped and the membrane is mounted on an Invar frame, completing the mask fabrication process [step (e)].

Figure 2.15 shows the final mask with typical dimensions of $20 \times 60\,mm$ mounted on an Invar frame and some representative SEM pictures of the microstructures present. This mask also includes alignment markers of the stencil type formed by selectively etching the titanium membrane under the markers creating suspended gold cross-hair pattern [4].

2.4.2
Silicon Nitride Mask Fabrication

Silicon nitride is widely used in MEMS as a dielectric material, as a passivation layer to protect from water and alkali metal ions and as a structural material. Because of its high etch resistance to chemicals, it is also used as an etch mask for silicon [76]. Its high strength and excellent radiation stability in addition to reasonable optical transmission make silicon nitride a desirable material for X-ray mask membranes, used extensively by several groups [77–81]. The fabrication method commonly used is illustrated in Figure 2.16.

In the first step (a), prime grade 100 mm diameter, 525 μm thick silicon wafers are used as carrier for the 1 μm thick silicon-rich nitride membrane, which is deposited on the front and back sides using a low-pressure chemical vapor deposition (LPCVD) process developed at the Delft Institute of Microsystems and Nanoelectronics (DIMES) [82–84]. The silicon nitride deposition is carried out at 850 °C

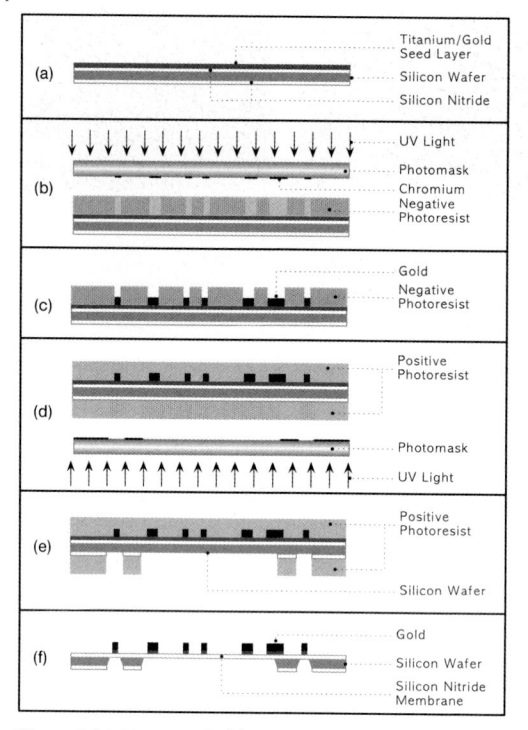

Figure 2.16 X-ray mask fabrication process on silicon nitride membrane mask. The membrane is deposited on a silicon wafer carrier which is selectively etched after forming the gold absorber releasing the silicon nitride membrane (see text for more details).

using a combination of dichlorosilane and ammonia gases; the reaction is summarized as

$$3SiCl_2H_2 + 4NH_3 \rightarrow Si_3N_4 + 6HCl + 6H_2 \qquad (6)$$

The silicon nitride films have a silicon-to-nitrogen ratio of 0.95 with no detectable hydrogen content [85]. The silicon nitride stress can be controlled from zero to several hundred MPa by varying the deposition conditions. The standard silicon-rich nitride deposited at DIMES for mask substrates has a tensile stress of 80 MPa. The maximum operating temperature of a silicon nitride membrane is estimated by rearranging Equation 4 to obtain

$$\Delta T_{max} = \frac{\sigma_{membrane}(1-v)}{\alpha E} \qquad (7)$$

Khakani and Chaker presented a range of typical values for the properties of a number of mask substrate materials including LPCVD silicon-rich nitride [86]. Using the highest v, α and E values, the 'lowest' ΔT_{max} is estimated to be 25 °C,

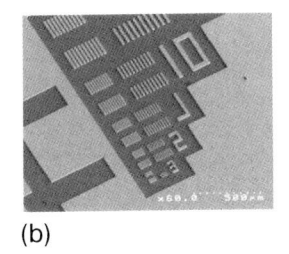

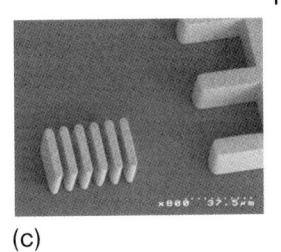

Figure 2.17 Silicon nitride membrane X-ray mask. (a) Back side of a mask with seven large etch windows and several smaller windows carrying alignment markers; (b) overview of gold absorber pattern down to 3 μm smallest structures; (c) close-up view of high aspect ratio gold absorber microstructure.

which is higher than the empirical ΔT_{max} of 10 °C reported by Vladimirsky et al. [87]. Because of its high resistivity (~$10^{16}\,\Omega$ cm), the silicon nitride layer is further coated with a thin bimetallic layer of 100 Å titanium and 300 Å gold by e-beam evaporation to serve as an electroplating seed layer.

In the second step (b), a chemically amplified negative tone photoresist, for example NFR-015 from JSR Microelectronic (Sunnyvale, CA, USA) is spin-coated on to the seed layer and soft baked, resulting in an up to 15 μm thick resist film. This photoresist is patterned by UV lithography, post-exposure baked and developed, forming a resist template for electroplating of the gold absorber structures, typically 8–10 μm thick [step (c)]. A thicker gold layer increases the chance of membrane failure due to the increased stress from the plated absorber structures. For exposures requiring a thicker gold absorber, alternative more robust mask substrates are used and will be discussed in Sections 4.3–4.5.

In order to minimize the open area in the silicon wafer and optimize membrane stability, the back side of the mask substrate is also covered with silicon nitride, which is patterned using front-to-back alignment UV lithography [step (d)]. A layer of thick positive photoresist is coated on the front side, protecting the gold pattern from damage during this process. The silicon nitride on the back side is reactive ion etched using a combination of SF_6 and He gases prior to KOH to etch the silicon wafer and release the silicon nitride membrane [step (e)]. Mounting the silicon wafer on a mask ring completes the fabrication process and results in X-ray masks as shown in Figure 2.17 [59, 79].

2.4.3
Graphite Mask Fabrication

The simplest mask fabrication method outlined in Figure 2.18 is realized by combining low-cost graphite mask substrates with thick resist UV lithography and gold electroplating. The use of graphite as the choice for LIGA X-ray mask substrate material was first published in 1998 by researchers at the Institute for Micromanufacturing (IfM) at Louisiana Tech University (Ruston, LA, USA) [6].

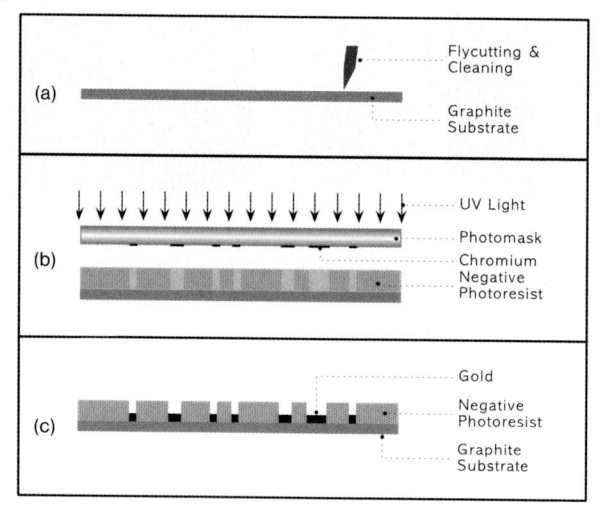

Figure 2.18 X-ray mask fabrication process for graphite-based working masks.

The fabrication of a *working* X-ray mask starts with the preparation of the urface of the graphite substrate [step (a)]. CAMD researchers use different graphite sheets made by POCO Specialty Materials (Decatur, TX, USA) [52]. The initial thicknesses of the graphite substrates range from 200 to 260 μm for a 10 cm disk with a density between 1.7 and 1.83 g cm^{-3}, a fairly high surface roughness (1.5 μm R_a, 5 μm R_z) and a low electrical resistivity between 1250 and 1930 μΩ cm. In order to improve surface quality and thickness uniformity, up to 50 μm of the graphite is flycut from each side of the substrate until a final thickness of ~150 μm is obtained with an improved surface roughness of $R_a \approx 800$ nm. Prior to spin-coating, the substrate is thoroughly cleaned with acetone and isopropyl alcohol and dried.

Considering the requirements for thick resist templates with nearly perfectly straight sidewalls patterned by UV lithography, SU-8 negative photoresist is an excellent choice using contact UV lithography in the wavelength range 350–450 nm [step (b)] [88]. Using different viscosity SU-8 products from MicroChem (Newton, MA, USA) [89], photoresist layer thicknesses ranging from 15 μm (SU-8 5) to 55 μm (SU-8 25) are realized reliably with optimized coating and baking parameters. It should be mentioned that the thickness of the SU-8 resist on graphite substrates is found to be consistently less than that on a silicon wafer due to the high porosity of the graphite requiring an adjustment of the spin-coating parameters. After exposure and completing the post-exposure bake and SU-8 development, typical SU-8 microstructures with CD of 5 μm for a maximum resist thickness of 15 μm and 10 μm for a 55 μm thick layer can be achieved. Gold electrodeposition of the absorber structures completes the X-ray mask fabrication process [step (c)]. Details of the gold electroplating process are discussed in Section 4.6. Due to the challenge of removing SU-8 crosslinked resist and considering the fact that SU-8 shows no radiation damage even for high exposure dose, the SU-8 resist typically remains on the X-ray mask. Recently, some successful SU-8 removal

Figure 2.19 Graphite X-ray working mask with mechanical microstructures (a), detail of SU-8 resist pattern after exposure and development (b) and electroplated gold absorber, ~10 μm thick after SU-8 removal (c).

efforts have been reported which will allow the SU-8 resist to be removed from very fine structures without risk of damaging the gold structures [90, 91]. Figure 2.19 illustrates some typical graphite mask structures that are routinely used for deep and ultra-deep X-ray lithography exposures [6].

In order to permit aligned exposures, Desta developed a modified fabrication process using small, up to 60 mm diameter, graphite sheets mounted on a glass ring as substrate [92]. During the optical lithography process, optical alignment markers present on the UV photomask will be transferred into resist patterned on the glass ring. After completing the fabrication process, these alignment markers are now located on the optically transparent glass ring and allow alignment of the absorber pattern on the graphite sheet relative to a patterned structure on the substrate.

2.4.4
Beryllium Mask Fabrication

One of the most promising but also challenging mask substrate materials is beryllium. Its high X-ray transparency allows the use of up to 1000 μm thick substrates, simplifying the fabrication process significantly. Beryllium does not produce any fluorescent radiation [27] and its high thermal conductivity in combination with high mechanical strength ensures excellent dimensional stability even for exposures at high-intensity sources such as the CAMD wiggler beamline. Beryllium substrates are commercially available in different qualities including passivation layer and metallic seed layer from Brush Wellman (Fremont, CA, USA) [93]. Due to its high cost, it is desirable to use beryllium substrates for high-resolution applications employing an intermediate X-ray mask. Researchers at CAMD have developed an alternative fabrication process amplifying the gold absorber thickness using a back-side exposure with soft X-rays as illustrated in Figure 2.20 [94].

The process starts with a commercial substrate which is rinsed with isopropyl alcohol and dried prior to application of a thin photoresist [step (a)]. An intermediate X-ray mask is fabricated by patterning a thin, typically 7 μm thick, photoresist using UV lithography [step (b)]. In this step, e-beam lithography in an even thinner resist layer can also be employed allowing even smaller CDs [4]. After development,

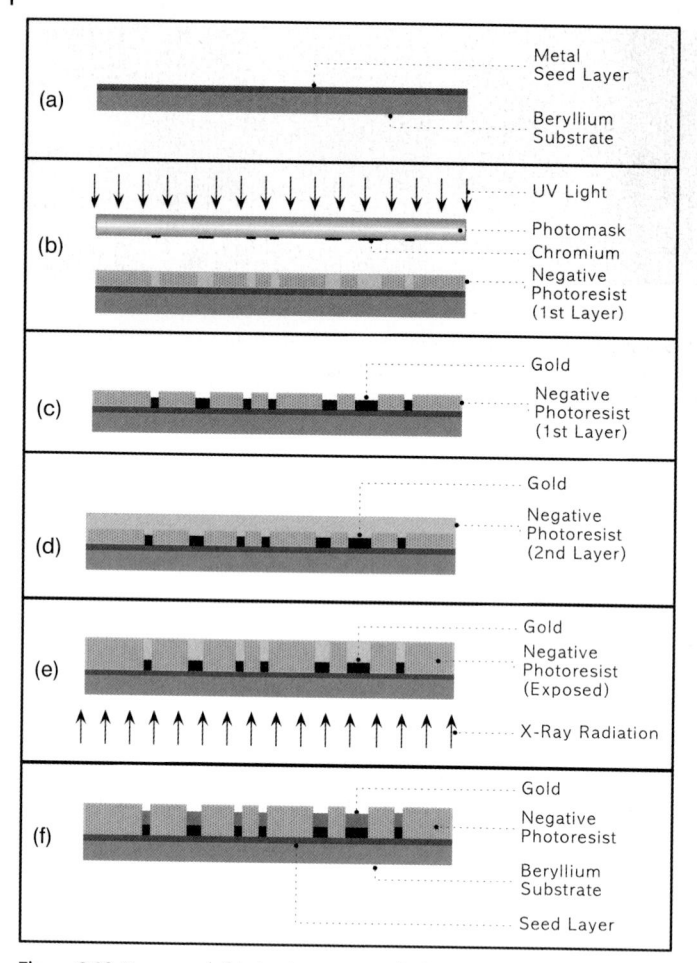

Figure 2.20 X-ray mask fabrication process for high resolution beryllium working masks.

an absorber structure of ~5 μm height is electroplated into the resist template and the photoresist is stripped [step (c)]. In the next step (d), a thick layer of SU-8 resist is applied on top of the intermediate mask pattern and prebaked. The thickness of this layer can range from 20 to 50 μm. Subsequently, an X-ray exposure is performed from the back side of the intermediate mask using soft X-rays [step (e)] followed by a post-exposure bake and development process. In the final step (f), additional gold is electrodeposited on top of the now open gold absorber, allowing its thickness to be increased to 30 μm or even greater. If needed, steps (c)–(f) can be repeated several times until the desired gold thickness is realized, providing a working mask well suited for ultra-deep X-ray exposures using a hard X-ray source.

Compared with a two-step, two-mask high-resolution fabrication process, this process achieves similar pattern quality but only uses one expensive beryllium substrate. Figure 2.21 shows some results achieved with this fabrication approach.

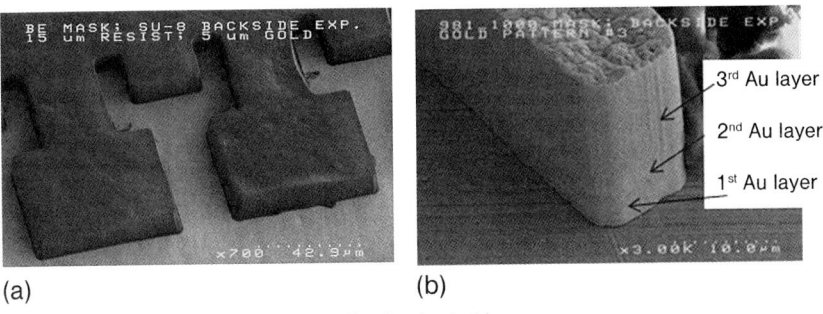

Figure 2.21 (a) SU-8 resist structures after first back-side exposure and development (b) 10 µm wide line approximately 30 µm tall fabricated in three gold deposition steps illustrating a clear and smooth transition between adjacent gold layers.

2.4.5
Glass Mask Fabrication

Borosilicate glass is a natural candidate for use as a mask-blank due to its optical transparency and relatively low density. Previously, Kupka *et al.* successfully produced X-ray masks on 25 µm thick glass membranes that were formed by etching borosilicate glass with an initial thickness of 80 µm using hydrogen fluoride [95]. Researchers at CAMD and BESSY used D263T borosilicate glass substrates (100 mm diameter, 30 µm thick) from Schott (Mainz, Germany) and developed a process to mount these in tension on Invar carrier frames [92, 96, 97]. The pattern formation is similar to the process described for graphite substrates (Figure 2.18). A 20 µm thick positive photoresist (SPR-220-7, Shipley, Marlborough, MA, USA) is spin-coated on a Ti/Au seed layer evaporated on to the glass substrate prior to the resist application. After lithographic patterning using UV light and structure development, the absorber structures are gold electroplated into the resist mold. This process requires careful control of the internal stress of the electroplated gold as the fragile glass substrate tends to break easily. After completion of the plating process and removal of the photoresist, the Ti/Au seed layer base has to be etched in the areas of the alignment markers to achieve optical transparency similar to the silicon nitride membrane. The images in Figure 2.22 demonstrate some results of this X-ray mask fabrication approach.

It should be noted that a number of suitable thick optical resists have been developed for MEMS applications, in particular to form molds for electroplating [98]. This provides researchers with the unique opportunity to choose from a variety of products taking into account substrate material properties and pattern resolution. However, it is always necessary to optimize the resist processing parameters in order to achieve high yield, to obtain minimum defect electroplating molds, which are a prerequisite for high-quality X-ray masks, and to ensure successful patterning of high aspect ratio microstructures using X-ray lithography.

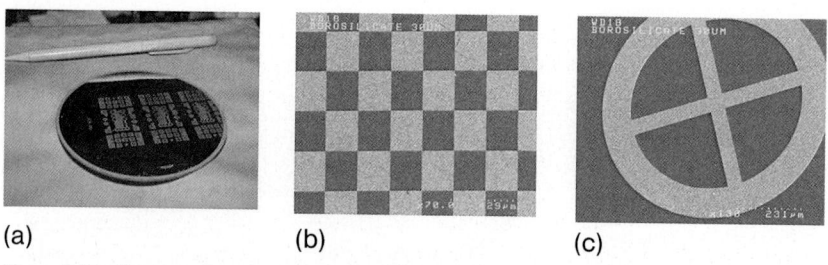

Figure 2.22 Glass membrane X-ray mask: (a) 30 μm thick glass substrate mounted on 100 mm diameter Invar ring with 5.5 μm thick gold absorber; (b) 200 μm by 200 μm checker gold pattern; (c) cross-hair alignment marker.

2.4.6
Alternative Approaches for X-ray Masks

The previous sections have addressed a number of *standard* materials and fabrication processes that are complemented by a number of alternative approaches often devised to reduce fabrication costs and improve processing time. The following section provides a brief overview of some of these techniques, which typically result in 'rapid prototyping' quality and will be replaced by one of the established techniques once a structure design has been finalized, backed by some preliminary experimental results. It also emphasizes the fact that researchers will use MEMS techniques available to them to build their own X-ray masks as they cannot purchase them commercially.

A very simple approach with limited feature size and sidewall roughness is possible using metal stencil masks; Utsumi *et al.* [99] described a stencil mask etched in a steel foil of 100 μm thickness and letter size. The flexible steel sheet is robust and easy to handle and can be directly clamped to the X-ray resist. Smallest feature sizes of 40 μm have been realized in an up to 1 mm thick resist. Due to the lack of a mask membrane, this mask has 100% transmission. However, designs are limited to mesh-type structures, requiring that the absorber structures are connected to each other. Instead of stencil masks with nearly vertically sidewalls, the same group used a mesh mask made from stainless-steel wire [100]. Due to the gradient thickness of the wire, there is a continuous increase in absorber thickness, resulting in a sloped resist sidewall. This effect is of interest especially for fabricating mold inserts with extremely tall microstructures where a sloped sidewall will support the demolding [101].

Flexible mask substrates can be realized with membranes made from polyimide [102, 103] or SU-8 [104]. Both polymers withstand a limited number of exposures and are therefore very suitable for prototyping. In the case of polyimide membranes, typically thin sheets between 25 and 50 μm thick are pre-stretched in a frame and metallized with a seed layer. After patterning of the photoresist structure and gold electroplating (these steps are similar to the corresponding steps in Figure 2.18), the final mask is mounted on a mask ring and used in X-ray exposures. Li and Sugiyama [102] used copper absorbers instead of gold and Kapton tape as flexible mask substrate. This tape can be wrapped round non-planar substrates, for example cylinders, allowing structures to be patterned on curved surfaces by X-ray lithography.

The use of thick silicon substrates for exposures with high-energy photons has been reported by Fischer *et al.* [105]. In this case, silicon wafers with thickness ranging between 100 and 500 μm were used as low-cost substrates and were processed similarly to X-ray masks made from graphite (Figure 2.18). The exposure results using a hard X-ray source (bending magnet at NSLS with a spectrum comparable to the CAMD wiggler beamline) indicated that the use of these masks is limited to hard exposure sources only. Silicon has also been used as both substrate and absorber pattern using a deep reactive ion etching (DRIE) process [106]. The authors calculated the silicon thickness for replacing the gold absorber, for example 6.7 μm of gold can be replaced with ~154 μm of silicon, and etched the corresponding pattern into the silicon using silicon dioxide as etch mask. The back side of the wafer was also etched, resulting in a thin, 20 μm thick silicone membrane. This mask has been used successfully for X-ray lithography exposures. They may be of interest for very small, sub-micrometer absorber structures in combination with soft X-ray lithography to fabricate sub-micrometer, high aspect ratio microstructures.

A final approach described here is the use of so-called conformal masks where the gold absorber is directly formed on top of the PMMA X-ray resist using SU-8 optical lithography and gold electroplating [107]. This mask offers the best performance in terms of proximity gap, relative shift between mask absorber and resist and mask substrate transmission (no substrate as the underlying X-ray resist forms the actual substrate). However, once an exposure is completed, the mask is of no use and has to be fabricated again for each new substrate.

2.4.7
Gold Electroplating for X-ray Masks

The primary absorber material for X-ray masks since the early days of LIGA has been electrodeposited gold. The requisite properties of the electroplated gold for X-ray masks are high density, fine grain and low stress. The electroplating solutions and operating conditions must also be compatible with a variety of photoresists and a multitude of mask substrate materials, as discussed previously.

Moderate pH, temperature and agitation are the preferred electroplating conditions and many commercial solutions compatible with the X-ray mask materials meet these requirements. The most common type of gold electroplating solution is the potassium gold(I) cyanide, $K[Au(CN)_2]$, bath, which was first developed in the 1840s [33]. However, this bath has poor compatibility with novolak photoresists. Since the addition of acid to cyanide solutions also produces hydrogen cyanide, a highly poisonous gas, inexperienced workers should avoid using it.

Sulfite-based electroplating solutions present safe alternatives to cyanide-based baths. Three types of commercially available sulfite-based $[(Au(SO_3)_2)^{3-}]$ electroplat-

Table 2.4 Commercial, sulfite-based gold electroplating solution used for X-ray mask fabrication.

	TECHNI-GOLD 25	TECHNI-GOLD 25 E	NEUTRONEX 309
Manufacturer	Technic, Inc.	Technic, Inc.	Enthone, Inc.
Gold concentration (g l^{-1})	8–16	8–16	8–12
pH	9.5–10.0	6.0–7.0 [6.5]	9.2–9.8
Specific gravity (Baumé)	10–30	10–30	10–30
Anode	Platinized electrode	Platinized electrode	Platinized electrode
Anode-to-cathode area ratio	≥1:1	≥1:1	≥2:1
Temperature (°C)	49–71	43–71	35–55
Agitation	Moderate to vigorous	Moderate to vigorous	Moderate
Current density (mA cm^{-2})	1.1–8.6	1.1–8.6	1.1–5.4

ing solutions and their baseline operating conditions are listed in Table 2.4. TECHNI-GOLD 25 (TG 25) and TECHNI-GOLD 25 E (TG 25E) are made and distributed by Technic (Cranston, RI, USA), and NEUTRONEX 309 is made by Enthone (West Haven, CT, USA) [108–110].

Gold electroplating systems typically use non-replenishing platinized anodes requiring periodic addition of gold salt to keep the concentration of the gold within the optimal range of operation. A simple method of determining the amount of gold plated out of the solution is to keep track of the total ampere-hours passed through the solution between gold replenishments. For the best results, the gold concentration needs to be kept in the range 90–100%. The pH of solutions is controlled by the addition of an appropriate base or acid solution. The concentration of sulfate in the baths increases over time, which is readily detected by an increase in the specific gravity. When the sulfate content exceeds specification, precipitation with calcium or barium acetate can be used to decrease the sulfate content [108] or the solution may simply be discarded.

The concentration of conductive salts must be monitored using titration and replenished to maintain chemical stability. The concentration of stabilizers in these solutions can be determined using UV absorption spectroscopy. Occasionally, organic contaminants, mainly from photoresist dissolution, cause a reduction in electroplating rate and a significant increase in stress. The solution can be cleaned by mixing activated carbon (100 g l^{-1}) into it, heating the mixture at 80 °C for 1 h and filtering it with a series of 5 and 0.5 μm pore PTFE filters after cooling it to room temperature. The activated carbon successfully absorbs organic contaminants and restores deposit characteristics to nominal.

In addition to high-density, highly pure gold deposits, a very uniform layer thickness ensuring sufficient contrast for every absorber structure is very important. Uniformity of gold thickness is influenced by a variety of factors, including the anode–cathode configuration, geometry and distribution of the features [111, 112], additives (leveling agents) [33] and electroplating mode (DC versus pulse; potentiostatic versus galvanostatic) [5, 113]. Although it is impractical to optimize all factors for every mask, varying the anode–cathode configuration and electroplating mode is sufficient to achieve good thickness uniformity.

Table 2.5 Factors affecting stress of gold electrodeposits, methods of measurement and control.

Parameter	Method of measurement	Method of control
Gold concentration	Track the total ampere-hours used	Replenish with gold salt
Conducting salt	Titration	Replenish with conducting salt
Additives	UV absorption, titration	Replenish with additives
Organic contaminants	Change in stress and/or plating rate	Filter with activated carbon
pH	Measure with pH meter	Adjust with base, acid or acid-adjusting salt
Specific gravity	Measure with hydrometer	Replace solution

The most important parameter for high-yield X-ray masks is internal stress in the gold absorber caused by several factors, including the mismatch in coefficient of thermal expansion between the absorber and mask substrate, inclusion of bath components including organic and inorganic contaminants in the deposits, phase transformation and growth dynamics. Excessively high stress may cause delamination of the absorber layer from the mask substrate, pattern distortion [114, 115] or even breakage of the mask substrate [97]. The methods of stress control include the use of additives, bath temperature, solution pH and plating mode (direct current versus pulse) [116]. Table 2.5 summarizes factors affecting the stress of gold electrodeposits and methods to measure and control it.

In 1909, Stoney published a method of stress measurement in electrodeposited thin films based on the deflection of beams on which the thin films have been deposited [117]. Stoney's method has resulted in what is commonly referred to as *Stoney's equation*:

$$\sigma_f = \frac{1}{6R} \frac{E_s t_s^2}{(1-v_s) t_f} \tag{8}$$

where σ_f is the stress in the deposited film, R is the radius of curvature, t_s is the thickness, E_s is the modulus of elasticity, v_s is Poisson's ratio of the substrate and t_f is the thickness of the deposited film [118]. Stoney's equation can also be expressed in terms of the tip deflection instead of the radius of curvature, which is the inverse of the second derivative of the deflection such that

$$R = \left(\frac{d^2 y}{dx^2}\right)^{-1} \Rightarrow y(x) = \frac{x^2}{2R} \tag{9}$$

For a beam of length L and small tip deflection δ, the radius of curvature is simply written as $R = L^2/2\delta$. Substituting for the radius of curvature in Equation 9 results in

$$\sigma_f = \frac{\delta}{3}\left(\frac{E_s t_s^2}{1-v_s}\right)\left(\frac{1}{t_f}\right) \tag{10}$$

For a cantilever beam of a given configuration, $E_s t_s^2/(1-v_s)$ does not vary, so the stress is a function of only the deflection of the beam and the thickness of the

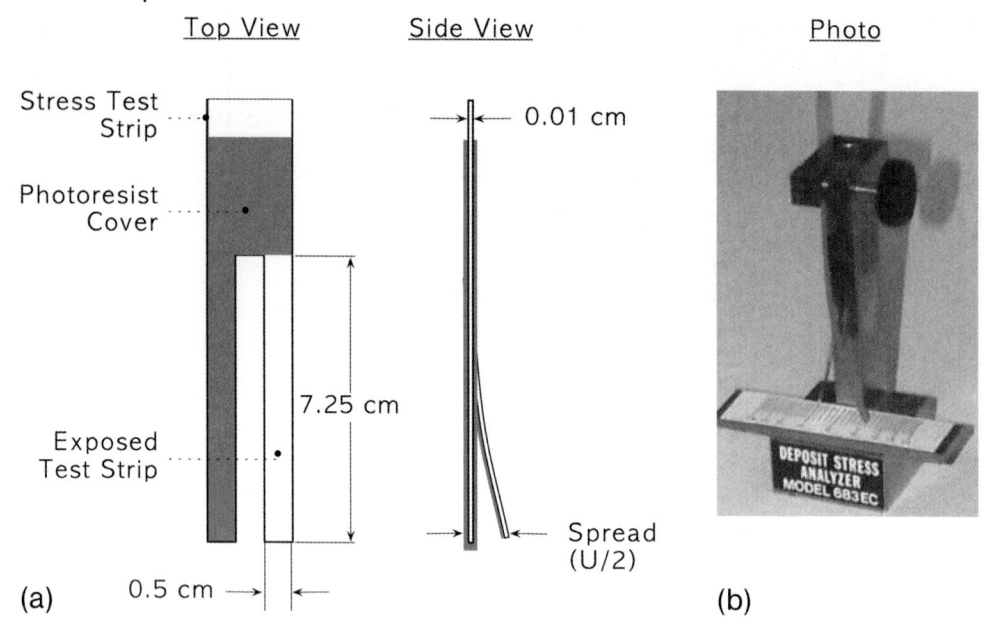

Figure 2.23 DSAS apparatus (Specialty Testing & Development, Jacobus, PA, USA): (a) schematic of beryllium–copper alloy test strip with only one side of one leg non-insulated; (b) Model 683 test stand used to measure the deflection of a test strip with a mounted strip.

film. This fact has been exploited by many to produce practical stress measurement techniques, one of which is the Stress Deposit Analyzer System (DSAS) (Specialty Testing & Development, Jacobus, PA, USA) – a simple and convenient method for measuring the average stress in an electrodeposited layer illustrated in Figure 2.23.

Standard test strips are insulated with photoresist everywhere except the alternating sides of the two legs on which thin films are deposited. Due to the stress, the strips deflect indicating compressive stress (away) when bending away and tensile stress when bending towards the electroplated film. The leg spread, U, is measured by use of the test stand and the reading is used to determine the stress. Stein et al. have shown that the resolution of the DSAS can range from 2 MPa for 8 μm thick films to 10 MPa for 4 μm films [116]. This simple device can be used to monitor stress in gold electrodeposited films. Studies of the solutions listed in Table 2.4 showed that internal stress in TG 25E (40 ppm arsenic) is very sensitive to current density ranging from −40 to 10 MPa. Stresses in NEUTRONEX 309 deposits range from only 2 to 12 MPa [92]. The stress for the NEUTRONEX 309 solution is shown as a function of current density and plating temperature in Figure 2.24 illustrating the importance of proper process parameter control. When deviations in the plating characteristics of the

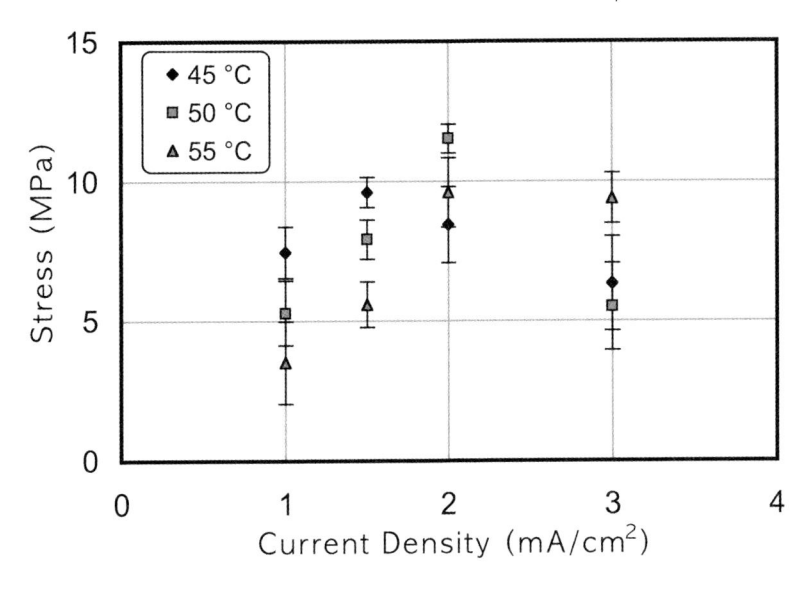

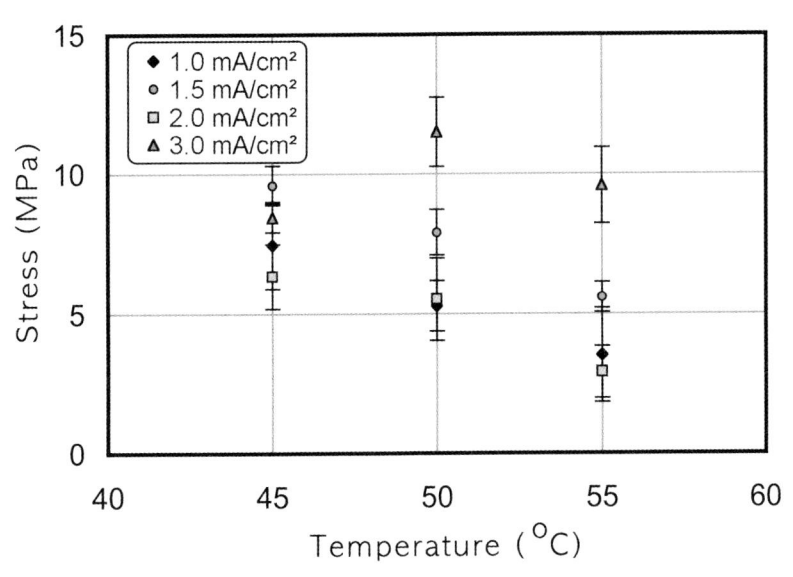

Figure 2.24 Internal stress in 2 μm thick gold electroplated using NEUTRONEX 309 with 30 ppm thallium grain refiner: (a) stress as a function of current density; (b) stress as a function of temperature.

solutions are encountered, restoring the original properties is often possible by cleaning of the organic contaminants using activated carbon filtration and replenishment of depleted components and typically allows bath operation for many months [92].

2.5
Summary and Conclusion

This chapter has described a number of different X-ray mask solutions, all of which are currently in use. The many existing X-ray mask material and fabrication options are a major reason why, despite great efforts within the LIGA community, commercialization of LIGA microparts is still accompanied by ongoing significant research and development efforts. In particular, high yield and reliable control of the critical dimensions of fabricated microparts is demanding and also suffers from a lack of 'standardized' X-ray masks [119, 120].

Fabrication processes range from simple, one-mask processes using thick UV lithography and gold electroplating, to complex two-mask processes combining e-beam and soft X-ray lithography with multiple electroplating steps. Combined with the substrate material choice, this determines the cost of X-ray masks, which can range between $3000 for a simple mask and up to $20 000 for a complex mask using e-beam writing. In addition to price, there is also a performance difference, ranging from low-resolution masks (CD of 10 µm) to high-resolution masks with sub-micrometer feature sizes. It is therefore suggested that LIGA users should clearly specify their designs, discuss their requirements with LIGA service providers and jointly determine the appropriate X-ray mask solution for their application.

Although masks made from beryllium promise the best performance, handling and costs are tremendous disadvantages and limit the use of these masks to demanding applications [6]. Masks made from titanium and silicon nitride membranes are typically used for high-quality applications and resist heights up to 500 µm. Design limiting factors are back-side etch windows which will reduce the open membrane area and also may cause deflection during X-ray exposure. Silicon nitride membranes offer the advantage of optical alignment windows without etching the membrane. Rigid graphite substrates are reasonable choices for low-cost X-ray masks but lack high-quality performance due to X-ray scattering in the substrate. Vitreous carbon is a promising carbon-based material but with higher costs and still ongoing research to improve the adhesion of the resist and gold absorber. Polymer-based substrate materials are another rapid prototyping approach which allow a limited number of exposures to make test samples. Although these materials are of low cost and many MEMS users are familiar with processing them, they are normally replaced with high-quality masks once a design has been finalized. Lastly, stencil masks can be easily made from wire meshes or by etching metal sheets, but are very limited in structure design. They do, however, promise large exposure areas that cannot be achieved with the commonly used X-ray mask substrates and are often used for rapid prototyping and optimization of process parameters.

Comparing lithographic performances of the different X-ray masks is not included in this chapter. A paper recently published by Aigeldinger et al. [121] provides a good summary of X-ray mask substrate materials and the effect on resist sidewall roughness in X-ray lithography exposure. From their Table 2.1, it can be concluded that vitreous carbon, beryllium and silicon demonstrate the best quality with mean average roughness (R_a) of around 20–30 nm. Similar values have also been reported for silicon nitride and titanium membrane masks. Rigid graphite has a significantly higher roughness (R_a = 300–400 nm) caused by X-ray scattering in the porous substrate. Another interesting result of their studies is the fact that low-z mask substrates, which provide higher X-ray transmission and require thinner gold absorber heights, show better process repeatability in the X-ray lithography process.

References

1 Becker, E.W., Ehrfeld, W., Münchmeyer, D., Betz, H., Heuberger, A., Pongratz, S., Glashauser, W., Michel, H.J. and Von Siemens, R. (1982) Production of separation-nozzle systems for uranium enrichment by a combination of X-ray lithography and galvanoplastics. *Naturwissenschaften (Historical Archive)*, **69**, 520–3.

2 Haghiri-Gosnet, A.M., Vieu, C., Simon, G., Carcenac, F., Madouri, A. and Chen, Y. (1995) Fabrication of sub-30 nm masks for X-ray nanolithography. *Journal of Vacuum Science Technology B*, **13**, 3066–9.

3 Ohki, S. and Yoshihara, H. (1990) High-precision X-ray mask technology. *Japanese Journal of Applied Physics*, **29**.

4 Schomburg, W.K., Baving, H. and Bley, P. (1991) Ti- and Be- X-ray masks with alignment windows for the LIGA process. *Microelectronic Engineering*, **13**, 323–6.

5 Desta, Y.M., Aigeldinger, G., Zanca, K.J., Coane, P.J., Goettert, J. and Murphy, M.C. (2000) Fabrication of graphite masks for deep and ultradeep X-ray lithography. *Proceedings of the SPIE – The International Society for Optical Engineering*, **4175**, 122–30.

6 Loechel, B., Goettert, J. and Desta, Y.M. (2007) Direct LIGA service for prototyping: status report. *Microsystem Technologies*, **13**, 327–34.

7 Coane, P., Giasolli, R., Carlo, F.D., Mancini, D., Desta, Y. and Goettert, J. (1998) Graphite-based X-ray masks for deep and ultra-deep X-ray lithography. *Journal of Vacuum Science and Technology*, **B16**, 3618–24.

8 Levinson, H.J. and Arnold, W.H. (1997) Optical lithography, in *SPIE Handbook of Microlithography, Micromachining and Microfabrication* (ed. P. Rai-Choudhury), Vol. **1**, Ch. 1, pp. 11–137.

9 (2007) A dedicated webpage addressing synchrotron radiation source issues and providing an overview and links to synchrotron laboratories worldwide is www.lightsources.org.

10 Koch, E.E. (ed.) (1983) *Handbook on Synchrotron Radiation*, North-Holland, Amsterdam, Vols **1–3**.

11 Anon. (1993) Summary of the history of X-ray lithography for VLSI application. *IBM Journal of Research and Development*, **37**, 287–474.

12 Cerrina, F. (1997) X-ray lithography, in *SPIE Handbook of Microlithography, Micromachining and Microfabrication* (ed. P. Rai-Choudhury), Vol. **1**, Ch. 3, pp. 251–319.

13 Kempson, V.C. et al. Experience of Routine Operation of Helios 1, Vol. **94**, EPAC.

14 Ehrfeld, W., Bley, P., Goetz, F., Mohr, J., Muenchmeyer, D. and Schelb, W. (1988) Progress in deep-etch synchrotron radiation lithography. *Journal of Vacuum Science Technology B*, **6**, 178–82.

15 Madou, M. (1997) In *Fundamentals of Microfabrication*, CRC Press, Boca Raton, FL, Ch. 6.

16 Christenson, T. (2005) X-ray based fabrication. *Micro/Nano R&D Magazine*, **10**.

17 BESSY Anwenderzentrum (2007) http://www.graphilox.de/azm/.

18 IMT at Forschungszentrum Karlsruhe (2007) http://www.anka-online.de/english/index.html.

19 IMT at Forschungszentrum Karlsruhe (2007) http://www.fzk.de/imt/.
20 (2007) Details of CAMD's LIGA beamlines are described at http://www.camd.lsu.edu/BEAMLINES.htm.
21 Kupka, R.K. (1996) Lithographie par rayons-X à trés haute résolution, PhD thesis, University of Paris XI.
22 Hubbell, J.H. and Seltzer, S.M. (2007) *Tables of X-Ray Mass Attenuation Coefficients and Mass Energy-Absorption Coefficients*, NIST, http://physics.nist.gov/PhysRefData/XRayMassCoef/cover.html.
23 Fischer, K., Chaudhuri, B., Stiers, E. and Guckel, H. (2000) Large area, cost effective X-ray masks for high energy photons. *Microsystem Technologies*, **6**, 117–20.
24 Maid, B. (1989) Anpassung der spektralen Verteilung der Synchrotronstrahlung fuer die Roentgentiefenlithographie, PhD thesis, University of Bonn.
25 CXRO (2007) *X-ray Optics Tools*, www-cxro.lbl.gov.
26 Lakowisc, J.R. (2006) *Principles of Fluorescence Spectrsocopy*, Springer, Berlin.
27 Pantenburg, F.J. and Mohr, J. (1995) Influence of seoncdary effects on the structure quality in deep X-ray lithography. *Nuclear Instruments and Methods in Physics Research B*, **97**, 551–6.
28 Pantenburg, F.J., Chlebek, J., El-Kholi, A., Huber, H.-L., Mohr, J., Oertel, H.K. and Schulz, J. (1994) Adhesion problems in deep-etch X-ray lithography by fluorescence radiation from the plating base. *Microelectronic Engineering*, **23**, 223–6.
29 Griffiths, S.K. and Ting, A. (2002) The influence of X-ray fluorescence on LIGA sidewall tolerances. *Microsystem Technologies*, **8**, 120–8.
30 Ehrfeld, W., Glashauser, W., Munchmeyer, D. and Schelb, W. (1986) Mask making for synchrotron radiation lithography. *Microelectronic Engineering*, **5**, 463–70.
31 Griffiths, S.K., Hruby, J.M. and Ting, A. (1999) The influence of feature sidewall tolerance on minimum absorber thickness for LIGA X-ray masks. *Journal of Micromechanics and Microengineering*, **9**, 353–61.
32 Fischer, K., Chaudhuri, B., Stiers, E. and Guckel, H. (2000) Large area, cost effective X-ray masks for high energy photons. *Microsystem Technologies*, **6**, 117–20.
33 Lowenheim, F.A. and Society, E. (1974) *Modern Electroplating*, 3rd edn, John Wiley & Sons, Inc., New York.
34 Ohki, S. and Yoshihara, H. (1990) High-precision X-ray mask technology. *Japanese Journal of Applied Physics*, **29**.
35 Sheu, J.T., Chiang, M.H. and Su, S. (1998) Fabrication of intermediate mask for deep x-ray lithography. *Microsystem Technologies*, **4**, 74–6.
36 Kadel, K., Schomburg, W.K. and Stern, G. (1993) X-ray masks with tungsten absorbers for use in the LIGA process. *Microelectronic Engineering*, **21**, 123–6.
37 Shew, B.Y., Cheng, Y., Shih, W.P., Lu, M. and Lee, W.H. (1998) High precision, low cost mask for deep X-ray lithography. *Microsystem Technologies*, **4**, 66–9.
38 Kebabi, B. and Malek, C.K. (1991) Fabrication of tensile stress minimized electroplated gold patterns for X-ray lithography masks. *Journal of Vacuum Science Technology B*, **9**, 154–61.
39 Honma, H. and Kagaya, Y. (1993) Gold plating using the disulfite–aurate complex. *Journal of the Electrochemical Society*, **140**, L135–7.
40 Horkans, J. and Romankiw, L.T. (1977) Pulsed potentiostatic deposition of gold from solutions of the Au(I) sulfite complex. *Journal of the Electrochemical Society*, **124**, 1499–505.
41 Smith, C.G. and Okinaka, Y. (1983) High speed gold plating: anodic bath degradation and search for stable low polarization anodes. *Journal of the Electrochemical Society*, **130**, 2149–57.
42 Li, Y.G. and Lasia, A. (1997) Electrodeposition of hard gold from acidic solution. *Journal of the Electrochemical Society*, **144**, 1979–87.
43 Wang, L., Christenson, T., Desta, Y.M., Fettig, R.K. and Goettert, J. (2004) High-resolution X-ray masks for high aspect ratio microelectromechanical systems applications. *Journal of Microlithography Microfabrication and Microsystems*, **3**, 423–8.

44 Desta, Y., Miller, H., Goettert, J., Stockhofe, C., Singh, V., Kizilkaya, O., Jin, Y., Johnson, D. and Webber, W. (2005) Deep X-ray lithography of SU-8 photoresist: influence of process parameters and conditions on microstructure quality. Proceedings of HARMST 2005, Gyeongju, South Korea.

45 Khounsary, A.M., Chojnowski, D., Mancini, D.C., Lai, B.P. and Dejus, R.J. (1997) Thermal management of masks for deep X-ray lithography. *Proceedings of the SPIE–The International Society for Optical Engineering*, **3151**, 92–101.

46 Synchrotron Radiation Center, Radiation Center providing more information about the lightsource, src.wisc.edu.

47 Vladimirsky, Y., Maldonado, J.R., Vladimirsky, O., Cerrina, F., Hanson, M., Nachman, R. and Wells, G.M. (1990) Thermoelastic effects in X-ray lithography masks during synchrotron storgae ring irradiation. *Microelectronic Engineering*, **11**, 287–93.

48 Zetter, T. (1994) *UETP–MEMS: Training in Microsystems. LIGA Technique*, Swiss Foundation for Research in Microtechnology. p. 140.

49 Khakani, M.A.E. and Chaker, M. (1993) Physical properties of the X-ray membrane materials. *Journal of Vacuum Science Technology B*, **11**, 2930–7.

50 Ehrfeld, W., Glashauser, W., Munchmeyer, D. and Schelb, W. (1986) Mask making for synchrotron radiation lithography. *Microelectronic Engineering*, **5**, 463–70.

51 Kouba, J., Ling, Z.-G., Wang, L., Desta, Y.M. and Goettert, J. (2003) Fabrication of large-area X-rays masks for UDXRL on beryllium using thin film UV lithography and X-ray backside exposure. *Proceedings of the SPIE–The International Society for Optical Engineering*, **5342**, 173–81.

52 POCO (2007) www.poco.com.

53 Harris, C., Desta, Y., Kelly, K.W. and Calderon, G. (1999) Inexpensive, quickly producable X-ray mask for LIGA. *Microsystem Technologie–Micro-and Nanosystems–Information. Storage and Processing Systems*, **5**, 189–93.

54 Desta, Y.M., Aigeldinger, G., Zanca, K.J., Coane, P.J., Goettert, J. and Murphy, M.C. (2000) Fabrication of graphite masks for deep and ultra-deep X-ray lithography. *Proceedings of the SPIE–The International Society for Optical Engineering*, **4175**, 122–30.

55 Divan, R., Mancini, D.C., Gallagher, S.M., Booske, J. and Van der Weide, D. (2004) Improvements in graphite-based X-ray mask fabrication for ultra-deep X-ray lithography. *Microsystem Technologies*, **10**, 728–34.

56 Scheunemann, H.U., Loechel, B., Jian, L., Schondelmaier, D., Desta, Y.-M. and Goettert, J. (2003) Cost effective masks for deep X-ray lithography. *Proceedings of the SPIE–The International Society for Optical Engineering*, **5116**, 775–81.

57 Jian, L., Loechel, B., Scheunemann, H.-U., Bednarzik, M. and Firsov, A. (2003) Vitreous carbon membrane X-ray masks for LIGA process. Proceedings of COMS 2003, Amsterdam.

58 Kouba, J., Scheunemann, H.-U., Waberski, C. and Rudolph, I. (2007) Advances in fabrication of X-ray masks based on vitreous carbon. Proceedings of HARMST 2007, Besancon, pp. 269–70.

59 Wang, L., Aristone, F., Goettert, J., Kong, J.R., Bradshaw, K., Christenson, T.R., Desta, Y.M. and Jin, Y. (2003) High resolution X-ray masks for high aspect ratio microelectromechanical systems (HARMS). *Proceedings of the SPIE–The International Society for Optical Engineering*, **4979**, 508–13.

60 Burbaum, C. *et al.* (1991) Fabrication of capacitive acceleration sensor by the LIGA technique. *Sensors and Actuators A*, **25**, 559–63.

61 Mohr, J., Burbaum, C., Bley, P., Menz, W. and Wallrabe, U. (1990) Movable microstructures manufactured by the LIGA process as basic elements for microsystems, in *Microsystem Technologies* (ed. H. Reichl), Springer, Berlin, pp. 529–37.

62 Desta, Y.M., Aigeldinger, G., Zanca, K.J., Coane, P.J., Goettert, J. and Murphy, M.C. (2000) Fabrication of graphite masks for deep and ultra-deep X-ray lithography. *Proceedings of the SPIE–The International*

Society for Optical Engineering, **4175**, 122–30.
63 Siddons, P., Johnson, E., Guckel, H. and Klein, J. (1997) Method and apparatus for micromachining using hard X-rays. US Patent 5 679 502.
64 For more details on the properties of Synchrotron radiation, see for example: Saile, V. (1983) Vorlesungsmanuskript of the 23. IFF-Ferienkurs, Forschungszentrum Jülich (1992), in *Handbook on Synchrotron Radiation* (ed. E.E. Koch), Vol. **1**, North-Holland, Amsterdam, pp. 11–28.
65 Ferguson, J.P. and Schoenfelder, S. (1999) Micromoulded spectrometers produced by the LIGA process. Colloquium on Microengineering in Optics and Optoelectronics, 1101–10.
66 Schulz, J., Bade, K., Guttmann, M., Hahn, L., Janssen, A., Koehler, U., Meyer, P. and Winkler, F. (2004) Ensuring repeatability in LIGA mold insert fabrication. *Microsystem Technologies*, **10**, 419.
67 Van Kan, J.A., Sanchez, J.L., Osipowicz, T. and Watt, F. (2000) Proton micromachining: a new technique for the production of three-dimensional microstructures. *Microsystem Technologies*, **6**, 82.
68 Ansari, K., Van Kan, J.A., Bettiol, A.A. and Watt, F. (2006) Stamps for nanoimprinting lithography fabricated by proton beam writing and nickel electroplating. *Journal of Micromechanics and Microengineering*, **16**, 1967–74.
69 Van Kan, J.A., Shao, P.G., Ansari, K., Bettiol, A.A., Osipowicz, T. and Watt, F. (2007) Proton beam writing: a tool for high-aspect ratio mask production. *Microsystem Technologies*, **13**, 431–4.
70 Gelorme, J.D., Cox, R.J. and Gutierrez, S.A.R. (1989) Photoresist composition and printed circuit boards and packages made therewith, US Patent 4 882 245.
71 Lorenz, H., Despont, M., Fahrni, N., LaBianca, N., Renaud, P. and Vettiger, P. (1997) SU-8: a low-cost negative resist for MEMS. *Journal of Micromechanics and Microengineering*, **7**, 121–4.
72 Rogers, A.K., Weber, K.E. and Hoffer, S.D. (1982) The alkaline peroxide prebon surface treatment of titanium: the development of a production process, in 27th National SAMPE Symposium, San Diego, pp. 63–72.
73 Mohr, J., Ehrfeld, W. and Muenchmeyer, D. (1988) Analyse der Defektursachen fuer die Genauigkeit der Strukturuebertragung bei der Roentgentiefenlithografie mit der Synchrotronstrahlung, KfK Report 4414, KfK, Karlsruhe.
74 Akkaraju, S. (1996) LIGA based fabrication of tips for scanning probe applications, Thesis, Louisiana State University, Baton Rouge, LA, p. 124.
75 Ehrfeld, W., Goetz, F., Muenchmeyer, D., Schelb, W. and Schmidt, D. (1998) LIGA process: sensor construction techniques via X-ray lithography, in IEEE Solid-State Sensor and Actuator Workshop, Hilton Head Island, pp. 1–4.
76 Kovacs, G.T.A. (1998) *Micromachined Transducers Sourcebook*, 1st edn, McGraw-Hill, New York.
77 Klein, J., Guckel, H., Siddons, D.P. and Johnson, E.D. (1998) X-ray masks for very deep X-ray lithography. *Microsystem Technologies*, **4**, 70–3.
78 Sheu, J.T., Chiang, M.H. and Su, S. (1998) Fabrication of intermediate mask for deep X-ray lithography. *Microsystem Technologies*, **4**, 74–6.
79 Wang, L., Zhang, M., Desta, Y., Melzak, J., Wu, C.H. and Peng, Z. (2006) A large format membrane-based X-ray mask for microfluidic chip fabrication. *Journal of Micromechanics and Microengineering*, **16**, 402–6.
80 Goettert, J., Datta, P., Desta, Y., Jin, Y., Ling, Z. and Singh, V. (2006) LIGA research and service at CAMD. *IOP Journal of Physics: Conference Series*, **34**, 912–18.
81 Sweatt, W.C. and Christenson, T. (2005) Method for the fabrication of three-dimensional microstructures by deep X-ray lithography. US Patent 6 875 544.
82 (2007) Silicon nitride membranes used at CAMD have been provided by DIMES, the Delft Institute of Microsystems and Nanoelectronics, http://www.tudelft.nl/live/pagina.jsp?id=9e21c86a-b526-41ae-bb68-8d548774acdb&lang=en.

References

83 Sarro, P.M., Van Hexwaarden, A.W. and Van der Vlist, W. (1994) A silicon–silicon nitride membrane fabrication process for smart thermal sensors. *Sensors and Actuators A*, **41–42**, 666–71.

84 Wu, M.-Y., Krapf, D., Zandbergen, M., Zandbergen, H. and Batson, P.E. (2005) Formation of nanopores in a SiN/SiO_2 membrane with an electron beam. *Applied Physics Letters*, **87** (113106), 1–3.

85 Kazinczi, R., Mollinger, J.R. and Bossche, A. (2000) 3-D resonator bridges as sensing elements, in SAFE-ProRisc-SeSens 2000, Veldhoven, The Netherlands.

86 Khakani, M.A.E. and Chaker, M. (1993) Physical properties of X-ray membrane materials. *Journal of Vacuum Science Technology B*, **11**, 2930–7.

87 Vladimirsky, Y., Maldonado, J.R., Vladimirsky, O., Cerrina, F., Hanson, M., Nachman, R. and Wells, G.M. (1990) Thermoelastic effects in X-ray lithography masks during synchrotron storgae ring irradiation. *Microelectronic Enigineering*, **11**, 287–93.

88 Ling, Z.-G., Lian, K. and Jian, L. (2000) Improved patterning quality of SU-8 microstructures by optimizing the exposure parameters. *Proceedings of the SPIE – The International Society for Optical Engineering*, **3999**, 1019–27.

89 (2007) Data sheets on different SU-8 products are available from MicroChem at http://www.microchem.com/products/su_eight.htm.

90 Engelke, R., Mathuni, J., Ahrens, G., Gruetzner, G. and Loechel, B. (2007) Investigations of SU-8 removal from metallic high aspect ratio microstructures with a novel plasma technique. Proceedings of HARMST 2007, Besancon, pp. 39–40.

91 Reznikova, E., Boerner, M., Jakobs, P.-J., Nazmov, V., Mohr, J. and Mappes, T. (2007) Soft X-ray lithography of high aspect ratio SU-8 submicron structures. Proceedings of HARMST 2007, Besancon, pp. 281–2.

92 Desta, Y.M. (2005) Fabrication of high aspect ratio vibrating cylinder microgyroscope structures by use of the LIGA process, PhD Thesis, Louisiana State University.

93 (2007) Details of beryllium products suitable for X-ray application are available from BrushWellman at www.berylliumproducts.com.

94 Kouba, J., Ling, Z.-G., Wang, L., Desta, Y.M. and Goettert, J. (2003) Fabrication of large-area X-rays masks for UDXRL on beryllium using thin film UV lithography and X-ray backside exposure. *Proceedings of the SPIE – The International Society for Optical Engineering*, **5342**, 173–81.

95 Kupka, R., Megtert, S., Roulliay, M. and Bouamrane, F. (1998) Tranparent masks for aligned deep X-ray lithography/LIGA: low-cost well-performing alternative using glass membranes. *Proceedings of the SPIE – The International Society for Optical Engineering*, **3512**, 271–6.

96 Jian, L.K., Loechel, B., Bednarzik, M., Scheunemann, H., Schondelmaier, D. and Firsov, A. (2003) Improvement in cost-effective X-Ray mask fabrication. Proceedings of HARMST 2003, Monterey, CA.

97 Desta, Y.M., Bednarzik, M., Bryant, M.D., Goettert, J., Jian, L., Jin, Y., Kim, D., Lee, S., Loechel, B., Scheunemann, H.-U. and Peng, Z. (2003) Borosilicate glass based x-ray masks for LIGA microfabrication. *Proceedings of the SPIE – The International Society for Optical Engineering*, **4979**, 514–22.

98 Koukharenko, E., Kraft, M., Ensell, G.J. and Hollinshead, N. (2005) A comparative study of different thick photoresists for MEMS applications. *Journal of Materials Science: Materials in Medicine*, **16**, 741–7.

99 Utsumi, Y., Kishimoto, T., Hattori, T. and Hara, H. (2007) Large area and wide dimensions X-ray lithography using energy variable synchrotron radiation. *Microsystem Technologies*, **13**, 417–23.

100 Mekaru, H., Utsumi, Y. and Hattori, T. (2002) Quasi-3D microstructure fabrication technique utilizing hard X-ray lithography of synchrotron radiation. *Microsystem Technologies*, **9**, 36–40.

101 Becnel, C., Desta, Y. and Kelly, K. (2005) Ultra-deep x-ray lithography of densely packed SU-8 features (Pt 1 and 2). *Journal of Micromechanics and Microengineering*, **6**, 1242.

102 Li, Y. and Sugiyama, S. (2007) Fabrication of microgratings on PMMA plate and curved surface by using copper mesh as X-ray lithography mask. *Microsystems Technologies*, **13**, 227–03.

103 Achenbach, S., Boerner, M., Kinuta, S., Bacher, W., Mohr, J., Saile, V. and Saotome, Y. (2007) Structure quality in deep X-ray lithography applying commercial polyimide-based masks. *Microsystem Technologies*, **13**, 349–53.

104 Cabrini, S., Perennes, S., Marmiroli, F., Olivo, B., Carpentiero, A., Kumar, A., Candeloro, R., Fabrizio, P. and Di, E. (2005) Low cost transparent SU-8 membrane mask for deep X-ray lithography. *Microsystem Technologies*, **11**, 370–3.

105 Fischer, K., Chaudhuri, B., Stiers, E. and Guckel, H. (2000) Large area, cost effective X-ray masks for high energy photons. *Microsystem Technologies*, **6**, 117–20.

106 Chen, D., Lei, W., Li, C., Guo, X., Mao, H., Zhang, D. and Yi, F. (2001) New type X-ray mask fabricated using inductively coupled plasma deep-etching. *Microsystem Technologies*, **7**, 71–4.

107 Lee, S., Kim, D., Jin, Y., Han, Y., Desta, Y.M., Bryant, M.D. and Goettert, J. (2004) A micro corona motor fabricated by an SU-8 built-on X-ray mask. *Microsystem Technologies*, **10**, 522–6.

108 (1996) TECHNI-GOLD 25 E Technical Data Sheet, Technic Inc., Cranston, RI.

109 (1995) TECHNI-GOLD 25 Technical Data Sheet, Technic, Inc., Cranston, RI.

110 (2003) NEUTRONEX® 309 Gold Electroplating Process Data Sheets, Enthone, Inc., West Haven, CT.

111 DeBecker, B. and West, A.C. (1996) Workpiece, pattern and feature scale current distributions. *Journal of the Electrochemical Society*, **143**, 486–92.

112 Dukovic, J.O. (1993) Feature-scale simulation of resist-patterned electrodepositon. *IBM Journal of Research and Development*, **37**, 125–41.

113 El-Sherik, A.M. and Erb, U. (1995) Synthesis of bulk nanocrystalline nickel by pulsed electrodeposition. *Journal of Materials Science*, **30**, 5743–9.

114 Muller, K.H., Toischer, P. and Windbrake, W. (1985) Influence of absorber stress on the precision of X-ray masks. Proceedings of the International EIPBN Conference, Portland, OR.

115 Acosta, R.E., Maldonado, J.R., Fair, R., Viswanathan, R. and Wilson, A.D. (1985) Distortion of masks for X-ray lithography. *Microelectronic Engineering*, **3**, 615–21.

116 Stein, B. (2000) Fast and accurate deposit internal stress determination. Surface Finishing 2000, Session N: Quality in Surface Finishing, Navy Pier, Chicago, IL.

117 Stoney, G.G. (1909) The tension of metallic films deposited by electrolysis. *Proceedings of the Royal Society of London*, **A82**, 172–5.

118 Ohring, M. (1991) *The Materials Science of Thin Films*, Vol. **20**, Academic Press, Boston.

119 Hahn, L., Meyer, P., Bade, K., Hein, H., Schulz, J., Loechel, B., Scheunemann, H.-U., Schondelmaier, D. and Singleton, L. (2005) MODULIGA: the LIGA process as a modular production method-current standardization status in Germany. *Microsystem Technologies*, **11**, 240–5.

120 Guckel, H., Fischer, K., Stiers, E., Chaudhuri, B., McNamara, S., Ramotowski, M., Johnson, E.D. and Kirk, C. (2000) Direct, high throughput LIGA for commercial applications: a progress report. *Microsystem Technologies*, **6**, 103–5.

121 Aigeldinger, G., Yang, C.-Y.P., Skala, C.-Y.P., Morse, D.M., Talin, D.H., Griffiths, A.A., Hachman, S.K. and Ceremuga, J.T. (2007) Influence of mask substrate materials on resist sidewall roughness in deep X-ray lithography. *Microsystem Technologies*, **14**, 277–86.

3
Innovative Exposure Techniques for 3D Microfabrication

Naoki Matsuzuka and Osamu Tabata

3.1 **Introduction** *51*
3.2 **Fundamentals of DXRL** *53*
3.2.1 X-ray Exposure Process *53*
3.2.2 Development Process *53*
3.3 **Innovative Exposure Techniques for 3-D Microfabrication** *55*
3.3.1 Technique Based on the Concept of Conventional DXRL *55*
3.3.1.1 Use of 3D Resist Substrates *55*
3.3.1.2 Inclined Exposure *57*
3.3.2 Technique to Yield Two-step Dose Distribution *60*
3.3.2.1 Use of an X-ray Mask with Low Contrast *61*
3.3.2.2 Double Exposure *62*
3.3.3 Technique to Yield a Multi-step or Arbitrary Dose Distribution *63*
3.3.3.1 Use of an X-ray Mask with Absorber Thickness Distribution *63*
3.3.3.2 Planar Movement of an X-ray Mask or a Resist Substrate *63*
3.3.4 Combination of Innovative Exposure Techniques *66*
3.4 **Conclusions** *67*
References *67*

3.1
Introduction

Microstructures several hundred micrometers or more in height have been demanded in order to progress microelectromechanical systems (MEMS). The LIGA process combined deep X-ray lithography (DXRL) based on synchrotron radiation with electroplating and molding technologies can mass-produce tall precision microstructures in various materials, for example, plastics, metals and ceramics [1]. In the LIGA process, DXRL is a primary technology to determine the structural dimensions and shapes of the microproducts.

DXRL is the most promising technology to fabricate tall resist microstructures with vertical sidewalls because the synchrotron-radiated X-rays are highly collimated and directly penetrate a very thick resist substrate. Resist microstructures 1 mm in height have already been realized due to these properties of the X-rays (e.g. [2, 3]). X-rays, in addition, are hard to diffract by absorber edges of an X-ray mask, which enables DXRL to realize very fine resist microstructures of sub-micrometer width (e.g. [4]). As a result, DXRL is good at fabricating precision resist microstructures with high aspect ratios. However, the sidewalls of the fabricated microstructures are always vertical to the resist surface, which can fabricate only the microstructures with a two-dimensional (2D) pattern with a thickness. The sidewall inclination and curvature of the cross-sectional shape cannot be controlled arbitrarily. This uncontrollability of the sidewalls has narrowed the application filed of DXRL and the LIGA process.

From the viewpoint of fabricating microstructures with high aspect ratio, deep reactive ion etching (DRIE) and UV lithography using a negative-tone resist named SU-8TM are also promising technologies. For DXRL, it was necessary to define its advantages compared with those of micromachining technologies. One of the measures is the demonstration of the fabrication of truly three-dimensional (3D) precision microstructures with inclined and freely shaped sidewalls because with DRIE and UV lithography it is difficult to control the sidewalls arbitrarily.

To realize truly 3D microstructures with inclined and freely shaped sidewalls, various innovative exposure techniques based on DXRL have been proposed and carried out by a number of research groups. These techniques can be roughly classified into following categories:

(1) Techniques based on the concept of conventional DXRL:
    - 'Use of 3D resist substrates' instead of 2D types [5–7].
    - 'Inclined exposure', tilting and rotating an assembly of an X-ray mask and a resist substrate to incident X-rays [8–10].
(2) Techniques to yield two-step dose distribution:
    - 'Use of an X-ray mask with low contrast' [11, 12].
    - 'Double exposure' with and without an X-ray mask [13].
(3) Techniques to yield multi-step or arbitrary dose distribution:
    - 'Use of an X-ray mask with distributed absorber thickness' [14, 15].
    - 'Plane movement of an X-ray mask or a resist substrate' during X-ray exposure [16–18].
(4) Combination of the innovative exposure techniques.

These techniques improved the flexibility of the fabricated structural shape of DXRL and the LIGA process. DXRL and the LIGA process can be expected to realize microcomponents and MEMS which have not easily been produced by the other micromachining technologies.

The important point of 3D microfabrication employing DXRL is an understanding of the development behavior of the exposed resist substrate. After describing the fundamentals, including the development behavior in DXRL, the innovative exposure techniques for 3D microfabrication are reviewed in this chapter.

3.2
Fundamentals of DXRL

Resist microstructures are fabricated by the X-ray exposure and the development processes in DXRL. In the X-ray exposure process, an X-ray mask divides an X-ray-sensitive resist into irradiated and non-irradiated regions. Only the irradiated regions change resist solubility to a solvent by the change in molecular weight. From the difference in solubility between irradiated and non-irradiated regions, the absorber pattern of the X-ray mask is transferred into the resist substrate. In the subsequent development process, either irradiated or non-irradiated regions are dissolved in a developer (solvent), depending on the polarity of the resist. The important point to note is that the solubility, namely the dissolution rate of the resist, depends on the absorbed dose.

3.2.1
X-ray Exposure Process

The absorbed dose D as a function of the depth z is expressed by following equation.

$$D(z) = E \int_\lambda D_S(\lambda) \exp[-\alpha(\lambda)z] d\lambda \qquad (1)$$

where λ = wavelength, E = exposure dose, $D_S(\lambda)$ = spectrum of the synchrotron radiation light on the resist surface per unit exposure dose and $\alpha(\lambda)$ = absorption coefficient of the resist. From Equation 1, it is obvious that the absorbed dose decrease in the z-direction.

In conventional DXRL, the primarily used X-ray-sensitive resist has been poly(methyl methacrylate) (PMMA) because of its high resolution. PMMA functions as a positive tone resist and the irradiated regions are dissolved in the subsequent development process. To dissolve the irradiated regions of PMMA completely, the exposure dose should be determined to be the absorbed dose at the bottom of the PMMA substrate sufficient to effect dissolution. On the other hand, an excess exposure dose results in damage to the PMMA such as bubble formation, which adversely affects the quality of the microstructures. It is important to determine the appropriate exposure dose to dissolve the irradiated region of PMMA without any damage in every DXRL experiment.

3.2.2
Development Process

Immersion of the exposed PMMA substrate in a developer dissolves the irradiated regions from the PMMA surface. The development behavior is shown schematically in Figure 3.1. As the absorbed dose increases, the PMMA is dissolved more rapidly. The propagation rate of the interface between the PMMA substrate and the developer decreases with increasing processed depth according to the Equation

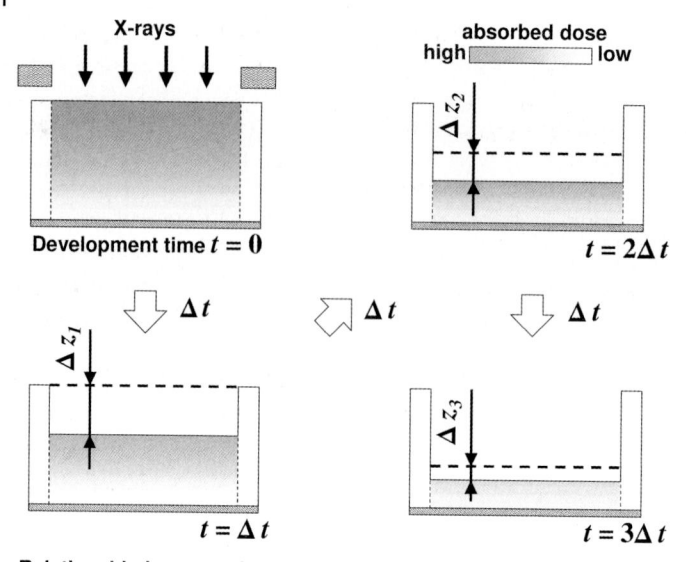

Figure 3.1 Development behavior of the PMMA substrate.

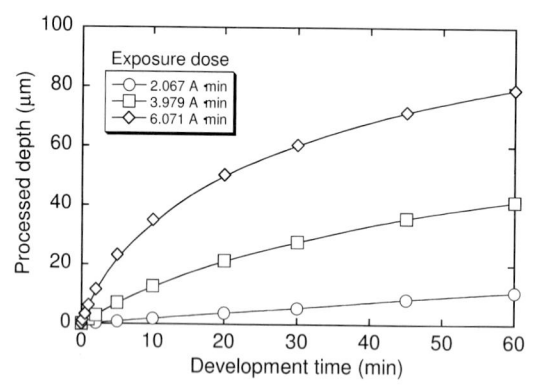

Figure 3.2 Relationship among the exposure dose, the development time and the processed depth.

1. In addition, the development time to process the required depth shortens with increase in the exposure dose under the same development conditions.

The experimental results for the relationship among the exposure dose, the development time and the processed depth are shown in Figure 3.2. These experiments were carried out at beamline 15 of the superconducting compact storage ring AURORA at Ritsumeikan University, Japan [19]. G-G developer, which is a mixture of three organic solvents and deionized water, is used primarily as a developer for DXRL because of the high quality of the developed PMMA microstructures [2, 20]. After exposing a PMMA sheet (CLAREX, Nitto Jushi Kogyo,

Tokyo, Japan) with a stencil mask, the exposed PMMA sheet was developed using G-G developer at 39 °C without stirring. These results clearly show the development behavior mentioned above.

The dissolution rate R as a function of the absorbed dose D is expressed by the following equation [21]:

$$R(D) = KD^{\beta} \tag{2}$$

The parameters K and β are characteristic constants depending on the polymer and the solvent. From Equations 1 and 2, the function $R(z)$ under an arbitrary exposure dose can be derived. The development behavior of conventional DXRL, which shows the relationship between the development time t and the processed depth z under the arbitrary exposure dose, is expressed by following equation:

$$t = \int_0^z R(z)^{-1} dz \tag{3}$$

This equation is useful for determining the experimental conditions, namely the exposure dose and the development time, in conventional DXRL. The development behavior is caused by the dependence of the dissolution rate on the absorbed dose, which has cleverly been applied to the 3D microfabrication of DXRL.

3.3 Innovative Exposure Techniques for 3-D Microfabrication

3.3.1 Technique Based on the Concept of Conventional DXRL

Conventional DXRL removes either irradiated or non-irradiated regions of the resist substrate. Two innovative exposure techniques applying this concept of conventional DXRL are described. One is a technique to use 3D resist substrates instead of planer ones. Another is a technique to control the angle of the incident X-rays to the resist substrate. The advantage of these techniques is predictability of the processed structural shape.

3.3.1.1 Use of 3D Resist Substrates

The X-ray-sensitive resist generally used is planar. Conventional DXRL realizes microstructures with a 2D absorber pattern with a thickness using the planar resist substrates. DXRL also has the advantage of transferring the absorbed pattern into 3D resist substrates with high accuracy because this technology has a wide focus range in the depth direction.

Utilizing this advantage, exposures of non-planar resist substrates can be carried out (e.g. [5–7]). In [5], a PMMA substrate with two-step height was fabricated using the LIGA process and then irradiated with X-rays as shown in Figure 3.3. The fabricated two-step PMMA microstructure is shown in Figure 3.4. In [6], the cylindrical PMMA formed on a core material was used to fabricate the cylindrical

3 Innovative Exposure Techniques for 3D Microfabrication

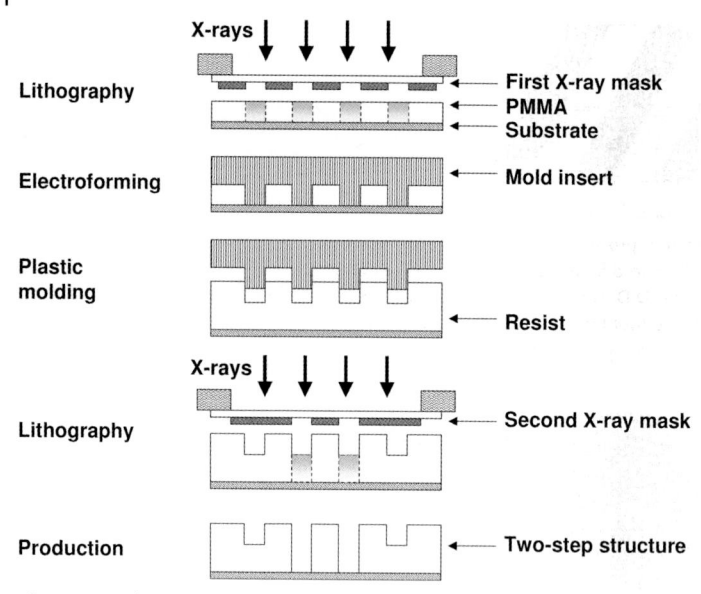

Figure 3.3 Fabrication process of the two-step microstructures.

Figure 3.4 SEM image of the two-step microstructure.
M. Harmening, W. Bacher, P. Bley, A. El-Kholi, H. Kalb, B. Kowanz, W. Menz, A. Michel, J. Mohr. Molding of three-dimensional microstructures by the LIGA process, Proc. of Micro Electro Mechanical Systems, © [1992] IEEE.

microstructure with two different diameters. The fabrication process is shown schematically in Figure 3.5. The cylindrical PMMA and the X-ray mask with line and space pattern were fixed to the apparatus and the PMMA substrate was rotated during the X-ray exposure, which uniformly irradiated the whole PMMA surface. The fabricated cylindrical microstructure with two different diameters is shown in Figure 3.6. As another complex microstructure, a helical microstructure of PMMA has been fabricated successfully [7].

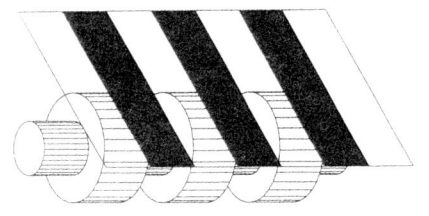

Figure 3.5 Fabrication process of the cylindrical microstructure with two different diameters. A.D. Feinerman, R.E. Lajors, V. White, D.D. Denton, X-ray lathe: an X-ray lithographic exposure tool for nonplanar objects, Journal of Microelectromechanical Systems, © [1996] IEEE.

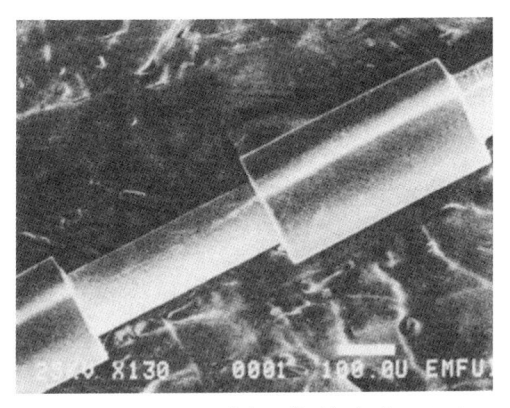

Figure 3.6 SEM image of the cylindrical microstructure with two different diameters. A.D. Feinerman, R.E. Lajors, V. White, D.D. Denton, X-ray lathe: an X-ray lithographic exposure tool for nonplanar objects, Journal of Microelectromechanical Systems, © [1996] IEEE.

The sidewalls of the microstructures are vertical as well as in conventional DXRL, and this technique can realize particular 3D microstructures which cannot be fabricated with the planar resist substrate.

3.3.1.2 Inclined Exposure

Incident X-rays in conventional DXRL are vertical to an assembly of the X-ray mask and resist substrate and go straight into the resist, which causes the sidewalls of the fabricated microstructures to be vertical. Put simply, the angle of the sidewalls is same as that of the incident X-rays. A technique to tilt and rotate the assembly of the mask and the resist to the X-rays as shown in Figure 3.7 can realize microstructures with inclined sidewalls.

In [8], the microstructure with sidewalls of multiple angles shown in Figure 3.8 was fabricated by multiple tilted exposures using a negative tone resist. It is expected that this microstructure can be applied as a photonic crystal in the optical MEMS field. In [9], the three-facet PMMA microstructure shown in Figure 3.9 was

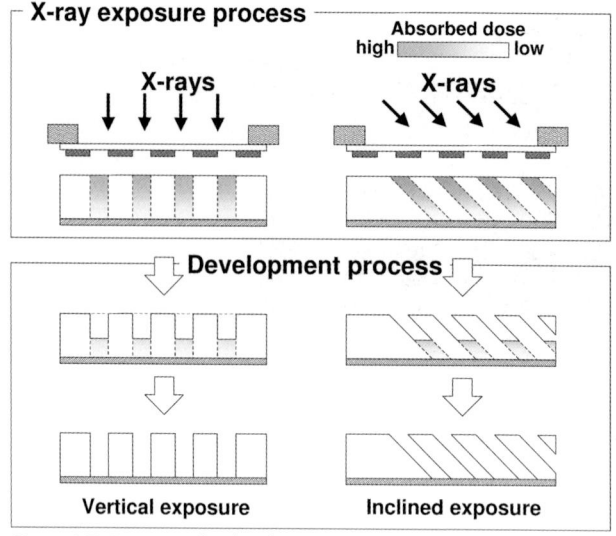

Figure 3.7 Concept of inclined exposure.

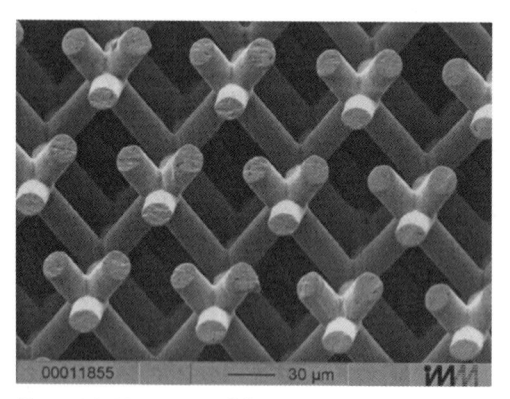

Figure 3.8 SEM image of the microstructure with sidewalls of multiple angles. Reused with permission from G. Feiertag, W. Ehrfeld, H. Freimuth, H. Kolle, H. Lehr, M. Schmidt, M.M. Sigalas, C.M. Souhoulis, G. Kiriakidis, T. Pedersen, J. Kuhl, and W. Koenig, Applied Physics Letters, 71, 1441 (1997). © [1997], American Institute of Physics.

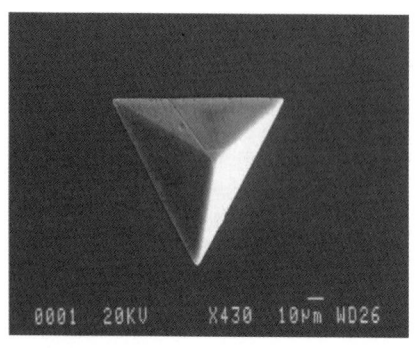

Figure 3.9 SEM image of the three-facet microstructure.

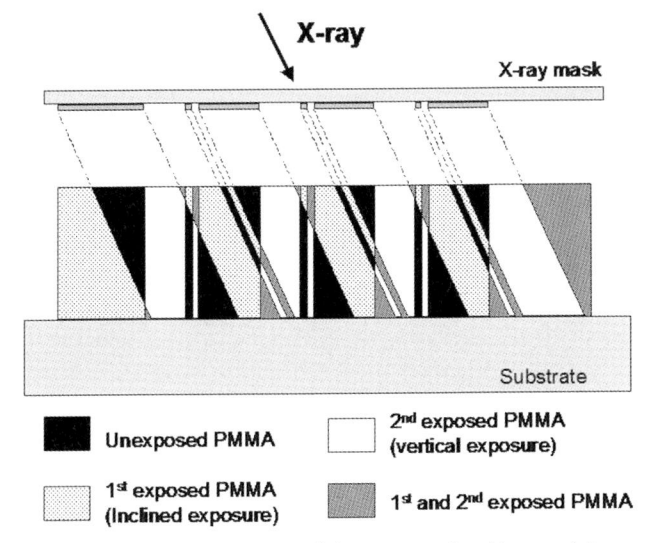

Figure 3.10 Fabrication process of the microneedle with a conduit.

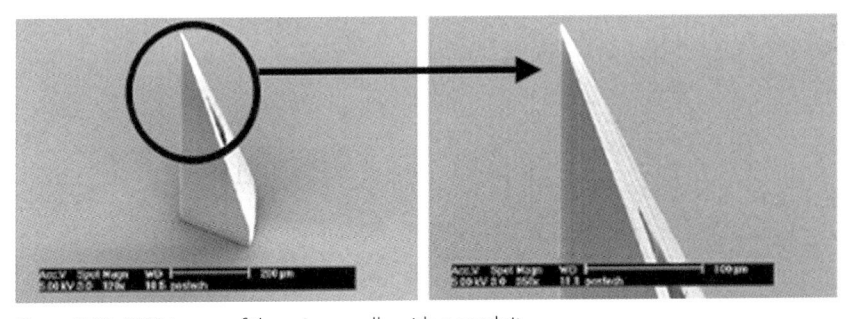

Figure 3.11 SEM image of the microneedle with a conduit.

fabricated by two exposures with an inclined angle of $\tan^{-1}(2°)$ and 45° using an X-ray mask with two equilateral triangular patterns. The surface flatness and roughness of the inclined sidewalls depend on the ability of X-rays to go straight and the line of the absorber edges, respectively. In this fabrication, the flatness and roughness were only a few hundred ångstroms. It is expected that this microstructure can be applied as a three-directional laser beam reflector. In [10], the fabrication of a microneedle with a conduit was carried out by vertical and successive inclined X-ray exposures using the X-ray mask which has a triangular pattern with a circular hole. The fabrication process is shown schematically in Figure 3.10. The successfully fabricated PMMA microneedle is shown in Figure 3.11. It is expected that the microneedle can be applied to blood extraction and drug delivery systems.

In this technique, the removable regions of the resist can be controlled with the parameters of the inclined angle, the gap between the mask and the resist, the absorber pattern and the polarity of the resist, and are easily predicted, which enables these parameters to be easily determined for the target 3D microstructure. The sidewalls of the fabricated microstructures are very flat and smooth.

3.3.2
Technique to Yield Two-step Dose Distribution

The development behavior of the PMMA substrate given a two-step dose distribution is described. This dose is the integrated spectrum on the PMMA surface with respect to the wavelength. The absorbed dose of the cross-section given a two-step dose distribution is shown schematically in Figure 3.12. For simplification of the description, the behavior is limited to two dimensions, namely vertical and lateral. Since the dissolution rate of PMMA depends on the absorbed dose, regions A and B in Figure 3.12 start to dissolve at different rates towards the depth direction by immersion in the developer. The difference in the dissolution rates between regions A and B leads to step formation at the boundary. By exposure of the sidewalls of the step to the developer, the interface between the PMMA substrate and the developer starts to propagate in the lateral direction from the sidewalls of the step. The upper part of the step dissolves faster than the lower part because of the difference in the absorbed dose and is exposed to the developer earlier. These effects form the inclined and curved sidewalls. The direction of propagation of the minute interface is normal to its interface. By the formation of the inclined and curved sidewalls, every point on the interface propagates in each direction.

The prediction method for the processed structural shape identifying the propagated interface is necessary to determine the experimental condition. The development behavior in this technique is more complex than that in conventional DXRL, which

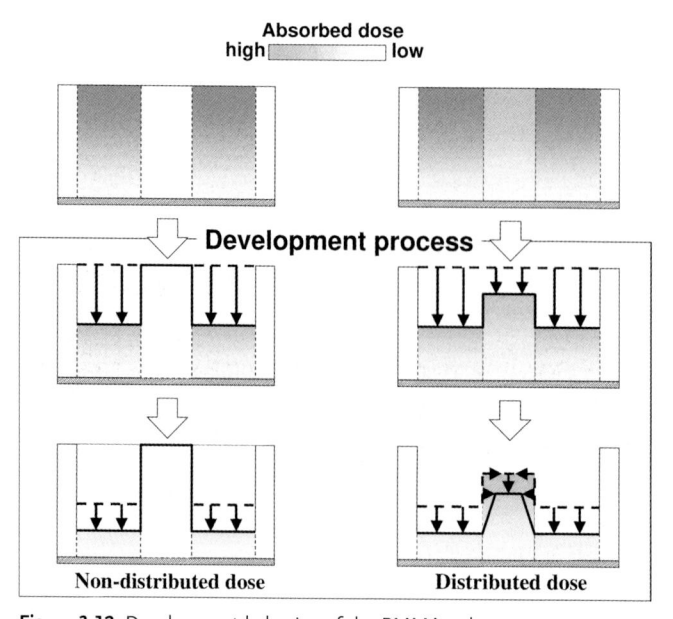

Figure 3.12 Development behavior of the PMMA substrate given a two-step dose distribution.

makes it difficult to predict precisely the processed structural shape. The calculation method for the shape by combining the calculated vertical and lateral directions can be adopted as a simple prediction method. The vertical and lateral propagations are expressed by the following equations using the functions $R_A(z)$ and $R_B(z)$ indicating the dependences of the dissolution rate on the depth at regions A and B:

- Vertical propagation:

$$t = \int_0^z R_A(z)^{-1} dz \quad \text{for region A}$$
$$= \int_0^z R_B(z)^{-1} dz \quad \text{for region B}$$
(4)

- Lateral propagation:

$$x = R_B(z)\left[t - \int_0^z R_A(z)^{-1} dz\right]$$
(5)

3.3.2.1 Use of an X-ray Mask with Low Contrast

A thin absorber of the X-ray mask realizes a two-step dose distribution on the resist surface owing to the transmission of X-rays. In one of the main processes for the fabrication of an X-ray mask, the absorbers of the X-ray mask are formed in the openings of the developed resist by an electroplating technique. The electroplating technique can control the absorber thickness, namely the contrast of the X-ray mask, without any difficulty. The 3D PMMA microstructures with inclined and curved sidewalls fabricated using X-ray masks with low contrast are shown in Figure 3.13 [11, 12]. The fabricated microstructure in [12] agreed well with the predicted structural shape using Equations 4 and 5.

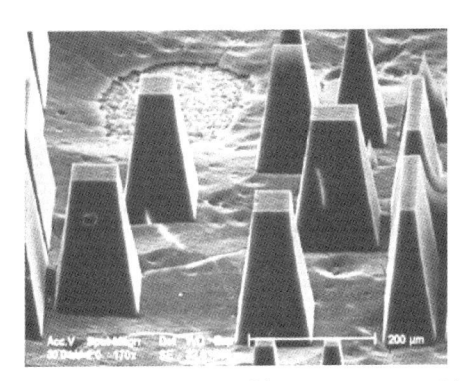

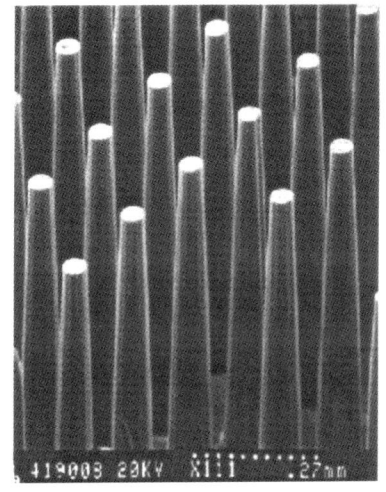

Figure 3.13 SEM image of the microstructure with inclined and curved sidewalls. Z. Liu, F. Bouamrane, M. Roulliay, P.K. Kupha, A. Labèque, and S. Megtert, Resist dissolution rate and inclined-wall structures in deep X-ray lithography, Journal of Micromechanics and Microengineering, 1998, Institute of Physics.

Although the X-ray mask suitable for the target structural shape is required, this technique can realize 3D microstructures with inclined and curved sidewalls by only a single exposure.

3.3.2.2 Double Exposure

Double exposure with and without an X-ray mask also realizes a two-step dose distribution [13]. This technique is shown schematically in Figure 3.14. The two-step dose distribution can be controlled by the combination of exposure doses with and without the X-ray mask. Two kinds of PMMA microprojections fabricated by two different combinations of exposure doses are shown in Figure 3.15. From the

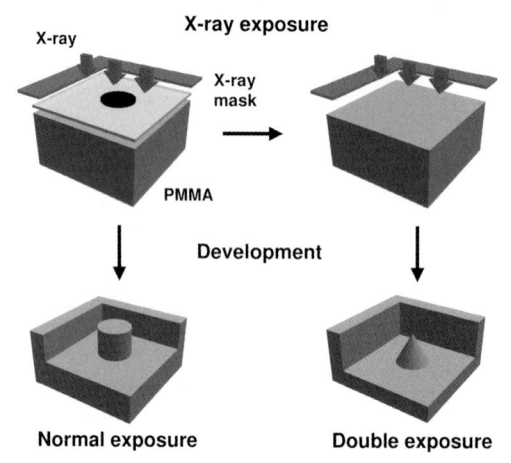

Figure 3.14 Concept of the double exposure technique.

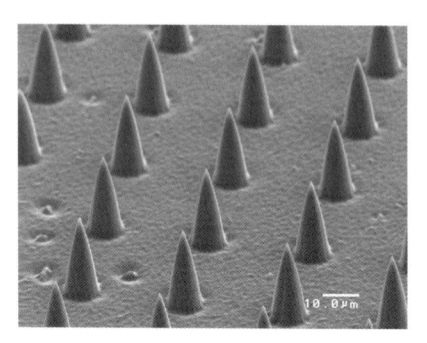

Exposure dose with mask
3.0 A·min
Exposure dose without mask
2.0 A·min
Height: 22.1 μm

Exposure dose with mask
3.0 A·min
Exposure dose without mask
3.0 A·min
Height: 15.9 μm

Figure 3.15 SEM images of microprojection arrays.

experimental results, it is obvious that the structural shape is controlled by the combination.

This technique requires two exposures, but variously shaped 3D microstructures with inclined and curved sidewalls can flexibly be fabricated by controlling the experimental parameters, with a normal X-ray mask.

3.3.3
Technique to Yield a Multi-step or Arbitrary Dose Distribution

The techniques to yield a multi-step or arbitrary dose distribution can control the structural shape more flexibly than the techniques mentioned in Section 3.2. The processed depth is determined by the minimum absorbed dose which can dissolve PMMA. The depth of the minimum absorbed dose in PMMA depends on the dose on the PMMA surface. The processed structural shape can be realized flexibly by the control of the dose distribution. However, it takes too much time to remove completely the PMMA region above the minimum absorbed dose in the development process. The prolonged development time leads to serious solvent attack. The structural shape, in addition, is not predicted easily with high accuracy before the interface between the PMMA substrate and the developer reaches the face formed by the minimum absorbed dose. As for this prediction method of 3D structural shape, see Chapter 5.

3.3.3.1 Use of an X-ray Mask with Absorber Thickness Distribution
The dose distribution can be realized by the absorber thickness distribution of the X-ray mask. The multi-step absorber thickness of the mask can be fabricated by the repetition of the lithography with the mask alignment and electroplating procedures [14]. An X-ray mask with a continuously distributed absorber thickness also can be fabricated by 3D micromachining of the resist in the lithographic procedure. The cone microstructure fabricated by this technique is shown in Figure 3.16. It is expected that the 3D microstructure can be applied as a lighting panel.

Although the X-ray mask suitable for the target structural shape has to be fabricated as well as the technique mentioned in Section 3.3.2.1, this technique can flexibly realize 3D microstructures with inclined and curved sidewalls by a single exposure. In this technique, precise control of the absorber thickness distribution is the important factor for realizing the target microstructures.

3.3.3.2 Planar Movement of an X-ray Mask or a Resist Substrate
A technique to move either the X-ray mask or the resist substrate shown in Figure 3.17 during the X-ray exposure can also realize the dose distribution [16–18]. In this technique, the dose distribution can be controlled by the mask pattern and the mask/resist movement patterns.

First, 3D microfabrication using the technique to control the mask movement pattern without the design of the mask pattern is described [16, 17]. The mask used has a pattern of a basic figure such as a circle or quadrilateral. A microlens array

Figure 3.16 SEM image of the cone microstructure.

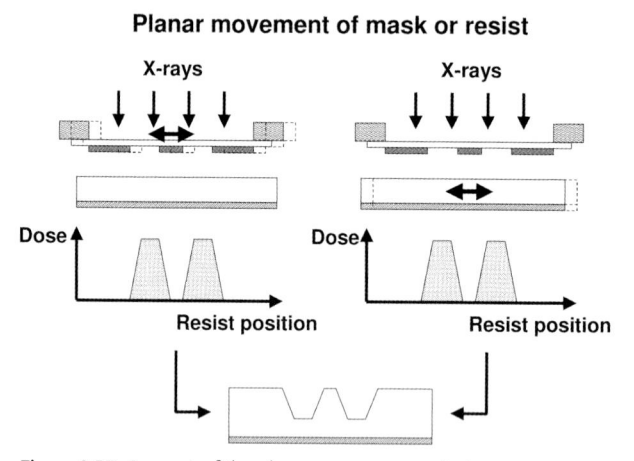

Figure 3.17 Concept of the planar movement technique.

has been fabricated by the 19 combined circular motions of the X-ray mask with a circular-arrayed pattern. The fabricated microlens array is shown in Figure 3.18a. The microstructure with V-shaped grooves shown in Figure 3.18b was successfully fabricated by non-constant speed reciprocation of the X-ray mask with line and space patterns. From these results, it is obvious that 3D microstructures with inclined and freely shaped sidewalls were successfully realized. The mask movement pattern to give an arbitrary dose distribution using the selected mask pattern can be derived mathematically [22], which helps to determine the exposure conditions.

(a)   (b)

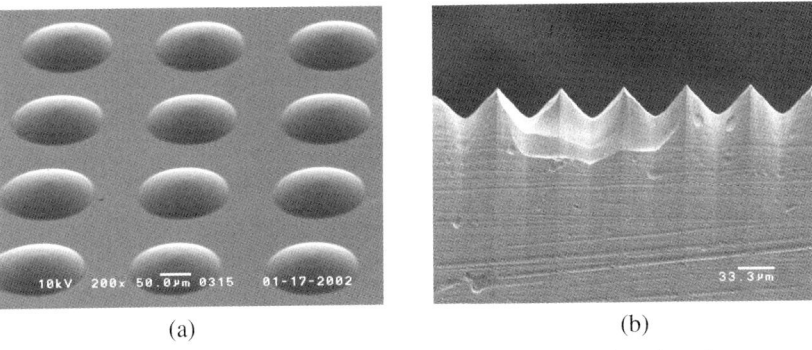

Figure 3.18 SEM images of the microlens array and microstructure with V-shaped groove.

(a)   (b)

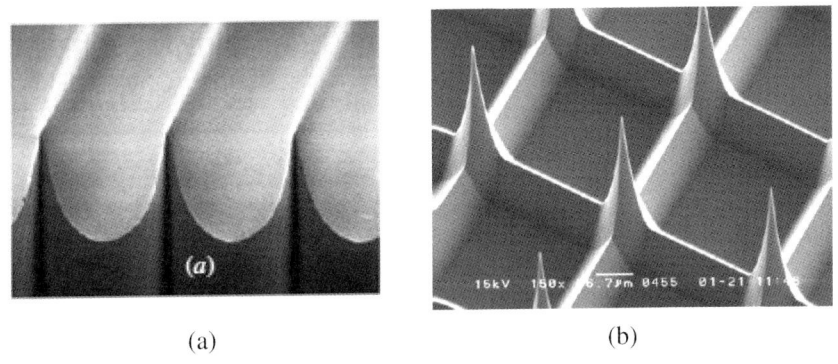

Figure 3.19 SEM images of the microstructure with sharp tips and microneedle array. S. Sugiyama, S. Khumpuang, G. Kawaguchi, Plain-pattern to cross-section transfer (PCT) technique for deep X-ray lithography and applications, Journal of Micromechanics and Microengineering, 2004, Institute of Physics.

Next, 3D microfabrication by moving a resist substrate using an X-ray mask with an arbitrarily designed pattern is described. This technique transfers a similar pattern of the mask to the cross-section of the PMMA. The microstructure with sharp tips shown in Figure 3.19a was fabricated successfully by fixing the mask with a triangular pattern and reciprocating the resist at constant speed. The microneedle array shown in Figure 3.19b was fabricated by half-rotating the PMMA substrate after the first reciprocation and second reciprocation. An arbitrary dose distribution can be realized easily by the design of the absorber.

Although it is difficult to fabricate sub-micrometer 3D microstructures with high accuracy because of Fresnel diffraction occurred by the gap between the mask and the resist, the arbitral control of the dose distribution on the resist surface fabricates 3D microstructures with inclined and curved sidewalls more flexibly than control of the two-step dose distribution.

(a)

(b)

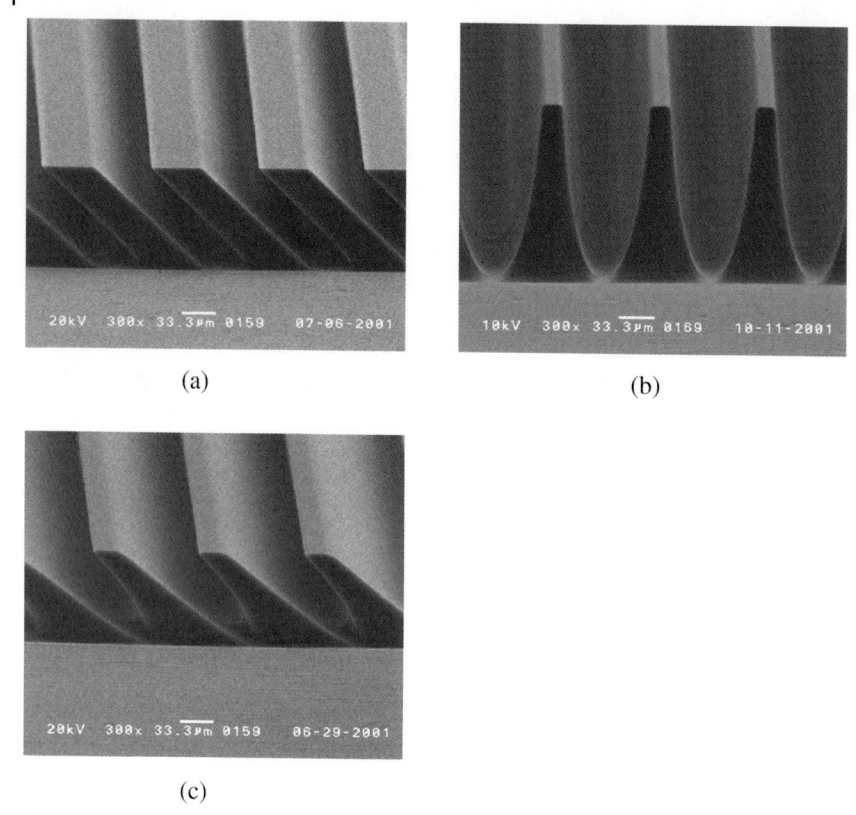

(c)

Figure 3.20 SEM images of the 3D microstructures fabricated by the innovative exposure techniques and its combination.

3.3.4
Combination of Innovative Exposure Techniques

The innovative exposure techniques have their own advantages and disadvantages. The inclined exposure technique described in Section 3.3.1.2 can easily realize 3D microstructures with flat and smooth inclined sidewalls as shown in Figure 3.20a but cannot realize microstructures with sidewalls of the cross-section. The technique to control the dose distribution described in Sections 3.3.2 and 3.3.3 can realize 3D microstructures with freely shaped sidewalls as shown in Figure 3.20b but cannot realize microstructures with both positive and negative slopes.

Combination of the innovative exposure techniques can realize 3D microstructures more flexibly than just a single technique. The microstructure with both positive and negative curved slopes shown in Figure 3.20c was fabricated successfully by the combination of the inclined and planar movement techniques. Combination techniques can be expected to achieve more flexible 3D microfabrication techniques based on DXRL.

3.4
Conclusions

Innovative exposure techniques for 3D microfabrication based on DXRL have been reviewed. Various techniques have been proposed and carried out. The technique based on the concept of conventional DXRL realizes multi-step and cylindrical microstructures with vertical sidewalls or microstructures with inclined sidewalls. The technique to yield the dose distribution on the resist surface realizes microstructures with freely shaped sidewalls. By combining these techniques, more complex 3D microstructures, for example, microstructures with both positive and negative curved sidewalls, can be realized. As a novel 3D fabrication process, the rolling sheet exposure technique was proposed recently [23]. The new aspect of this technique is that the resist sheet is exposed towards not the resist surface but the cross-section of the resist. For this technique, see Chapter 4. These techniques are expected to enable DXRL to play an important role as promising technology for the development of MEMS.

References

1 Becker, E.W., Ehrfeld, W. and Münchmeyer, D. (1982) Production of separation-nozzle systems for uranium enrichment by a combination of X-ray lithography and galvanoplastics. *Natuwissenschaften*, **69**, 520–3.

2 Pantenburg, E.J., Achenbach, S. and Mohr, J. (1998) Characterisation of defects in very deep-etch X-ray lithography microstructures. *Microsystems Technologies*, **4**, 89–93.

3 Bednarzik, M., Jian, L., Kouba, J., Herzog, J., Loechel, B., Scheunemann, H.-U., Firsov, A. and Haase, D. (2005) Optimization of the fabrication process of ultra thick, high aspect ratio SU-8 structures using X-ray lithography, in Proceedings of High Aspect Ratio Microstructure Technology (HARMST 2005), Gyeongju, South Korea, pp. 136–7.

4 Deguchi, K., Miyoshi, K., Oda, M., Matsuda, T., Ozawa, A. and Yoshida, H. (1996) Extendibility of synchrotron radiation lithography to the sub-100 nm region, *Journal of Vacuum Science Technology B*, **14**, 4294–7.

5 Harmening, M., Bacher, W., Bley, P., El-Kholi, A., Kalb, H., Kowanz, B., Menz, W., Michel, A. and Mohr, J. (1992) Molding of three-dimensional microstructures by the LIGA process, in Proceedings of the Micro Electro Mechanical System (MEMS1992), Travemünde, pp. 202–7.

6 Feinerman, A.D., Lajors, R.E., White, V. and Denton, D.D. (1996) X-ray lathe: an X-ray lithographic exposure tool for nonplaner objects. *Journal of Microelectromechanical Systems*, **5**, 250–5.

7 Mekaru, H., Kusumi, S., Sato, N., Yamashita, M., Shimada, O. and Hattori, T. (2004) Fabrication of mold master for spiral microcoil utilizing X-ray lithography of synchrotron radiation. *Japanese Journal of Applied Physics*, **43**, 4036–40.

8 Feiertag, G., Ehrfeld, W., Freimuth, H., Kolle, H., Lehr, H., Schmidt, M., Kiriakidis, G., Pedersen, T., Kuhl, J. and Koenig, W. (1997) Fabrication of photonic crystals by deep X-ray lithography. *Applied Physics Letters*, **71**, 1441–3.

9 Oh, D.-Y., Gil, K., Chang, S.S., Jung, D.K., Park, N.Y. and Lee, S.S. (2001) A tetrahedral three-facet micro mirror with inclined deep X-ray process, *Sensors and Actuators A*, **93**, 157–61.

10 Moon, S.J. and Lee, S.S. (2005) A novel fabrication method of a microneedle array using inclined deep X-ray exposure.

Journal of Micromechanics and Microengineering, **15**, 903–11.

11 Klein, J., Guckel, H., Slddons, D.P. and Johnson, E.D. (1998) X-ray masks for very deep X-ray lithography. *Microsystems Technologies*, **4**, 70–3.

12 Liu, Z., Bouamrane, F., Roulliay, M., Kupka, R.K., Labèque, A. and Megtert, S. (1998) Resist dissolution rate and inclined-wall structures in deep X-ray lithography. *Journal of Micromechanics and Microengineering*, **8**, 293–300.

13 Matsuzuka, N., Hirai, Y. and Tabata, O. (2005) A novel fabrication process of 3D microstructures by double exposure in deep X-ray lithography (D^2XRL). *Journal of Micromechanics and Microengineering*, **15**, 2056–62.

14 Cabrini, S., Gentili, M., Fabrizio, E.D., Gerardino, A., Nottola, A., Leonard, Q. and Mastrogiacomo, L. (2000) 3D microstructures fabricated by partially opaque X-ray lithography masks. *Microelectronic Engineering*, **53**, 599–602.

15 Utsumi, Y., Minamitani, M. and Hattori, T. (2004) Dot array microoptics for lighting panel using synchrotron radiation lithography. *Japanese Journal of Applied Physics*, **43**, 3872–6.

16 Tabata, O., Terasoma, K., Agawa, N. and K. (2000) Yamamoto, 3-dimensional microstructure fabrication using multiple moving mask deep X-ray lithography process. *IEEJ Trans. SM*, **120**, 321–6.

17 Lee, K.-C. and Lee, S.S. (2003) Deep X-ray mask with integrated actuator for 3D microfabrication. *Sensors and Actuators A*, **108**, 121–7.

18 Sugiyama, S., Khumpuang, S. and Kawaguchi, G. (2004) Plain-pattern to cross-section transfer (PCT) technique for deep X-ray lithography and applications. *Journal of Micromechanics and Microengineering*, **14**, 1399–404.

19 You, H., Matsuzuka, N., Yamaji, T. and Tabata, O. (2003) Deep X-ray exposure system with multistage for 3D microfabrication. *Journal of Micromechatronics*, **2**, 1–11.

20 Glashauser, W. and Ghica, G.-V. (1982) German Patent 3 039 110.

21 Meyer, P., El-Kholi, A., Mohr, J., Cremers, C., Bouamrane, F. and Megtert, S. (1999) Study of the development behavior of irradiated foils and microstructure. *Proceedings of the SPIE – The International Society for Optical Engineering*, **3874**, 312–20.

22 Matsuzuka, N. and Tabata, O. (2005) Algorithm to derive optimal mask and movement pattern in moving mask deep X-ray lithography (M2DXL). *IEEJ Trans. SM*, **125**, 222–8.

23 Oh, H.-S., Kim, J.-S., Lee, Y.-W., Chung, S.-I. and Choi, Y.-J. (2005) Innovative roll to roll patterning system with flexible substrate for the large scale high aspect ratio micro-structure, in Proceedings of High Aspect Ratio Microstructure Technology (HARMST 2005), Gyeongju, South Korea, pp. 72–3.

4
Hot Embossing of LIGA Microstructures

Mathias Heckele and Matthias Worgull

4.1   **Introduction**   *69*
4.2   **Hot Embossing**   *72*
4.3   **Simulation**   *76*
4.3.1   Software Tools   *76*
4.3.2   Material Properties as a Basis of Process Simulation   *76*
4.3.3   Mold Filling   *77*
4.3.4   Demolding   *79*
4.3.5   Use of Simulation in Practice   *82*
4.4   **Applications**   *83*
4.4.1   Simple Technologies   *83*
4.4.1.1 Optical Microstructures   *83*
4.4.1.2 High Aspect Ratio Replication   *86*
4.4.1.3 Large Area Replication   *86*
4.4.2   Advanced Technologies   *88*
4.4.2.1 Replication on a Substrate   *88*
4.4.2.2 Diced Microstructures   *90*
4.4.2.3 Multilayer Replication   *91*
4.4.2.4 Polymer Mold Inserts   *93*
4.5   **Plant Technology**   *94*
4.5.1   Tools   *94*
4.5.2   Hot Embossing Facilities   *96*
4.6   **Summary and Outlook**   *98*
        References   *100*

4.1
Introduction

Initial application of the LIGA technique was aimed at fabricating separation nozzle elements for uranium enrichment. The nozzles had been produced by precision technologies until they were replaced by electrochemically deposited nickel components. Hence the German acronym LIGA originally comprised lithography (Li) and subsequent electrodeposition (Ga) [1] only.

Plastic components were produced by lithography, followed by a subsequent development step. However, direct lithography soon turned out not to be an economically efficient alternative in the plastics sector. In spite of the excellent properties, for example, the high aspect ratios and extremely smooth lateral walls, high irradiation costs in the fabrication cycle caused the cost limits expected for a Plexiglas component to be exceeded considerably. The success of plastics as a basis of mass production and for replacing traditional materials is due to the fact that they are deemed to be inexpensive.

Mass production of plastics by molding with a negative mold or a molding die appeared reasonable. Attempts were made to develop further the electrodeposited nickel component into such a molding tool. For this purpose, the component was subjected to an additional electrodeposition step in order to produce a monolithic nickel tool. As an alternative, attempts were to use directly the nickel structure grown on the substrate as a tool.

Of course, separation of the production of plastic components from lithography and the use of molding tools for replication mean a change of paradigms: away from the mostly parallel lithography step that is typically used in the semiconductor industry towards a series production of plastic consumer goods.

Today's key technologies for the fabrication of most of our products, silicon batch production and plastics series production, cannot be more different. Microelectronics require constantly larger formats of masks and substrates and increasingly precise lithography techniques for structure replication. Hence more complex and larger plants and production facilities are needed. Plastics mass production does not so much require larger formats, but rather shorter cycle times. These are reached by more rapid machines that produce enormous numbers of pieces for the market.

At the beginning, when the first LIGA mold inserts were electrodeposited, experts were not sure about whether typical LIGA geometries could be filled up with plastics and demolded from the mold insert. To prevent the filigree metal structures from being damaged, a reaction casting method was applied first. A low-viscous, short-chain polymer was mixed with an initiator and injected into the microstructure. The water-like fluid consistency of this mixture ensured easy filling of the evacuated mold. Polymerization was started by heating, and a stable plastic microstructure was obtained, which could be demolded from the base plate. Construction of a filling and distribution system, however, proved to be problematic. Filling of cavities that were not connected with each other had to be ensured. In addition, a high dwell pressure had to be applied to compensate for shrinkage due to polymerization [2]. Hence a very large expenditure was associated with a small laboratory facility (Figure 4.1).

In principle, the process worked and microstructures of poly(methyl methacrylate) (PMMA) and polyamide (PA) were molded successfully. As reaction injection molding was a well-established technique for the production of mainly duroplastic

Figure 4.1 A unit for mixing polymer and initiator, a vacuum chamber to evacuate the microstructure, a heater system for polymerization and a hydraulic cylinder to apply the dwell pressure required make up the facility for reaction injection molding (RIM). The photograph shows the prototype of 1986.

products, implementation on a large technical scale was planned in cooperation with a specialized company.

For various microengineering products, however, thermoplastic polymers such as polycarbonate (PC) or polyoxymethylene (POM) were also of interest. Use of the known injection molding process on the microscale was therefore also tried. To test the pressing of thermoplastics into microcavities, first experiments were carried out with a laboratory facility for reaction injection molding. By means of the heating system, the polymer was heated above its softening point. With a hydraulic cylinder, it was pressed into the evacuated mold. Since vast experience had been gained from the use of PMMA as an X-ray resist, this material was used for the first tests, as it can also be processed thermoplastically (Figure 4.2).

Success of the first tests with thermoplastic material might have been due to the fact that small shear forces only were encountered when pressing the polymer into the mold. Moreover, all webs in the mold insert were embedded homogeneously and stabilized immediately by filling from above. These experiments in

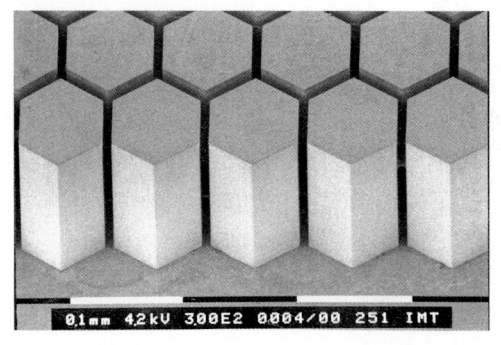

Figure 4.2 One of the first LIGA structures replicated by hot embossing in PMMA. Pillars of 125 µm height from a honey comb mold insert. The PMMA sheet was polymerized at the Institute to achieve a very low viscosity at the glass transition temperature T_g.

1989 marked the start of the use of micro hot embossing for the production of LIGA structures [3].

4.2
Hot Embossing

The most attractive feature of hot embossing is the very simple process procedure that allows for the production of several replicas of a master with relatively small expenditure (Figure 4.3). In a simple way, identical structures are obtained for experiments or applications. In principle, it is sufficient to place a plastic foil between two heated metal plates. After the thermoplastic material has been softened sufficiently, the two plates are pressed against each other. Either one of the plates or both plates may be provided with microstructured surfaces. As soon as the plastic material has cooled, the ready component can be removed from the tool [4].

This description of the process procedure reveals several advantages. Due to the use of a semi-finished product, it is no longer necessary to process granules. Hence the two processes of 'injection' and 'molding' do not have to be synchronized. In the experimental stage or when highly filigree structures are required, the requirements of these two processes often conflict with each other. The much broader process window of hot embossing is suited in particular for large parts with microstructured details. On the one hand, a large area will be filled up homogeneously. On the other hand, locally optimum conditions have to prevail for the precise molding of details. Especially in the fabrication of optical components with minimum internal stresses, it is reasonable to separate microforming and macroforming from each other. Insertion of a semi-finished product limits the variety of materials for use. Compared with the number of polymer types and modifications supplied in the form of granules, that of foils and plate materials is strongly reduced.

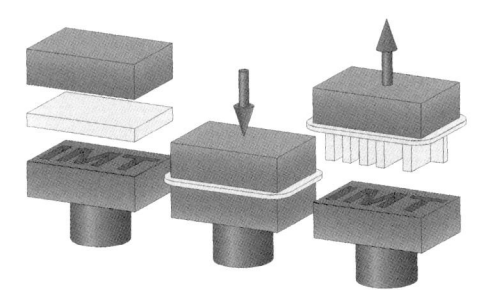

Figure 4.3 Basic principle of hot embossing: Insertion of the semi-finished foil, heating between the embossing die and countertool, closing of the tool and filling of the microstructures: The residual layer ensures maintenance of the internal pressure of the tool also during cooling. Demolding by vertical removal from the microstructured mold insert as close to the solidification temperature as possible.

The main difference between hot embossing and injection molding lies in the fact that heating and molding are carried out at the same place. The resulting disadvantages, such as long cycle times, however, are compensated by several advantages. On the microscale, melt temperatures often have to be very high, as a minimum viscosity is needed to fill the finest cavities. However, the melt must not be kept too long at such high temperatures. Consequently, tool and injection cycles have to be synchronized precisely in the injection molding process. Here, hot embossing can be conducted much more stably. Although hot embossing machines are no longer equipped with inert oil-heated heating plates, but with electrical heating elements that reach very high heating powers, thermostating technology still offers a high potential for improvement. The high cyclic loads and the problem of having to supply and remove large heat volumes represent challenges that have been only partly solved.

As soon as the polymer melt has reached the correct temperature, the tool can be closed for embossing proper (Figure 4.4). The embossing process depends strongly on the geometry and the height of the cavities. Molding or primary shaping? – that is the question. Is the embossing die pressed into the polymer mass and is excessive material displaced laterally? Or does the thin plastic material flow into the deep microstructured bag holes? This question cannot be answered clearly in hot embossing. For low structures, such as Fresnel optics or nanostructured surfaces, molding certainly will predominate. In areas, where high LIGA structures are located close to each other, however, the primary shaping character will prevail. This behavior is confirmed by simulation calculations. Such calculations show that the filling of the cavity largely takes place in the low-pressure range. High embossing forces for mold filling are required for very low structures and extreme aspect ratios only.

Structural heights of a few hundred micrometers compared with the heights of several millimeters and centimeters of classical cast molds initially appeared to be analogous to a die that is pressed into the heated, soft polymer. This concept gave rise to the term hot stamping or hot embossing. Later, when the process was

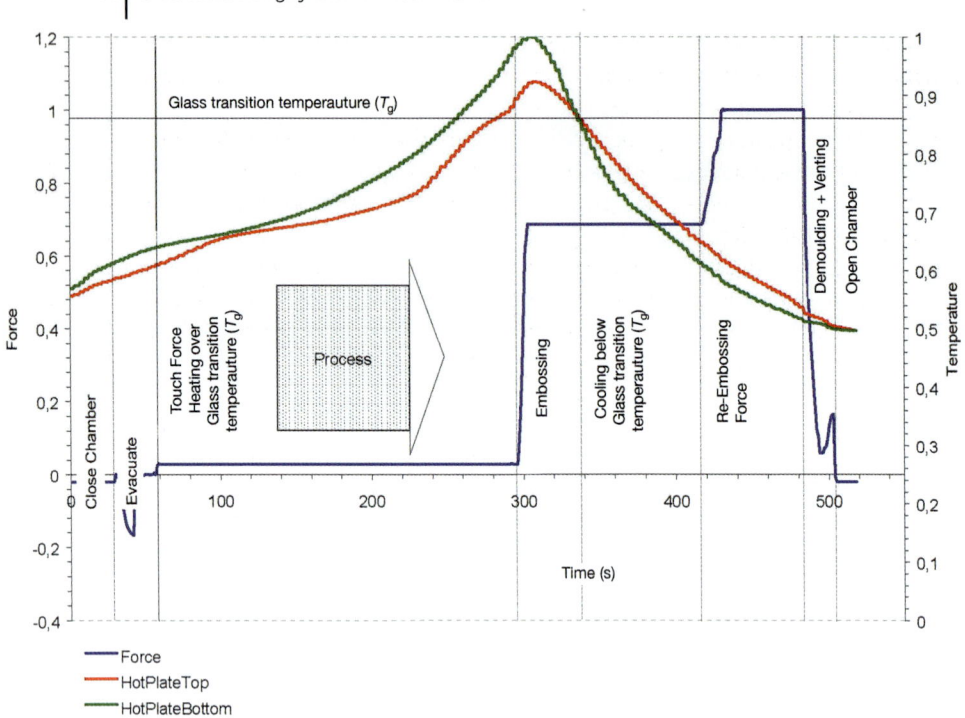

Figure 4.4 Diagram of most relevant parameters for a typical hot embossing cycle. Temperature and force are given in arbitrary units.

understood in more detail, other terms such as compression molding [5] were introduced. Today, various terms are used.

The embossing time proper is very short, but, of course, cannot be compared with the injection times of less than 1 s. As the material mostly flows in the embossing direction only and hardly in the lateral direction, the flow length is in the range of the structural height. Internal stresses or material modifications due to an accelerated flow through narrow gaps do not occur in hot embossing.

In the heating phase, the tool is controlled such that a constant preliminary force is applied. On reaching the target temperature, the tool starts to close. With the lower speed limit of a few micrometers per second, this process step takes a maximum of a few minutes. As soon as the cavities are filled, a dynamic force equilibrium develops. Flow resistance of the polymer displaced through the embossing gap counteracts the embossing force. This situation is typical of hot embossing. Both embossing plates form an open tool and can be moved towards each other with a minimum gap remaining. The material located in the embossing gap is referred to as the residual or carrier layer. It cannot be reduced to zero, as it is not possible to produce embossing plates with a surface roughness of less than 1 μm. On the other hand, the coherent residual layer ensures that the embossing force is maintained in all cavities. This is necessary during the complete cooling process, as free polymer shrinkage would cause strong deviations of dimensions and shapes. This embossing pressure, which is comparable to the dwell pressure phase of injection molding, requires high pressures and, hence, a very sophisticated plant technology.

When the temperature drops below the glass transition point, demolding may take place. This demolding step is decisive for the quality of a microstructure. In this last process step, the surface generated by replication must not be damaged or destroyed again. Separation of a polymer from the tool must be clean and homogeneous. For this purpose, the temperature must be very high in order to prevent the plastic structures from shrinking on to the mold insert. In this case, additional lateral forces would make demolding impossible or induce damage. Demolding is a critical process step for structures with high aspect ratios in particular. The polymer–metal contact area is very large. This also applies to the degree of interlock of material and tool. This phenomenon is especially pronounced when using tools with vertical side walls produced by X-ray deep-etch lithography (LIGA). The differences between a modified laboratory press and a professional hot embossing machine become obvious at this point at the latest. Cooling to room temperature and lateral withdrawal of the plastic foil from the mold insert inevitably generate defects. In the case of one-sided molding, the plates may be moved apart in the closed and evacuated chamber, with the foil adhering to the unstructured side of the countertool (Figure 4.5). In this way, the plastic part can be removed vertically from the tool.

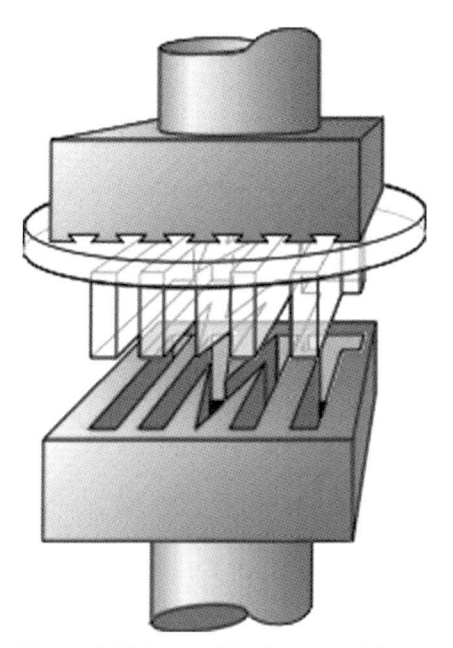

Figure 4.5 High demolding forces result from the large contact area between the mold insert and the plastic component. To transmit these demolding forces, the residual layer has to be closely interlocked with the tool counterplate.

4.3
Simulation

4.3.1
Software Tools

With the further development of the LIGA technique towards higher structures, smaller details and increasing aspect ratios, it became necessary to determine in advance optimum molding parameters by a process simulation based on a solid theoretical knowledge [6–8]. Today, process simulations are part of many plastic molding processes. In the field of microreplication, however, its development potential still is very high [9].

FEM simulation tools are state of the art in plastic molding. However, no simulation tool exists that can reproduce satisfactorily the complete process chain of hot embossing – from the filling of the cavities to the demolding of the microstructures [10, 11]. Furthermore, complete FEM modeling of a typical LIGA mold insert is impossible, due to the high number of nodes with PC-oriented FE systems [12].

The analyses performed were aimed at studying the individual process steps of hot embossing on the basis of simply structured tools, at understanding relationships and at deriving improvement potentials. Theoretical and practical analyses today focus on demolding, as this process step is associated with the highest risk of destroying the microstructures (Figure 4.6 shows typical damage patterns). Based on simulation models, parameter studies are carried out to derive optimum process parameters, in particular for reducing demolding forces.

The individual process steps of molding and cooling and demolding were studied separately with two different simulation tools (Figure 4.7). The established injection molding simulation tool MOLDFLOW had been applied to simulate molding by injection embossing. Since the embossing step of injection embossing and of hot embossing is based on the displacement- and force-controlled embossing of a polymer melt, MOLDFLOW can also be used to describe molding by hot embossing. Process-specific properties, such as the use of substrates and demolding, cannot be modeled with MOLDFLOW. Hence cooling and demolding were analyzed in detail using the simulation software ANSYS. The tool, polymer and substrate composite were modeled to analyze influences of the tool in a manner that is very close to reality. In parallel with the simulations, demolding forces were measured systematically with varying process parameters. The relationships derived were compared with the simulations. By dividing the process into two groups, the process chain can now be described nearly completely [13].

4.3.2
Material Properties as a Basis of Process Simulation

By way of example, the description of the demolding behavior and the MOLDFLOW simulation referred to the PMMA Lucryl G77Q11 of BASF. The MOLDFLOW material data, however, could not be transferred directly to ANSYS. Hence

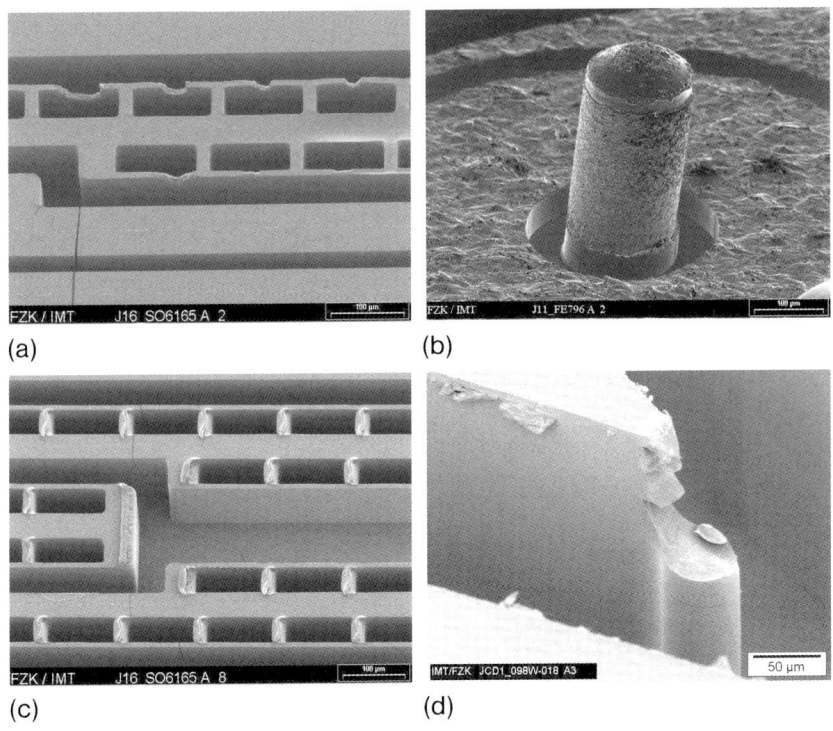

Figure 4.6 Typical deformations of molded microstructures. Incomplete filling of the microcavities (a) results from for example, a small filling pressure. The distortion of a structure (b) is caused by shrinkage differences in the molded part. Overdrawn edges (c) and torn structures or fragments (d) are due to shrinkage effects during demolding.

the viscoelastic material behavior was described in ANSYS by a generalized Maxwell model with 10 elements. The necessary material parameters were determined by measurements.

By means of the dynamic thermomechanical analysis (DMA) method, the Young's modulus was measured as a function of time and temperature. At a reference temperature of $T = 100\,°C$, the time–temperature shift was determined and the constants C_1 and C_2 were calculated for the WLF shift function. Figure 4.8 shows Young's modulus as a function of time and temperature. The shift function was interpolated for $T < 100\,°C$ in order to consider the temperature range of demolding in the simulation.

4.3.3
Mold Filling

The mold filling process simulated for a rectangular component in the LIGA format with individual free-standing, symmetrically arranged, rectangular

4 Hot Embossing of LIGA Microstructures

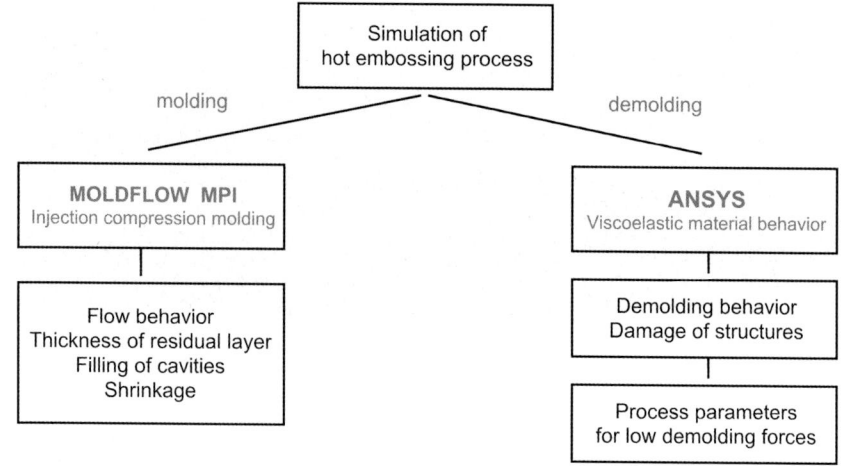

Figure 4.7 Schematic representation of the analysis of hot embossing. Software to describe the hot embossing process completely is not available. Hence molding was simulated using the commercial software MOLDFLOW MPI with the injection embossing option. It yielded results with respect to the flow behavior of the plastic melt during embossing. Work was focused on the simulation of demolding with the FEM software ANSYS. In ANSYS, the material behavior of the plastics was described by a viscoelastic material model. A model of the demolding process was developed, by means of which statements could be made with respect to the demolding behavior of microstructures. It also allowed the determination of process parameters to reduce demolding forces.

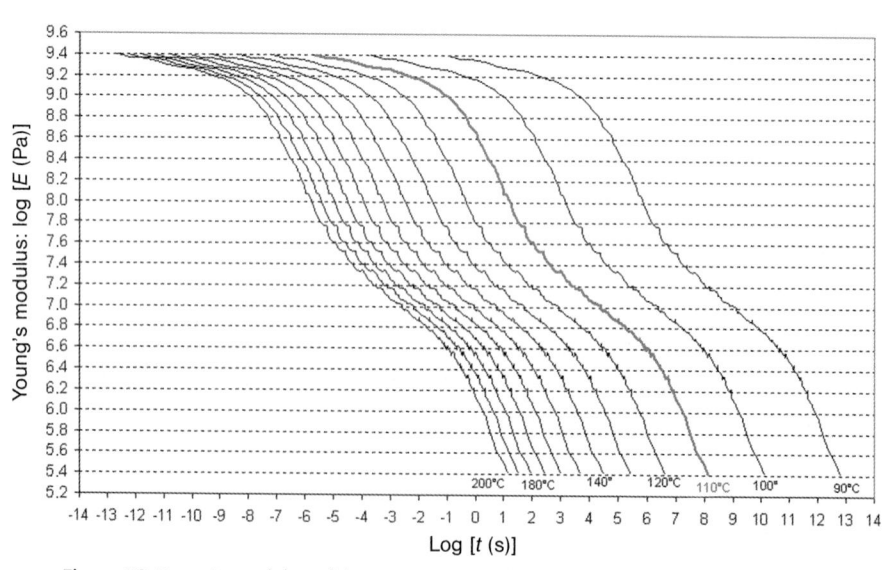

Figure 4.8 Young's modulus of the PMMA Lucryl G77Q11 as a function of time and temperature. The decreasing Young's modulus is caused by relaxation processes. At a temperature of about 170 °C, relaxation times are in the range of typical process times. Above this temperature, stress reduction due to relaxation processes has to be expected in the embossed components. Below 170 °C, relaxation times exceed typical process times. Consequently, stresses in the molded part are reduced to a far smaller extent.

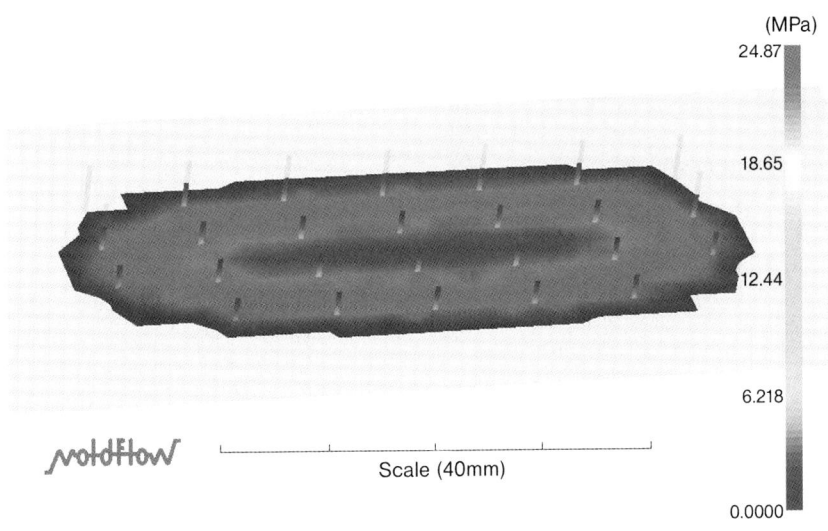

Figure 4.9 Pressure distribution during mold filling, with a rectangular component in the LIGA format being used as an example. Free-standing microstructures are arranged systematically on the component. Pressure distribution at the end of the displacement-controlled embossing phase is shown. According to the pressure distribution in the residual layer, the cavities are filled from the inside to the outside.

structures is shown in Figure 4.9. Filling is based on the representation of the pressure distribution of the melt in the component. The pressure peak at the end of the displacement-controlled embossing process is generated along the central line of the component. According to the pressure gradient towards the edge of the component, the cavities are filled from the inside to the outside. In the free-standing structures, a pressure gradient develops, starting at the foot of the structures with the pressure of the residual layer.

4.3.4
Demolding

Using a microstructured mold insert as an example, Figure 4.10 illustrates the characteristic steps of the demolding process in a sequence. To assess the load of the microstructure, the von Mises stresses were determined. They may be compared with the temperature-dependent stress–strain curves measured for the material.

Before starting demolding, the pressure load of the molded part is reduced. Due to the compressibility of the polymer, the molded part is relieved and thickness increases. The point of highest load is located at the bottom of the shrinkage of the residual layer directly at the foot of the structure in the direction of the center of the molded part. With the starting demolding movement, the highest component load moves to the lower quarter of the structure. The side facing the center of the molded part is subjected to a higher load. As demolding proceeds, the

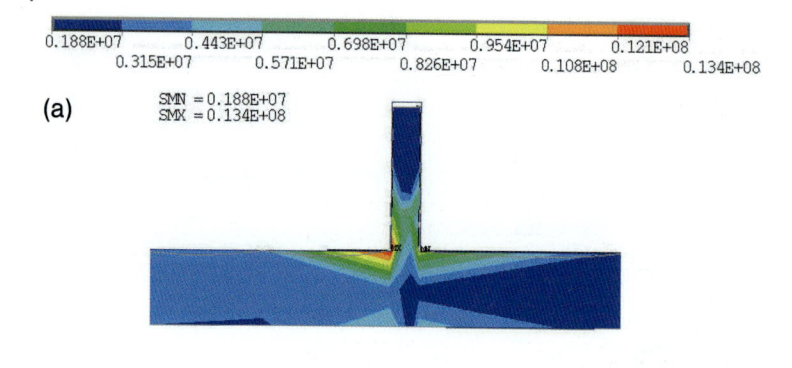

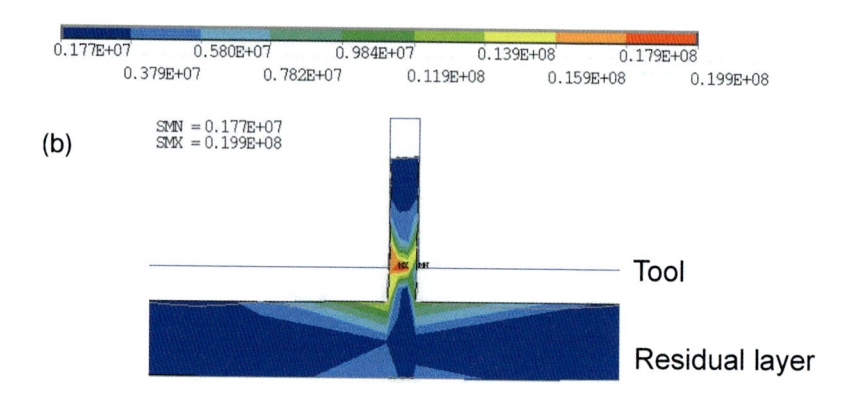

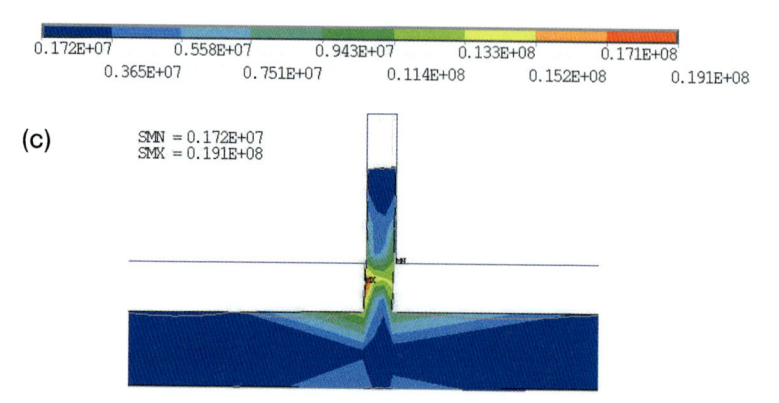

Figure 4.10 Demolding of a 20 μm wide and 100 μm high microstructure. The stresses in the component (von Mises reference stresses) at various times are shown. Stress distributions are marked by colors. SMX stands for the maximum stress value and SMN for the minimum stress. The highest load of a typical structure is encountered at the beginning of demolding.

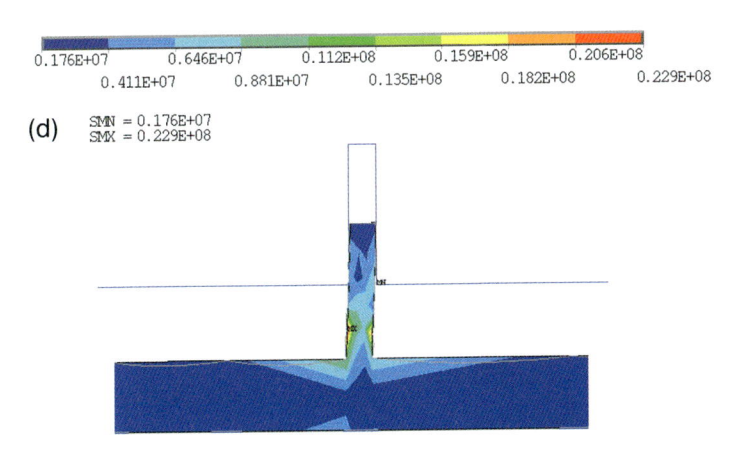

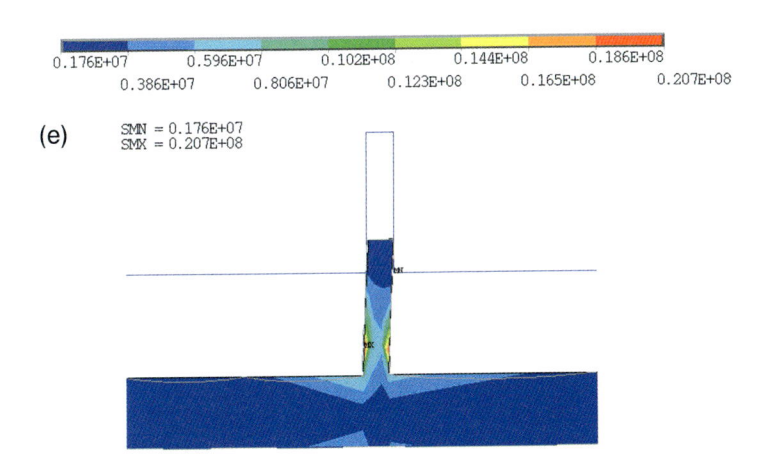

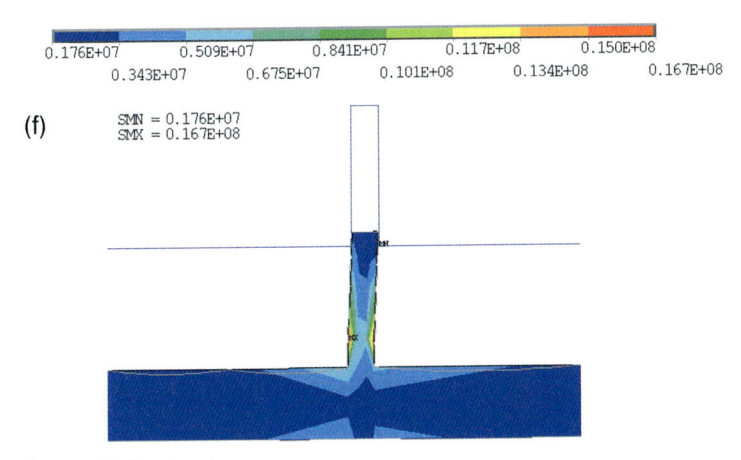

Figure 4.10 *Continued*

highest load is shifted to the outer areas of the structures. During the first half of the demolding path, the load of the structure is highest. The risk of a plastic deformation of the side walls or of rupture is highest in this phase.

4.3.5
Use of Simulation in Practice

A major criterion of process simulation is its usability in practice. As a complete microstructured tool can be modeled only with high expenditure, modeling was restricted to characteristic features of a LIGA tool. Hence simulation results have to be interpreted and transferred to real microstructured mold inserts. In this case, the simulation results with respect to optimization of the process parameters were verified by a practical example, a mold insert of a miniaturized Fourier transform spectrometer. Structures reached a height of 412 μm and the aspect ratio varied between 5 and 8, depending on the structure. Interest focused on a 12 μm wide gap between two 412 μm high walls (Figure 4.11). With typical molding para-

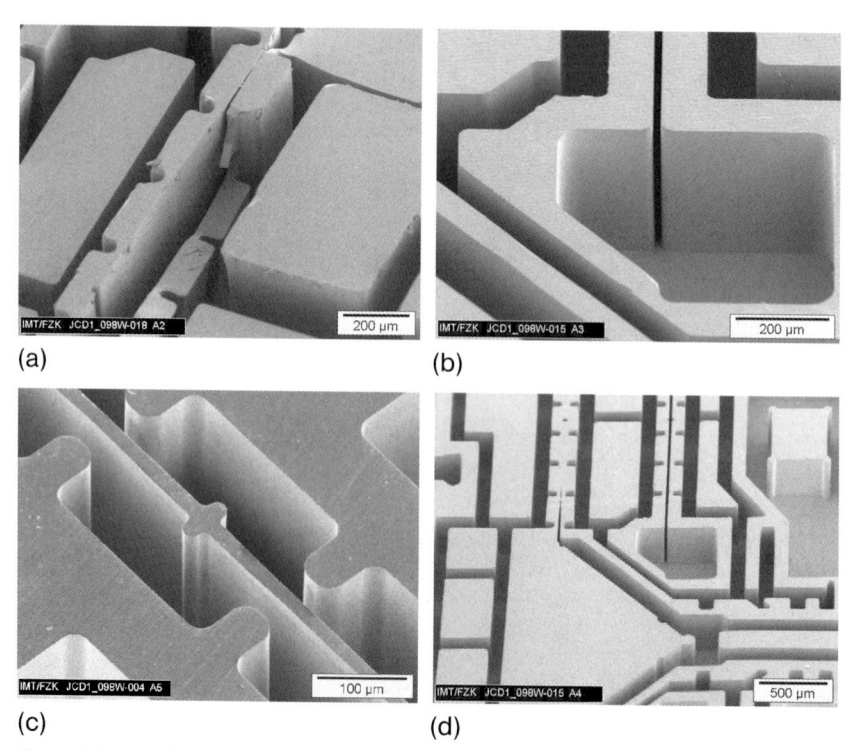

Figure 4.11 Detailed view of a Fourier transform spectrometer. Structural height amounts to 412 μm. Critical structures are 12 μm wide trenches. Typical damage prior to the optimization of the process parameters (a) and the result after optimization (b–d).

meters, the structures could not be demolded. The structures were destroyed by demolding.

In the next step, the process parameters were adapted according to the simulation results. At an increased molding temperature and a higher embossing force, the structures were filled completely. At the same time, a thin residual layer was generated. As a result, low shrinkage of the molded part was ensured and the demolding force was reduced. The demolding temperature was close to the glass transition point.

4.4 Applications

4.4.1 Simple Technologies

4.4.1.1 Optical Microstructures

Most of the optical structures made by molding of polymers are diffractive structures with lateral dimensions in the range of some hundreds of nanometers and low heights. Low structures molded with LIGA mold inserts are monomode waveguides, for example [14]. Trenches of some micrometers in depth and an aspect ratio of 1 can be molded easily. The reason why LIGA mold inserts are used in this case is the precision required. Multimode applications require higher structures. Typical examples are coupling elements (Figure 4.12). The critical area is the edge that chops the light as its quality determines the distribution of light at the exit. With the LIGA process, curves with radii of <1 µm at structural heights of up to 1000 µm can be fabricated. Such parts are needed in great numbers for optical networks; they are therefore preferably made by injection molding.

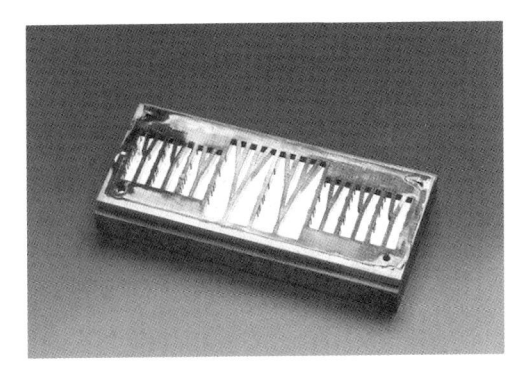

Figure 4.12 Mold insert for coupling elements (structure height 1050 µm). In the center of the mold insert, two elements are arranged for splitting the intensity into four equal parts.

Figure 4.13 SEM of a micro-optical bench completely assembled with a beamsplitter and two microlenses.

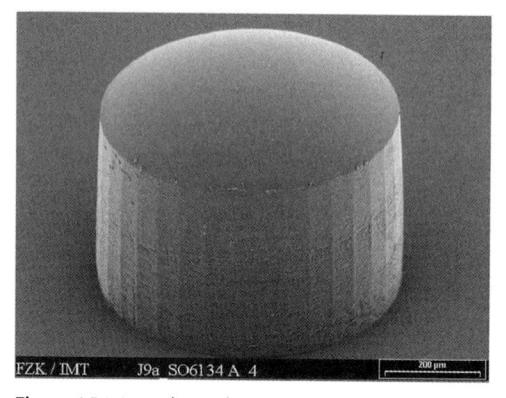

Figure 4.14 Microlens (diameter: 600 μm) produced by molding in a LIGA mold insert.

When different micro-optical components are to be combined, for example, spherical lenses with waveguides, micro-optical benches (Figure 4.13) are indispensable. They require extreme dimensional precision [15, 16]. This is a challenge for polymer technology with its materials of comparably high thermal expansion. Having optimized the design and molding parameters, modern high-performance polymers, such as PEEK, yield good results.

Lenses are other important optical components. Their dimensions are large compared with those of diffractive elements. Again, the requirements to be met in surface quality call for the highest precision. Various approaches exist to fabricating the curved lens surface in mold inserts. One way is by remelting cylindrical LIGA microcolumns and further processing them to mold inserts [17]. Molding by hot embossing in different thermoplastics produces microlenses of excellent quality (Figure 4.14), which are used in micro-objectives because of their imaging accuracy [18]. An advantage of the LIGA mold inserts is parallel fabrication, which yields identical components needed for arrays, for example, the array of microlenses shown in Figure 4.15.

Figure 4.15 Array of microlenses produced by LIGA structuring and subsequent X-ray exposure to create a bended optical surface.

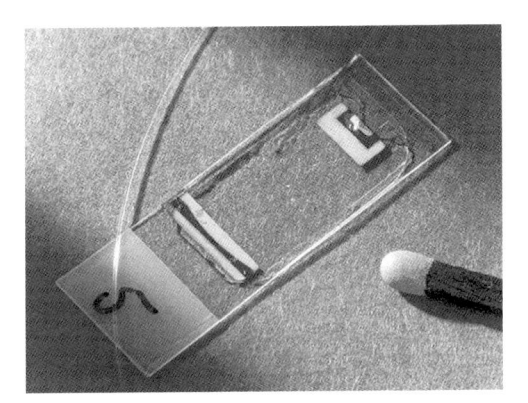

Figure 4.16 UV–VIS microspectrometer, complete with fiber (spectral range: 380–780 nm).

A well-known LIGA product and a key application of polymer replication in high numbers is the LIGA UV–VIS microspectrometer (Figure 4.16). In principle, this is a waveguide structure, but as far as function is concerned, a microspectrometer already is a complete microsystem [19]. In addition to the beam-guiding system in the waveguide, a self-focusing diffraction grating, a launching shaft and a sloped edge for coupling out the diffracted light on to a diode array are integrated. A system of this type never needs to be adjusted; this feature, in addition to production by plastics molding, is the reason for the low price. From the point of view of production technology, two types of spectrometers are distinguished. Initially, the spectrometer was built as a three-layer waveguide and structured by X-ray lithography. The main problem of fabrication by hot embossing was to eliminate the residual layer in order to obtain a sharp transition from the structured core layer to the unstructured cladding layer, where loss of light must be avoided. Using a soft aluminum countertool, this problem could be solved. The additional step of welding a cladding on to the core layer made this technology too sophisticated.

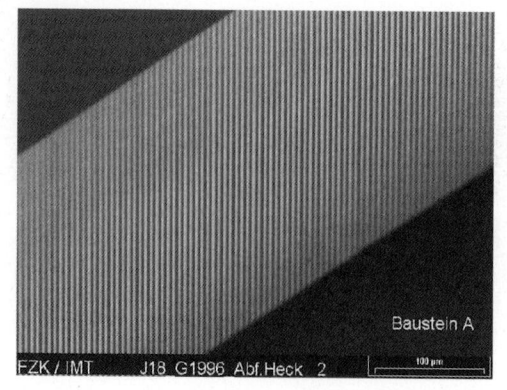

Figure 4.17 Near-infrared spectrometer. Detail of the grating (structure height 340 μm). The optical grating has a step height of 1 μm.

Therefore, the next generation is designed as a hollow waveguide component, where light is conducted in the air (Figure 4.17). Microstructures are integrated in the support and light conduction structures. The advantage of this technique lies in only one material being employed, which also allows for the use of injection molding [20].

4.4.1.2 High Aspect Ratio Replication

Apart from optical structures, typical LIGA structures are characterized by large structure heights and high aspect ratios. Such structures are applied where large lateral wall areas are required or in solid joints and mechanical frameworks. If metallized, such structures possess high electrical capacities that are needed for electrostatic drives or detectors [21].

Other examples of the use of LIGA mold inserts are the structures shown in Figure 4.18. Although they have a height of 500 μm and a lateral width of 100 μm, that is, an aspect ratio of only 5, they are extremely difficult to mold in plastics. The reason is the large number of parallel standing walls with a length of several centimeters. Pressure differences in the structure that would damage the free-standing walls are prevented by filling the mold from above during hot embossing.

4.4.1.3 Large Area Replication

Examples of large area replication are diffracted optics or waveguides and microfluidic components. For large area replication, UV-LIGA mold inserts or micromechanically structured mold inserts are preferred. The expenditure associated with the production of large format LIGA mold inserts with the respective masks and irradiation technology still limits the possibilities of this type of plastic molding. The potential of this technique, however, is illustrated by the microtiter plates. In medical engineering, these plates with a standard format of

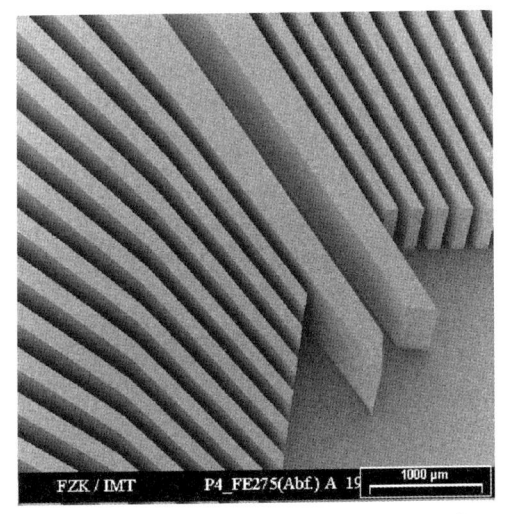

Figure 4.18 Lamellar structures for lead frame application. Structural height 500 µm with 100 µm grooves.

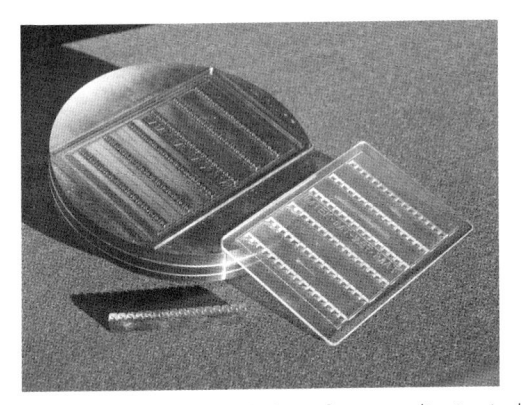

Figure 4.19 An example of a large format application is the dispensing well plate. The classical format of 85.5 × 127.8 mm is provided with 96 individual structures with micronozzles.

85.5 × 127.8 mm are applied for analysis purposes. Initially, these plates were provided with 96 wells. As a result of miniaturization, 384 or even 1536 wells are now accommodated on the same surface area [22]. Instead of increasing the number of wells, microtechnical methods also allow their complexity to be enhanced. Simple wells can be replaced by a complete CE analysis system or a dosage unit with a reservoir [23]. Larger mold inserts (Figure 4.19), however, result in higher requirements on the machine and the tool in which the mold insert is installed (cf. Section 4.5.1).

4.4.2
Advanced Technologies

4.4.2.1 Replication on a Substrate

Hot embossing is also of interest, as the relatively slow flow speeds and the wide process window allow for a number of process variations. Moreover, embossing does not have to be synchronized with a cyclic injection process.

As a result of hot embossing, any microstructure is located on a residual layer that is mostly used as a carrier layer, such that entire fields of microstructures on a common substrate can be further processed or used in this fixed, well-defined arrangement. In some cases, however, additional or different material properties of this carrier layer are required. A metal substrate may be needed for a conductive carrier or a ceramic material or silicon wafer may be required due to its more favorable thermal properties. An example is the micro-optical bench used in a micro-optical distance sensor (Figure 4.20). By means of hot embossing, the structure is directly replicated on a polymer-coated silicon wafer. This results in a silicon–plastic composite with optimal adhesion properties. The residual layer is kept so small that it no longer has any mechanical stability. This means that the complete thermal expansion behavior is determined by the silicon wafer. In this way, the embossed optical components are decoupled from each other.

The substrate used for replication may assume additional functions. It may be applied as a plating base for filling up with metal the trenches embossed in the plastic material. For this purpose, however, a conductive base is required at the bottom of the structure. This may be a metal base plate or metallization on a wafer (Figure 4.21). However, here also, the residual layer formed by hot embossing cannot be avoided. At first, the bottom of the structure will not be conductive and the separating polymer layer will have to be removed. This can be done by reactive ion etching. Although this process possesses high anisotropy, polymer is also

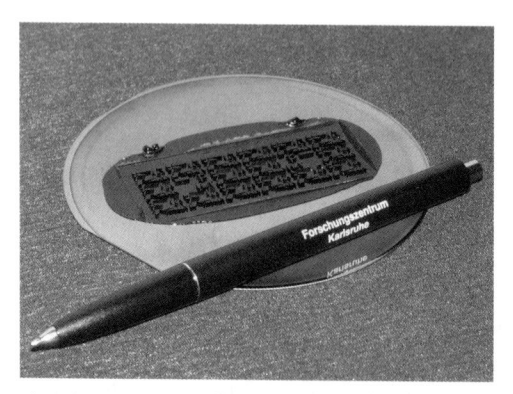

Figure 4.20 An array of optical benches for an distance sensor working by triangulation. To decrease thermal expansion, the replication is made on top of a silicon wafer with minimized residual layer, Material: filled PC.

(a) (b)

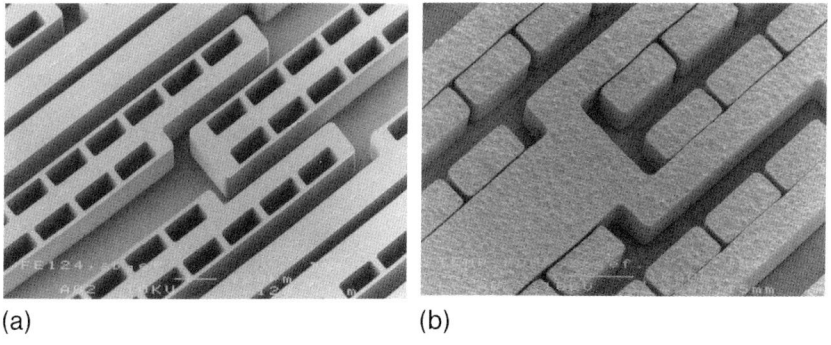

Figure 4.21 An embossed plastic microstructure (a) serves as a lost mold for a metallic element (b) on a substrate. Detail of an acceleration sensor made of electroplated nickel (height 200 µm).

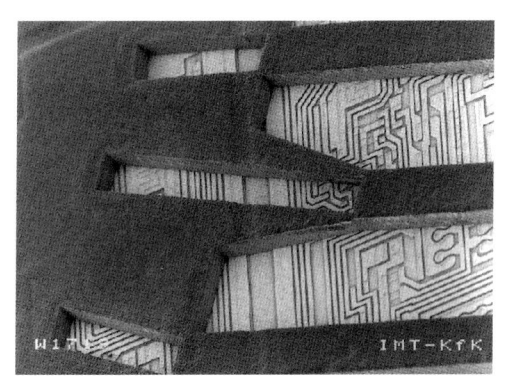

Figure 4.22 A nickel microstructure on a CMOS wafer. Hot embossing and electrodeposition are compatible with the CMOS process. The electronic functions are not affected adversely by subsequent process steps.

removed from lateral walls. This limits the aspect ratio achieved with the LIGA technique [24]. Alternative technologies, such as later metallization, can be used for high LIGA structures with certain reservations only, since contacting no longer takes place via the bottom of the structure and only connected areas lie on a common electrical potential. Honeycomb structures are typical examples of such problem structures [25].

Replication on substrates allows for hybrid integration. Apart from the combination of polymer with metal, plastic material may also be combined with silicon [26]. This leads to a considerable increase in applications. Silicon technology can make use of the complete variety of material properties of plastics and, by adding semiconductor components to the plastic microstructures, complete microsystems are obtained (Figure 4.22). By means of electrodeposition, metals can also be integrated in this kit [27].

4.4.2.2 Diced Microstructures

A crucial problem with hot embossing is the characteristic residual layer, which represents an obstacle to the production of diced and hole structures. Several attempts have been made to eliminate this drawback compared with technologies with closed tools.

The production of microspectrometers has demonstrated that embossing without residual layers is feasible, but subsequent demolding of the microstructures is no longer possible without an ejector. In a second embossing step, the microspectrometers are provided with a cladding that has an optical function and also serves to demold the microstructured components. This approach was further extended to embossing of composite foils. A combination of POM and acetate is embossed jointly. The embossing tool completely penetrates the POM layer, but only partly enters the AC carrier layer. Subsequently, this foil composite can be demolded as usual via its residual layer. The material properties of POM and AC ensure that adhesion is sufficient for demolding, but subsequent dicing is still possible [28]. Apart from through-holes, diced structures can also be produced in this way (Figure 4.23).

Another possibility for structuring holes in hot embossing results from the precise orientation of the upper and lower tools of the micro hot embossing machines (Figure 4.24). In a first embossing step, a high-temperature plastic material (PEEK) is structured with the mold insert. At the same time, this material is permanently welded on to the lower counterplate. Thus, the upper tool and the lower tool are combined and act like a puncher. The mold insert precisely engages with the depressions of the countertool. Now, a polymer foil (PMMA, PC) can be provided with holes at reduced deformation temperatures.

Through-holes may also be produced by the most obvious technology. The microstructured mold insert is closed down to the countertool, such that the polymer is displaced completely. This technology only works, however, when the area of the

(a) (b)

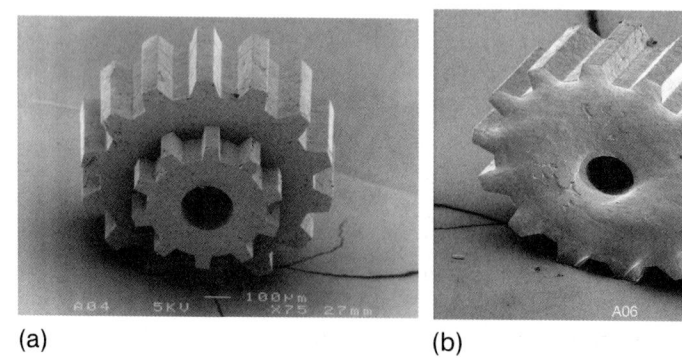

Figure 4.23 Diced LIGA gearwheel made of POM (a). The composite foil technique prevents a sharp replication of the rear (b).

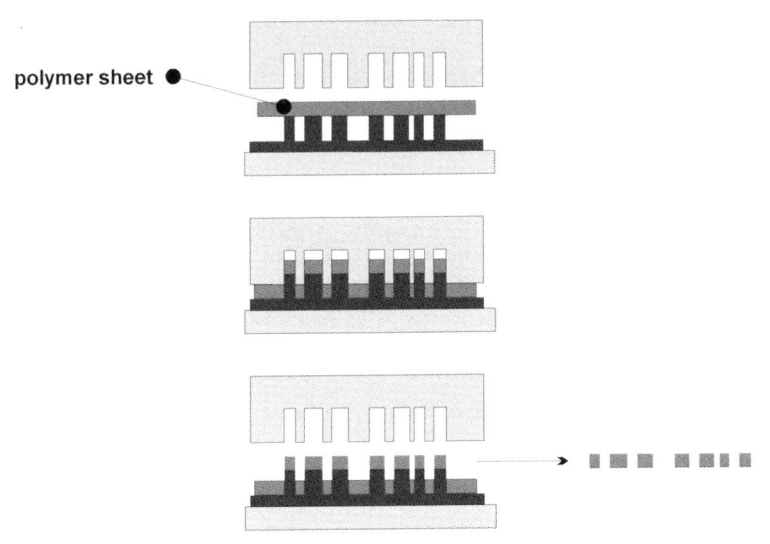

Figure 4.24 Principle of hot punching. Low-melting polymers sheets (e.g. PMMA, PC) can be structured with holes. The countertool (replicated from the same mold insert) is made of a high-temperature polymer (PEEK).

holes is much smaller than the complete area, such that sufficient space is available for the displaced polymer [29]. Moreover, the raised sections that come into contact with the countertool have to be hardened to prevent damage due to mechanical loading. At the same time, the countertool has to be designed for wear, which means that it can be replaced easily. To compensate for deviations from the ideal dimensions, that is, tool roughnesses, the counterplate is not designed in a rigid manner. As a result, the expenditure associated with tool construction is increased, as known from injection molding. In this way, nozzle holes were produced for a dispensing well plate (Figure 4.25).

4.4.2.3 Multilayer Replication

Another type of hot embossing, which has already been mentioned in Section 4.4.1.1, will now be dealt within more detail. By embossing multilayer foils, various material properties may be combined to increase the applicability of polymers. The embossing of through-holes makes use of various chemical affinities, such that the adhesion between the foils is located precisely in the transition range desired. In the production of waveguides, materials with various refractive indices are combined. In a similar way, microstructures may be provided with various mechanical properties, for example, hard/soft or brittle/elastic. It is also possible to combine variable surface properties, for example, hydrophilic/hydrophobic. All these modifications can be produced by hot embossing. As a transition technology, it may have the character of melt molding or the

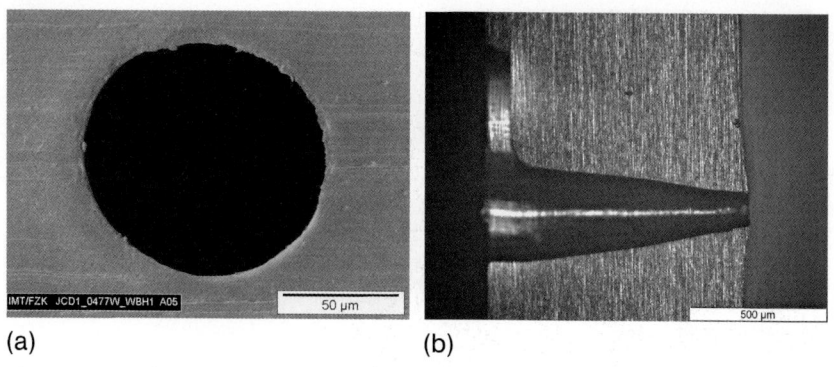

Figure 4.25 Hard tool–soft countertool combination to open through holes. (a) Diameter 100 µm (view from above); (b) height 1000 µm (transverse section).

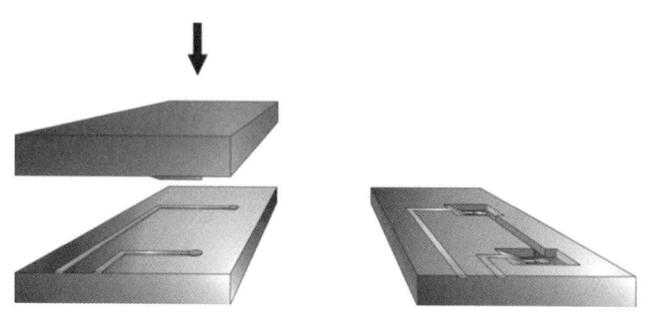

Figure 4.26 Principle of embossing of conduction paths. Metal on polymer is an extension of the multilayer technology.

forming character may predominate. The more forming is applied, the more the originally used foil composite is maintained and the more clearly are the properties separated.

It is largely irrelevant which material combinations are used. Hence combination is not restricted to various polymers. It is also possible to combine polymers with metals, as will be shown in the next example [30].

A structured gold layer is applied to a plane polymer semi-finished product. Through a mask, the polymer surface is sputtered with gold and subsequently reinforced by electrodeposition. This composite can then be embossed with a microstructured tool (Figure 4.26). Thus, the conduction path is adapted to the embossed topology, that is, the conduction path extends from the surface to the lateral walls to the bottom of the structure. On the other hand, the conduction path is pressed into the plastic material, such that a plane surface is obtained. In this way, conduction paths a few micrometers thick are embedded completely in the plastic material and do not cause any problem when the surface of the structure is to be covered by a lid.

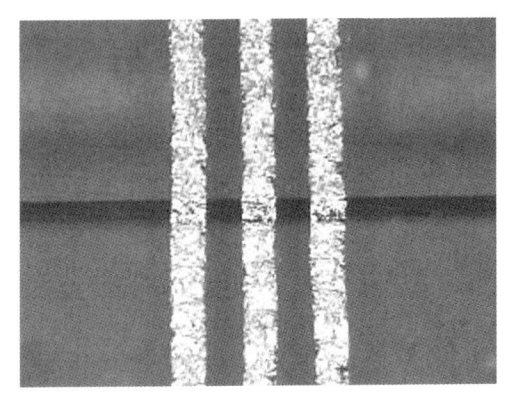

Figure 4.27 Conduction path (width 300 µm) via oblique LIGA edge (height 135 µm).

This technology works not only for sloped lateral walls (Figure 4.27), but also for vertical edges. In this case, the conduction path no longer follows the topology and forms a sharp corner of 90°, but it is completely immersed in the polymer and gradually disappears from the surface. The conduction path is not interrupted, such that the contact from the structure surface down to the trench is maintained.

4.4.2.4 Polymer Mold Inserts

It is possible to vary the materials not only of molded structures, but also of mold inserts. Apart from electrodeposited metals or alloys [31], non-metal materials may also be used. An elegant way of producing mold inserts is embossing in plastics. In principle, any embossed microstructure can be used as a mold. It does not matter whether it will be used as a lost mold in a casting process or for a second electrodeposition step. It is also possible to use such a structure for hot embossing, provided that the requirements are met in terms of mechanical stability and temperature resistance [32, 33]. A suitable mold insert material is PEEK (Figure 4.28). This partly crystalline high-temperature material has a very high deformation temperature and a very narrow softening range. As a result, it can be used in a hot embossing tool up to temperatures above 200 °C. This is sufficient for conventional technical plastics such as PMMA and PC. An advantage of this technology lies in the fact that many identical mold inserts can be fabricated from a single parent structure. Although plastic mold inserts mostly cannot compete with metal mold inserts in terms of stability, they are suited for use as disposable mold inserts for material tests or for work with strongly abrasive, filled polymers. Having the inverted structure of the original mold insert, such mold inserts supply duplicates of the mold insert structure. This property can also be made use of when such a mold insert cannot be fabricated in a conventional way for technical reasons.

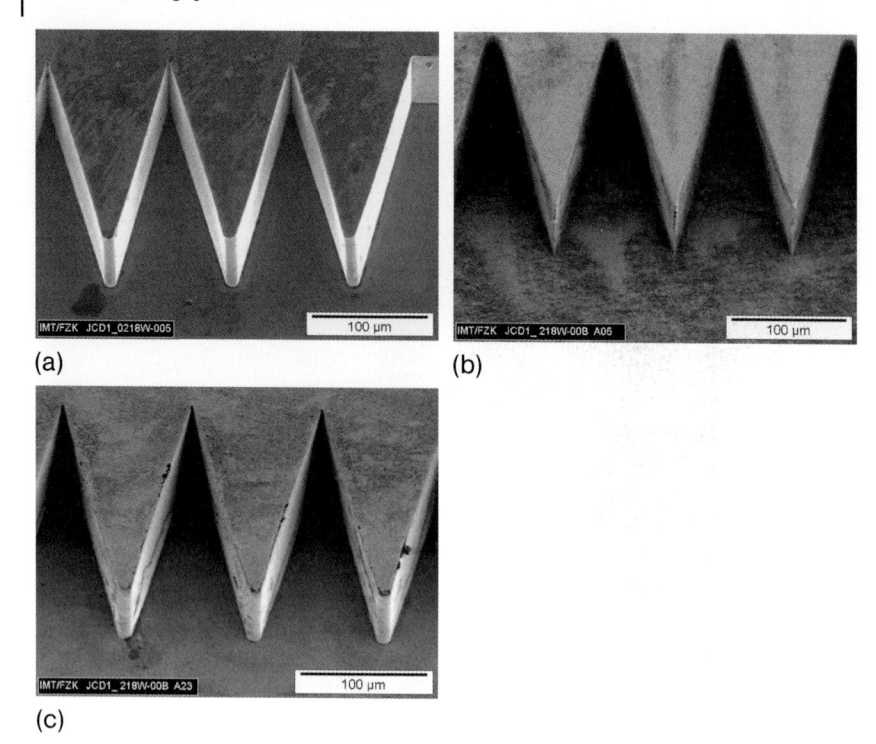

Figure 4.28 Mold insert inversion. From a LIGA mold insert (a) is replicated a PEEK structure (b). The PEEK structure is used again as mold insert to create a PMMA pattern (c).

4.5
Plant Technology

4.5.1
Tools

Although hot embossing is intended to ensure quality rather than high numbers of pieces, process capacity is always of importance in commercial application, as it significantly affects the costs. Increasingly large mold inserts and shorter cycling times are required. However, they can only be guaranteed by more sophisticated plant technology. When analyzing the durations of the different process steps of hot embossing, three characteristic phases can be distinguished: heating, cooling and handling. The drawback associated with heating is that the plastic material is heated via the tool, which is why high tool temperatures are required. Today, compact electric heating cartridges allow high thermal powers to be transferred to the tool. The times needed to insert semi-finished products and remove the embossed parts cover most of the time needed for the entire process. In laboratory operation, the expenditure of automated operation is by no means justified by the

Figure 4.29 Exploded view of a high-end hot embossing tool to illustrate the complexity by means of which the highest requirements in terms of economic efficiency and precision can be fulfilled.

benefit. When formats are fixed and numbers of pieces are higher, however, the expenditure associated with a handling system is worthwhile. For the fabrication of optical prototypes in the chip format, a process chain was presented, which consists of a magazine with the semi-finished products, a fivefold tool and a removal robot.

The most difficult problem is the long cooling time. Because of the high heating powers, large heat volumes have to be removed from the tool. Heat volumes also are enormous, because the stable construction requires a massive setup. Hence large masses have to be heated. By reducing the thermal mass, heating and cooling powers can be reduced. Such a new construction is based on decoupling the heating process from the embossing process. High mechanical stability, parallel conduct and exact opening movement are required during the dwell pressure and demolding phases only (Figure 4.29). The cavities can be heated and filled using low force and tool halves with a floating bearing. This means that the mechanical expenditure needed is low in this process phase and that thermal masses can be reduced decisively. This also accelerates heating. The task now consists in combining these two requirements, small and light for heating and stable and heavy for cooling. In a tool designed within the project mikroFEMOS, the heating and cooling blocks were separated by an air gap [34]. During heating, the heating plate with the microstructured mold insert is separated from the cooling plate. A low contact pressure ensures thermal contact between the mold insert, semi-finished product and countertool. After filling the tool, a high embossing force is applied. At the same time, the gap between the hot embossing die and the cold cooling block is closed. Now, mechanical stability is high for a precise demolding movement and the small thermal mass of the embossing die can be cooled efficiently by the cooling reservoir. This thermal insulation additionally prevents the entire machine from being subjected to a temperature drift during longer phases of operation.

The design of a test tool for hot embossing technology goes a step further [35]. Whereas low tool costs were at the focus when developing the above mikroFEMOS tool, this test tool was conceived such that the surface area was large, cycle times were short and microstructures were embossed on both sides and removed automatically from the mold inserts. A highly precise piezodrive is responsible for the orientation of both tool halves. A complex multi-cycle thermostating unit and an additional tool drive allow for the symmetric demolding of the microstructured front and rear sides.

LIGA tool construction is mainly based on a linear embossing lift, as this is a prerequisite for embossing and demolding high microstructures with high aspect ratios. Continuous processes, such as roll-to-roll embossing, are restricted to low aspect ratios.

4.5.2
Hot Embossing Facilities

Progress has also been achieved in mechanical engineering for hot embossing. Development started from a simple laboratory facility for reaction injection molding. Similar table-based facilities have been built in many application laboratories. In the early 1990s, industry did not see any market for hot embossing facilities. Attention was focused on injection molding, which was given priority in the mass production of plastic microstructures. Manufacturers of presses and embossing machines did not risk entering the microstructurization sector. Jenoptik Mikrotechnik, a manufacturer of lithography facilities, eventually built the first commercial micro hot embossing machine. The facility was based on a modular approach developed at the Forschungszentrum Karlsruhe [36]. The core of the facility was a universal testing machine from Zwick with a torsion-proof frame, a highly precise electric servodrive and a high-quality electronic displacement/force control unit, complemented by a vacuum chamber with a fixing plate for a microstructured mold insert and a heating and cooling unit. An open computer architecture and a number of points to measure the process parameters allowed for the execution of research-oriented work. In the second stage of development, the machine was provided with an adjustment table and a movable microscope for adjusted embossing (HEX03). A high-performance image processing system

Figure 4.30 HEX03, the flagship of the Hex series by Jenoptik Mikrotechnik. The vacuum chamber and a sophisticated demolding system are the basis of high-quality molding.

allowed one to identify and evaluate adjustment marks at various structural heights for the automatic table control system to move to the right position (Figure 4.30).

Following the commercialization of the HEX facilities by Jenoptik, the development potential was exhausted for the time being. With more than 50 facilities sold, hot embossing technology today is represented in a variety of laboratories worldwide.

The facilities of EVG, Obducat and Molecular Imprint that have been commercialized in parallel with the advent of nanoimprint lithography (NIL) are suited for the molding of LIGA mold inserts with only certain limitations. Using wafer bonders or mask aligners, these machines not only allow thermoplastic polymers to be processed, but can also structure UV-sensitive resist systems. The machines are designed for large surfaces (>8 inch) and the thinnest polymer layers. Maximum embossing forces are in the range of 20–25 kN. In addition, the machines are equipped with a simple demolding system (Figure 4.31).

Although these machines can be used for microstructuring with certain limitations, the development does not fit to the above-mentioned approaches to tool construction focusing on injection molding plants and presses. Plant technology for these tools was developed further at the Forschungszentrum Karlsruhe in cooperation with Wickert Maschinenbau. The increase in the embossing area to 8 inch requires an increase in embossing force from a maximum of 250 kN in the past to 1000 kN. In terms of construction, entirely new approaches have to be pursued (Figure 4.32). A two-step hydraulic drive ensuring a high maximum force and smallest touch forces and a large clearance between the four guiding columns for an unrestricted use of a tool handling system characterize the new machine [37]. It is controlled by a classical storage-programmable control system (SPS). In this way, integration of various tool controls in machine control is guaranteed. Moreover, the use of a known user interface with a touch screen is important. It

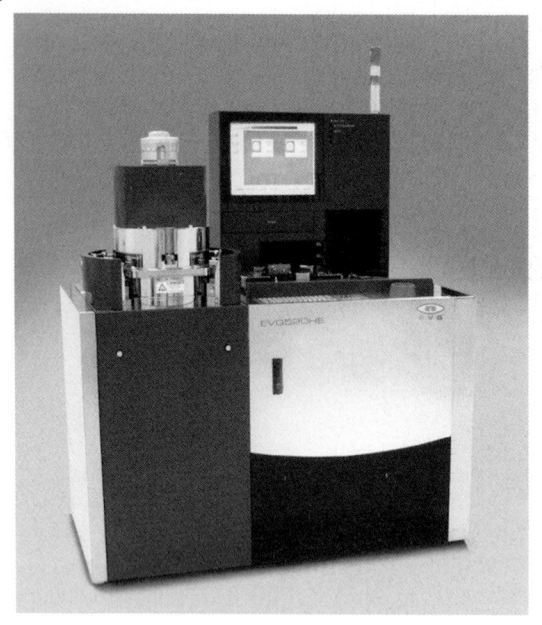

Figure 4.31 Trendsetter in nanoimprint lithography. The EVG 520 is suited for molding LIGA structures with only certain limitations. Its strengths lie in the field of smaller dimensions with smaller aspect ratios.

gives the largely unknown micro hot embossing technology in classical plastics production a familiar appearance. This approach to production is promising, as reflected by the fact that Jenoptik has announced a HEX4 machine with similar equipment.

4.6
Summary and Outlook

Micro hot embossing is of high relevance to molding high structures with large aspect ratios in plastics. Although this replication technique does not play any role on the macroscopic scale, where forming techniques such as thermoforming or blow molding and injection molding methods are applied, it has become an established microstructuring method. It is a flexible process for producing small numbers of prototypes within relatively short periods of time and components of the highest quality for the optics sector in particular.

Hot embossing initially was a mainly empirical laboratory technique. The development of simulation tools and in-depth understanding of the process resulted in the theoretical background knowledge required for its systematic development.

Figure 4.32 Hot embossing machines for production. The Wickert WMP 1000 was the first facility on the market for large surface areas and short cycle times. A maximum embossing force of 1000 kN ensures a sufficient pressure for the embossing of surface areas of up to 8 inch.

In parallel, progress has been achieved in tool and plant construction. Although many newly developed microcomponents are still produced on simple laboratory presses, the research facilities commercialized by Jenoptik and the production plant by Wickert will allow for further, also commercial, developments in this sector.

Of course, nanoimprint lithography has also pushed hot embossing, although it is conducted in other dimensions and at smaller aspect ratios. Today, the situation with plastic replication techniques on the microscale can be described as follows. Of paramount importance is injection molding, which has shifted its focus from the macroscale to smaller dimensions and details. It even allows for the molding of certain LIGA structures. Micro injection molding today represents an established method for production on the microscale. Other plastic molding techniques are applied for special purposes.

The LIGA technique based on X-ray deep-etch lithography has given rise to a variety of similar developments for the production of microstructured mold inserts. UV lithography, laser irradiation and other types of irradiation provide for a large number of alternatives. Hot embossing has turned out to be an ideal partner of these techniques in the development sector. Of course, it will continue to show its strengths at this interface in the future. With an increasing number of plastic microcomponents on the market, however, niche techniques such as micro hot embossing will have to fulfill higher commercial requirements.

References

1 Becker, E.W., Betz, H., Ehrfeld, W., Glashauser, W., Heuberger, A., Michel, H.J., Münchmeyer, D., Pongartz, S. and Siemens, R.V. (1982) Production of separation nozzle systems for uranium enrichment by a combination of x-ray lithography and galvanoplastics. *Naturwissenschaften*, **69**, 520–3.

2 Becker, E.W., Ehrfeld, W., Hagmann, P., Maner, A. and Münchmeyer, D. (1986) Fabrication of microstructures with high aspect ratios and great structural heights by synchrotron radiation lithography, galvanoforming and plastic molding (LIGA process). *Microelectronic Engineering*, **4**, 35–56.

3 Harmening, M., Bacher, W., Bley, P., El-Kholi, A., Kalb, H., Kowanz, B., Menz, W., Michel, A. and Mohr, J. (1992) Molding of threedimensional microstructures by the LIGA process, in Proceedings of MEMS 1992, Travemünde, February 1992, IEEE.

4 Heckele, M., Bacher, W. and Müller, K.D. (1998) Hot embossing-the molding technique for plastic microstructures. *Microsystem Technologies*, **4**, 122–4.

5 Moon, S., Lee, N. and Kang, S. (2003) Fabrication of microlens array using micro-compression molding with electroformed mold insert. *Journal of Micromechanics and Microengineering*, **13**, 98–103.

6 Young, W.-B. (2005) Analysis of the nanoimprint lithography with a viscous model. *Microelectronic Engineering Issues*, **77**, 405–411.

7 Hirai, Y., Konishi, T., Yoshikawa, T. and Yoshida, S. (2004) Simulation and experimental study of polymer deformation in nanoimprint lithography. *Journal of Vacuum Science and Technology B – Microelectronics and Nanometer Structures*, **22**, 3288–93.

8 Sirotkin, V., Svintsov, A., Zaitsev, S. and Schift, H. (2006) Viscous flow simulation in nanoimprint using coarse-grain method. *Microelectronic Engineering*, **83** (4–9), Micro- and Nano-Engineering, MNE, 880–3.

9 Rowland, H.D. *et al.* (2005) Impact of polymer film thickness and cavity size on polymer flow during embossing: toward process design rules for nanoimprint lithography. *Journal of Micromechanics and Microengineering*, **15**, 2414.

10 Juang, Y., Lee, L.J. and Koelling, K.W. (2002) Hot embossing in microfabrication. Part 1: experimental. *Polymer Engineering and Science*, **42**, 539–50.

11 Juang, Y., Lee, L.J. and Koelling, K.W. (2002) Hot embossing in microfabrication. Part 2: rheological characterization and process analysis. *Polymer Engineering and Science*, **42**, 551–6.

12 Worgull, M., Hetu, J.-F., Kabanemi, K.K. and Heckele, M. (2006) Modeling and optimization of the hot embossing process for micro- and nanocomponent fabrication. *Microsystem Technologies*, **12**, 947–52.

13 Worgull, M., Heckele, M. and Schomburg, W.K. (2003) Analysis of the Micro Hot Embossing Process. FZKA-Bericht, 6922, Forschungszentrum Karlsruhe, Karlsruhe.

14 Göttert, J., Marth, M., Mohr, J., Patterson, B. and Söchtig, J. (1996) Integration of III/V-devices on polymer micro-optical benches with single mode waveguides. OSA Conference on Lasers and Electro-Opticcs (CLEO 96), Anaheim, CA.

15 Mohr, J., Göttert, J. and Müller, A. (1996) Microoptical devices based on free space optics with LIGA micro-optical benches: examples and perspectives. *Proceedings of the SPIE – The International Society for Optical Engineering*, **2783**, 48–54.

16 Müller, A., Göttert, J., Mohr, J. and Rogner, A. (1996) Fabrication of stepped micro-optical benches for fibre and free space application. *Microsystem Technologies*, **2**, 40–5.

17 Ossmann, C., Göttert, J., Ilie, M., Mohr, J. and Ruther, P. (1997) Fabrication of PMMA based microlenses using the LIGA-process. EOS Topical Meeting on Microlens Arrays, Teddington.

18 Ruther, P., Gerlach, B., Gottert, J., Ilie, M., Mohr, J., Muller, A. and Obmann, C. (1997) Fabrication and characterization of microlenses realized by a modified LIGA process. *Journal of Optics A – Pure and Applied Optics*, **6**, 643–53.

19 Müller, C. and Mohr, J. (1993) Microspectrometer fabricated by the LIGA process. *Interdisciplinary Science Reviews*, **18**, 273.

20 Krippner, P., Mohr, J., Müller, C. and van der Sel, C. (1996) Microspectrometer for the infrared range, in Proceedings of Microoptical Technologies for Measurement, Sensors and Microsystems, Besançon, pp. 277–82.

21 Wallrabe, U., Solf, C., Mohr, J. and Korvink, J. (2005) Miniaturized Fourier transform spectrometer for the near infrared wavelength regime incorporating an electromagnetic linear actuator. *Sensors and Actuators A*, **123–124**, 459–67.

22 Gerlach, A., Knebel, G., Guber, A., Heckele, M., Herrmann, D., Muslija, A. and Schaller, T. (2002) Microfabrication of single-use plastic microfluidic devices for high-throughput screening and DNA analysis. *Microsystem Technologies*, **7**, 265–8.

23 Steger, R., Mehne, C., Wangler, N., Heckele, M., Zengerle, R. and Koltay, P. (2006) Drop in drop nanoliter kinase assay made with hot embossed disposable multi channel dispenser, in The 10th International Conference on Miniaturized Systems for Chemistry and Life Sciences (µTAS2006), 5–9 November, Tokyo, pp. 999–1001.

24 Müller, K.-D., Bacher, W. and Heckele, M. (1998) *Flexible integration of non-silicon microstructures on microelectronic circuits*, in MEMS 1998: Proceedings of the 11th annual Workshop on Micro Electro Mechanical Systems, Heidelberg, pp. 263–7.

25 Harmening, M., Bacher, W., Bley, P., El-Kholi, A., Kalb, H., Kowanz, B., Menz, W., Michel, A. and Mohr, J. (1992) Molding of three-dimensional microstructures by the LIGA process, in Micro Electro Mechanical Systems 1992, Travemünde, pp. 202–7.

26 Michel, A., Ruprecht, R., Harmening, M. and Bacher, W. (1993) Abformung von Mikrostrukturen auf prozessierten Wafern. Bericht des Kernforschungszentrums Karlsruhe, KfK 5171, Kernforschungszentrum Karlsruhe, Karlsruhe.

27 Both, A., Bacher, W., Heckele, M., Müller, K.D., Ruprecht, R. and Strohrmann, M. (1996) Fabrication of LIGA-Acceleration sensors by aligned molding. *Microsystem Technologies*, **2**, 104–8.

28 Heckele, M. and Durand, A. (2001) Microstructured through-holes in plastic films by hot embossing. in Proceedings of the 2nd EUSPEN International Conference, Turin, pp. 196–8.

29 Mehne, C., Heckele, M., Steger, R. and Warkentin, D. (2006) Hot embossing of large area microfluidic devices with through holes. 6th International Conference of the European Society for Precision Engineering and Nanotechnology (EUSPEN), Baden.

30 Heckele, M. and Anna, F. (2002) Hot embossing of microstructures with integrated conduction paths for the production of lab-on-chip systems. *Proceedings of the SPIE – The International Society for Optical Engineering*, **4755**, 670–4.

31 Guttmann, M., Schulz, J. and Saile, V. (2005) Lithographic fabrication of mold inserts, in *Advanced Micro and Nanosystems*. Vol. **3**, *Microengineering of Metals and Ceramics* (eds H. Baltes, O. Brand, G.K. Fedder, C. Hierold, J.G. Korvink, O. Tabata), Wiley-VCH Verlag GmbH, Weinheim, pp. 187–219.

32 Schröer, M., Heckele, M. and Saile, V. (2004) Polymer mold inserts for hot embossing, MNE 2004, Rotterdam, pp. 362–3.

33 Khan Malek, C., Coudevylle, J.R., Jeannot, J.C. and Duffait, R. (2007) Revisiting micro hot-embossing with moulds in non-

conventional materials. *Microsystem Technologies*, **13**, 475–81.

34 Heckele, M., Dittrich, H., Mehne, C. and Wissmann, M. (2005) Heissprägen von mikrooptischen Komponenten in grossen Stückzahlen, in *µFEMOS – Mikro-Fertigungstechnik für Hybride Mikrooptische Sensoren* (ed. M. Bär) Universitätsverlag Karlsruhe, Karlsruhe, pp. 33–40.

35 Dittrich, H., Heckele, M. and Schomburg, W. (2004) Forschungszentrum Karlsruhe; Werkzeugentwicklung fuer das Heisspraegen beidseitig mikrostrukturierter Formteile, Wissenschaftliche Berichte Forschungszentrum Karlsruhe, FZKA-7058, Forschungszentrum Karlsruhe, Karlsruhe.

36 Heckele, M. and Bacher, W. (1997) Modular hot embossing equipment for MEMS. *Micromachine Devices*, **2** (*2*), 1–3.

37 Dittrich, H., Heckele, M. and Mehne, C. (2005) Double sided large area hot embossing for polymer microstructures with high aspect ratio, in Proceedings of High Aspect Ratio Microstructure Technology (HARMST 2005), Gyeongju, South Korea, pp. 226–7.

5
Exposure and Development Simulation for Deep X-ray LIGA

Jan G. Korvink, Sadik Hafizovic, Yoshikazu Hirai, and Pascal Meyer

5.1	**Modeling and Simulation Needs**	*104*
5.2	**Beam-Line Modeling and Calculation**	*106*
5.2.1	Beamline Data	*109*
5.2.1.1	The Synchrotron Source ANKA	*109*
5.2.1.2	AURORA and BL-15 at Ritsumeikan University	*110*
5.2.2	Object-oriented Data Structures for Beamline and Exposure Modeling	*111*
5.2.2.1	Utility Classes	*114*
5.2.2.2	Optical Element Family	*115*
5.3	**Summary**	*118*
5.4	**Development Simulation**	*118*
5.4.1	Photoresist Damage Theory	*119*
5.4.1.1	PMMA Degradation	*120*
5.4.1.2	Development of Irradiated PMMA	*120*
5.4.2	Degradation	*121*
5.4.2.1	Measurement Methods of PMMA Dissolution Rate	*121*
5.4.2.2	PMMA Dissolution Rate Results	*122*
5.4.3	Advancing Front Methods and Data Structures	*122*
5.4.4	Mesh-free Topology Representation	*124*
5.4.4.1	Fast Marching Topology Representation	*125*
5.4.5	Level Set Topology Representation	*126*
5.4.5.1	Mathematical Geometry Toolkit	*127*
5.4.5.2	Front Normal	*128*
5.4.5.3	Propagation Speed	*128*
5.4.5.4	Surface Curvature	*128*
5.4.5.5	Mean Surface Curvature	*128*
5.4.5.6	Fast Marching Propagation	*128*
5.4.5.7	Eikonal Partial Differential Equation	*129*
5.4.5.8	Front Propagation	*129*
5.4.5.9	The Algorithm	*130*
5.4.5.10	Heap	*130*
5.4.5.11	Update Scheme for Regular Grids	*131*

Advanced Micro & Nanosystems Vol. 7. LIGA and Its Applications.
Edited by Volker Saile, Ulrike Wallrabe, Osamu Tabata and Jan G. Korvink
Copyright © 2009 WILEY-VCH Verlag GmbH & Co. KGaA, Weinheim
ISBN: 978-3-527-31698-4

5.5	**Available Software Packages**	132
5.5.1	Simulation of the Development Time with DoseSim	133
5.5.2	X3D	134
5.6	**Conclusion and Outlook**	139
	References	140

5.1
Modeling and Simulation Needs

When developing complex engineering systems, the entire process from component specification to drawing preparation can be handled by the designer, but the specific requirements demanded by the fabrication techniques employed must be taken into account when optimizing the performance of the final system. For example, microsystem structures close to the optimal should be taken into account in the optical design stage as the optical performance can be influenced by these structures. The final design should meet the requirements of the fabrication process. A concept [1] for a general MEMS process simulation capability, that addresses key processes in the X-ray LIGA process, is shown in Figure 5.1. The process simulation tasks include:

- the mask manufacturing proces
- the resist application technology
- resist irradiation and development steps
- electroplating
- rework of designs
- estimation of the cost of manufacture
- design rule checking and the assessment of manufacturability
- generating work plans.

Design and simulation tasks include:

- verification of the physical function of each component
- extraction of process design rules.

Since the design and fabrication software tools currently available for the LIGA process are still inadequate or unavailable [2], the central idea here is to determine a process flow representation that can be used to organize the tasks mentioned above, namely:

- process design
- process prediction through simulation.

The exposure and development steps of deep X-ray LIGA are unique to the process; the different LIGA steps (UV lithography, proton lithography, e-beam lithography) are mainly differentiated by the type of exposure. We will focus attention on these two steps.

Calculations of the radiation dose deposited in the resist are needed. The calculations required for synchrotron beamline design are related to the spectral

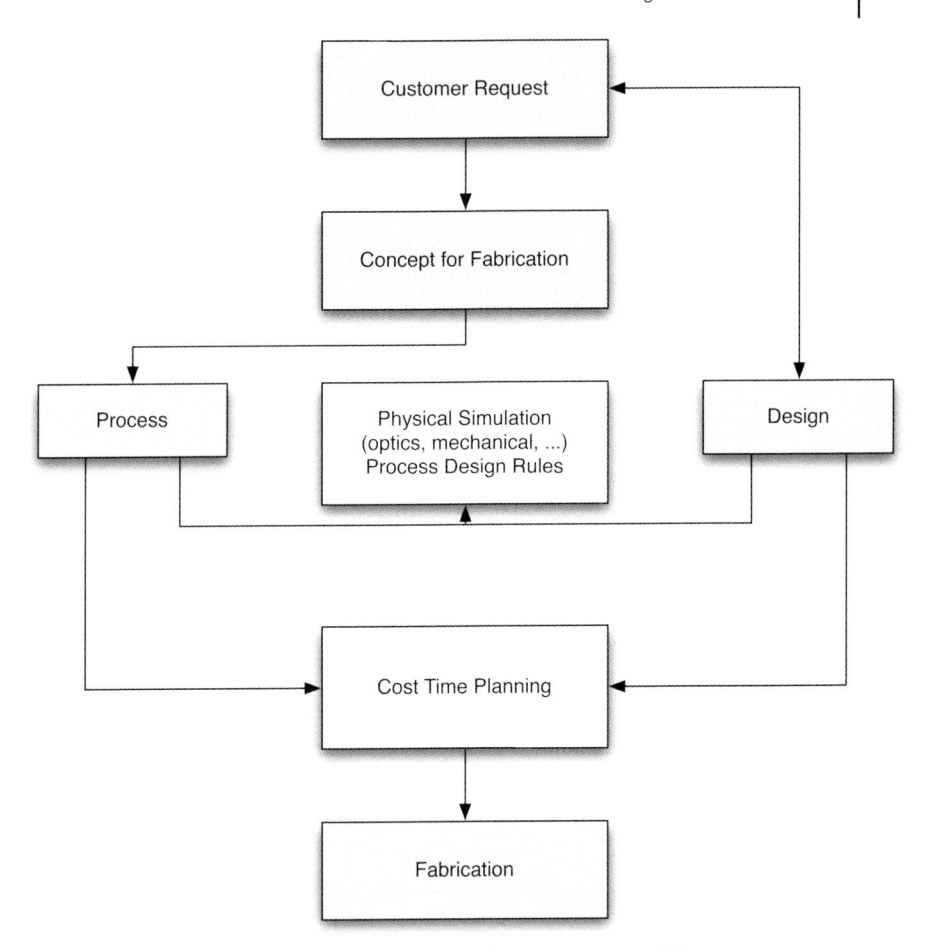

Figure 5.1 Concept for a general MEMS process physics simulation capability.

characteristics and to the modeling of the beamline optical elements (mirrors, filters, beam-stop). A number of programs to calculate the steps already exist, for example:

- XOP [3, 4], for calculation of spectral properties of synchrotron radiation sources and for interaction of X-rays with optical elements;
- Shadow [5, 6], which simulates the photon flux in a SR beamline;
- Transmit [7], which models the properties of the radiation as it propagates through a beamline and subsequently through the mask and resist;
- X-Ray [8], which predicts the cumulative dose delivered to any given layer of a semiconductor process involving X-ray lithography (XRL) steps.

None of the programs listed is specifically dedicated to deep X-ray lithography. In addition to these calculations, further parameters or data may be important for LIGA process users. These are, for example:

- the dose rate;
- the deposited dose profile through the thickness of the resist;
- the exposure dose required (the parameter which should be given to the scanner which moves the sample);
- the time required (or recommended) for optimal development of an irradiated resist sample.

For the LIGA community, the above programs have limited and incomplete applicability. Due to the high complexity involved in successfully interfacing the currently available programs, and also enforcing requirements specific to the LIGA process, the concept then arose of creating an all-inclusive beamline and development program dedicated to deep X-ray lithography. A diagrammatic representation of the modeling of the irradiation and development steps is presented in Figure 5.2. It consists of four tasks:

- Calculation of the spectrum of the source. The light comes from an insertion device such as a bending magnet or wiggler.
- Calculation of the spectrum modification due to the optics before the scanner. Optical elements include filters, mirrors, beam-stops and slits.
- Calculation of the deposited dose in the resist.
- Simulation of the development of the exposed resist.

The beamline modeling approach is presented in Section 5.2 and the exposure modeling in Section 5.3. The resulting program that implements these ideas, called X3D, is presented in Section 5.4.

5.2
Beam-Line Modeling and Calculation

The simulation of resist exposure can be separated into the following interdependent sub-tasks:

- Calculation of the primary dose.

- Simulation of the secondary dose due to secondary effects of the irradiation. Secondary radiation during resist exposure adversely affects feature definition, side-wall taper and overall side-wall offset. Additionally, it can degrade the resist adjacent to the substrate, leading to the loss of free-standing features through undercutting during resist development or through mechanical failure of the degraded material. The source of this radiation includes photoelectrons, Auger electrons and fluorescence photons.

- Effect of heating during the irradiation. Heating of the resist causes thermal expansion, which leads to loss of pattern fidelity.

- The effect of proximity (Fresnel effects at the feature edges) and of the source (e-divergence of the beam radiation).

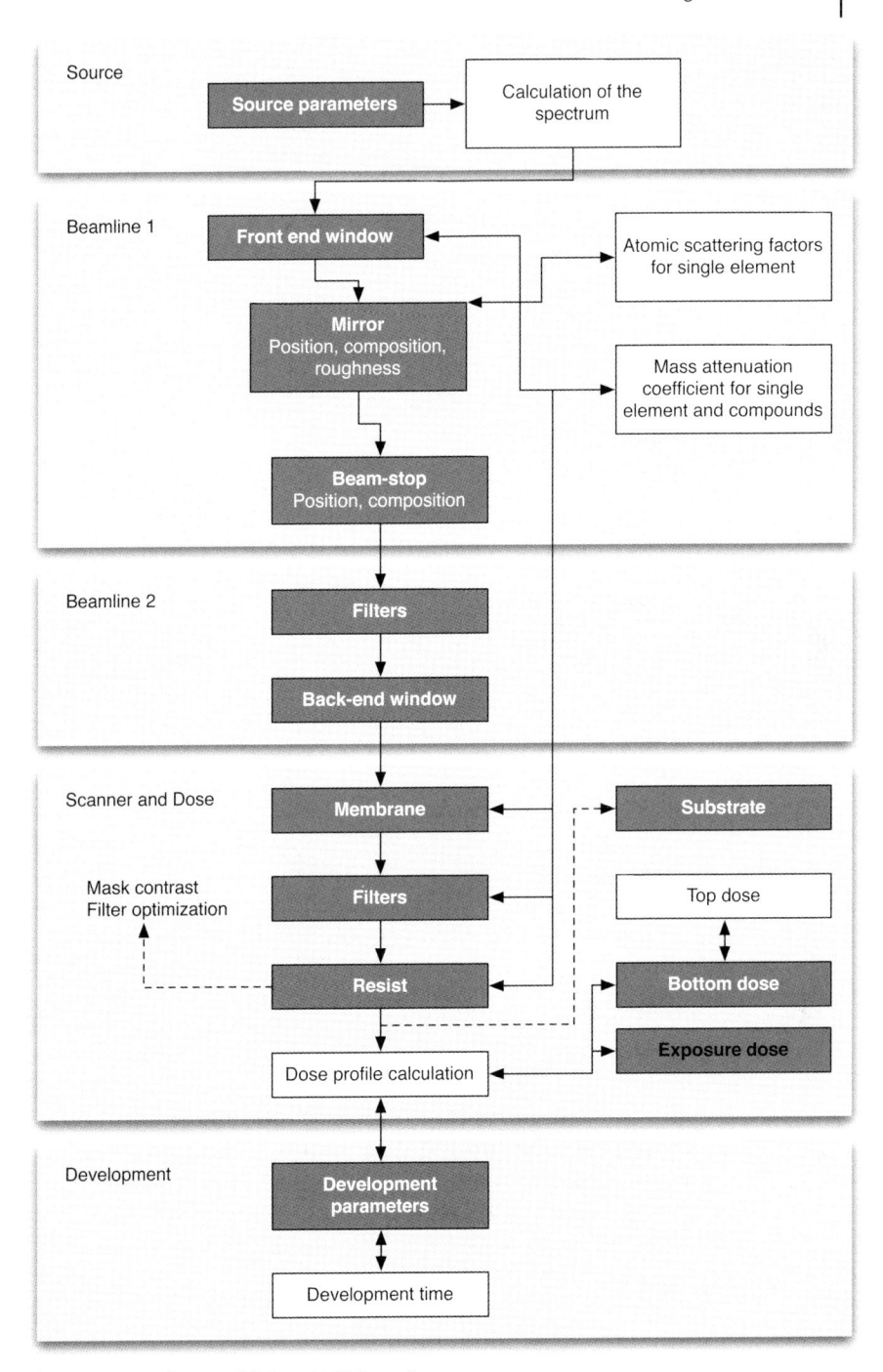

Figure 5.2 Data flow model for a LIGA beamline.

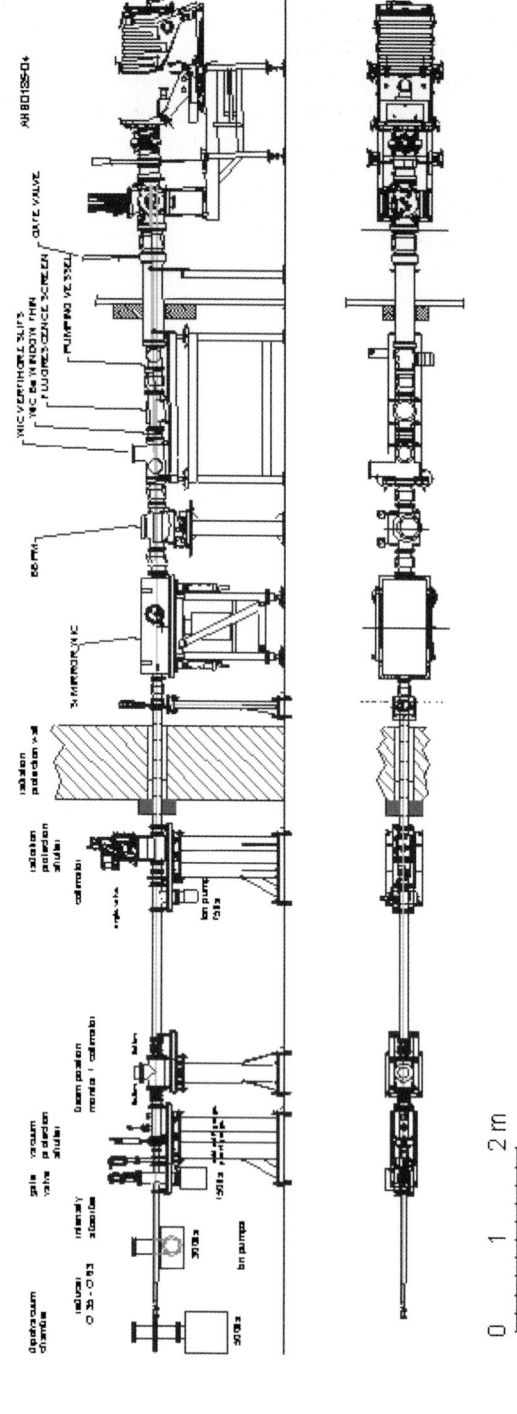

Figure 5.3 Schematic drawing of Lito 2 of the ANKA at the Forschungszentrum Karlsruhe.

As can be deduced, the creation of a simulation tool that takes into account all parameters influencing the exposure and development represents a huge task that needs to be well organized. As noted previously, a number of LIGA exposure and development software tools already exist, each being different in the number of parameters that they take into account and also in program architecture. The next section emphasizes comprehensive exposure program design by taking a closer look at the design decisions employed in creating the X-ray lithography simulator X3D. X3D is a program developed in a joint project between Ritsumeikan University, Japan, and IMTEK, Germany.

5.2.1
Beamline Data

As inspiration for the structure of a LIGA optical path, we provide beamline data in Table 5.1 for the following two synchrotron sources:

- ANKA, located at the Forschungszentrum Karlsruhe (Germany)
- AURORA, located at Ritsumeikan University (Japan).

Due to the capital investment involved, these parameters will not change very often. However, we note that the parameters vary between different beamlines even at a single site. For program design, it will be more important to make a collection of the different optical elements that can arise and their parameters and features, so that these can be modeled in a fairly general manner in the program.

5.2.1.1 The Synchrotron Source ANKA
ANKA is located in Karlsruhe (Germany). It operates a primary source at 2.5 GeV with an initial current of 200 mA. It belongs to the third generation of synchrotron sources. The microstructure laboratory at ANKA consists of three beamlines (LIGA1, LIGA2 and LIGA3) sited on a bending magnet with different optics for deep X-ray lithography, each dedicated to a specific task in the LIGA process. The specification of the lines for the resist PMMA are given in Table 5.2, the corre-

Table 5.1 Outline parameters of two synchrotron light sources actively used for LIGA.

Parameter	ANKA	AURORA
Operating electron energy (GeV)	2.5	0.575
Electron beam current (mA)	200	300
Critical wavelength of emitted X-rays (nm)	0.2	1.5
Critical energy (keV)	6	0.8
Bending magnetic field (T)	1.5	3.8
Circumference (m)	100.4	3.14
Typical beam size Horizontal (mm):	0.45	1.3
Vertical (mm):	0.2	0.14

Table 5.2 The three different LIGA beamlines at the synchrotron ANKA (Germany).

Beamline	LIGA1	LIGA2	LIGA3
Dedicated to	High resolution X-ray mask making and thin-film X-ray lithography (XRL). Typical structural height is up to 100 μm with a smallest width of 5 μm	Deep X-ray lithography (DXRL). Typical structural height ranges from 100 to about 400 μm with a smallest width of 5–20 μm	Ultra-deep X-ray lithography (UDXRL). Typical structural height ranges from 400 to 2500 μm with a smallest width of 8–40 μm
Structure	2.5 D	2.5 D	2.5 D
Optics	Be window thickness 175 μm in total. Grazing incidence mirror Si/200 nm Cr at 15.4 mrad. Optional low-Z filters: C, Al	Be window thickness 225 μm in total. Grazing incidence mirror Si/200 nm Ni at 8.65 mrad. Optional low-Z filters: C, Al	Be window thickness 225 μm in total. Optional low-Z filters: C, Al

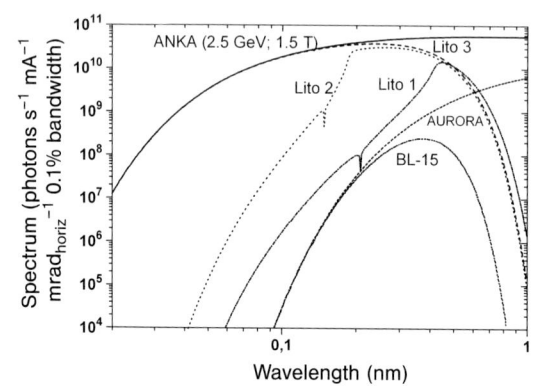

Figure 5.4 Synchrotron radiation spectra for the ANKA ring and the beamline Lito1, 2 and 3 and the AURORA storage ring and BL-15.

sponding spectra are shown in Figure 5.4. A schematic drawing of the LIGA2 beamline is presented in Figure 5.3. The three beamlines are equipped with scanners supplied by Jenoptik (see Figure 5.5); the maximum exposure surface area corresponds approximately to that of a 6 inch wafer.

5.2.1.2 AURORA and BL-15 at Ritsumeikan University

The compact storage ring AURORA, designed and manufactured by Sumitomo Heavy Industries, is installed at Ritsumeikan University, Japan [9]. AURORA consists of 16 beamlines for various applications and four of them (BL-5, -6, -13 and -15) are dedicated to the LIGA process. Each beamline has its own unique specification for deep X-ray lithography and 3D X-ray lithography methods, for

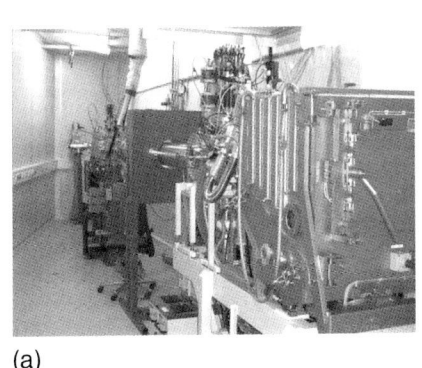

Figure 5.5 Construction of the exposure system: the scanner (a) overview and (b) the mask and sample holder.

example moving mask deep X-ray lithography (M2DXL) [10] with BL-15 and plane-pattern to cross-section transfer method [11] with BL-13.

BL-15 was built based on the M2DXL concept, which is planned to be a precise 3D microfabrication method. It is composed of two beryllium windows (each of thickness 200 μm) as a bandpass filter and isolation between the vacuum of the AURORA and the exposure chamber, as shown in Figure 5.6. The distance between source and substrate is 3.88 m. Figure 5.4 shows the spectral intensity of AURORA and BL-15. The exposure chamber with a size of $0.6 \times 0.6 \times 0.6$ m can offer two kinds of exposure environment: vacuum (below 1×10^{-4} Pa) and helium atmosphere (up to 0.1 MPa), which serves as a coolant for the substrate and mask during the X-ray exposure process.

Figure 5.7 shows the multistage exposure system in the exposure chamber, which is mainly made up of six stages. At the end of them is a precision piezo scanning stage (P-731.10, Physik Instrumente, with a stroke of 200 μm with 10 nm step resolution in closed-loop mode) for the M2DXL technique. By combining these stages, various 3D microstructures such as microprisms, microneedle arrays, microcavity arrays and microlens arrays have been realized by this exposure system [10]. Unlike the standard X-ray lithography process, it is more difficult to control method-specific parameters of the 3D lithography process, such as optimum stage movements and exposure energy due to inhomogeneous dose distributions in the exposed resist, realized by this exposure system. Hence a dedicated process simulation tool to design and support the 3D lithography process was deemed necessary to advance the capability of this free-form 3D X-ray lithography technique.

5.2.2
Object-oriented Data Structures for Beamline and Exposure Modeling

In this section, we emphasize design rather than implementation. We will comment on implementation aspects also, to do justice to the inherent connection of implementation to performance.

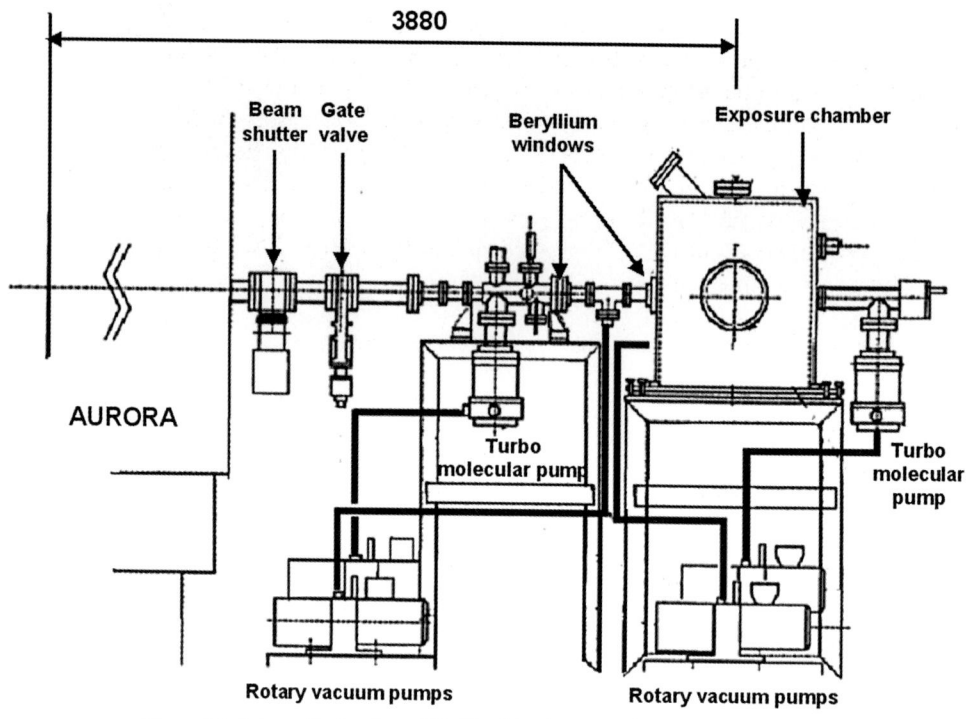

Figure 5.6 Schematic drawing of BL-15 of the AURORA facility at Ritsumeikan University.

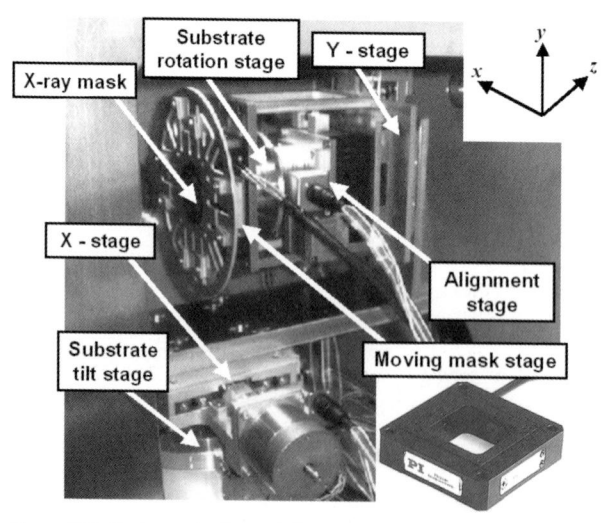

Figure 5.7 Photograph of the multi-exposure stage system. The exposure system is mainly made up of six stages: an X-stage, a Y-stage, a substrate tilt stage, a substrate rotation stage, an X–Y moving mask stage and a manual precision X–Y stage for the alignment between mask and substrate.

In designing a software system that models the exposure process, object-oriented techniques prove extremely useful. They allow coherence of the software model to the real world setup to a degree which is otherwise out of reach. The physical state of the exposure system is projected to a group of classes. In doing so, it becomes apparent that all elements involved in the system share the property that they may cast light on a successive element. Therefore, an abstract class named Optical_Element (the classes are specified in the Z-notation; see http://en.wikipedia.org/wiki/Z_notation for a good introduction where this feature is defined) (see also Figure 5.8):

```
_Optical_Element_____
≻ Transient
↾beam : R² → Beam              [beam cast on successive element at (x,y)]
⨯back_trace : R² → R²
                    [ trace beam from (x,y) on successor element back to self ]
distance_to_successor : R
_____
```

This forms the backbone of the exposure simulation. Note that it is a child of class Transient which provides access to global time. This is needed to allow an optical element to perform a scan movement at a precise global time instant. Details are given below and in Figure 5.9.

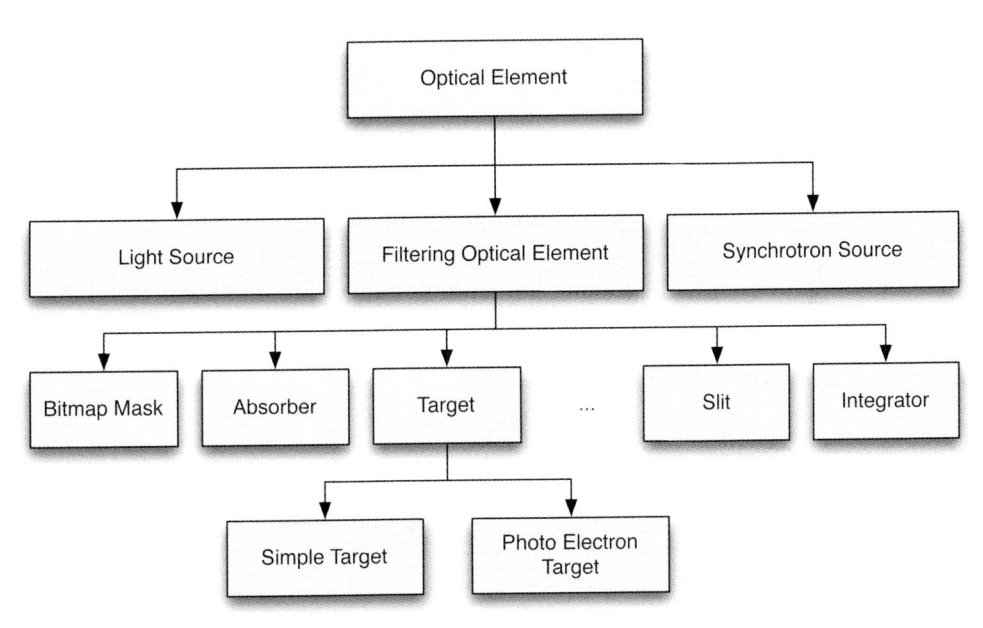

Figure 5.8 Subset of the class family hierarchy Optical Element. Omission is marked by '...'. Note the position of target – it allows additional use as a filter. Therefore, multiple targets may be stacked.

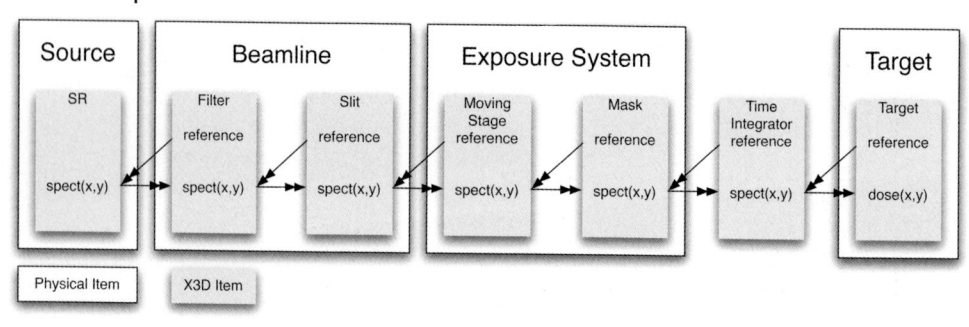

Figure 5.9 Optical elements chain. The white boxes depict physical units whereas the gray are classes in X3D. The left arrows denote references that pass coordinates, making the chain similar to a single linked list. The right arrows denote the flow of spectral data.

5.2.2.1 Utility Classes

The class Transient is the global time keeper. All optical elements inherit from it and thus are given the means to set and to obtain the time to adjust their state accordingly. For example, the synchrotron's beam current intensity depends on a global clock just as the position of the moving stage does.

$$
\begin{array}{l}
\underline{\quad Transient\quad}\\
\quad\underline{\Delta set}\\
\quad ?newtime : \mathbb{R}\\
\quad time' = newtime\\
\\
\quad \underline{\Xi get}\\
\quad time! : \mathbb{R}\\
\quad time! = time\\
\\
\quad static\ *\ time : \mathbb{R} \qquad\qquad \text{[time is stored in units of seconds]}
\end{array}
$$

Another utility is the Path class family. The basic property to implement an arbitrary geometric mapping is given with the following abstract class Path:

$$
\begin{array}{l}
\underline{\quad Path\ [TYPE]\quad}\\
\quad pos : Time \to TYPE
\end{array}
$$

Incarnations of path are constant, exponential and linear interpolating paths.

$$
\begin{array}{l}
\underline{\quad Constant\_Path[TYPE]\quad}\\
\succ Path\\
\\
pos = \_ \mapsto constant\_value\\
constant\_value : \mathbb{R}
\end{array}
$$

5.2 Beam-Line Modeling and Calculation

```
┌─ ExpPath[TYPE] ─────────────────────────────────
│ ≻ Path
│ c : TYPE
│ λ : ℝ
│ ┌─ Ξpos ──────────────────────────────────────
│ │ time? : ℝ
│ │ ret! : TYPE
│ │
│ │ ret! = c e^{λ·t}
└─┴──────────────────────────────────────────────
```

```
┌─ LinearPath[TYPE] ──────────────────────────────────────────────
│ ≻ Path
│ ┌─ Ξpos ────────────────────────────────────────────────────
│ │ time? : ℝ
│ │ ret! : TYPE                    [linear interpolation based on list]
│ │
│ │ list : ⟨< ℝ, TYPE >⟩           [ ascendingly ordered w.r.t. first ele. ]
└─┴───────────────────────────────────────────────────────────────
```

5.2.2.2 Optical Element Family

All optical elements inherit from the abstract class Optical_Element as depicted in Figure 5.8. Hence all optical elements have in common that they may be queried for a beam at an out-of-axis coordinate x,y at the axis coordinate of an immediately successive optical element. Therefore, every optical element needs and knows the distance to its successor optical element. Also, every optical element has access to the global time through the inheritance of Transient.

An object of class Beam is passed through the chain of filtering optical elements, each processing it. It carries the spectrum, angular orientation and beam divergence along the length of the beamline.

```
┌─ Beam ──────────────────────────────────────────────────────────
│ intensity : ℝ → ℝ                              [histogram]
│ ─────────────────────────────────────────────
│ ∀freq : ℝ • intensity(freq) = spectral intensity
│ divergence : ℝ                                  [needed for propagation]
│ rot_x, rot_y : rad                              [needed for propagation]
└─────────────────────────────────────────────────────────────────
```

A direct, instantiable descendant of Optical_Element is a light source, slightly more sophisticated is the synchrotron source:

```
┌─ Light_Source ──────────────────────────────────────────────────
│ ≻ Optical_Element ≻ Transient
│ ─────────────────────────────────────────────
│ The most basic light source possible: constant intensity.
│ beam : ℝ² → Beam
└─────────────────────────────────────────────────────────────────
```

```
┌─ Synchrotron_Source ──────────────────────────────────────┐
│ ≻ Optical_Element ≻ Transient                             │
│ beam : $\mathbb{R}^2 \to$ Beam                            │
│ ∗electron_energy : $\mathbb{R}$          [[e⁻ energy]=eV] │
│ ∗radius : $\mathbb{R}$                   [[SR-radius]=m]  │
│ ∗currentPath : Path              [[electron beam current]=A] │
└───────────────────────────────────────────────────────────┘
```

Another branch of the family of optical elements shares the property that it always has a preceding optical element. We could think of these as being a type of filtering optical element. The abstract class Filtering_Optical_Element enriches the basic optical element by additional properties.

```
┌─ Filtering_Optical_Element ───────────────────────────────┐
│ ≻ Optical_Element ≻ Transient                             │
│                                                            │
│ All classes inheriting from Filtering_Optical_Element construct with an object
│ of Optical_Element. In the sequel, whenever the filtering element will be asked
│ for the intensity at coordinates x,y, it will in turn ask its previous_element for
│ input, and then return the filtered and propagated beam.  │
│ beam : $\mathbb{R}^2 \to$ Beam   [implementation of previously abstract method]
│ apply : $\mathbb{R}^2 \times$ Beam $\to$ Beam   [virtual function called by beam]
│ ∗previous_element : Optical_Element                       │
│ ∗propagate : Beam $\to$ Beam                              │
└───────────────────────────────────────────────────────────┘
```

Any element inheriting Filtering_Optical_Element, may rely on the beam method implemented therein. It handles all ray tracing and propagation and calls the application (apply) method which is supplemented by the inheriting class. Here are two useful variants:

```
┌─ Slit ────────────────────────────────────────────────────┐
│ ≻ Filtering_Optical_Element ≻ Optical_Element ≻ Transient │
│ apply : $\mathbb{R}^2 \times$ Beam $\to$ Beam   [the most basic filtering]
│ width, height, x_center, y_center : $\mathbb{R}$          │
└───────────────────────────────────────────────────────────┘
```

```
┌─ Bitmap_Mask ─────────────────────────────────────────────┐
│ ≻ Filtering_Optical_Element ≻ Optical_Element ≻ Transient │
│                                                            │
│ Uses a bitmap to filter the beam.                         │
│ A gray scale tiff file is read and multiplied with the in beam. Black
│ (value=0) and white (value=255) correspond to block completely and
│ pass completely, respectively.                            │
│  ┌─ INIT ─────────────────────────────────────────────┐   │
│  │ tiff_file_name? : String                           │   │
│  │ width?, height?, x_offset?, y_offset? : $\mathbb{R}$ │ │
│  └────────────────────────────────────────────────────┘   │
│ apply : $\mathbb{R}^2 \times$ Beam $\to$ Beam             │
│  ┌──────────────────────────────────────────────────────┐ │
│  │ ∗domain : $\mathbb{PR}^2$                            │ │
│  │ $x,y \in$ domain $\Leftrightarrow x \in$ (x_offset, x_offset + width)
│  │              $\wedge\ y \in$ (y_offset, y_offset + height)) │
│  └──────────────────────────────────────────────────────┘ │
│ $[x,y] \in$ domain $\Rightarrow$ apply$(x,y,\_) = 0$     │
└───────────────────────────────────────────────────────────┘
```

The Absorber_Mask filters the beam either through the absorber and membrane or the membrane only.

```
┌─ Absorber_Mask ─────────────────────────────
│ ≻ Filtering_Optical_Element ≻ Optical_Element
│ apply : R² × Beam → Beam
│ membrane_attenuation : Path
│ absorber_attenuation : Path
└─────────────────────────────────────────────
```

```
┌─ Absorber_Mask ─────────────────────────────
│ ≻ Filtering_Optical_Element ≻ Optical_Element
│ apply : R² × Beam → Beam
│ membrane_attenuation : Path
│ absorber_attenuation : Path
└─────────────────────────────────────────────
```

A somewhat special class is the Integrator. Since the light sources in the general case emit intensities rather than energies, time integration is required. The Integrator class uses the mechanism provided by the inherited Transient class to set the global time and then to evaluate a beam. The integration scheme is an adaptive Gauss integration. However, its implementation is non-trivial because moving masks generate non-smooth, ill-behaving integrands: a very small time step might illuminate a spot previously shadowed by the moving mask.

```
┌─ Integrator ────────────────────────────────────────────
│ ≻ Filtering_Optical_Element ≻ Optical_Element ≻ Transient
│ implements an operator: $\int_{start\_time}^{end\_time} dt$
│ beam : R² → Beam
│              [overwrites the definition in Filtering_Optical_Element ]
│ start_time, end_time : R
└─────────────────────────────────────────────────────────
```

The abstract class Target introduces the virtual member dose. A call to dose will trigger the respective time-integration. A simple target simply requires integration.

```
┌─ Target ────────────────────────────────────────────────
│ ≻ Filtering_Optical_Element ≻ Optical_Element ≻ Transient
│ dose : R³ → R
│ dump_geometry : Stream → void
└─────────────────────────────────────────────────────────
```

```
┌─ SimpleTarget ──────────────────────────────────────────
│ ≻ Target ≻ Filtering_Optical_Element ≻ ...
│ dose : R³ → R
│ absorption_coefficient : Path
│ thickness : R                       [a target also is a filter!]
└─────────────────────────────────────────────────────────
```

5.3
Summary

The object-oriented structure given above enables us to model the physical beamline in a fairly general manner. Optical elements of a variety of types can be cascaded to form a model of the physical beamline. In fact, it is fairly straightforward to parse a user specification of a beamline design into an instantiated structure using the above definitions.

The advantage of object orientation is also made clear. Abstract classes define the generic behavior of the objects and specialized classes introduce more detail. The inheritance mechanism is used to integrate the timekeeping function and functions associated with control instrumentation. Should more detail be required of the behavior of a single optical element, such as the incorporation of secondary effects on the distribution of beamline intensity, this is now easily added without making modifications to the overall structure.

A program that executes its task based on the presented data structures will compute a three-dimensional delivered radiation dose profile in the target material as depicted in Figure 5.10. These data are necessary for the resulting development simulation described next.

5.4
Development Simulation

Development simulation aims at precisely predicting the final 3D structure for a given irradiation profile and choice of developer and resist. A prerequisite for such a simulation process is the availability of a resist damage model, which predicts

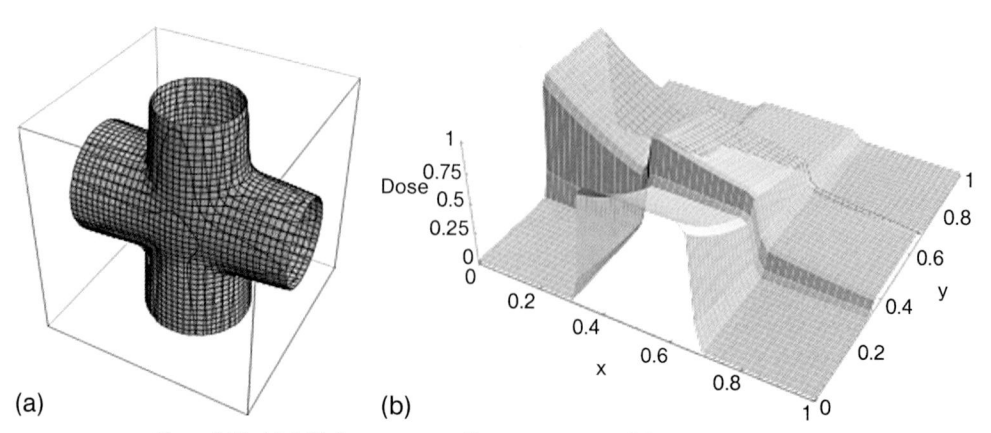

Figure 5.10 (a) 0.05 dose contour; (b) cross-section of dose profile. These dose profiles are used for the example calculations in this chapter.

the conversion process in a resist due to irradiation and a dissolution model, which predicts the 'etch' rate as a function of the level of damage in a polymer resist. In this section, we discuss these two issues and then present data structures and a numerical method for the advancing front simulation.

5.4.1
Photoresist Damage Theory

In X-ray lithography, X-rays typically in the 0.5–10 range are used, which interact with matter by the photoelectric effect, the Compton effect and through Raleigh scattering. The net effect depends on the cross-section of the different possibilities of interaction. In deep X-ray lithography, about 95% of the deposited dose is due to the photoelectric effect in the resist. Secondary effects, such as fluorescence radiation, generation of photoelectrons in the mask membrane and the plating base are negligible during exposure.

A distinction should be made between positive and negative resists as summarized in Table 5.3. In the first case, the radiation will damage the polymer by reducing its molecular weight. The most often used resist material is poly(methyl methacrylate) (PMMA). The damaged parts become soluble in a developer. In the case of a negative resist, the radiation will 'damage' the polymer by increasing its molecular weight (for example, due to curing or crosslinking). In contrast, the irradiated volume is now insoluble in a developer. The most often used resist material is an epoxy-based resin such as SU-8^{TM} [12].

In one case, we need to know the dose to apply to the resist so that it becomes soluble in a developer; the developer having a negligible influence on the non-irradiated parts; in the second case, we need to know the dose to apply to the resist so that the exposed part becomes insoluble in the developer.

The development (or etching) of the irradiated resist concerns only the positive resist since all of the resist can be attacked. The negative resist is also developed but only the unexposed part is removed, the exposed part being practically insoluble. This makes simulation considerably simpler. The next section will discuss the positive resist PMMA. A study of its dissolution rate as a function of the dose applied will be provided.

Table 5.3 Comparison of characteristics of 'positive' and 'negative' resists.

Type of resist	Example	Effect of X-ray	Development
Positive	PMMA	Chain scission–decrease in the molecular weight	Function of the dose applied and the kind of PMMA
Negative	SU-8	Crosslinking–increase in the molecular weight	Function of the type of SU-8 applied

5.4.1.1 PMMA Degradation

During X-ray irradiation of PMMA, synchrotron light is absorbed in the exposed PMMA area, which results in a chemical modification; scission of the polymer chain leads to radiation-induced degradation of the molecular weight and it becomes soluble in an organic developer. By increasing the dose of radiation, the average molecular weight decreases from an initial value $M_{W(D_0)}$ to a minimum limiting value between 2500 and 3000 g mol^{-1} at very high doses of radiation. The degradation mechanism of radiation-excited PMMA depends on the chemical structure of the resist and the exposure energy. The radiochemistry of PMMA is a complex mixture of consecutive reactions including excitations, fission, crosslinking, recombination, disproportion, rearrangements and transfer reactions. The most important step of degradation is the scission of the methyl ester group, which is responsible for the major amount of the gases evolved. The remaining polymer chain stabilizes after hydrogen abstraction by formation of a double bond or by chain scission.

Radiochemical degradation of PMMA is categorized into two schemes [13–15]. About 80% of the main-chain scissions have been observed after preceding side-chain degradation. The radiation-excited polymer molecule splits off an ester side-chain. The remaining chain radical stabilizes after hydrogen abstraction by formation of a double bond or reacts by way of a main-chain scission. In the case of stabilization by hydrogen abstraction the molecule has one ester side-chain less than before irradiation. The remaining 20% of the main-chain scissions are due to a direct decomposition of the polymer into two macromolecules. Recombination of these fragments results in the primary polymer molecule.

The effect of X-rays on the molecular weight can be characterized by the G value, defined as:

$$G = G[s] - 4G[x] \tag{1}$$

where $G(s)$ is the scission coefficient and $G(x)$ is the crosslink coefficient. The G value is defined as the number of effective scission events per 100 eV. The molecular weight $M_{W(D)}$ after X-ray irradiation is calculated by the following expression:

$$\frac{1}{M_{W(D)}} = \frac{1}{M_{W(D_0)}} + G \frac{D}{100 \times 1.6 \times 10^{-19} \times N_A \times \rho} \tag{2}$$

where $M_{W(D_0)}$ represents the molecular weight before the exposure, N_A is Avogadro's number, D is the dose in J cm^{-3}, ρ is the density of PMMA and 1.6×10^{-19} is the conversion factor from eV to J [16, 17].

5.4.1.2 Development of Irradiated PMMA

The dissolution rate is a function of molecular weight, which is related to the initial PMMA molecular weight, dose and the main-chain scission yield. This lowering of the average molecular weight causes the solubility of the resist in the developer to increase dramatically. A developer suitable for PMMA in X-ray lithography, commonly referred to as the GG developer [18], is composed of 15 vol.% deionized water, 60 vol.% 2-(2-butoxyethoxy)ethanol, 20 vol.% tetrahydro-1,4-oxazine and 5 vol.% 2-aminoethanol.

In chemistry, many models [19–21] concerning polymer dissolution may be found. As a dose profile is deposited during X-ray lithography and the GG developer consists of four components, these models cannot be applied easily.

Many investigations in X-ray lithography show that the dissolution rate is a bare function of dose. A dissolution rate obtained from a specific depth h_1 in a thick PMMA sheet at dose $D(h_1)$ agrees with the dissolution rate in another sheet with a top dose value $D(h_0 = 0)$. The dissolution rate is only controlled by the absorbed X-ray dose value.

The dissolution rate R of a positive resist irradiated with dose D [16, 19–24] is described by the following equation (all other parameters being constant):

$$R(D) = R_0 + C \left[M_{W(D)} \right]^\beta \tag{3}$$

where R_0 correspond to the dissolution rate of an infinite MW and C and β are characteristic constants of the polymer and solvent. R_0 is negligible in the case of the PMMA–GG developer system. Based on Equations 2 and 3, one obtains

$$R(D) = C \times \left[\frac{1 + \frac{G M_{W(D_0)} D}{100 \times 1.6 \times 10^{-19} \times N_A \times \rho}}{M_{W(D_0)}} \right]^\beta \approx C \times \left(\frac{G}{100 \times 1.6 \times 10^{-19} \times N_A \times \rho} \right)^\beta \times D^\beta \tag{4}$$

Finally, the dissolution rate as a function of the dose is approximated by the following equation:

$$R(D) = kD^\beta \tag{5}$$

with the constant

$$K = C \times \left(\frac{G}{100 \times 1.6 \times 10^{-19} \times N_A \times \rho} \right)^\beta$$

based on the assumption that the G value is constant. The dissolution rate R is usually given in $\mu m\, min^{-1}$ and the dose in $kJ\, cm^{-3}$ [19, 20].

5.4.2
Dissolution

5.4.2.1 Measurement Methods of PMMA Dissolution Rate

In order to calculate the necessary development time or the resist development profile of 2.5/3D microstructures, the resist dissolution rate is primary input data and must be determined experimentally. As the X-ray lithography process simulation needs to calculate high aspect ratio microstructure and realistic MEMS device geometries with sizes on the order of millimeters, a macroscopic resist dissolution and easily measurable approach are required rather than describing the very complicated problem of microscopic resist dissolution. Then a dedicated experimental method, that takes into account the non-linear dependence of dissolution rate on dose, is required to determine the dissolution rate with high precision. Many authors have investigated the dissolution rate of PMMA as a function of:

- synchrotron radiation source
- radiation dose
- development temperature (room temperature or higher temperature)
- development method (with agitation, without agitation, kind of agitation and so on)
- other factors influencing dissolution rate (type of PMMA, aspect ratio of the structure, development directly or not after the irradiation and so on).

The dissolution rate as a function of dose can be obtained with a single exposure by calculating the dose profile as a function of PMMA depth and measuring the dissolution rate at various depths. The major approach to measure the dissolution rate involves the following steps. The exposed PMMA samples are immersed in the developer. At fixed time intervals, the samples are taken out of the developer, rinsed with deionized water, dried and the development depth is measured. Then, the samples are immersed in the developer again and this process is repeated until the exposed PMMA is completely developed or until the thickness change is small. In another approach to measure the development depth with high precision, a mechanical *in situ* measuring instrument with a probe tip [16], and using stacked PMMA slabs that are exposed from the slab side (that is parallel to the slab surface) in order to measure for wide dose energy ranges with a single exposure (see Figure 5.11) [22], have been proposed.

5.4.2.2 PMMA Dissolution Rate Results
The different PMMA types, methods of measurements, models and their results found in the literature are summarized in Table 5.4 [16, 19, 21, 23, 25] and some results are plotted in Figure 5.12. The dissolution rate is strongly dependent on the development temperature, the kind of PMMA and the type of agitation used. As a first approximation, the dissolution rate can be defined as a function of dose only. The relationship of dissolution rate and dose is more complicated; the analysis of other factors (influence of the dose rate and of the spectrum and their effects) is a key issue for a complete and precise development process simulation tool.

5.4.3
Advancing Front Methods and Data Structures

Following an introduction to the representation of topological data, this section describes a method to construct topologies from speed functions, the so-called fast marching method.

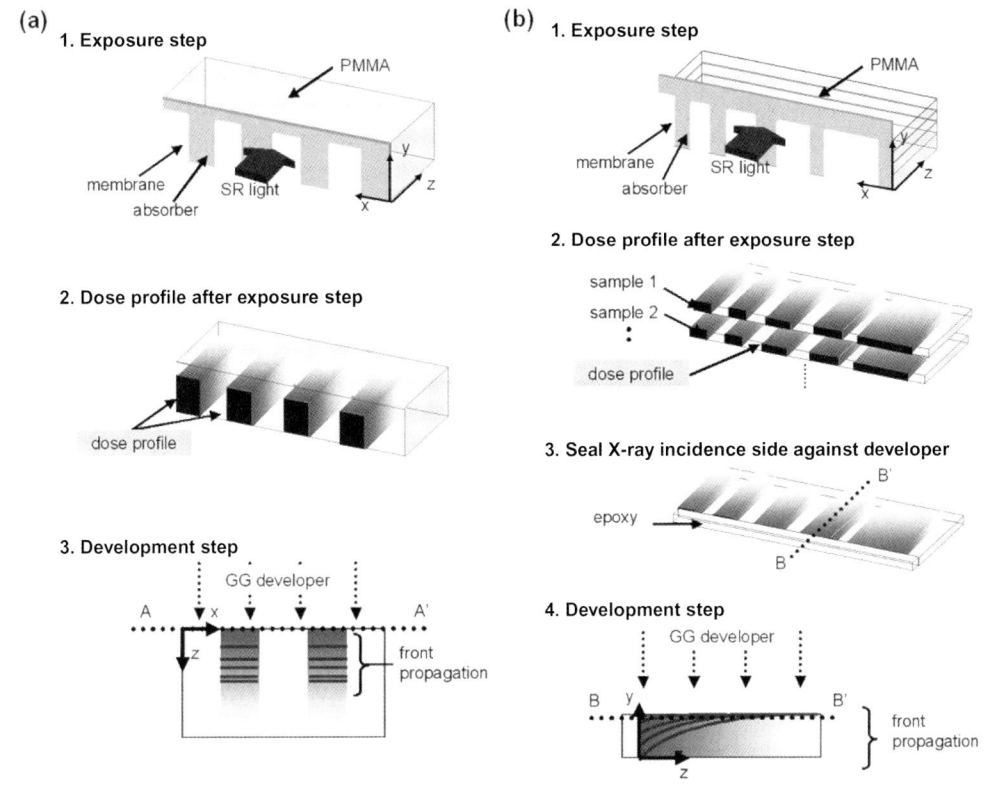

Figure 5.11 Comparison of measurement method for dissolution rate.

In describing and working on the spatial structure of objects, two aspects become apparent: *geometry* and *topology*. Geometry ascribes distances whereas topology expresses connectedness. The traditional approach to encoding spatial structure is to combine both geometric and topological information directly into a polygonal mesh. We call this an explicit representation, since the surface is directly given by the coordinates of the mesh's nodes. This strategy is suitable provided that only the geometry changes but the topology remains invariant. Once this precondition is no longer valid and surfaces join, break up, emerge and vanish, intricate complications arise from the problem of connecting and disconnecting nodes of the mesh. Notwithstanding these drawbacks, parts of this strategy have been developed and successfully applied to dynamic boundary tracking. Above all, so-called string methods have been used in microlithography simulations [26].

Implicit spatial structure encoding and tracking techniques such as level set methods, first introduced by Sethian [27] and subsequently refined in many respects and extensively reported [28], are potent replacements for explicit surface tracking techniques. An example of a level set is shown in Figure 5.13. The main idea in implicit methods is not to rely on nodes to represent the described surface. This connection between the mesh and the surface to be described is given up. Hence a unique separation of geometric data and topological data is achieved:

Table 5.4 Results of studies concerning the PMMA dissolution rate.

Reference	Type of PMMA; MW	Development conditions	Type of samples	Type of measurements	Relation of rate vs dose
Liu et al., 1998 [19]	Cast MW = 2.75 × 10^6	Stirring, 21°C, 37°C	Dose profiles Foils; 205–750 μm	Ex situ, micrometer	
Pantenburg et al., 1998 [23]	Foils GS 233, Rohm MW = 1.2 × 10^6	Stirring; 21°C, 37°C	Homogeneous dose 100 μm foils	Ex situ, weight loss measurement	
	Cast MW = 6.0 × 10^6	Stirring; 21°C, 37°C	Homogeneous dose 100 μm foils	Ex situ, weight loss measurement	
Griffiths et al., 1999 [25]	Foils Perspex CQ-grade; Goodfellow MW = 2.75 × 10^6	Stirring; 35°C	Dose profiles 1 mm foils	Ex situ, micrometer	a, b, c are coefficients for synchrotron sources
Meyer et al., 1999 [16]	Foils GS 233, Rohm MW = 1.2 × 10^6	No agitation; 21°C	Dose profiles 500 and 100 μm foils	In situ, mechanical measuring system	
	Cast MW = 6.0× 10^6	No agitation; 21°C	Dose profiles 500 μm microstructures	In situ, mechanical measuring system	
Meyer et al., 2002 [21]	Foils GS 233, Rohm MW = 1.2 × 10^6	No agitation; 21°C	Dose profiles 100, 400 and 500 μm microstructures	In situ, mechanical measuring system	
		Megasonic supported 21°C	Dose profiles 100, 400 and 500 μm microstructures	Ex situ, mechanical measuring system	

whereas the described surfaces may join or break up, emerge or vanish, the structure of the underlying mesh is not affected.

5.4.4
Mesh-free Topology Representation

The topological interface set Γ is represented by a level-set of a scalar function $f : \mathbb{R}^m \rightarrow \mathbb{R}$. Mathematically, a level-set Γ may be defined as

$$\Gamma(level) = \{\mathbf{x} : \mathbb{R}^m | f(\mathbf{x}) = level\} \tag{6}$$

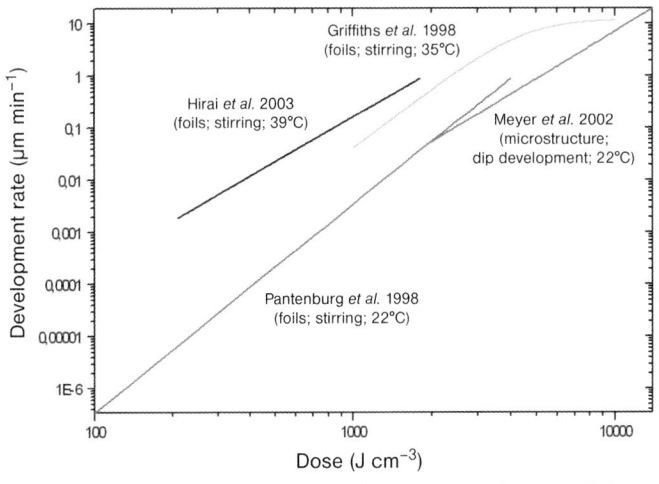

Figure 5.12 Comparison of PMMA dissolution rate as a function of dose with GG developer.

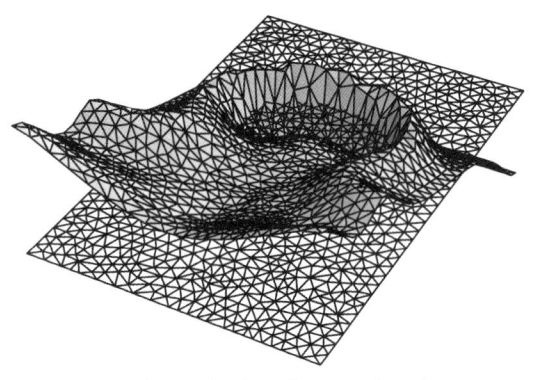

Figure 5.13 A discrete level set function plotted over a triangle mesh. The intersection of the surface with the horizontal plane implicitly encodes a 2D boundary curve of arbitrary topological complexity in R^2, in this case a liquid–gas interface. Image courtesy of Oliver Rübenkönig, IMTEK.

Furthermore, to capture evolving topographies, a time variable t is introduced. There are two ways to achieve this: the *fast marching* approach is more straightforward and computationally cheaper but is limited to strictly monotone moving or 'marching' interfaces. The *level set* approach overcomes this limitation at the cost of adding one dimension (the time variable) to the domain of f.

5.4.4.1 Fast Marching Topology Representation

All topological information is contained in an arrival time field:

$$T : \mathbb{R}^m \to \mathbb{R} \tag{7}$$

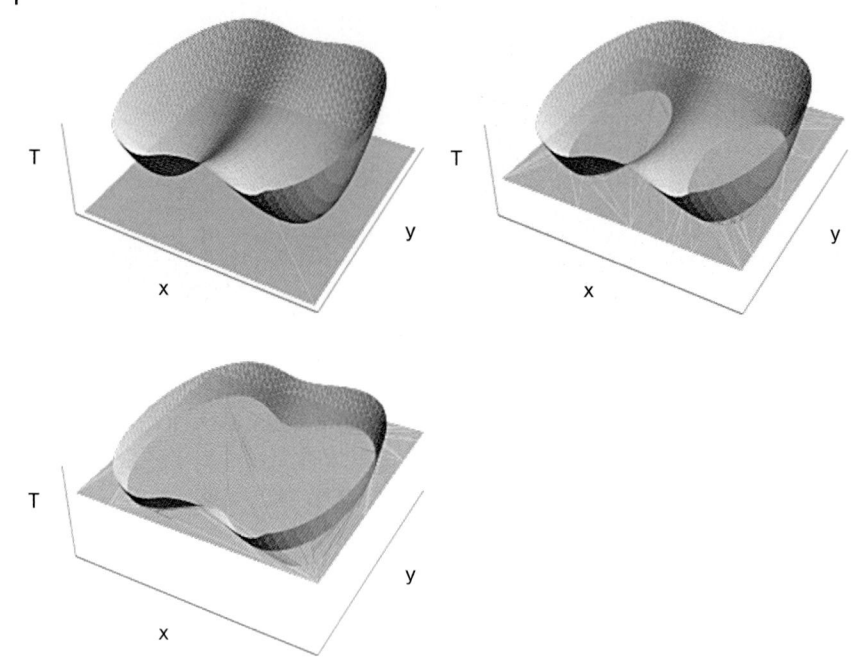

Figure 5.14 Fast marching formulation (2D). Evolving 2D spatial interface embedded in an arrival time function $T:\Re^2 \circledR \Re$. Plotted $T(\mathbf{x})$ together with a plane marking a time t. The intersection of $T(\mathbf{x})$ with t yields the interface $\Gamma(T)$ which divides Ω_{world} into the two subsets $\Omega_{sol} = \{\mathbf{x}:\Omega T(\mathbf{x}) > t\}$ and $\Omega_{liq} = \{\mathbf{x}:\Omega T(\mathbf{x}) < t\}$. In the left scene, we have $\Gamma = \Omega_{liq} = \emptyset$. In the middle, two separate domains are visible. The right example shows the previously disjoint domains have merged to form a single domain.

and Figure 5.14 contains a demonstration thereof. The interface $\Gamma(t)$ is defined by

$$\Gamma(t) = \{\mathbf{x}: \mathbb{R}^m | T(\mathbf{x}) = t\} \tag{8}$$

All topological information is contained in an arrival time field $T:\Re^m$. Note that each coordinate $\mathbf{x}$ is mapped to exactly one arrival time. This implies that the front passes $\mathbf{x}$ either not at all or exactly once, which is equivalent to restricting the propagation speed of a front to positive values. In terms of fabrication processes, it means that we can represent etching or deposition, but not both at the same time. We are able exploit this restriction such that it becomes possible to implement highly efficient front propagation algorithms such as the fast marching method.

5.4.5
Level Set Topology Representation

Although the original level set representation will not be further utilized in this chapter, the underlying idea is briefly portrayed for the sake of completeness.

5.4 Development Simulation

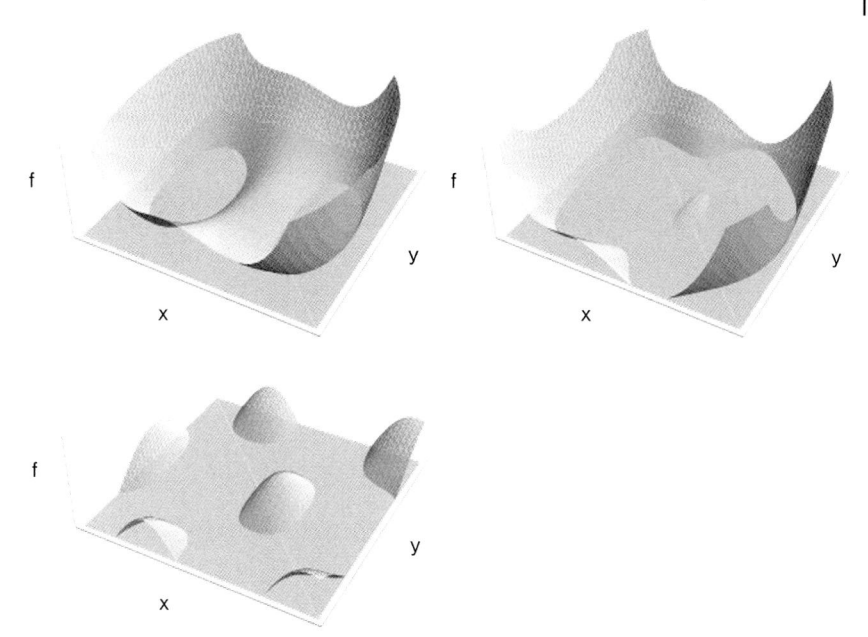

Figure 5.15 Level_set: Level Set method formulation. The evolving 2D spatial interface is embedded in a zero-level-set of an extended, $n + 1$-dimensional, arrival time function $\phi: \Re^{2+1} \circledR \Re$. Plotted is $\phi(\mathbf{x},t)$ with t advancing from the left to the right plot together with the base plane marking the zero-level. Their intersection yields the interface $\Gamma(T)$ which divides Ω_{world} into the two subsets $\Omega_{sol} = \{\mathbf{x}: \Re^n \phi(\mathbf{x},t) > 0\}$ and $\Omega_{liq} = \{\mathbf{x}: \Re^n \phi(\mathbf{x},t) < 0\}$.

To drop the limitation to uniformly moving interfaces, the arrival time function T is replaced by a version extended by one dimension, namely the time t:

$$\phi(\mathbf{x}, t): \mathbb{R}^{n+1} \to \mathbb{R} \tag{9}$$

Furthermore, the Interface Γ is now always the zero-level-set (see Figure 5.15):

$$\Gamma(t) = \{\mathbf{x}: \mathbb{R}^n | \phi(\mathbf{x}, t) = 0\} \tag{10}$$

In this sense, ϕ behaves like T-t.

5.4.5.1 Mathematical Geometry Toolkit

To extract properties and measures from topologies encoded in arrival time fields, the gradient operator takes a central role. It is the source for the front propagation direction and also for the front speed. It needs to be implemented carefully and with maximum available accuracy.

5.4.5.2 Front Normal

The front normal is a unit vector pointing towards the propagation direction. It is used in the calculation of many other properties and needs to be efficiently evaluated.

$$\mathbf{n} = \frac{\nabla T}{|\nabla T|} \tag{11}$$

Difficulties may arise when ∇T is not a smooth function and measures should be taken in this case.

5.4.5.3 Propagation Speed

The propagation speed is easily extracted from T. The absolute value or magnitude is

$$v = |\nabla T|^{-1} \tag{12}$$

whereas its vectorial value is available through

$$\mathbf{v} = \frac{\mathbf{n}}{|\nabla T|} \tag{13}$$

5.4.5.4 Surface Curvature

Curvature in the direction of $\mathbf{e}_i$, κ_i, is given by

$$\kappa_i = \partial_{e_i} n_i = \partial_{e_i} \frac{\partial_{e_i} T}{|\nabla T|} \tag{14}$$

Note that the calculation of ∇T might not be required, since $|\nabla T|$ is often exactly provided by the inverse of the speed function v^{-1}.

5.4.5.5 Mean Surface Curvature

Another more common measure for curvature is the mean curvature, since its value is more versatile:

$$\kappa_m = \nabla \cdot \mathbf{n} = \nabla \cdot \frac{\nabla T}{|\nabla T|} \tag{15}$$

5.4.5.6 Fast Marching Propagation

Since the fast marching (FM) method works with a front representation, it is forced to use speed functions F restricted to $F \geq 0$. With regard to development simulation, this means that re-deposition may principally not be implemented. However, with this price paid, we may state two powerful principles:

- **Entropy Principle:** Once a point is crossed by the front, it remains crossed. This allows us to categorize the FM method to the class of the fastest possible algorithms, since every point will be visited only once and afterwards its arrival time is fixed.

- **Causality Principle (upwind):** Front dynamics depend only on previous and not on future states. Similarly, an arrival time T_0 depends solely on upwind values. An upwind value T with respect to T_0 is defined by $T < T_0$.

Based on these two principles, we can specify the order of how arrival times are calculated. The initialization step sets all arrival times T to infinity (or a sufficiently large value), except for nodes lying on the initial front, whose arrival times T are set to zero. After the initialization step, we can take the smallest of all calculated T and at once accept it as fixed, since it cannot possibly be influenced by points that have not seen the front at time T yet. This permits a transformation of the transient problem into a stationary problem. As a consequence, time- and geometry-dependent causal speed functions, taking into account, for example, the concentration field, are applicable. In this context, 'causal' means to depend only on past and current states.

The computational cost of the pure propagation algorithm expressed in terms of the number of voxels in each space direction, N, without considering the speed up achievable by employing a tree-like spatial data structure for the voxels, is in $O[N^3]$ with respect to resolution and in $O[N^2]$ with respect to domain size. Basically, this means that the fast marching propagation algorithm scales linearly with the number of voxels involved and linearly with the front area. Taking into account the ordered spatial data structure, the respective computational costs become $O(N^3 \log N)$ and $O(N^2 \log N)$.

5.4.5.7 Eikonal Partial Differential Equation

Given the scalar speed function $F : \mathbb{R}^n \to \mathbb{R}$, the eikonal partial differential equation, which describes the propagation of the development front, follows directly from

$$|\nabla T|^{-1} = F \qquad (16)$$
$$T = const \text{ on } \Gamma_0$$

Here, Γ_0 is the set containing the initial front. Note that the speed function will only be evaluated on the front Γ and therefore a version F_{local} only defined on Γ suffices: $F_{local} : \Gamma \to \mathbb{R}$. This renders construction of extension velocities, as required e.g., by the narrow band method, unnecessary. Clearly, F_{local} is determined by the dissolution rate or etch rate of the resist in the developer solution.

5.4.5.8 Front Propagation

The propagation algorithm classifies mesh nodes into three groups.

- *Known*: for these nodes the arrival time is known ($t <$ active.smallest).
- *Active*: calculation for these nodes is in progress.
- *Unknown*: not considered yet ($t =$ BIG).

During the simulation, only the active class is explicitly kept accessible in a container. Thus nodes pass through the active nodes container once only and on their way through they obtain an arrival time less than the BIG value. The container is usually a spatial data structure that allows the rapid evaluation of neighbor relations.

5.4.5.9 The Algorithm

The basic algorithm of the fast marching method is independent of the grid and update scheme employed. It has the following steps:

1. **Initialize**
   Given start time t_{start} and the set of voxels, add all voxels that have at least one neighbor with $T < t_{start}$ and at least one neighbor with $T > t_{start}$ to the sorted container active.

2. **Propagate**
   While $T_{earliest}$ is smaller than a given threshold and *active* is not empty:
   (a) Remove the first voxel from the sorted sequence active and name it earliest voxel. It meets the predicate:
   for all voxel:active, earliest voxel.$T \leq$ voxel.T.
   It is accepted and accounted for as known.
   (b) Update all downwind neighbors of *earliest voxel*.

Figure 5.16 shows snapshots of a front propagated by this algorithm. Critical sub-steps are the sorted container, a heap and the update procedure, both treated in the following.

5.4.5.10 Heap

The sorted container is implemented by a heap data structure [29]. It stores elements according to a less-than ordering and allows addition and removal at the cost of $O(N \log N)$ and $O(\log N)$, respectively, where N is the number of elements in the container. In order to obtain a well-defined ordering, a more sophisticated version than $a.T < b.T$ is needed to sort voxels, since the cases where $a.T = b.T$ need to be covered also.

Definition voxel_ordering(a, b):

1. If $a.T\ != b.T$ return ($a.T < b.T$).
2. If $a.coord_0 < b.coord_0$ return true.
3. If $b.coord_1 < b.coord_1$ return true.
4. If $b.coord_2 < b.coord_2$ return true.
5. Return false.

Another advantage arising from this total ordering is that it may be used to identify voxels, namely voxel_ordering(a, b) = voxel_ordering(b, a) $\Leftrightarrow a = b$. This enables voxel retrieval in $O(\log N)$ using a binary search.

All of the mentioned features are readily available in C++ by providing the voxel data type together with the less-than ordering to the generic STL type set [30].

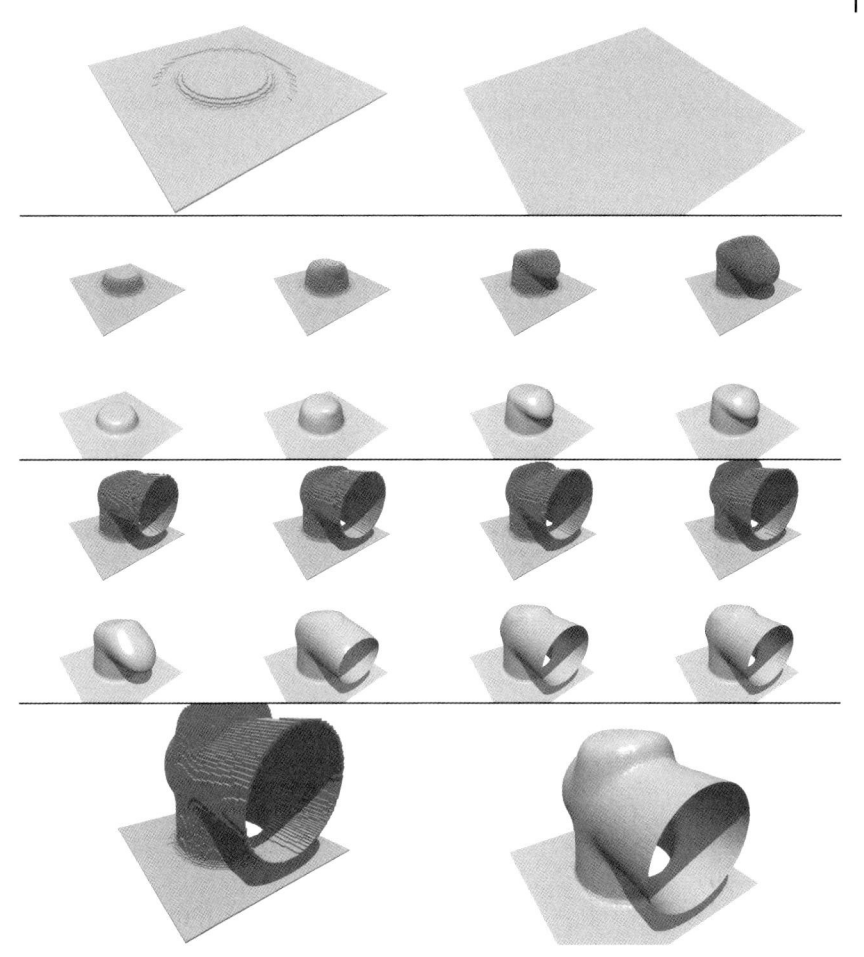

Figure 5.16 Recording of active voxels and last complete front. Plotted are snapshots equally spread in time of all active voxels together with the corresponding extracted front for speed function from Figure 5.10. Note the first frame's planar surface and the corresponding enclosing active voxels.

5.4.5.11 Update Scheme for Regular Grids

An update of the point with value T_0 consists of adjusting T_0 to satisfy $|\nabla T|^{-1} = F$. Consideration of the causality principle dictates use of one-sided differential operators, for example, forward and backward

$$\partial_- T = \frac{T - T_{-\Delta}}{\Delta}, \partial_+ T = \frac{T - T_{+\Delta}}{\Delta}$$

for the first-order case. We introduce the upwind differential operator tagged with an upward facing arrow '↑' to indicate upwind:

$$\partial_i^\uparrow T = \frac{T - T_{+\Delta_i \times \mathbf{e}_i}}{\Delta_i} \qquad (17)$$

where Δ_i takes positive or negative values as to preserve causality and $\mathbf{e}_i$ is the unit base vector. Expanding the norm in the eikonal and raising the equation by the power of two, which we may do since $F \geq 0$, provides

$$\sum_{i=1}^{n} \left(\partial_i^\uparrow T\right)^2 = F^{-2} \qquad (18)$$

Substituting the differential operator leaves us with the expression which T_0 has to satisfy:

$$\sum_{i=1}^{n} \left(\frac{T - T_{+\Delta_i \times \mathbf{e}_i}}{\Delta_i}\right)^2 - F^{-2} = 0 \qquad (19)$$

With the assumption of $F \geq 0$, we use Mathematica® to find the general solution:

$$T = \frac{BF \pm \sqrt{A + B^2 F^2 - ACF^2}}{AF} \cdot A \qquad (20)$$

with

$$A = \sum_{i=1}^{n} \frac{1}{\Delta_i^2},\ B = \sum_{i=1}^{n} \frac{T_{+\Delta_i \times \mathbf{e}_i}}{\Delta_i^2},\ C = \sum_{i=1}^{n} \frac{T_{+\Delta_i \times \mathbf{e}_i}}{\Delta_i^2} \qquad (21)$$

Setting the dimension n to 2 or 3 yields the schemes reported by Sethian. A feature of typical LIGA structures, and lithography in general, is that at local spots, near side-walls, a very high spatial resolution is required, whereas for the largest planar part, either exposed or unexposed, a coarse resolution suffices. Using a uniform grid seems inappropriate since locally, near side-walls, high-resolution representation is needed, whereas large areas are planar and should be coarsely represented. Not only is storage/memory consumption an issue, also computation time is reduced with fewer data points.

5.5
Available Software Packages

In Table 5.5 we list some available software packages, together with their freeware/commercial status, their platforms, their goals and a short summary of their features. The most popular program is the package DoseSim, due to its friendly graphical user interface and its many possibilities.

Table 5.5 Available software packages for deep X-ray LIGA.

Program	Platform	Status	Possibility	Used by	Contact
DoseSim v3.1	Windows	Freeware	Source: bending magnet Optics: mirror, double mirror, beam-stop Dose: primary dose Development: one dimension	Bessy (Berlin, Germany) CAMD (Baton Rouge, LA, USA, USA) IMT (Karlsruhe, Germany) MiniFAB (Melbourne, Australia) National University of Singapore (Singapore) Elettra (Trieste, Italy)	http://www.fzk.de/imt
LEX-D	Dos	Commercial	Source: bending magnet Optics: mirror, double mirror, beam-stop Dose: primary and secondary Development: three dimensions (profile, time)	Sandia National Laboratory (Livermore, CA, USA)	http://www.ca.sandia.gov/liga/
X3D	Linux	laboratory-level (will be freeware)	Source: bending magnet Optics: beam-stop Dose: primary dose Development: three dimensions (profile, time)		

5.5.1
Simulation of the Development Time with DoseSim

In the popular simulation software DoseSim [31], the development time and its corresponding development depth are calculated by integration of Equation 6 using an increment of 1 µm:

$$t(h) = \int_0^h \frac{1}{R(D,z)} dh \approx \sum_{i=0}^{i=h/increment - 1} \frac{increment}{R\left(D, \frac{increment}{2} + i \times increment\right)} \tag{22}$$

The model of dissolution rate and its parameter are chosen by the user. As an example, we give a comparison between experimental and simulation results in Figure 5.17. The development is done at 21 °C and is supported with megasonic (1 MHz; 10 W cm^{-2}), the model used is the $R(D) = KD^\beta$ model ($K = 0.0067$ and $\beta = 3.03$ [21]).

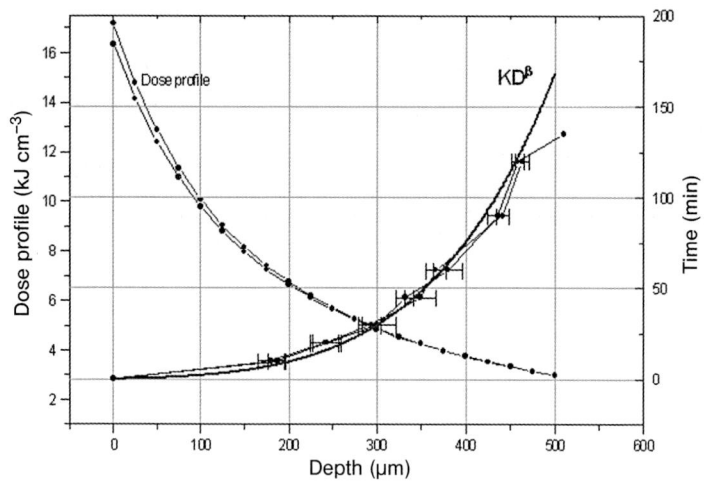

Figure 5.17 Comparison of experimental and calculated results. The development are done at 21 °C and is supported with megasonic (1 MHz; 10 W cm^{-2}).

5.5.2
X3D

There are many complicated process parameters and difficulties in determining 3D X-ray lithography parameters, such as the optimum exposure energy and development time, not only for moving mask X-ray lithography but also for other methods. Therefore, a resist profile simulation tool and an analytical method to determine process parameters is of major interest among 3D X-ray lithography research groups and was the major motivation in ensuing with the specification and development of X3D.

The X3D simulation program combines exposure and development simulation. Its organization is illustrated in Figure 5.18. It enables the user to model completely the X-ray source, all static optical elements and a freely selectable number of movable masks and a movable stage containing the target. In a first step, the geometry of each mask is discretized into a two-dimensional bitmap pattern so as to facilitate faster processing. Movable stages are specified with time displacement specifications. During a second phase, the exposure at each coordinate in the target is integrated by tracing a ray bundle from the synchrotron source through all optical elements into the target and integrating the intensity over the exposure time. Since optical elements may move, the modulation of the ray bundle is taken into account. The result of this phase is an absorbed dosage per unit volume at each point in the photoresist. This value is then converted into a dissolution rate for the resist in a specified developer. The final phase computes the result of developing the target for a specified number of minutes. This last aspect involves some special considerations which we briefly address.

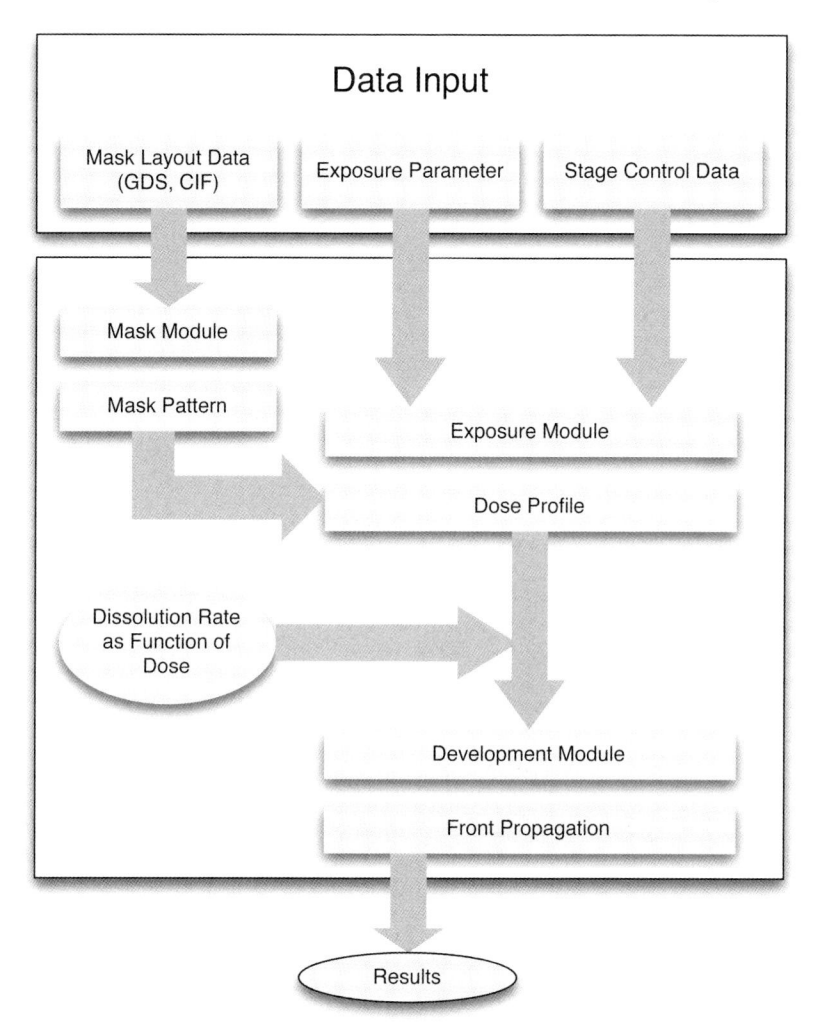

Figure 5.18 Data flow in X3D.

The most important aspect governing the accurate modeling of the 3D X-ray lithography process is that the resist profile simulation tool should take into account a dissolution speed vector in addition to the resultant dose distribution. As we have noted previously, the photoresist dissolution rate is eventually defined as a function of depth (position) in the exposed photoresist and the dissolution front propagation is dominated by this dissolution rate. In the case of 3D X-ray lithography, there is a dissolution rate gradient along the depth of the photoresist as well as along the lateral directions. Thus the dissolution rate vector at the photoresist–developer interface does not point in a direction perpendicular to the initial photoresist surface once the development process starts, namely, the

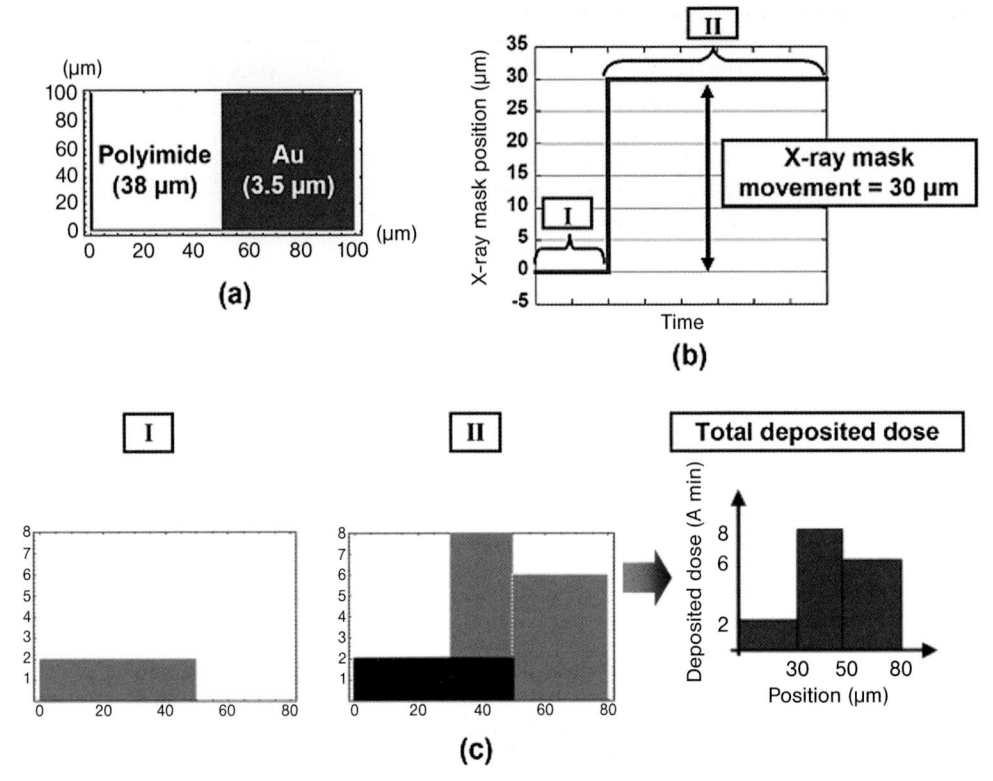

Figure 5.19 Simulation parameters for case study: (a) X-ray mask pattern; (b) X-ray mask movement pattern of the 'stairs-like' dose deposition pattern; (c) deposited dose given by M2DXL. In (c), I and II show the amount of dose deposited by the corresponding phases I and II in (b).

dissolution front tends to move faster towards the higher dissolution direction [22]. In order to address the complicated situation and consider the dissolution vector in the development process, an algorithm based on the fast marching method as discussed above is suitable for handling this problem and was implemented in the X3D code.

To demonstrate a system validation of X3D [32, 33], a 'stair-like' dose deposition pattern shown in Figure 5.19 and Table 5.6 was chosen as a case study. Despite the dose deposition pattern's geometric simplicity, both the necessity and the validity of X3D, which takes into account a dissolution vector in addition to the resultant dose distribution, are clearly shown. Figure 5.20 and 5.22 shows the simulation result, resulting from dose distribution pattern shown in Figure 5.19. It should be emphasized that the side-wall shape did not end up being vertical and the corner was rounded off. As mentioned above, the 3D dissolution process (that is increasing in edge roundness) was observed both in the experiments and in the simulation results.

Table 5.6 Parameters of the X3D simulation example.

Category	Parameter		Value
SR (AURORA)	Operating electron energy (GeV)	ε	0.575
	Critical wavelength (nm)	λc	1.5
	Typical source size (vertical) (mm)	σy	0.14
	Distance between source and mask (m)	D	3.88
Exposure process	Filter (Be) (µm)		200×2
	Scan (mm)		20
	X-ray mask movement (µm)		30
X-ray mask	Absorber (Au) (µm)		3.5
	Membrane (polyimide) (µm)		38
Development process	Developer		GG developer
	Temperature (°C)		39

Development time (min)	SEM	X3D
30		
60		
90		
120		

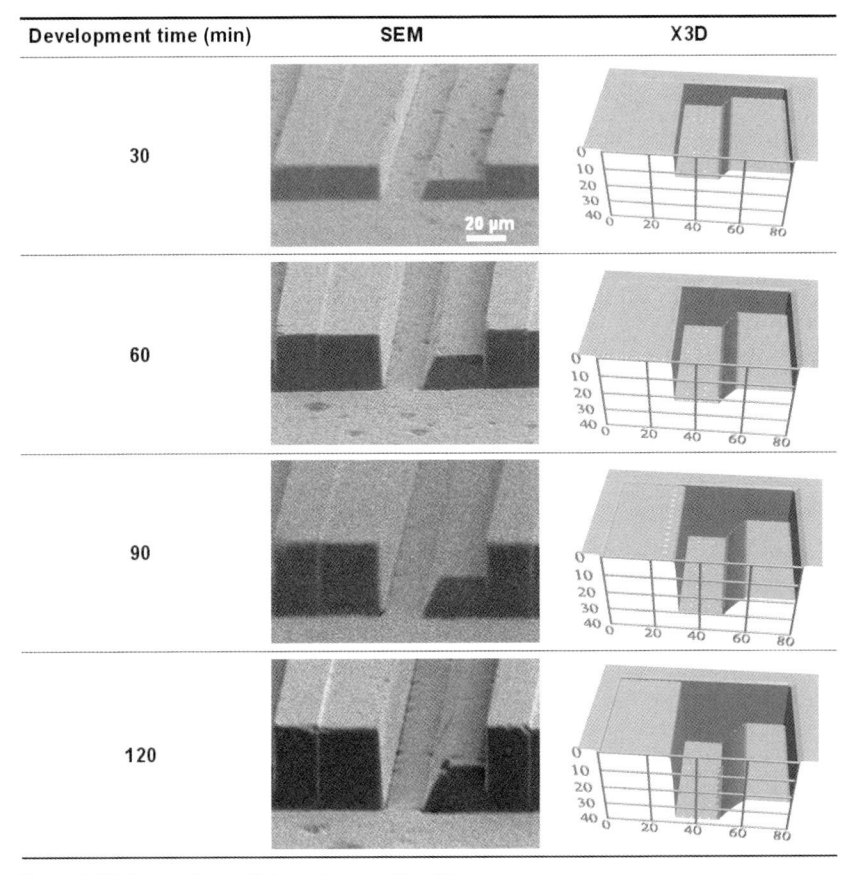

Figure 5.20 Comparison of simulation results with measurements with development times of 30, 60, 90 and 120 min.

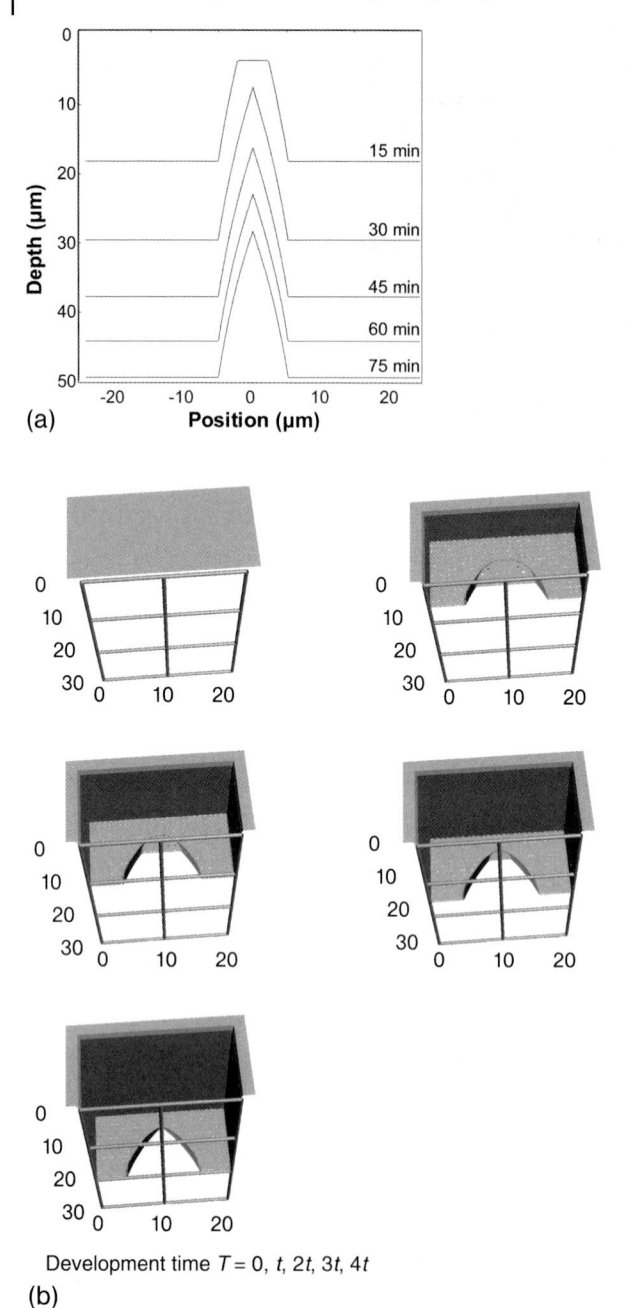

Figure 5.21 Examples of resist profiles simulation tool for D2XRL: (a) fast marching method; (b) X3D.

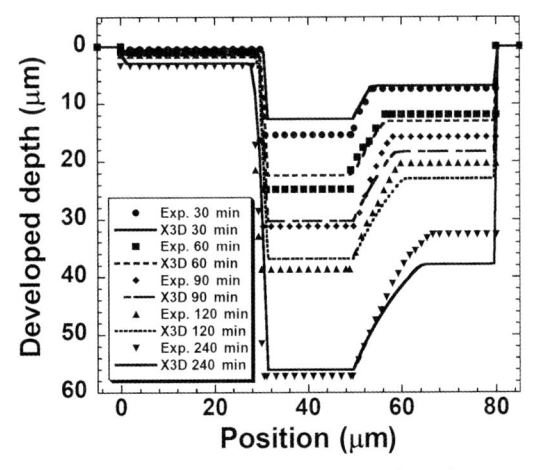

Figure 5.22 Comparison of simulation results with measurements with development times of 30, 60, 90, 120 and 240 min.

The double X-ray exposure technique is a newly devised technique to realize 3D microstructures without M2DXL [34, 35]. The feasibility of this technique was confirmed through experiments and simulation using the X3D as shown in Figure 5.21 and 5.22.

5.6
Conclusion and Outlook

Modern approaches in LIGA manufacturing are exploring the possibility of creating free-form 3D microstructures, in addition to using novel, less cumbersome resist materials. Both of these exploration activities have profound consequences for the establishment of robust, easy to use predictive CAE tools. For one, 3D exposure systems require the absorption dose to be known point-wise throughout the resist. High-resolution yet large-area (volume) exposures imply also a dramatic growth in the number of simulation 'voxels' and, of course, demand more sophisticated approaches for the simulation tools. In addition, where previously optical elements remained stationary with respect to the beam, 3D exposure requires relative movement of all optical components and the subsequent point-wise dose accumulation in time. Second, the exploration of different resist systems has implications both for exposure simulation and for subsequent development calculations. The conversion mechanisms vary considerably across polymers, from 'mere' damage to the chain lengths to photo-polymerization. The development mechanism also varies, from dissolution to sublimation. These effects must be accounted for in the simulation tool and require a sophisticated experimental program of characterization and fine tuning.

We are still a long way from a complete and satisfactory CAE or TCAD capability for the LIGA process. Nevertheless, important advances have been made in capturing the details of individual processing steps and these activities have helped to more precisely define the requirements and set the research agenda for the future.

References

1 Meyer, P., Schulz, J. and Solf, C. (2004) Concept of a MEMS process sofware specially dedicated to the deep x-ray lithography process, in Proceedings of Industrial Simulation Conference 2004, pp. 195–7.
2 Hahn, K., Brueck, R., Priebe, A. and Schneider, C. (2000) Cost estimation for LIGA fabrication flows using process design methods. *Microsystem Technologies*, **6**, 145–8.
3 Sanchez del Rio, M. and Dejus, R.J. (1996) XOP: a graphical user interface for spectral calculations and X-ray optics utilities. *Review of Scientific Instruments*, **67**, 1–4.
4 Sanchez del Rio, M. and Dejus, R.J. (1997) XOP: a multiplatform graphical user interface for synchrotron radiation spectral and optics calculations. *Proceedings of the SPIE – The International Society for Optical Engineering*, **3152**, 148–57.
5 Cerrina, F. and Lai, B. (1986) Shadow: a synchrotron radiation ray tracing program. Nuclear Instruments and Methods in Physics Research Section A, Volume 246, Issue 1–3, pp. 337–41.
6 Welnak, C., Chen, G.J. and Cerrina, F. (1994) Shadow: a synchrotron radiation and X-ray optics simulation tool. *Nuclear Instruments and Methods*, **347**, 344–7.
7 Cerrina, F., Baszler, F., Turner, S. and Khan, M. (1993) Transmit: a beamline modelling program. *Microelectronic Engineering*, **21**, 103–6.
8 Baszler, F., Khan, M. and Cerrina, F. (1992) X-ray dose density: a new radiation damage modeling tool. *Proceedings of the SPIE – The International Society for Optical Engineering*, **1671**, 451–60.
9 Iwasaki, H., Nakayama, Y., Ozutsumi, K., Yamamoto, Y., Tokunaga, Y., Saisho, H., Matsubara, T. and Ikeda, S. (1998) Compact superconducting ring at ritsumeikan university. *Journal of Synchrotron Radiation*, **5**, 1162–5.
10 Tabata, O., You, H., Matsuzuka, N., Yamaji, T., Uemura, S. and Dama, I. (2002) Moving mask deep X-ray lithography system with multi stage for 3D microfabrication. *Microsystems Technologies*, **8**, 93–8.
11 Sugiyama, S., Khumpuang, S. and Kawaguchi, G. (2004) Plain-pattern to cross-section transfer (PCT) technique for deep X-ray lithography and applications. *Journal of Micromechanics and Microengineering*, **14**, 1399–404.
12 Cremers, C., Bouamrane, F., Singleton, L. and Schenk, R. (2001) SU-8 as resist material for deep X-ray lithography. *Microsystem Technologies*, **7**, 11–16.
13 Gad-el-Hak, M. (2001) *The MEMS Handbook*, CRC Press, New York.
14 Schmalz, O., Hess, M. and Kosfeld, R. (1996) Structural changes in poly(methyl methacrylate) during deep-etch X-ray synchroton radiation litography. Part I: degradation of the molar mass. *Die Angewandte makromolekulare Chemie*, **239**, 63–77.
15 Schmalz, O., Hess, M. and Kosfeld, R. (1996) Structural changes in poly(methyl methacrylate) during deep-etch X-ray synchroton radiation litography. Part II: radiation effects on PMMA. *Die Angewandte makromolekulare Chemie*, **239**, 79–91.
16 Meyer, P., El-Kholi, A., Mohr, J., Cremers, C., Bouamrane, F. and Megtert, S. (1999) Study of the development behavior of irradiated foils and microstructure. *Proceedings of the SPIE – The International*

Society for Optical Engineering, **3874**, 312–20.

17 Charlesby, A. (1960) *Atomic Radiation and Polymers – International Series of Monographs on Radiation Effects in Materials*, Vol. **1**, Pergamon Press, Oxford.

18 Glashauser, W. and Ghica, G.V. (1982) German Patent, 3 039 110.

19 Liu, Z., Bouamrane, F., Roulliay, M., Kupka, R.K., Labèque, A. and Megtert, S. (1998) Resist dissolution rate and inclined-wall structures in deep X-ray lithography. *Journal of Micromechanics and Microengineering*, **8**, 293–300.

20 El-Kholi, A., Bley, P., Göttert, J. and Mohr, J. (1993) Examination of the solubility and the molecular weight distribution of PMMA in view of an optimised resist system in deep etch X-ray lithography. *Microelectronic Engineering*, **21**, 271–4.

21 Meyer, P., El-Kholi, A. and Schulz, J. (2002) Investigations of the development rate of irradiated PMMA microstructures in deep x-ray lithography. *Microelectronic Engineering*, **63**, 319–28.

22 Hirai, Y., Hafizovic, S., Korvink, J.G. and Tabata, O. (2003) Measurement of PMMA dissolution rate and system calibration for predictive 3D simulation of moving mask deep X-ray lithography. Proceedings of the 5th International Workshop on High Aspect Ratio Micro-Structure Technology, Monterey, CA, pp. 37–8.

23 Pantenburg, F.J., Achenbach, S. and Mohr, J. (1998) Influence of developer temperature and resist material on the structure quality in deep x-ray lithography. *Journal of Vacuum Science and Technology B* **16**, 3547–51.

24 Greeneich, J.S. (1975) Developer characteristics of poly-(methyl electron resist methacrylate). *Journal of the Electrochemical Society*, **122**, 970–6.

25 Griffiths, S.K., Hruby, J.M. and Ting, A. (1999) The influence of feature sidewall tolerance on minimum absorber thickness for LIGA X-ray masks. *Journal of Micromechanics and Microengineering*, **9**, 353–61.

26 Sefler, J.F. (1995) 3D surface modeling utilities for use in TCAD, PhD thesis, University of California, Berkeley, Berkeley, CA.

27 Sethian, J.A. (1998) Adaptive fast marching and level set methods for propagating interfaces, *Acta Mathematica Universitatis Comenianae*, **67**, 3–15.

28 Sethian, J.A. (1999) *Level Set Methods and Fast Marching Methods*, 2nd edn, Cambridge University Press, Cambridge.

29 Knuth, D.E. (1998) *The Art of Computer Programming*, Addison-Wesley.

30 Austern, M.H. (1999) Generic programming and the STL: using and extending the C++ standard template library, in *Professional Computing Series*, Addison-Wesley.

31 Meyer, P., Schulz, J. and Hahn, L. (2003) DoseSim: MS-Windows graphical user interface for using synchrotron X-ray exposure and subsequent development in the LIGA process. *Review of Scientific Instruments*, **74**, 1113–19.

32 Hafizovic, S., Hirai, Y., Tabata, O. and Korvink, J.G. (2003) X3D: 3D X-ray lithography and development simulation for MEMS, in Proceedings of the 12th International Conference on Solid-State Sensors Actuators and Microsystems (Transducers 2003), Boston, MA, pp. 1570–3.

33 Hirai, Y., Hafizovic, S., Matsuzuka, N., Korvink, J.G. and Tabata, O. (2006) Validation of X-ray lithography and development simulation system for moving mask deep X-ray lithography. *Journal of Microelectromechanical Systems*, **15**, 159–68.

34 Matsuzuka, N., Hirai, Y. and Tabata, O. (2005) A novel fabrication process of 3D microstructures by double exposure in deep X-ray lithography (D2XRL). *Journal of Micromechanics and Microengineering*, **15**, 2056–62.

35 Matsuzuka, N., Hirai, Y. and Tabata, O. (2006) Prediction method of 3D shape fabricated by double exposure technique in deep X-ray lithography (D2XRL), in 19th IEEE International Conference on Micro Electro Mechanical Systems (MEMS2006), Istanbul, pp. 186–9.

36 Adalsteinsson, D. and Sethian, J.A. (1995) A level set approach to a unified model for etching, deposition and lithography I: algorithms and two-dimensional

simulations. *Journal of Computational Physics*, **120**, 128–44.

37 Adalsteinsson, D. and Sethian, J.A. (1995) A level set approach to a unified model for etching, deposition and lithography II: algorithms and three-dimensional simulations. *Journal of Computational Physics*, **122**, 348–66.

38 Osher, S. and Sethian, J.A. (1988) Fronts propagating with curvature-dependent speed: algorithms based on Hamilton–Jacobi formulations. *Journal of Computational Physics*, **79**, 12–49.

39 Sethian, J.A. and Adalsteinsson, D. (1997) An overview of level set methods for etching, deposition and lithography development. *IEEE Transactions of Semiconductor Manufacturing*, **10**, 167–184.

40 Spivey, J.M. (1998) *The Z Notation: a Reference Manual*, Oriel College, Oxford.

6
Design for LIGA and Safe Manufacturing

Ulrich Gengenbach, Ingo Sieber, and Ulrike Wallrabe

6.1	**General Aspects of Design for Safe Manufacturing**	*144*
6.1.1	Basic Rules of Embodiment Design	*146*
6.1.2	General Design Principles	*147*
6.1.2.1	Principles of Force Transmission	*147*
6.1.2.2	Principle of the Division of Tasks	*147*
6.1.2.3	Principles of Self-help	*148*
6.1.2.4	Principle of Self-protection	*149*
6.1.2.5	Principle of Stability and Bi-stability	*149*
6.1.2.6	Principles of Fault-free Design	*150*
6.2	**General Microsystems Design Rules**	*150*
6.2.1	Design to Allow for Expansion	*151*
6.2.2	Design Against Wear	*151*
6.2.3	Design for Recycling	*152*
6.2.4	Design for Fabrication, i.e. What Does 'Design for LIGA' Mean	*152*
6.2.4.1	Structure Inversion	*152*
6.2.4.2	Dimensional Deviations in the LIGA Process	*152*
6.2.4.3	Loads and Stresses in the LIGA Process	*153*
6.3	**Design for LIGA Manufacturing**	*154*
6.3.1	Introduction	*154*
6.3.2	Design Rules for Deep X-ray Lithography	*154*
6.3.2.1	Avoidance of Cracks	*154*
6.3.2.2	Mechanical Stabilization with Regard to Thermal Expansion and Swelling of the Resist	*160*
6.3.3	Design Rules for Electroplating	*163*
6.3.4	Design Rules for Molding	*167*
6.4	**Design for the Assembly of LIGA Parts**	*168*
6.4.1	General Aspects of the Assembly of Microsystems	*169*
6.4.2	Design for Feeding	*170*
6.4.3	Design for Gripping	*171*
6.4.4	Design for Positioning	*172*
6.4.5	Design for Joining	*174*
6.4.5.1	Adhesive Bonding	*174*

Advanced Micro & Nanosystems Vol. 7. LIGA and Its Applications.
Edited by Volker Saile, Ulrike Wallrabe, Osamu Tabata and Jan G. Korvink
Copyright © 2009 WILEY-VCH Verlag GmbH & Co. KGaA, Weinheim
ISBN: 978-3-527-31698-4

6.4.5.2 Mechanical Microclamping 176
6.5 Functional Design for LIGA 178
6.5.1 Introduction 178
6.5.2 Design Requirements 179
6.5.3 Comprehensive Modeling 179
6.5.4 Modeling and Simulation of a Micro-optical Distance Sensor 180
6.5.4.1 Adjustment of the Model to the Manufacturing Chain 181
6.5.4.2 Robust Design 184
6.5.4.3 Thermal Analysis 186
6.6 Conclusions 187
References 188

An old rule in engineering says that '80% of the fabrication costs of a product are determined in the design phase'. This rule holds true for microsystems in the same way as it does for macroscopic products. Key elements for keeping these costs as low as possible are following a proper design methodology and applying design rules that enable reliable and cost efficient fabrication.

These two issues are the focus of this chapter. Figure 6.1 gives an overview of the chapter's structure. It begins with general aspects of design for safe manufacturing, which illustrate how 'macroscopic' design methodology can be applied to LIGA. Then general, which are also 'macroscopic', design rules are rephrased for microsystems technology. In the next section, design rules are elaborated for the most important LIGA process steps. This is followed by a discussion of design rules which result from the assembly of LIGA parts. Finally, a section on functional design for LIGA outlines the application of the design methodology and design rules using the example of a micro-optical distance sensor.

6.1
General Aspects of Design for Safe Manufacturing

Pahl and Beitz's groundbreaking book [1] is one of the cornerstones of design methodology and design rule development for mechanical and precision engineer-

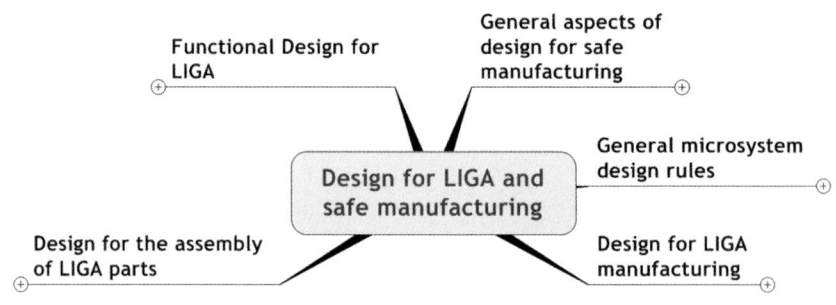

Figure 6.1 Structure of the chapter.

ing. It is the intention in the following section to demonstrate that this approach can also be applied to LIGA technology. Pahl and Beitz subdivide the product development process into four phases (Figure 6.2):

- The first phase focuses on the clarification of the task and elaborating the list of requirements.
- The main task of the second phase is conceptual design, that is, the development of the principle solution and the establishment of functional structures. The resulting concept variants are evaluated against technical and economic criteria.

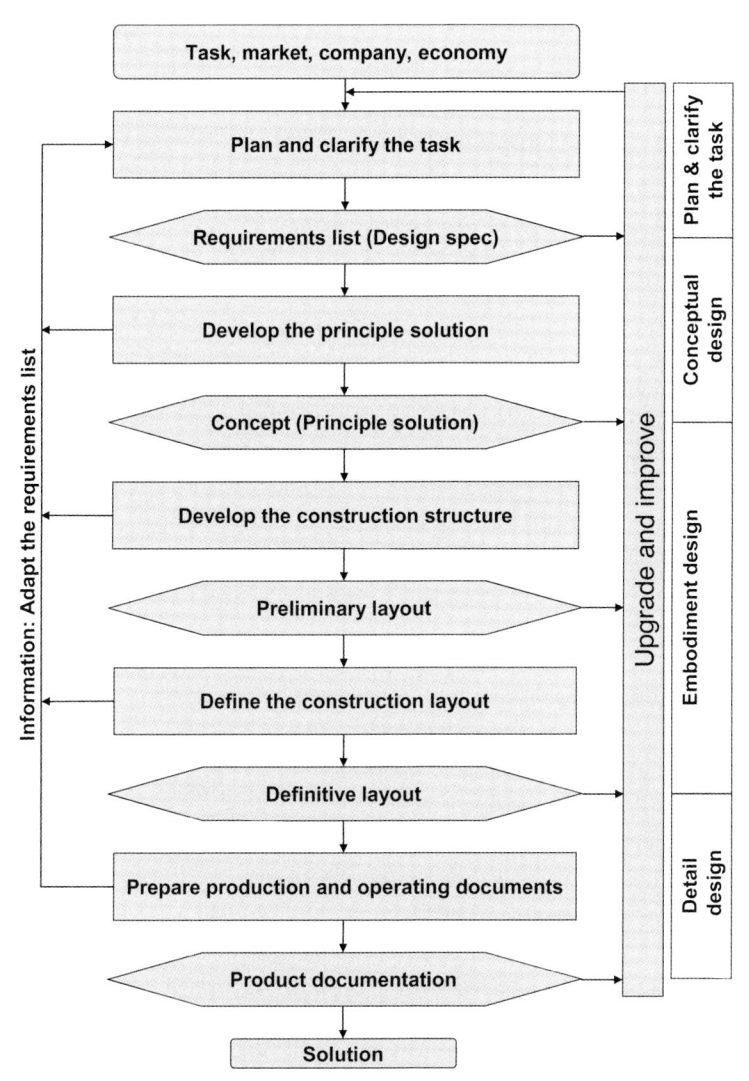

Figure 6.2 Design methodology (derived from [1]).

- The solution chosen is the input of the third phase, the embodiment design. Here, materials are selected, first shapes are determined and analyzed by means of analytical calculations or computer simulation. This leads to preliminary layouts, which are evaluated against economic and technical criteria. The chosen layout is further elaborated with parts lists, production and assembly documents.

- In the fourth phase, detail design, this definite layout is finalized with detailed drawings, parts list, production, assembly, handling/transport and operating instructions.

Apparently, the level of detail and effort increases from phase to phase. While the emphasis in the first two phases is on clarification of the task and on finding a solution, the last two phases represent the design process proper. Under the assumption that the product requirements have been specified to a complete extent and that a product concept has been selected, it is in the embodiment design where important decisions on product design have to be taken. This decision-making process can be supported by a set of basic rules.

6.1.1
Basic Rules of Embodiment Design

The most elementary rules of embodiment design are

- clear
- simple
- safe.

At first glance, this seems to be a trivial statement. Each of these rules, however, can be translated into tangible requirements on design with respect to function, active principle, safety, ergonomics, fabrication and control, assembly and transport, usage, maintenance and recycling. Violation of any of these fundamental rules may lead to serious consequences in product function. Depending on the perspective in the design process, each fundamental rule may be applied in multiple ways for different aspects of the product.

- '*Clear*' means unambiguous assignment of subfunctions to dedicated components or subsystems. On the system level, it may also mean, for example, unambiguous assignment of functions to keys or menu points on the user interface of a device.

- '*Simple*' on the component level may imply the use of elementary geometric shapes to allow for straightforward analytical solutions in dimensioning and the simple fabrication of the part. On the system level, it may mean implementation of just the functionality required and no extras that complicate handling and usage.

- '*Safe*' on the component level may be translated into proper design and dimensioning to ensure that the part performs its function without failure throughout its lifetime. On the system level, it implies that the system must not become a hazard to the operator and environment both during operation and after disposal.

Unfortunately, these basic rules are not disjoint and their priority may differ from task to task. It may be necessary to violate one rule to some extent in order to fulfill completely another one of higher priority.

There are numerous (macroscopic) devices on the market that clearly violate these rules. Hence, during product design, it is a good idea to step back once in a while and reconsider the design with respect to the basic rules of embodiment design.

6.1.2
General Design Principles

The basic rules above can be transformed into a set of design principles that should be applied in a product. Although the design methodology was initially developed for macroscopic devices, it will become evident that they can equally be applied to the design of microsystems.

6.1.2.1 Principles of Force Transmission

For mechanical microparts, force transmission is as much of an issue as for macroscopic parts. While other microtechnologies are limited by crystal orientations, LIGA can capitalize on one of its main strengths, the freedom of design in the plane.

The principle of direct and short force transfer can be implemented by designing LIGA microstructures for optimum force transmission, such as low inertia, high stiffness scaffold structures or flexure hinges optimized for high cycle counts. This approach is illustrated by the design of the LIGA gyroscope (see Figure 6.15).

Another important aspect in microtechnology is the principle of matched deformations. The possibility of combining different materials, another strength of LIGA, may lead to differences in deformation, if such a composite part is under mechanical or thermal stress. This may cause delamination at the interface and, thus, component failure. This principle influences both component design and the selection of joining processes.

6.1.2.2 Principle of the Division of Tasks

In macroscopic engineering, it is good practice to subdivide tasks into subtasks and define specific subfunctions that fulfill these tasks. A standard example taken from mechanical engineering is the precise joining of two parts. The task 'joining of two parts' can be subdivided into the subtasks 'alignment' and 'fixation'. Each of these subtasks can be fulfilled by different functions, which are either integrated in the same part or realized as separate parts. In the example above, integration of the 'alignment function' could mean that stop faces are integrated into the part, while centering pins would fulfill the same function as separate parts. In the same way, the fixation subtask can be implemented by integrated fixation elements (e.g. clips) or separate fixation elements (e.g. screws).

The same approach can be used in microtechnology. Here, the LIGA process, with its wide range of materials, has proven to be particularly favorable. It allows one to assign subfunctions to different parts made of materials optimally suited to the functional requirements. As is apparent from the example above, the division of tasks and their execution influence the assembly process considerably.

The principle of division of tasks is important for a clear design with parts optimally designed for their specific requirements. In microtechnology, it should be applied wisely, since it may need to be brought into line with the objectives of function integration and miniaturization.

6.1.2.3 Principles of Self-help

The principle of self-help means designing the system in such a way that its design and/or the arrangement of its components either improve its function or prevent malfunction. There are several principles of self-help:

Self-enforcement means that the function is implemented by an initial effect, which is enforced by an additional effect resulting from the operation of the device. This principle can be illustrated by the passive membrane valves in the thermo-pneumatic AMANDA micropump [2].

In the case of the membrane valve, the initial effect is a pre-tension that pushes the valve membrane lightly against the valve seat. Figure 6.3 shows the expulsion stroke of the pump, whereby the pump pressure presses the membrane of the inlet valve further against the valve seat, thus increasing the sealing effect. At the same time, the outlet valve is pushed open. In the intake cycle, the situation is reversed.

Self-compensation stands for a design that compensates adverse external influences. The LIGA acceleration sensor represents a fine example of this principle. The measuring principle is the change in capacitance with a change in the distance between two capacitor plates. The plate capacitor consists of a set of fixed plates (rectangular rods in Figure 6.4) and a movable seismic mass (fork-shaped structure in Figure 6.4). Upon acceleration, the seismic mass moves and the distance to the fixed plates changes. This effect in turn changes the capacitance and yields the desired signal [3].

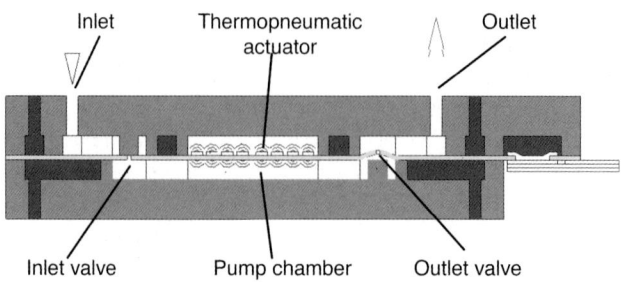

Figure 6.3 Self-enforced valves in the AMANDA micropump [4].

Figure 6.4 Detail of a LIGA temperature-compensated acceleration sensor.

Without any external acceleration, temperature changes and the resulting thermal expansion also lead to a change in the distance of the plates and thus result in measuring errors. The design shown in Figure 6.4 compensates for this effect. The fixed plate is attached to a ceramic substrate with a low thermal expansion coefficient. Consequently, the problem that one has to cope with is the thermal expansion of the seismic mass made of nickel. In the lower half of Figure 6.4 the seismic mass is the handle of the fork and located between the fixed plates. Here, thermal expansion leads to a decrease in the gap. In the upper half, the fixed plates are located between the tines of the fork. Here, thermal expansion leads to an increase in the gap. Since the surface areas of handle and tines are matched, these two effects compensate each other.

6.1.2.4 Principle of Self-protection

In the case of overload, the component should not be destroyed. For mechanical parts, this implies, for example, that excess force is being taken up by some additional force transmission path. In microsystems, this is sometimes difficult to achieve, and sometimes it comes almost for free. In the example above of the acceleration sensor, the fixed plates serve as a catch against too large motions in the seismic mass under excess acceleration. Obviously, in this situation the capacitor plates come into contact, which is why the electronics have to be protected against such a short-circuit.

6.1.2.5 Principle of Stability and Bi-stability

Usually, stable behavior of a component or a system is paramount and the design should be made accordingly. Sometimes, however, a bi-stable behavior is a welcome feature. This is illustrated by the bi-stable membrane microvalve [4] (see Figure 6.5). The valve consists of a polyimide membrane with two stable states. In the closed state, the membrane bulges upwards and, thus, seals the inlet. In the open state, the membrane bulges downwards and a free flow channel exists between the inlet and outlet. The membrane is flipped between these two states by means of a pressure or vacuum pulse that is applied via an actuation channel on the rear of the membrane.

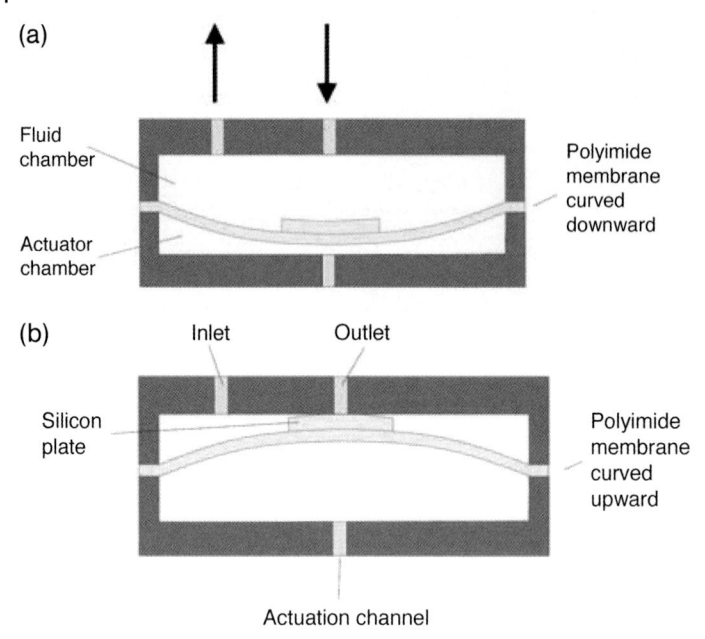

Figure 6.5 Bi-stable microvalve: (a) open state; (b) closed state.

6.1.2.6 Principles of Fault-free Design

The principle of fault-free design also sounds like a trivial statement, because any design is intended to be fault-free in the first place. The idea of this principle, however, is to consider explicitly already in the design phase what faults might arise from either fabrication problems or proper/improper usage of the component. Measures to achieve a fault-free design are:

- simple components and structures
- design that minimizes the influence of errors
- selection of active principles that are invariant to disturbances
- mutual compensation of disturbances.

It is evident that these measures revert to the basic rules and to some of the principles outlined above. The fork-shaped seismic mass in the acceleration sensor above, for instance, is an example of mutual compensation of disturbances.

6.2
General Microsystems Design Rules

The above basic rules and principles refer to the conceptual and embodiment design. Although the examples were derived from macroscopic mechanical design theory, they are equally applicable to microsystems technology. There are, however, general process aspects that have to be taken into account.

6.2.1
Design to Allow for Expansion

The variety of materials used in the LIGA process implies a wide range of thermal expansion coefficients, ranging from 2.8×10^{-6} K^{-1} for silicon substrates to 10×10^{-6} K^{-1} for nickel up to 70×10^{-6} K^{-1} for poly(methyl methacrylate) (PMMA) (see also Table 6.2). For composite parts that are intended to be used in a wider temperature range, this diversity needs to be taken into account. An example is the micro-optical distance sensor, where the micro-optical board, a silicon–PMMA composite, has been optimized from both the design and fabrication points of view for robustness over a wider temperature range. This will be illustrated in the section on functional design for LIGA.

6.2.2
Design Against Wear

Wear is frequently related to friction. Wherever possible, friction should be avoided, in particular in microtechnology, where forces and torques usually are so small that the losses resulting from friction cannot be tolerated. Hence, for small linear or angular motions, mechanical joints should not be scaled down, but rather replaced by flexure hinges.

Where contact between mating surfaces cannot be avoided, for example in gear trains (Figure 6.6), the LIGA process should be used to design the shape of the mating surfaces such that only normal forces and no shear forces are transmitted. Moreover, friction should be minimized by taking advantage of the high surface quality in the order of some 10 nm Ra that can be reached by LIGA.

Figure 6.6 Nickel gears and watch parts of optimized tooth geometry and high surface quality for reduced wear.

6.2.3
Design for Recycling

Due to the small size and high degree of integration, designing for recycling is barely an issue for microproducts. One of the few examples of relevance is biodisposables (microfluidic chips, labs on a chip, etc.). These polymer parts are fabricated worldwide as polymer disposables in large volumes and yield tons of waste. For such applications, the use of biodegradable polymers, such as polylactates, should be considered.

6.2.4
Design for Fabrication, i.e. What Does 'Design for LIGA' Mean

Nowadays, design of macroscopic devices usually begins with emphasis on device function. In a second step, if at all done explicitly, the design is reworked with respect to fabrication issues. In microsystems technology, this approach is not feasible, since device function and fabrication frequently cannot be treated independently. Moreover, knowledge about macrofabrication processes is fairly commonplace for designers, but it is not that widespread for microtechnologies. In order to bridge this gap, design rules are provided for the various microfabrication processes and microsystems designers are already well advised to take them into account at the functional design level.

Another major difference between macro- and microfabrication processes is the interdependence of process steps [5]. Whereas this interdependence can often be neglected in macrofabrication, it plays an important role in microsystems technology. Hence it does not suffice to take into account a single- or few-process steps, and it is necessary to consider the entire process chain required to fabricate a microdevice and also to take account of the following interdependences between its sub-processes:

6.2.4.1 Structure Inversion
An obvious interdependence is the inversion of the structure from one process step to the next. This is common sense in other lithography-based microfabrication technologies. However, inversion has a special place in LIGA technology, since it may also show up in process steps other than lithography. Depending on the desired output, that is, primary structured polymer microparts (X-ray lithography), metallic microparts (electroforming) or secondary structured polymer microparts (replication), and the process sequence, in the design the geometric shape has to be inverted, in order to obtain the desired final part.

6.2.4.2 Dimensional Deviations in the LIGA Process
Systematic dimensional deviations occur in several LIGA sub-processes. Their extent depends on various factors, such as sub-process properties, substrate material, micropart design and micropart material. Dimensional deviations of a few

hundred nanometers in intermediate masks result from electron scattering during the irradiation of the resist and are basically independent of part-specific influences. The dimensional deviations of metal microstructures fabricated by electroforming are also in the range of a few hundred nanometers and are largely independent of part geometry. For polymer parts, the situation is different. Material shrinkage, for example 0.8% for PMMA, significantly influences the dimensional accuracy of the final part.

The overall dimensional deviation of a part results from its process chain and the material. In the detail design phase, the LIGA part designer copes with these systematic deviations by amending the part dimensions with the help of sub-process-specific correction factors.

6.2.4.3 Loads and Stresses in the LIGA Process

Other important factors that influence the design of LIGA parts in addition to material selection are loads and stresses in the fabrication process. These influence both important tooling such as masks and mold inserts and the resulting parts.

For the main sub-processes, these loads and stresses can be broadly categorized as shown in Table 6.1.

It is important to consider completely the loads and stresses that both the part and key fabrication tools (masks, mold inserts) will be subjected to while passing through the process chain. These influences require adaptations to the design and influence material selection. The corresponding issues will be covered in more detail in the following sections on sub-process-specific design rules.

Table 6.1 Loads and stresses in the LIGA sub-processes.

Sub-process	Loads and stresses on masks, mold inserts and parts			
	Mechanical	Thermal	Chemical	Irradiation
Mask fabrication	×	×	×	×
Primary structuring by deep X-ray lithography	×	×	×	×
Electroforming	(×)		×	
Secondary structuring by replication	×	×	(×)	
Handling	×			
Joining	×	(×)	(×)	(×)

6.3
Design for LIGA Manufacturing

6.3.1
Introduction

This part of the chapter is concerned with design for safe manufacturing from the pure LIGA process point of view, namely for X-ray lithography, electroplating and molding. Most of the design rules that will be presented here have not been published before and are based either on internal reports of the Forschungszentrum Karlsruhe or on the experience of a number of people working daily on the process in the laboratory. Wherever scientific material has been published, a reference is cited; all other rules have to be understood simply as hints for the designer, based on the personal knowledge of the authors, gathered through years of developing designs and subsequently observing the resulting structures under the microscope. One reference, however, must be mentioned in advance, as it is an almost complete selection of rules which were published in a German PhD thesis [5]. For the large majority of readers who are probably not familiar with German, this thesis is nevertheless helpful, since it is illustrated with a large number of diagrams, which make the task clear.

For the lithography section, this chapter will focus on the

- avoidance of cracks
- mechanical stabilization of structures in terms of thermal expansion and swelling of the resist, namely, of PMMA
- rules for aligned exposure.

For the electroplating section, it will concentrate on

- auxiliary structures for homogeneous plating rates
- avoidance of failure due to gas bubbles.

For the molding section, hints will be given for

- maximum aspect ratio
- features for easy demolding.

6.3.2
Design Rules for Deep X-ray Lithography

6.3.2.1 Avoidance of Cracks
The most important issue for successful deep X-ray lithography is the avoidance of cracks. Cracks occur already during exposure, but most often after the exposure, during the development and even after the development. They appear horizontally on all level heights of the LIGA structures, and also vertically at narrow resist lines and sharp corners. In many cases, the microscopic cracks become visible on the side-walls of the developed structures. Internal cracks are hard to see and do not

Table 6.2 Thermal expansion coefficients of substrate and resist materials.

Material	Coefficient of thermal expansion α (10^{-6} K^{-1})
Silicon (substrate)	2.3–2.6
Al$_2$O$_3$ ceramics (substrate)	5.4–8.1
Copper (substrate for mold inserts)	16.8–17.0
PMMA (X-ray resist)	70–80

influence the ongoing processes of electroplating and molding. In the case of optical applications of purely lithographic structures, however, optical transparency may be affected adversely by internal cracks.

There are two main reasons for the generation of cracks. The most important one is tension stress in the resist layer, typically in the PMMA. The amount of tension stress in the PMMA is crucially dependent on the procedure of resist deposition on the substrate. In the case of casting the liquid PMMA on to a substrate, the stress is imposed by volume shrinkage during the polymerization process. Even annealing after polymerization cannot reduce the stress to zero, because the thermal expansion coefficients of the substrate and the polymer differ by a factor of 10 or even more. Table 6.2 gives an overview of the thermal expansion coefficients of PMMA and various substrate materials.

When bonding a PMMA platelet to a substrate, which has been pre-polymerized and annealed before bonding, stress is drastically reduced compared with the casting procedure. Here, the stress is imposed during bonding in the adhesive layer and once again later during exposure and development due to thermal expansion.

This residual stress has to be dealt with and the risk of crack generation can be reduced drastically by following some very simple design rules, which will be presented later in this section.

Figures 6.7 and 6.8 show some typical examples of cracks related to tension stress [6].

The second cause of cracks are microscopic hydrogen bubbles, which are formed during exposure and within the hours following exposure. The hydrogen is released when the PMMA chains are dissected into shorter pieces by the X-rays. It diffuses through the PMMA and eventually forms microscopic bubbles in the exposed resist. With increase in gas pressure, the risk of cracks also increases. It might be argued that cracks in the exposed regions will disappear with the development and, hence, do not need to be considered. However, the pressure may also damage the unexposed neighboring structures. When the developer penetrates a gas bubble, the bubble explodes and, thus, causes a pressure pulse into the unexposed structures. Figures 6.9 and 6.10 show strongly affected side-walls of unexposed structures, on which a network of very fine line cracks is spread across the surface [6]. This underlines the necessity for choosing the process parameters in such a

Figure 6.7 Vertical crack at a very narrow resist line [7].

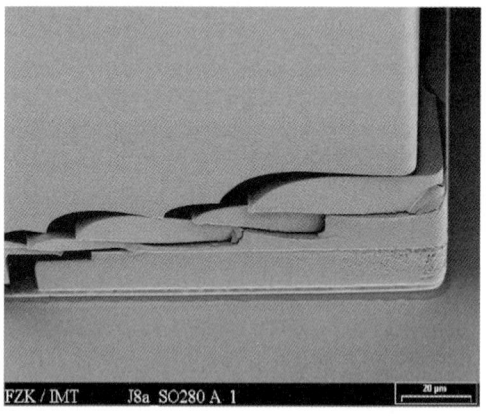

Figure 6.8 Horizontal cracks at the bottom of the PMMA structure due to thermal stress [7].

Figure 6.9 Completely destroyed PMMA structures showing cracks in all directions [7].

Figure 6.10 Less severe crack pattern with cracks following the symmetry of the structure [7].

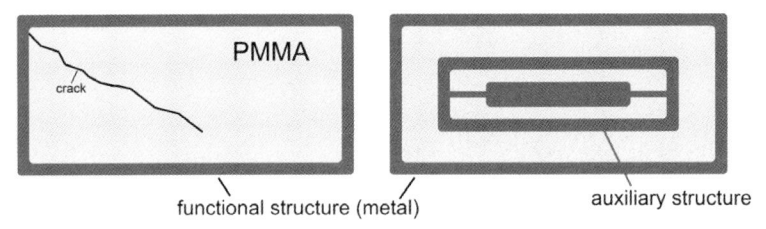

Figure 6.11 Auxiliary structures interrupt large PMMA areas, where possible.

way as to reduce the generation of hydrogen bubbles to a minimum. This can be achieved by applying the minimum exposure dose necessary for the development and by starting the development immediately after the exposure.

In order to support the minimization of cracks, a couple of design rules have proved to be really helpful:

Lithography design rule 1: avoid large resist areas! Dissect the resist pattern wherever possible. To this end, one has several options. First, one can avoid large resist areas within the functional microstructure in such a way that the function is not influenced, which is illustrated in Figure 6.11. Here, we assume a rectangular frame of narrow beams to be the required functional microstructure. The beams will be metallic, so they need to be electroplated. Actually, a simple frame formed by its four side lines would fulfill this easy function. The frame would leave a similar simple rectangle of PMMA in its center. In order to destroy a possible path for a crack to run across the whole PMMA rectangle, it is much safer to fill the rectangle with auxiliary structures that interrupt all possible travel paths for cracks. If the auxiliary structures disturb its function, they can be placed on a sacrificial layer and are released from the substrate in the end (Figure 6.12).

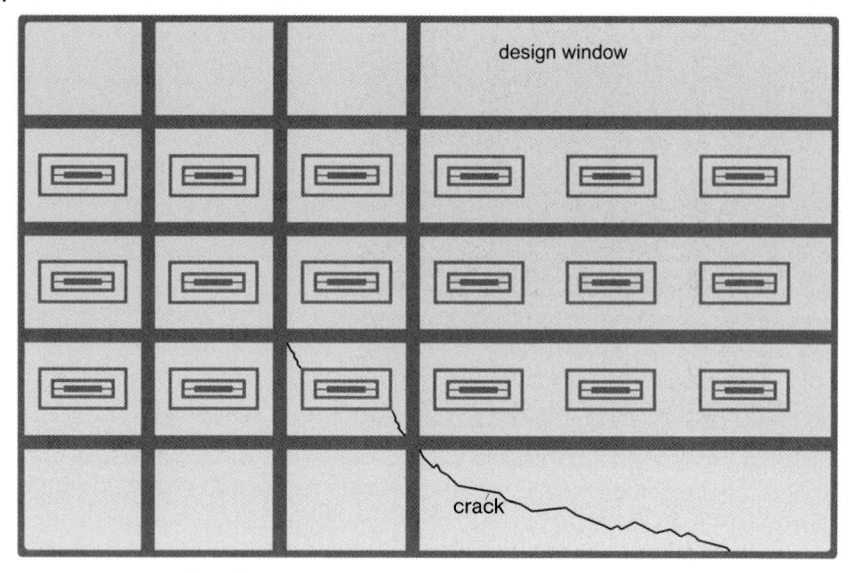

Figure 6.12 Exposed mesh surrounding the functional microstructures. In the case of metal structures, they are to be electroplated.

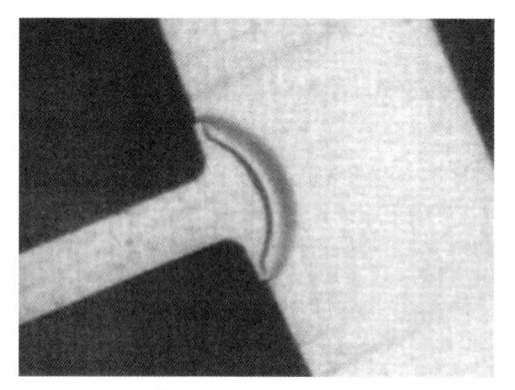

Figure 6.13 Crack at anchor structures.

A mesh of exposed and developed lines surrounds the functional microstructures. Again, the lines interrupt a continuous path for a crack along the whole diagonal of the pattern window.

Lithography design rule 2: avoid thin and long resist lines! Of course, a long resist line is a design feature that typically arises from the required function. Hence, the hint to avoid thin and long resist lines in many cases seems useless. Two metallic capacitor plates separated from each other by a resist line with a thickness of a few micrometers is a good example. Figures 6.13 and 6.14 show two typical cracks in thin resist lines. In Figure 6.13 the crack occurs at the anchor

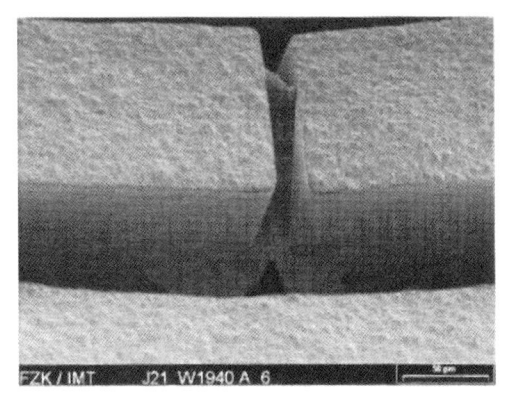

Figure 6.14 Metal structure with former crack filled with metal bridging the gap.

Figure 6.15 Detail of a gyroscope: the gap is 2 μm wide.

structure of the resist line. Figure 6.14 shows a metal structure, where a tiny metal line can be noticed which bridges the gap between the two larger rectangles. Before electroplating, this gap was a PMMA line that suffered a continuous crack. The crack was then filled with metal and appears as a bridge after stripping the PMMA.

In some cases, however, the thin resist line is allowed either to be very short or to be tapered. Figure 6.15 shows a dead stop as a detail of a gyroscope [7]. The dead stop features an air gap of 2 μm. Hence the PMMA line is also 2 μm wide, but only along a very short length, as can be seen from the broadening of the air gap. The tapered structures stabilize the PMMA itself and, additionally, increase the adhesion to the substrates, which helps avoiding cracks.

Lithography design rule 3: avoid sharp corners! Avoid them, if possible, in general. Concave corners are much more crack sensitive than convex corners. Give

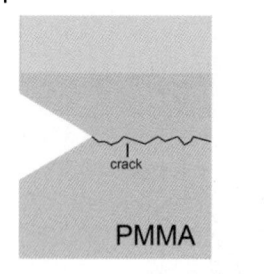

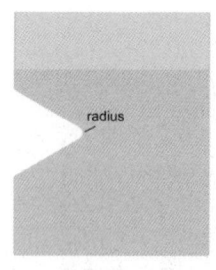

Figure 6.16 Crack generation at concave corners and implemented radii to suppress it.

all corners a radius of at least 2 μm; even better would be 5 μm. If radii are not crucial to the design, make them as large as one can or wishes. Figure 6.16 illustrates the generation of cracks at concave corners and how the added radii help to suppress them.

6.3.2.2 Mechanical Stabilization with Regard to Thermal Expansion and Swelling of the Resist

Thermal expansion and swelling of the resist most severely influence structure quality during electroplating. Many electroplated metals are most optimally deposited from aqueous electrolytes at elevated temperatures around 50 °C. Consequently, one has to deal with thermal expansion and swelling in the same way as tension stress. All design rules derived above are also valid for the compensation of thermal expansion and swelling, since both effects also impose stress and may cause cracks. However, some additional effort is recommended to improve mechanical stabilization.

A good example was published by Ruzzu and Mathis [8]. The authors were working on electrostatic motors for an optical switch matrix (see Chapter 8), which showed a design-dependent strong influence on both expansion and swelling. Figure 6.17 presents a section of the initial design. The black structures are the functional structures that are to be electroplated in metal, the dark gray structures are auxiliary structures, which have already been implemented in the design and will also be electroplated, and the light gray background indicates the PMMA that remains after development. For this design, large dimensional errors were observed, especially at the outer radius of the larger black ring, which is crucial for proper function of the device.

Figure 6.18 shows the deviations in the outer radius of this ring as a function of the annular position at two electroplating temperatures, namely 52 °C and room temperature of 23 °C. At 52 °C, the radius shows very large errors up to 12 μm; at 23 °C it is reduced to a residual value of 4 μm. The remaining error at 23 °C can be attributed to swelling of the neighboring PMMA structures due to water uptake in the electrolyte. The outer radius is influenced by the surrounding PMMA pattern that is not rotationally symmetric. The increasing PMMA expands towards the auxiliary structures (region A), but remains almost in place between them (region B), where PMMA bridges stabilize the inner ring. This explains the strong

6.3 Design for LIGA Manufacturing | 161

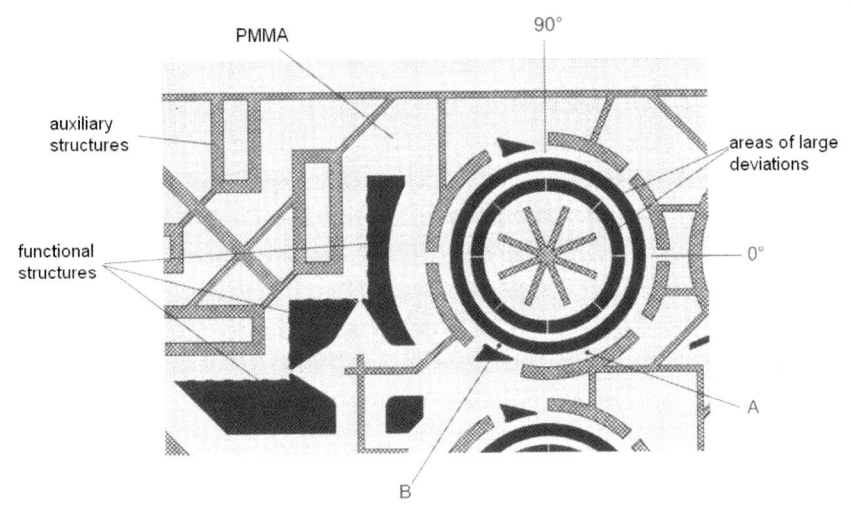

Figure 6.17 Design detail, including auxiliary structures: the outer black ring suffered large dimensional errors due to resist swelling.

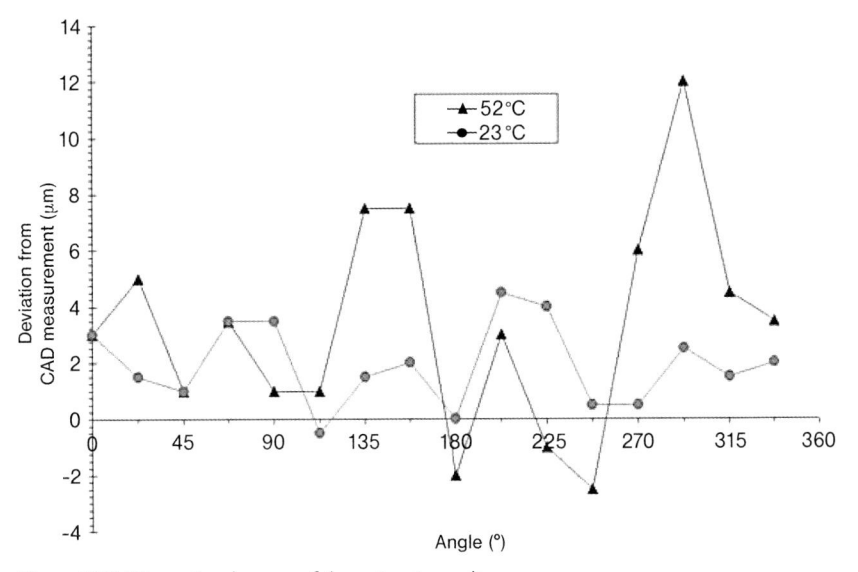

Figure 6.18 Dimensional errors of the outer ring radius electroplated at 52 °C and room temperature (cf. Figure 6.8).

variation in errors along the circumference. Figure 6.19 illustrates the geometric consequences of the volume increase by the bent PMMA ring.

The remaining errors can be further reduced via better mechanical stabilization based on an improved design, which is shown in Figure 6.20. Compared with Figure 6.19, the outer black ring is dissected into several segments. The dissecting

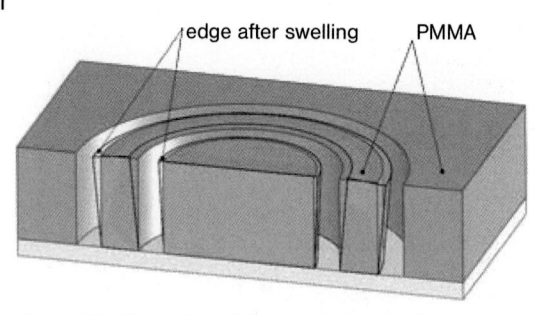

Figure 6.19 Illustration of dimensional errors due to water uptake and swelling.

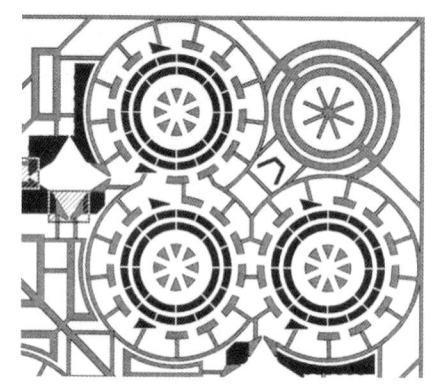

Figure 6.20 Improved design for better mechanical stabilization.

resist lines help to hold the surrounding PMMA in place, because they provide a connection to the center of the circles. The similarly segmented surrounding auxiliary structures have been arranged as symmetrically as possible. Their base length opposing the black circle is equal to the base length of the small triangles that are part of the set of functional structures. Finally, an almost complete outer ring surrounds the whole device in order to separate it from the rest of the mask pattern. Figure 6.21 compares the remaining errors in the second layout with the original errors in the first layout, both electroplated at 23 °C. The errors could be reduced to an average of 2 µm.

Of course, the swelling behavior could be perfectly predicted by simulation, but it is not presented here. The results presented above can be summarized by two additional design rules.

Lithography design rule 4: avoid long and contiguous electroplated structures! Dissect electroplated structures like the ring above in order to connect the resist pattern for stabilization. Note and consider why this design rule completely contradicts the rules for crack avoidance discussed above. In the end, it depends on the designer's experience as to what he or she considers to be more important in a specific case and the decision is finally driven by the expected function of the device.

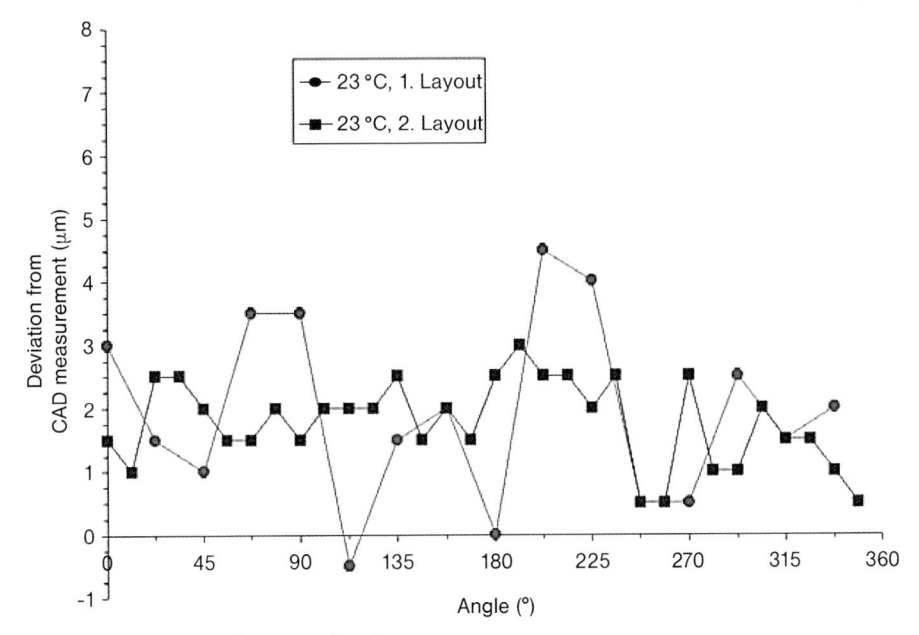

Figure 6.21 Dimensional errors reduced to 2 µm.

Lithography design rule 5: look for a symmetric arrangement of stabilizing structures! This does not really help to suppress errors related to swelling and thermal expansion, but makes them at least homogeneous and easier to predict.

6.3.3
Design Rules for Electroplating

Electroplated layers typically possess design-dependent inhomogeneities in thickness or layer growth, namely in the deposition rate. This happens across the whole substrate on a more macroscopic scale as well as on a microscopic scale, that is, in each individual microstructure. The different plating rates are also related to an inhomogeneous electric field. In the macroscopic case, this field is generated by the fact that a typically large anodic electrode opposes a fairly small substrate, the cathode, of which only a limited area is foreseen for electroplating. This is illustrated in Figure 6.22.

In the microscopic case, electrically conductive surface areas (the structures of which are to be electroplated) lie in the direct neighborhood of electrically insulating surfaces (the PMMA structures), which also cause the electrical field lines to bend and result in an inhomogeneous field and inhomogeneous layer thickness, as can be seen in Figure 6.23.

The macroscopic deformation can be compensated by microstructured frames, which surround the actual design window (Figure 6.24). The strong deformation of the electrical field then concentrates on the frames, whereas the field is more

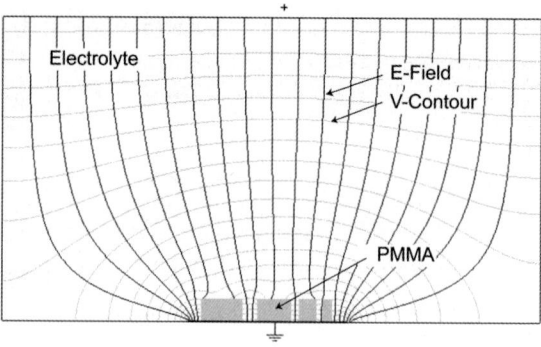

Figure 6.22 Macroscopic field deformation resulting in deviations of plating rates across the substrate.

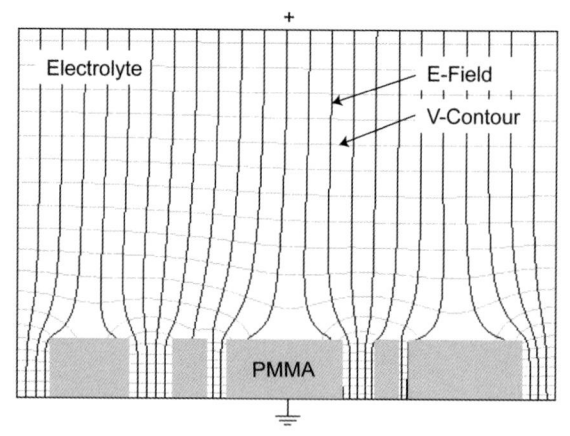

Figure 6.23 Microscopic field deformation resulting in deviations of plating rates even in separate microstructures.

homogeneous inside the frames, where the functional structures are located. An additional measure is to use dielectric apertures between the anode and substrate. This topic, however, does not belong to a design chapter and will not be discussed here.

Electroplating design rule 1: let a set of electroplated frames surround the functional design! One should pattern at least three frames; even better would be more, depending on the space one has left for it. The width of the frames should increase from inner to outer, starting with 100 µm and increasing upwards to 500 µm or even more.

Microscopic field deformation is a crucial design-dependent aspect and a perfect task for simulation. However, function (mechanical, optical, electrical, etc.) and field deformation result in a very complex coupled problem, which is not limited to the single microstructure, but includes all others in the neighborhood. So yet again, auxiliary structures are the measure of choice. The rules are very similar to the rules derived for X-ray lithography.

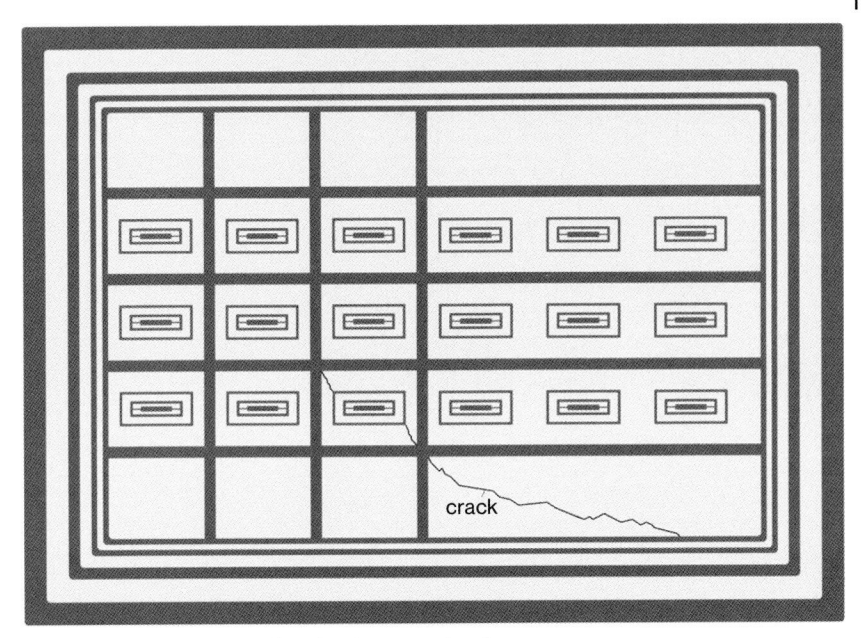

Figure 6.24 Frames surrounding the design window for reducing macroscopic deviations in layer growth during electroplating.

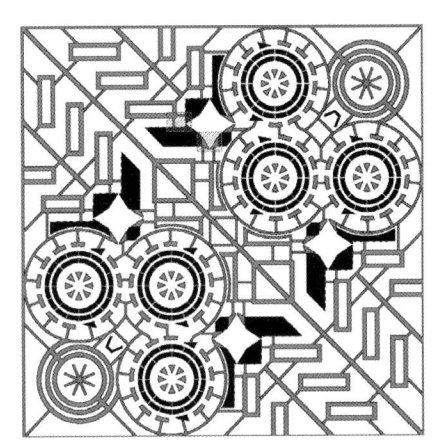

Figure 6.25 Connected set of auxiliary structures for constant layer growth during electroplating.

Electroplating design rule 2: avoid large contiguous areas! Dissect large areas, wherever possible, and implement auxiliary structures wherever space is available and function is not disturbed, no matter whether it is resist or electroplated metal. In taking the design from Figure 6.20 and printing a larger section of it (Figure 6.25),

Figure 6.26 Chip with functional structures in the right claw and the linked and released auxiliary structures in the left claw.

one finds many additional auxiliary structures, which have been designed primarily for electroplating purposes. These structures should be of similar size as the functional structures and if they are very close they should have a similar geometry.

Electroplating design rule 3: try to connect all auxiliary structures that will be electroplated! If one wishes to eliminate them at the end of the process, it is necessary to perform sacrificial layer etching. In the case of many tiny and non-linked auxiliary structures, there is the risk of having many sand-like pieces in the etchand which may fall back into the functional microstructures, where they disturb or even destroy the function. On the other hand, they can be picked out with a pair of tweezers, if necessary, when they are connected to form a single larger contiguous piece. An example is shown in Figure 6.26, representing the design of Figure 6.20. The scorpion holds a chip with the remaining functional structures in its right claw and the connected auxiliary structures that have been released in the other claw. Another example can be found in [9].

A very specific error exists in electroplating in simple circles, which are frequently not filled completely. Metal deposition starts and then stops after a while, although all neighboring structures are filled nicely. Hydrogen bubbles, which are generated during the process, are thought to cause this problem. They are captured in the developed cylindrical holes and cannot escape, because the substrate is oriented vertically and the cylinder axis horizontally. Once the diameter of the bubble reaches the diameter of the cylinder, it seals the cylinder perfectly and no electrolyte can penetrate it any more, so that the deposition stops. The resulting design rule in order to avoid this sealing is very easy:

Electroplating design rule 4: avoid cylindrical holes for electroplating! Instead of a cylinder, one may use a honeycomb or place an auxiliary structure in the center, such as a cross or similar. Figure 6.27 shows a typical defect and Figure

Figure 6.27 Typical electroplating defect at a cylindrical hole due to a hydrogen bubble.

Figure 6.28 Design which prevents defects due to hydrogen bubbles.

6.28 a cylinder with a cross. If one cannot avoid cylindrical holes, it can be risked up to an aspect ratio of approximately 1.5.

6.3.4
Design Rules for Molding

The design rules for molding concern the following main aspects: again, the avoidance of cracks, de-molding after hot embossing or injection molding in order to remove the polymer structure from the mold without destroying the structure and the shrinkage of the polymer (see Chapter 4). Figure 6.29 shows a de-molded structure with typical defects, namely with a deformation of the thin polymer structures.

Figure 6.29 Stretched polymer structures after de-molding. The aspect ratio is ~20.

In order to prevent cracks, almost the same rules hold true as those derived for X-ray lithography. Dissect the design and avoid sharp corners! The first part of this rule, however, completely contradicts the recommendation for de-molding. The larger the total surface area of the microstructures is, the higher the de-molding forces are. Hence the dissection of the large areas into many smaller pieces will generate additional surface area and increase the force. On the other hand, the effect of shrinkage once again calls for smaller polymer sections, because the more polymer that is available to shrink on to the mold insert, the larger the de-molding forces are. This is a complex dilemma and it is difficult to provide general design rules. If we nevertheless attempt to do so, it could be:

Molding design rule 1: enlarge radii to a minimum of 5 µm to avoid cracks at sharp corners!

Molding design rule 2: make the design as simple as possible! Try to avoid fancy and tiny details, which are not really necessary for the function in order to keep the surface small.

Molding design rule 3: dissect polymer sections into fairly small pieces! If one inserts auxiliary structures, try to keep the aspect ratio between 1 and 5.

Molding design rule 4: try to avoid very high aspect ratios in general! This is recommended not only because of the increasing surface area, but also because of stability in general.

6.4
Design for the Assembly of LIGA Parts

This section focuses on assembly aspects. As pointed out in the first part of this chapter, design principles and design rules should play a major role in a state-of-the-art design process. Applying such design rules reduces the effort required to develop reliable and efficient assembly-friendly designs.

6.4.1
General Aspects of the Assembly of Microsystems

Miniaturization leads to the following effects that considerably influence micro-assembly:

- reduction of volume
- reduction of mass
- reduction of handling surfaces on the part
- tremendous change in the volume to surface area ratio
- part sizes on the same scale as their tolerances, etc.

As a result, 'parasitic forces' caused by these scaling effects, such as electrostatic and/or adhesion forces, can have a considerable impact on the microassembly processes.

For this reason, the most general design rule reads:

Fundamental assembly design rule: avoid assembly! Design devices as monolithically as possible! This rule should be followed as much as possible. It saves time, money and avoids quality issues.

In some instances, however, different materials are required to achieve the desired function or the fabrication processes do not allow for a monolithic design. In these cases, it is important to consider the entire assembly process and its implications on front-end processes for micropart fabrication.

Figure 6.30 depicts the main steps of the assembly process:

- feeding
- gripping
- positioning
- joining.

Numerous interrelations exist between these steps and with upstream and downstream processes, such as manufacturing and packaging. As every process

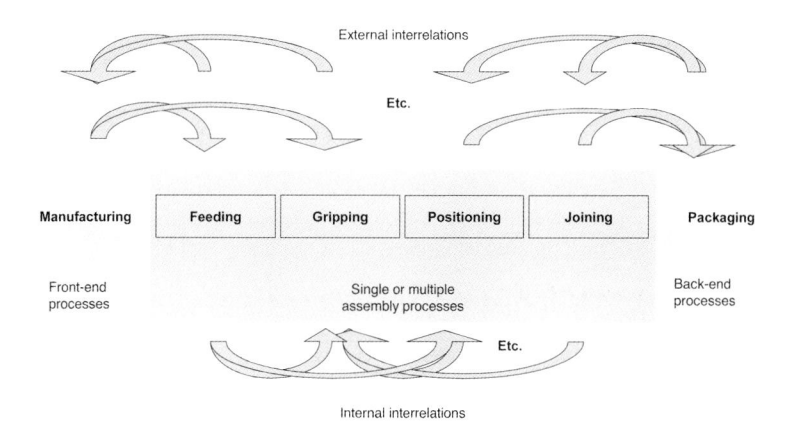

Figure 6.30 Internal and external interrelations of microassembly processes.

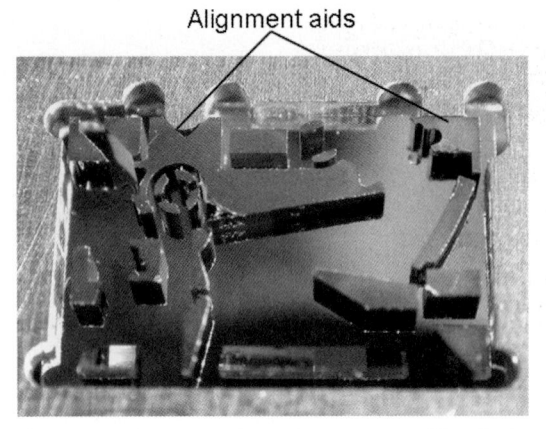

Figure 6.31 Alignment aids on the micro-optical bench, the base part of the micro-optical distance sensor.

step and its interrelations result in particular requirements on the design of the LIGA parts, these will now be discussed one by one.

6.4.2
Design for Feeding

Feeding means transport of the parts to be assembled into the work space of the assembly system. It is good practice to transport the base part of the machine in an ordered manner, on a substrate or clamped on to a tray. Stop faces and alignment aids should be incorporated into the design of the base part. This is illustrated by Figure 6.31, where two alignment aids have been implemented at the rim of the micro-optical bench.

The major issue frequently lies in how to feed the other part(s) that has/have to be mounted on the base part. While bulk feeding and sorting are fairly common in macroscopic assembly, it should be avoided in microassembly. The parasitic effects outlined above, such as electrostatics and adhesion, make separation and sorting rather impractical.

If the fabrication process of the part already provides an order of parts in line or array form, this order should be maintained and used at any cost. Thus, the part arrangement obtained from injection molding or hot embossing is the basis for developing a solution for feeding. This arrangement should be maintained during dicing by sawing the array attached to a blue tape.

Assembly design rule 1: avoid bulk! Maintain part order created in upstream fabrication processes.

The LIGA process itself can be applied favorably by using the sacrificial layer process entirely to not under-etch parts, but to do so only so far that a small patch of the layer remains to keep each part in place (Figure 6.32).

Figure 6.32 LIGA cylinder lenses attached to the substrate by the sacrificial layer.

Care needs to be taken that the parts are not subjected to excessive mechanical stress during feeding. In the example in Figure 6.31, the clamping mechanism in the tray was optimized to avoid excessive shear forces on the micro-optical bench structures.

Assembly design rule 2: Avoid undue stresses on microparts during the feeding process.

6.4.3
Design for Gripping

The gripping process can be subdivided into the pick-up and the release process. In the same way as feeding, both sub-processes are heavily influenced in the micro-domain by surface-bound effects, such as electrostatics and adhesion. Provided that the feeding has been properly designed, the influence of these effects is considered to be negligible in the part pick-up process. In the release process, however, such effects may pose a real challenge. After switching off the gripping force, parts may stick to the gripper, such that they cannot be set down and released. In this case, the release of the part must be coordinated with the joining process.

Assembly design rule 3: take into account that releasing parts from the gripper may be difficult, because they stick to the gripper due to electrostatics or adhesion. Coordinate part release with the joining process.

The mass of the microcomponents is not usually the factor that determines the gripping force. More relevant are forces required for picking up the part from a feeding device (e.g. shearing it off from an adhesion layer; see Figure 6.33) and forces in some joining processes where the part is pressed into the base part. Both the gripper and the part have to be adapted to withstand these forces. Moreover, the part design has to be adapted accordingly. This means that surfaces for gripping and surfaces for force exertion during pick-up have to be included in the part design. Unless the grippers are very sensitive, these surfaces should be designed outside of the functional surfaces of the microcomponent.

Figure 6.33 Vacuum gripper to shear off and pick up cylinder lenses from the substrate.

Assembly design rule 4: avoid undue stress on parts during gripping; in particular avoid gripping on functional surfaces. Instead, include dedicated gripping surfaces in the part design.

6.4.4
Design for Positioning

Positioning means moving the microcomponent on to or into the base part and aligning it at the target position. In this process, the part is again subjected to forces along one and along up to three axes and potentially to friction. As in the gripping process, these stresses have to be considered in the design in order to avoid damage to the functional surfaces. Moreover, if more complex insertion and alignment motions are necessary, the required free insertion space has to be reserved in the base part in an early stage of the design process.

Assembly design rule 5: reserve free space in the base part to allow for insertion and alignment motions.

Usually, the absolute positioning tolerance of an assembly machine is too inexact for proper alignment of microcomponents and/or the position of the parts is not well defined. Hence there are three main strategies for aligning two parts relative to each other:

- Active alignment, with the two parts being moved relative to each other, while measuring some functional parameter (e.g. intensity). Here, the design has to ensure that the gripper does not interfere with the measurement and related equipment.

Assembly design rule 6: optimize both part design and gripper for active alignment.

- Alignment with digital image processing is the most frequently used technology. Here, the designer should make sure that there are clear geometric features that allow for the measurement of the relative position and orientation of the two

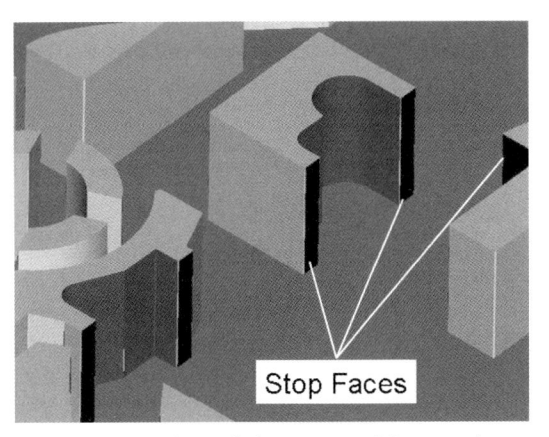

Figure 6.34 Stop faces of a lens mount of the LIGA distance sensor.

parts. This may mean structuring special alignment marks on to the parts or to make sure that the already existing features (corners, edges) of the parts are of sufficient quality to be used for this purpose. Both approaches have implications for the upstream fabrication processes.

Assembly design rule 7: for alignment with digital image processing, either design the part with proper geometric features, good edge quality and contrast, or provide dedicated alignment marks.

- Passive alignment implies using entirely mechanical means to position the part at the target position, that is, (gently) pushing the part against stop faces. To this end, the stop faces have to be integrated into the design of the base part and the mating faces of the microcomponent have to be designed accordingly.

Figure 6.34 illustrates the stop faces designed in the lens mounts of the LIGA distance sensor. A three-point contact is implemented for a statically determinate alignment of the lens. The stop faces have to be sufficiently wide and stiff enough to withstand the contact forces during alignment.

Assembly design rule 8: design stop faces and mating faces sufficiently robust so that they can withstand the alignment forces. The aim is a statically determinate alignment.

The stop faces, however, must not impair the function of the part, for example, by obstructing functional surfaces. In the example above (Figure 6.34) of the micro-optical distance sensor, the width of the rightmost stop face was determined by optical simulation to avoid obstruction of the optical path.

Assembly design rule 9: design stop faces that do not obstruct functional surfaces or functional ranges (e.g. ranges of motion, beam paths) of the part.

The aforementioned aspect of a sufficient insertion space is particularly important in passive alignment, since the alignment motions are frequently larger in

Figure 6.35 LIGA micro-optical bench with conical guide at the beginning of the fiber channel (left hand side).

this process. Although it may be straightforward to leave free space in the design in order to insert the component and push it against the stop face, there are cases where it may be reasonable to implement a shape that guides the component into the target position. Figure 6.35 shows a conical guide that facilitates insertion of an optical fiber into a channel on an optical bench.

Assembly design rule 10: use guiding structures to facilitate insertion and to reduce degrees of freedom during alignment.

6.4.5
Design for Joining

The ultimate goal of any assembly process is to align two parts relative to each other and permanently join them. In the discussion of the gripping sub-process, it was already pointed out that the part may stick to the gripper due to surface forces, so that releasing it may be difficult. In addition to fixing the part, the joining process has to provide means to release the part from the gripper. To achieve this, a force that is larger than the retention force of the gripper has to be generated in the joining process.

Assembly design rule 11: if parts stick to the gripper, a retention force can be created by proper selection of joining process and part design.

In the joining process itself, generation of the retention force must not alter the position achieved in the previous step.

Assembly design rule 12: neither the joining process itself nor generation of the retention force may alter the alignment of the joined parts.

The main joining processes used in the LIGA process are adhesive bonding and mechanical microclamping. Both are suitable for directly generating the forces which release the parts from the gripper.

6.4.5.1 **Adhesive Bonding**
Adhesive bonding uses an additional material, the adhesive, to join the two parts. Obviously, the chemistry of the adhesive must fulfill chemical and mechanical requirements. From the chemical point of view, it has to be compatible with the materials of the parts to be bonded, for example, not induce stress cracks in polymer parts.

Assembly design rule 13: take care that the chemical properties of the adhesive (e.g. solvents) are compatible with the materials of the microcomponents.

The adhesive bonding process can be broken down into two sub-processes, dispensing and curing. Adhesive dispensing frequently requires particular adaptation to the product's design. In microdimensions, there are two main challenges in adhesive dispensing:

1. Put the adhesive at the intended position.
   Dispensing small quantities of adhesive precisely is sometimes a difficult task. The quantity has to be controlled properly and the medium has to be brought to the correct position. Proper part design solves both problems at the same time. In the lens mount in Figure 6.34, a large pocket has been structured which leaves enough space to position the dispenser needle. On the other hand, the volume is large enough to accommodate over-dispensing due to dosage tolerances.

   Assembly design rule 14: take dosing tolerances into account and design sufficiently large areas for adhesives or provide overflow areas.

   Assembly design rule 15: unless contactless dispensing, such as the ink jet principle, is used, reserve touch-down areas in the structure for dispensing devices, such as needles or tips.

   Assembly design rule 16: if the structure is too narrow to allow for touch-down areas for dispensing devices, dispense on to the gripped part (for example, by dipping it into an adhesive bed).

2. Prevent adhesive from creeping into places where it may be harmful to the function of the part. Due to capillary forces, low-viscosity adhesives may creep into unforeseen places in a microstructure. This effect may lead to the blocking of structures that should be movable or to the obstruction of optical surfaces. In order to prevent such effects, capillary obstacles, such as stopping grooves, should be designed in the structure.

Assembly design rule 17: prevent adhesive from creeping into unwanted places in the microstructure by means of overflow areas (see rule 14) or by means of stop grooves.

On the other hand, the capillary effect can also be used productively in adhesive bonding. In the above LIGA micropump, a capillary network has been implemented that precisely transports the adhesive from a reservoir to the joining positions.

Assembly design rule 18: take advantage of capillary forces by designing a capillary network that transports adhesive to the joining positions.

The curing process has to be compatible with the material; that is, thermal curing temperatures have to remain below the glass transition temperatures of polymers and UV curing must not damage the polymer chains. Moreover, outgassing of the adhesive during curing must not impair the component function (e.g. optical surfaces).

Assembly design rule 19: take care that the curing process (e.g. temperature regime, UV irradiation, outgassing) is compatible with the materials and the function of the microcomponents.

Moreover, adhesive shrinkage during the curing process should not impair precise assembly. Ideally, the part design and the adhesive bonding strategy are optimized in such a way that adhesive shrinkage pulls the part against the stop faces.

Assembly design rule 20: design for adhesive shrinkage. To avoid impairing precise assembly, make sure that adhesive shrinkage pulls the part against stop faces.

6.4.5.2 Mechanical Microclamping

Mechanical microclamping means using the elastic or plastic deformation of the material in the base part or of an intermediate part to fix the microcomponent. Depending on the material and the temperature range to which the device is subjected, material relaxation or creep may be of relevance. In such a case, the joint should be secured with an additional bonding process, such as adhesive bonding.

Assembly design rule 21: mechanical clamping joints that are subjected to temperature cycling may slacken with time due to creep and relaxation. This in particular applies to polymer material. If in doubt, secure the joint additionally (for example, with adhesive).

The main categories of mechanical microclamping are press fits into the bulk material and dedicated clamping elements.

- Press fits

    Use of press fits leads to elastic and potentially to plastic deformation in the material. This is illustrated in Figure 6.36, where a glass lens is pressed into a nickel lens mount. Plastic deformation can be distinguished clearly along the upper rim. The resulting plastic deformation (see Figure 6.36) has to be

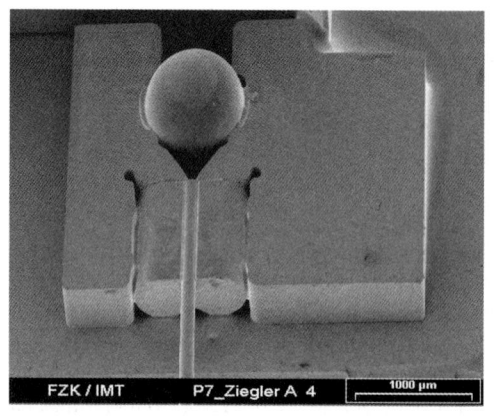

Figure 6.36 Press fit of a glass lens into the bulk material of a nickel lens mount.

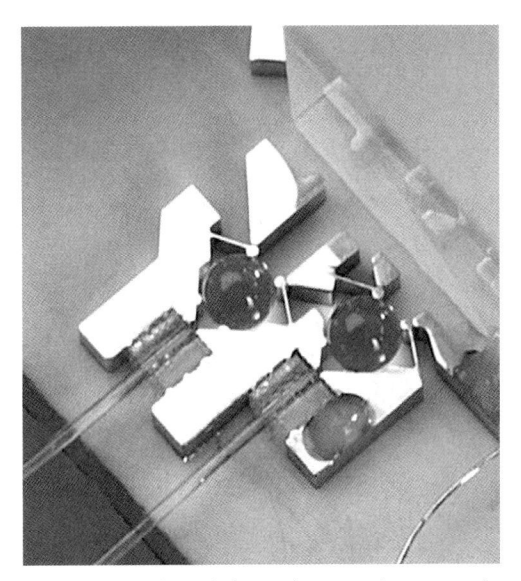

Figure 6.37 Dedicated elastic clamping elements in the LIGA FTIR [9].

compatible with the functional requirements of the part. The designer has to ensure that the base layer of the substrate is stiff enough to avoid warpage or shearing off of the structure walls.

Assembly design rule 22: when designing press fits, avoid deflection and shearing off of the side-walls from the substrate as well as substrate warpage.

- Dedicated clamping elements

    Dedicated clamping elements are more difficult to design than press fits and may require additional fabrication steps (e.g. sacrificial layer processes). On the other hand, they are a means to control precisely joining and fixation forces. The LIGA process in particular provides the designer with all the freedom needed in order to develop elegant and durable solutions. Figure 6.37 shows another nickel lens mount, this time with leaf spring-shaped elastic clamping elements that keep the spherical lenses at a statically determined place [9].

Assembly design rule 23: for precisely controllable mechanical clamping, design dedicated clamping elements that deform only in the elastic range.

A similar leaf spring-shaped elastic clamping element with a hook at its end is used to assemble the yoke of the electromagnetic chopper (Figure 6.38). This approach represents a function separation, as was outlined at the beginning of this chapter. With analytical methods of engineering mechanics or with FEM, the dedicated clamping elements can be dimensioned straightforwardly.

The rules outlined in the description of the stop faces above apply to both press fits and dedicated clamping elements: The design of the microclamping joint must be compatible with the function of the part.

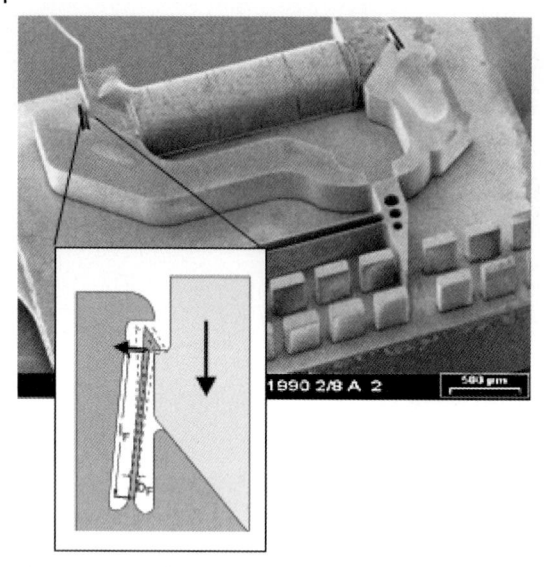

Figure 6.38 Metal hook joint in the LIGA chopper.

Assembly design rule 24: the design and function of the mechanical clamping elements must be compatible with the function of the part.

Treatment of joining processes has so far been limited to the two classes most frequently used in LIGA, adhesive joining and micromechanical clamping. Apart from these two main classes, other additional joining processes are applied, such as ultrasonic welding, laser joining and soldering (for metal LIGA parts). A description of their influence on LIGA part design exceeds the scope of this section.

6.5
Functional Design for LIGA

6.5.1
Introduction

Target-oriented application of computer-based tools for the design and simulation of microsystems shortens the time needed for development and increases the reliability of the systems in operation. To ensure the system's functionality in operation, it is not sufficient to model the microsystem on an ideal model environment. As stated in Section 6.1.2.6, all relevant requirements and constraints defined by the manufacturing processes and the application of the system in a real environment must already be considered in the design stage. Furthermore, every individual manufacturing step adds its own tolerances to the system. An adapted functional design is able to compensate for the tolerances and leads to a robust design. In addition, ambient influences, such as ambient temperature changes, may disturb the functionality of the microsystems and must hence be analyzed.

6.5 Functional Design for LIGA

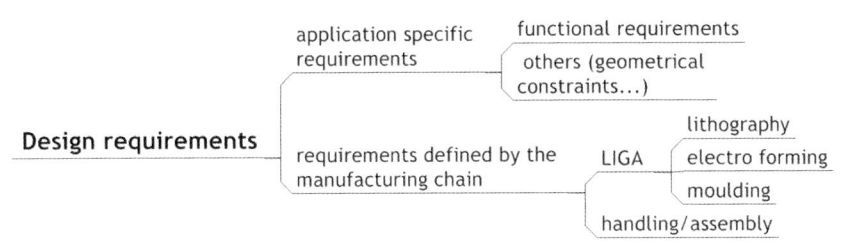

Figure 6.39 Diagram of the design requirements.

6.5.2
Design Requirements

Design requirements can be divided into application-specific requirements and requirements defined by the manufacturing processes (see Figure 6.39). Application-specific requirements can be further subdivided into functional requirements (requirements which ensure the functioning of the system) and more general requirements, for example geometric constraints.

Apart from the application-specific requirements, the designer has to be aware of the requirements defined by the manufacturing processes. Every individual process step has its own requirements, which must be considered by the designer. A detailed discussion of design rules for the LIGA process and for the assembly process can be found in the previous two sections. The following list only presents those requirements that are considered to be most important for the case study presented [10]:

- Requirements of the LIGA process steps:
  - Lithography: optically effective structures may only be curved around one axis. The result of this process step are so-called 2.5D structures. Furthermore, the design rules defined in the previous section have to be considered.
  - Molding: the hot embossing process, for example, requires certain clearances between individual structures and certain thicknesses of the structures themselves in order to allow for de-molding.
  - Molding material: material properties have to be considered in modeling (for example, the thermal expansion coefficient).
- Requirements of the assembly process:
  - Free space for collision-free manipulation of the components to be assembled (see Assembly design rule 5).
  - Integration of structures needed for assembling the components (see Assembly design rules positioning).

6.5.3
Comprehensive Modeling

Comprehensive modeling and simulation must consider the following three aspects:

- adjustment of the model to the production chain/design rules (design for manufacturing)
- compensation of tolerance effects by means of functional design (robust design)
- analysis of ambient and operational effects.

Implementation of these aspects in the modeling environment will be described in this section.

- **Adjustment of the design to the production chain:** knowledge of the requirements of the individual process steps is essential to design for manufacturing. The designer has to plan his layout carefully in order to fulfill all these requirements. Compliance with all manufacturing requirements by the system's model will save time and money and help to avoid additional re-design steps.

- **Adjustment of the design to tolerances:** consideration of tolerance effects in the processing steps not only requires detailed knowledge of the process chain, but also data on the material and the shape of the components under consideration. Hence tolerancing can be used not only to analyze the system's performance subject to the individual tolerances or in interaction with the tolerances of other components, but also for an adaptation of the surface shape of the optical components to compensate for the tolerances. Therefore, an optimization over the complete tolerance range, weighted with the tolerance distribution, if known, must be carried out to find the most robust solution with respect to manufacturing tolerances.

- **Analysis of ambient influences:** another important issue to be considered while modeling is ambient and operational effects. Ambient effects, for example thermal variations (daily or seasonally), may have an impact on the proper alignment and shape of the individual components and hence on the performance of the microsystem. Knowledge of the impact of thermal variations on the positioning and shape of the individual functional components allows one to calculate the influence on the system's performance and eventually to adjust the layout, such that these thermal effects are compensated.

The functional design for LIGA will now be discussed exemplarily in the field of micro-optics; the case study will be based on a micro-optical distance sensor.

6.5.4
Modeling and Simulation of a Micro-optical Distance Sensor

The comprehensive modeling approach described is applied to a micro-optical distance sensor developed under the BMBF-funded joint project μFEMOS. The working principle of the micro-optical distance sensor is the principle of triangulation.

Considering the 'Principles of division of tasks' described above, the micro-optical distance sensor is set up in a modular manner: that is, the passive optical unit is separated from the electronics [11]. This approach results in two

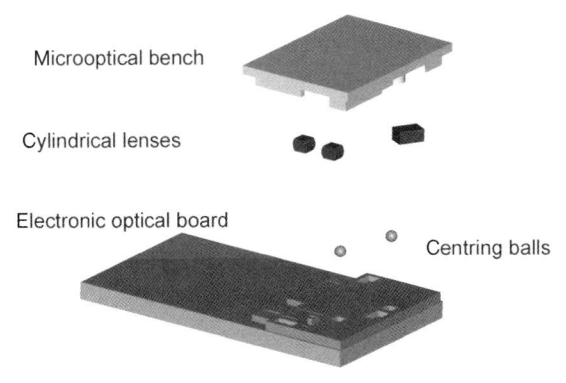

Figure 6.40 Illustration of the flip chip-like approach to manufacturing the micro-optical distance sensor [12].

subsystems (see Figure 6.40): the first subsystem is called the micro-optical bench (MOB) and contains the microstructured positioning and fixation elements and the passive micro-optical components; the second one contains the active optical components and the electronics and is referred to as the electronic–optical board (EOB).

For optical modeling of the micro-optical bench, the MOB is divided into two independent functional modules: the optical function components of the emission optical path will be modeled, simulated and optimized separately from the components of the detection optical path (see Figure 6.41). The emission optical path consists of the beam path from the laser diode up to the target; the detection optical path consists of the beam paths of different target positions to the detecting element.

The combination of the two models results in the optical model of the micro-optical bench. The combination takes place only after the separate optimization of the beam-forming and beam-manipulating components (Figure 6.42) [12].

6.5.4.1 Adjustment of the Model to the Manufacturing Chain

Comprehensive modeling of the micro-optical bench must consider the requirements imposed by the manufacturing chain. Table 6.3 lists some of these requirements, which result from the application of design rules defined by the individual process steps and the technologies involved, for example the design rules described in Sections 6.3 and 6.4. The requirements listed concern the fields of *electronics* (e), *function and reliability* (fr), *handling and assembly* (ha) and *molding* (m).

With respect to these requirements, the mechanical structures for positioning and fixation are designed and placed in the optical model. Thus, the optical model of the micro-optical bench is extended beyond pure optical functions by adding the mechanical functions.

For process technology reasons, all structures are metallized, with the only exception of cylindrical lenses, which will be assembled in a later process step. This means that the mechanical structures (positioning and fixation structures)

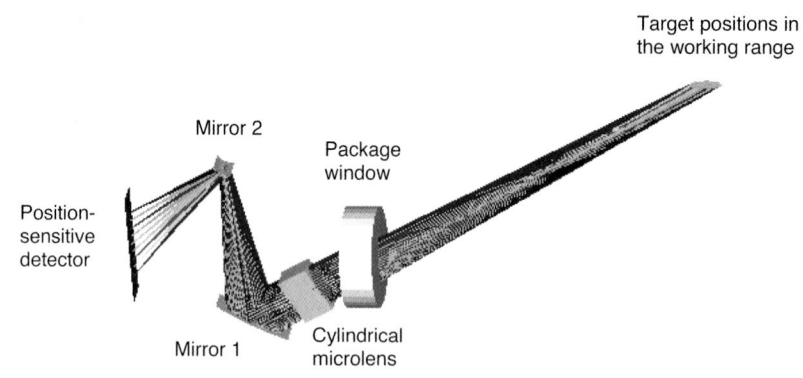

Figure 6.41 Emission optical path (a), detection optical path (b) of the micro-optical distance sensor.

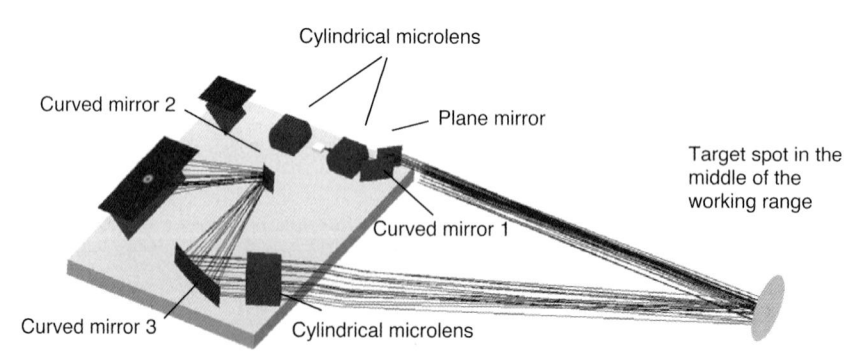

Figure 6.42 Optical design of the micro-optical sub-system. The optical functional model with some arbitrary rays is displayed.

are as reflective as the optical structures (mirrors). Consequently, in optical modeling both the mechanical structures and the optical structures must be considered in order to find possible interactions between the optical beam path and the mechanical structures.

For this purpose, an element coupling has to be performed as a first step [13]. This means that the optical function structures and the mechanical structures have

6.5 Functional Design for LIGA

Table 6.3 Some requirements for the design of the LIGA micro-optical distance sensor [10].

Requestor[a]	Requirements	Motivation
fr	Free space for adhesive cavities for joining of EOP and MOB: 1.3×0.8 mm (see Assembly design rule 14)	To ensure adhesion
fr	Edge thicknesses of adhesive cavities: $300\,\mu m$	To advance marginal rigidity
fr	Release of the mirror structures (see Lithography design rule 1) (see Electroplating design rule 2) (see Molding design rule 3)	To prevent warpage of the optical structures
fr	Closed structures of the PMMA–boundary around the detection unit	Reduction of stray light
fr	Slanted wall structures (perpendicular to the substrate)	Suppression of parasitic light
e	Open edge structures of the LD	Enabling visual control of the bonding wire
e	Openings in the rear edge structures	Free space for the PSD wires
m	Minimum distance between PMMA structures: $200\,\mu m$	To ensure de-molding of the structures
m	Minimum wall thickness: $200\,\mu m$	To ensure stability of the structures
m	Rounding of the PMMA edges (see Molding design rule 1)	To ensure the de-molding of the structures/to avoid cracks at sharp corners
ha	Free space for assembly paths (see Assembly design rule 5)	Enabling collision-free assembly
ha	Free space for mounting/alignment structures	Enabling the mounting of the optical components

[a] e, Electronics; fr, function and reliability; ha, handling/assembly; m, molding.

to be combined. This coupling allows for the analysis of the interaction between the optical beam path and the mechanical structures.

The properties of the surfaces of the mechanical structures can be adapted to reality by means of simulation software. As a result, it is possible to simulate influences on the mechanical structures. Such influences could be light backscattered at the mechanical structures (see Figure 6.43). Thus, the effects of this stray

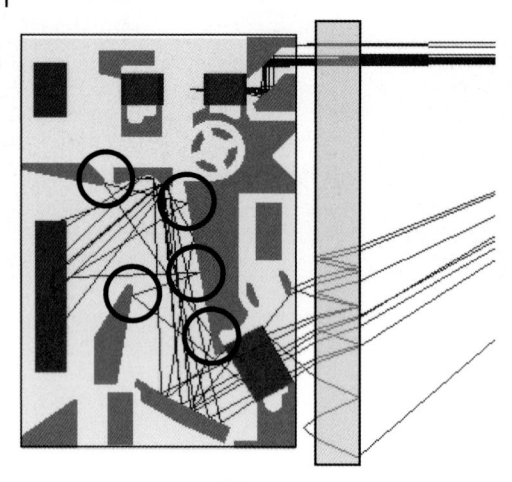

Figure 6.43 Top view of the entire model. Rays scattered at the mechanical structures are shown. The interactions between the scattered rays and the mechanical structures are marked by circles.

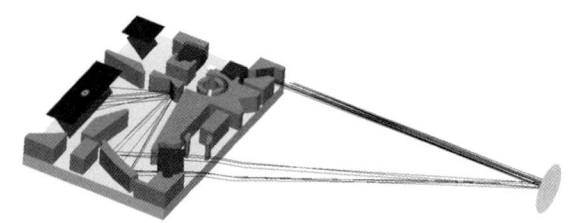

Figure 6.44 Representation of the entire model.

light on optical performance can be analyzed and, if necessary, reduced by additional structures or an adapted design [14].

Figure 6.44 shows the entire model with the reflected/refracted beam path in the case of a target position in the middle of the measuring range. The substrate is shown in light gray, mechanical structures in medium gray and the optical components in dark gray.

The entire model also serves as basis for the compensation for tolerance effects by means of the functional design (robust design) and for the analysis of ambient effects.

6.5.4.2 Robust Design

The next step in a comprehensive design approach is compensating for the effects of manufacturing tolerances on the individual functional components and structures by means of functional design. This approach follows very closely the design principles of a fault-free design, as described in Section 6.1. Every single process step adds its own specific tolerances to the microsystem. These tolerances cannot

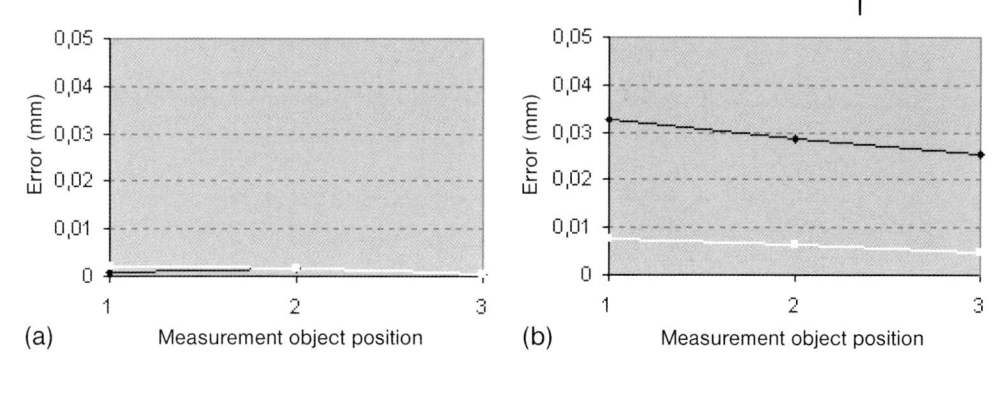

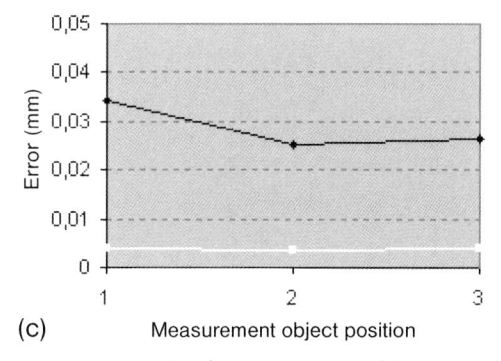

Figure 6.45 Results of the optimization of the shape of the curved mirror 2 as denoted in Figure 6.42.

be avoided. By means of a comprehensive design approach, the attempt is made to compensate for the effects of these tolerances by means of functional design. In this section, this approach is illustrated using the example of the mirror structure of the detection optical path of the micro-optical distance sensor (curved mirror 2 in Figure 6.42). The LIGA technique, with its process steps lithography, electroforming and hot embossing for replication, adds a tolerance of about ±0.5 μm to the ideal position of the carrier structure of the mirror. The cross-section of the 2.5D mirror has to be optimized for the complete tolerance range. This means that the mirror element can take every position in the tolerance range with the same likelihood. An optimization criterion is to find the best mirror design for the complete tolerance range: the robust design. This approach compensates for the effects of the position tolerance on the system's performance by means of functional design of the mirror element. As a result, Figure 6.45 shows the deflection of the spot position on the detector element as a function of the measurement object position. Position 1 corresponds to the near point of the measurement object, position 2 to the central position and position 3 to the far point in the measurement range of the distance sensor.

The black curves are the result of the original mirror cross-section and the white curves represent the optimized design. Figure 6.45a shows the deviation of the spot position on the PSD as a function of the measurement object position for the ideal position of the mirror element 2, which is zero tolerance. As expected, the result obtained with the original mirror cross-section is better than that of the tolerance-compensating design, since this is the case for which the original mirror was optimized. Nevertheless, the difference between the two curves is very small. The situation is different when looking at the curves at the lower threshold of the tolerance range, which means a mis-positioning of the mirror element by −0.5 µm (Figure 6.45b). Over the complete measurement range, the deviation of the spot position on the PSD is far smaller for the optimized design than for the original mirror. The same applies to a mirror position at the upper tolerance threshold, which means a mis-positioning by +0.5 µm (Figure 6.45c).

The results discussed show that it is worthwhile to model and optimize the functional structures not only for the ideal position, but also in the context of the fabrication process. Tolerances caused by different fabrication processes have to be considered in the design phase. The result is a robust design with respect to fabrication tolerances.

6.5.4.3 Thermal Analysis

Thermal loads on the structural components cannot be calculated by means of optical simulation tools. Data exchange has to be established between the optical tool and a finite element method (FEM) tool capable of solving such problems. Results of the FEM calculations are geometric changes in the optical components and their mounting structures due to a change in the thermal load. Figure 6.46 displays the micro-optical bench in the case of a constant ambient temperature of

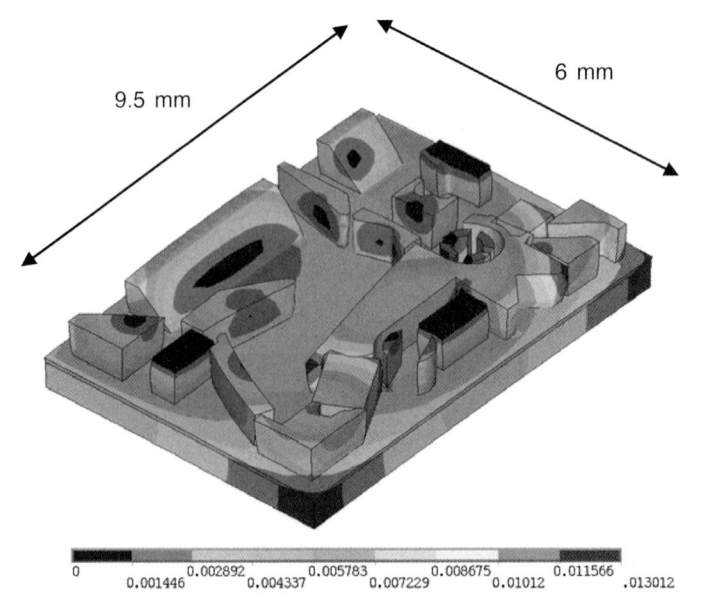

Figure 6.46 MOB (PMMA) under a temperature load of 85 °C (constant ambient temperature). The displacement values are given in millimeters.

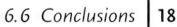

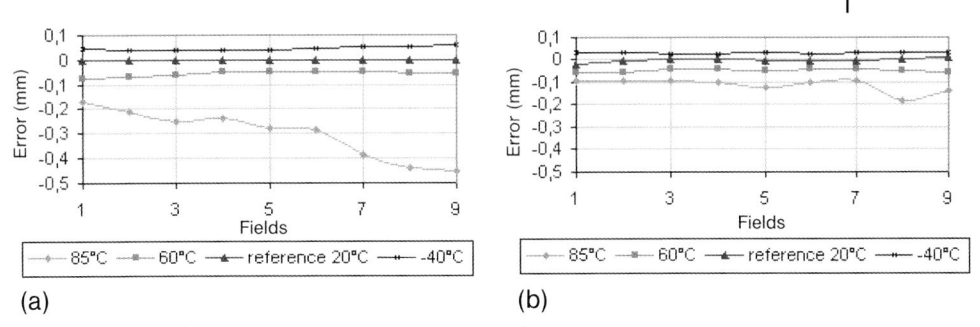

Figure 6.47 Error of the spot position on the PSD versus the fields, representing equally spaced distances spanning the measuring range for the two materials PMMA (a) and PC (b). Graphs are displayed for the temperatures −40, 20 (reference temperature), 60 and 85 °C.

85 °C and PMMA as material. The displacement values of the structures are given in millimeters.

To determine the effects of the temperature load on the optical performance of the micro-optical subsystem, the resulting FEM simulation data need to be re-transferred to the optical simulation tool. In the next step, optical analysis of the potentially deformed and misaligned optical function structures can be carried out. For example, the thermal effect on the optical performance of the micro-optical distance sensor will be shown subject to different materials and different constant ambient temperatures. Both PMMA and polycarbonate (PC) are considered as materials. Both materials can be used for replication by a hot embossing process. Four different constant temperatures [−40, 20 (reference temperature), 60 and 85 °C] are used for the simulations.

Figure 6.47 shows the results of the optical analysis of an MOB manufactured in PMMA (a) and of an MOB manufactured in PC (b) at the different temperatures given above. It can be clearly seen that the error in the spot placement on the PSD is smaller for PC than PMMA for the whole temperature range.

This approach allows conclusions to be drawn with respect to the system's performance for different materials in the case of ambient temperature variation and also in the case of influences resulting directly from the system's operation. With this knowledge, the right material can be chosen for the right field of operation and the desired performance of the system.

6.6 Conclusions

This chapter has introduced the application of 'macro' design methodology to the design of microsystems based on LIGA technology. Moreover, sets of design rules for different LIGA process steps and for the assembly of LIGA parts have been defined. The application of this toolbox and the benefits of a systematic approach in modeling and simulation in functional design of LIGA-based microsystems have been illustrated in the case study of a micro-optical distance sensor.

The set of design rules covers a substantial number of common issues, but is by no means complete. Still, application of the overall methodology described here should enable designers of microsystems based on LIGA to design robust devices that can be fabricated with high yield at competitive cost.

References

1 Pahl, G. and Beitz, W. (1995) *Engineering Design*, Springer, Berlin.
2 Schomburg, W.K., Ahrens, R., Bacher, W., Goll, C., Meinzer, S. and Quinte, A. (1998) AMANDA – low-cost production of microfluidic devices. *Sensors and Actuators*, **A70**, 153–8.
3 Strohrmann, M., Eberle, F., Fromhein, O., Keller, W., Krömer, O., Kühner, T., Lindemann, K., Mohr, J. and Schulz, J. (1994) Smart acceleration sensor systems based on LIGA micromechanics, in Microsystem Technologies 94, 4th International Conference on Micro Electro, Opto, Mechanical Systems and Components (ed. H. Reichl), Springer, Berlin, pp. 753–76.
4 Schomburg, W.K. and Goll, C. (1997) Design optimization of bistable microdiaphragm valves. *Sensors and Actuators*, **A64**, 259–64.
5 Leßmöllmann, C. (1991) Fertigungsgerechte Gestaltung von Mikrostrukturen für die LIGA-Technik, PhD thesis, Universität Karlsruhe.
6 Achenbach, S., Pantenburg, F.J. and Mohr, J. (2000) Optimization of the Process Conditions for the Fabrication of Microstructures by Ultra Deep X-Ray Lithography (UDXRL). FZKA Wissenschaftliche Berichte 6576, Karlsruhe.
7 Schumacher, K., Wallrabe, U., Mohr, J. (1999) Design, Herstellung und Charakterisierung eines mikromechanischen Gyrometers auf der Basis der LIGA-Technik. *FZKA wissenschaftliche Berichte* 6361, Karlsruhe.
8 Ruzzu, A. and Matthis, B. (2002) Swelling of PMMA structures in aqueous solutions and room temperature Ni-electroforming. *Microsystem Technologies*, **8**, 116–19.
9 Solf, C., Janssen, A., Mohr, J., Ruzzu, A. and Wallrabe, U. (2004) Incorporating design rules into the LIGA technology applied to a fourier transformation spectrometer. *Microsystem Technologies*, **10**, 706–10.
10 Hollenbach, U., Heckele, M., Hofmann, A. and Mohr, J. (••) Fertigungsgerechtes Design für einen mikrooptischen Abstandssensor, in *μFEMOS Mikro-Fertigungstechniken für hybride mikrooptische Sensoren* (ed. M. Bär), Universitätsverlag Karlsruhe, Karlsruhe, pp. 9–26.
11 Mohr, J., Last, A., Hollenbach, U., Oka, T. and Wallrabe, U. (2003) A modular fabrication concept for micro-optical systems. *Journal of Lightwave Technology*, **21**, 643–7.
12 Hofmann, A., Gengenbach, U., Scharnowell, R. and Skupin, H. (••) Montagekonzept für einen mikrooptischen Abstandssensor, in *μFEMOS Mikro-Fertigungstechniken für hybride mikrooptische Sensoren* (ed. M. Bär), Universitätsverlag Karlsruhe, Karlsruhe, pp. 41–60.
13 Hollenbach, U., Mohr, J., Oka, T., Wallrabe, U. and Sieber, I. (2004) Novel design and prototyping of a micro optical distance sensor considering a distributed fabrication *Optical sensing*, Proc. of SPIE, Vol. 5459.
14 Sieber, I., Hofmann, A. and Hollenbach, U. (2003) Application of the micro-optical construction kit to a micro-optical distance sensor, in Proceedings of MICRO.tec, Munich, pp. 381–6.

7
Commercialization of LIGA

Ron A. Lawes

7.1 Introduction *189*
7.2 Competing Technologies *190*
7.3 The Commercial Cost of LIGA *192*
7.3.1 Cost Aspects of LIGA *192*
7.3.2 A Cost Model for MEMS Technologies *194*
7.3.3 Comparative Costs *196*
7.3.4 Device Costs *198*
7.4 Foundry Operation *199*
7.5 Conclusions *201*
   References *202*

7.1
Introduction

Almost since its earliest R&D stage, LIGA, using X-rays from synchrotron radiation, has been considered to be too expensive and too inconvenient to access by industry and too difficult to use by non-specialist scientific staff. The absence of an appropriate infrastructure (for example, mask making), a mature resist and exposure technology and a number of specialized foundries have added to the difficulties of exploiting LIGA commercially.

As MEMS device applications have grown over the last 20 years, these perceived problems have driven industry to seek alternative manufacturing techniques, even at the expense of not fully realizing best-quality microstructures for some designs. Commercialization of MEMS has therefore been on two fronts. First, essential high aspect ratio microstructures have been manufactured by alternative techniques such as deep reactive ion etching (DRIE), UV LIGA and excimer laser ablation, and second, design methodologies have been adapted to use the thin-film techniques readily available from the CMOS semiconductor industry.

The emphasis on high aspect ratio micromachining techniques for microsystems/MEMS has been used mainly to achieve novel devices with, for example, high sensing or actuation performance. Often these utilize deep structures (depth $H = 100–1000\,\mu m$) with vertical wall layers but with relatively modest spatial resolution (1–10 μm). As these techniques have moved from research to industrial manufacture, the capital cost of the equipment and the cost of device manufacture have become important. The quality of the microstructures produced varies depending on the technique used. Often more than one micromachining technique can meet the performance requirements and a range of materials, such as silicon, polymers, metals, quartz and glass, can be used. The cheapest manufacturing technique is likely to be adopted.

CMOS thin-film techniques are currently being used to fabricate various sensors and actuators, which, although often limited in performance, offer closely coupled, integrated signal processing electronics and have the significant advantages of using well-established and cost-effective manufacturing techniques readily available in both in-house and foundry facilities. As the semiconductor industry investment cycle seeks improvements in performance and functionality (through reduced feature size and larger wafers) to spread the fabrication costs over more devices, high-quality equipment and often whole CMOS facilities become obsolescent. Yet such facilities are more than adequate for MEMS device manufacture. For example, as the semiconductor industry has moved from 0.35/0.25 μm processes on 150 mm diameter wafers to sub-100 μm processes on 300 mm wafers, the MEMS industry has developed mixed MEMS–CMOS devices using modest, sub-micron processes on 150 and 200 mm wafers. However, these techniques are still effectively planar and many advanced sensor designs require thicker structures to achieve adequate sensor sensitivity or actuator force.

This is the competitive scenario that the commercialization of LIGA must face if it is to play a significant role in industrial manufacture, where its unique advantages can be exploited. To successfully comercialize LIGA, synchrotron access has to be simplified, design rules and standard processes have to be available from more than one synchrotron center and an industry-based infrastructure must be established for mask-making and resist materials. Above all, the cost structure of LIGA (and competing technologies) must be better understood and efforts made to reduce costs, for example through automation.

7.2
Competing Technologies

Each of the high aspect ratio microfabrication technologies operates by a different physical mechanism, offer a different quality of micromachining (such as wall angle and roughness) and may expose thick resist or machine material directly. Some techniques are more suitable for, or only applicable to, certain materials. Silicon is the dominant material.

The main techniques, which compete with X-ray LIGA, are shown in Figure 7.1, along with the main features that affect the cost of manufacture. High aspect ratio processes such as DRIE and bulk micromachining can produce structure in the wafer layer itself, whereas X-ray LIGA and UV LIGA require an intermediate, thick

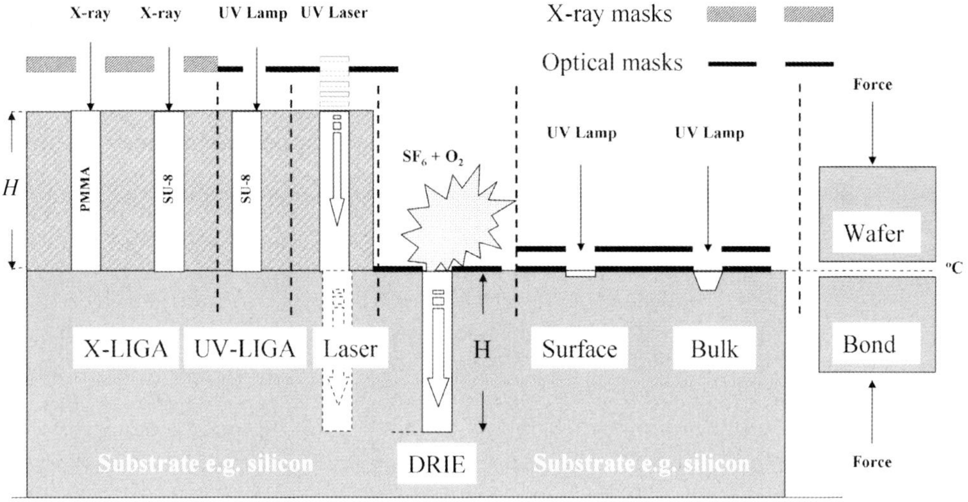

Figure 7.1 High aspect ratio microfabrication techniques.

resist, process. Excimer laser technology might machine material directly, for example a polymer, or use a resist stage.

The difference in resist thickness can be of the order of 1000:1 and the resist handling and development times (and costs) will vary accordingly. The maximum layer thickness that each technique can process will be determined by the wall angle, which may be non-vertical by a small but significant fraction of a degree, and the tolerances required for the top and bottom dimensions of the microstructure. In respect of wall verticality and roughness, X-ray LIGA is a superior technique.

All these processes require some form of mask to define the device's design features. The time to process a wafer layer to the required depth depends on the machine technology and the degree of automation in substrate handling. Many of the machines process a whole wafer simultaneously, as sub-micron resolution and alignment of individual die are rarely required. The exposure time algorithm depends on the time taken for the exposing radiation to interact with either the resist or the substrate material.

Bulk micromachining relies on the wet chemical etching of materials with preferential crystallographic planes [1], such as silicon, and can produce relatively simple devices at very low cost. Inflexibility due to the few fixed planes and low resolution (50–100 μm) are the main limitations. Cost of ownership is low, as the equipment required is simple mask aligners, resist handling tools and wet etching baths.

Laser micromachining and DRIE are two well-established techniques that do not require deep resist exposure and development to form a high aspect ratio structure.

An excimer laser produces a high-flux, demagnified optical image of a conventional chrome-on-quartz mask, which normally structures the layer material directly by ablation. Each laser pulse removes an image of part of the mask, to a depth of a fraction of a micron, depending on the material being machined [2]. The area of the mask can then be covered by a synchronized mask–wafer motion to expose the whole wafer surface. The wall angles and roughness in the substrate can be controlled by the illumination flux. However, the maximum exposure rate and flux are set by the maximum energy [3] and power that the mask can absorb ($0.1\,J\,cm^{-2}$ and $20\,W\,cm^{-2}$). Note that laser micromachining can be used both to manufacture deep resist molds and to micromachine directly a variety of materials.

DRIE is a form of plasma etching that requires a conventional lithography step to define features in a masking layer on the surface of the wafer, followed by an alternating series of vertical, sub-micron etching and side-wall passivation steps (known as the Bosch process) to the required layer depth. The reactive etchant is a fluorine radical, typically derived from an $SF_6 + O_2$ gas mixture [4] followed by a passivation plasma from C_4F_8 to deposit a protective polymer coating on the previously etched steps.

Surface microengineering [5] is similar to conventional semiconductor fabrication in that thin films (1–5 μm) of polysilicon are deposited on a silicon dioxide sacrificial layer that can be selectively removed to leave a free-standing structure. Commercial foundry processes exist with up to 3–5 layers of polysilicon. It should be noted that relatively modest, sub-micron semiconductor plants can be adapted or built to produce large MEMS device volumes at minimal cost. Strictly, surface microengineering does not produce the high aspect ratio structure typical of LIGA so that MEMS device designs must be limited to near planar options. However, it is useful to compare the costs with those of the high aspect ratio techniques, as surface microengineering can be compatible with CMOS processing and offers significant cost competition.

It was soon realized that the SU-8 resist was sensitive to UV radiation (g-line, h-line and i-line) and hence usable with a conventional mask aligner and conventional, commercially available chrome-on-quartz masks [6]. This has given rise to a cost-competitive technique, commonly known as UV LIGA, which is not only ideal for R&D applications but has also found its way into industrial production. The physical processes and manufacturing performance of UV LIGA [6–8] and X-ray LIGA [9] are now reasonably well understood.

Wafer bonding [10] is included, as it is often used to join several layers of a device. Wafer bonding, using pressure and high temperature, is probably the cheapest way in which two structured layers, typically manufactured in silicon and/or quartz, can be bonded together, for example, to form a buried microchannel as part of a microfluidics device.

7.3
The Commercial Cost of LIGA

7.3.1
Cost Aspects of LIGA

LIGA using X-rays from a synchrotron [7] was one of the first high aspect ratio techniques developed. LIGA utilizes the ability of intermediate energy X-rays (3–

10 keV) to penetrate deep resist layers without significant absorption or scattering. While the performance is excellent, offering very high aspect ratios (e.g. 100:1) and capable of exposing deep resists (500–2000 μm) with good resolution (1–5 μm) and tolerances (<0.5 μm), the cost to date has been perceived generally as prohibitive. The cost of X-ray LIGA has been assumed to depend on the time to expose a given resist, which is resist sensitivity and thickness dependent and, most crucially, on the cost per hour of using a synchrotron. The time to expose a given thickness of resist and the resist type, has been extensively analyzed through beam simulation programs such as DoseSIM [12] and RALBEAM [13] but the cost per hour has received less attention, not only for LIGA but also for other competing high aspect ratio techniques.

In the early days, the only LIGA resist used was a positive resist poly(methyl methacrylate) (PMMA), which required 5–10 h of exposure at a synchrotron. This time has been much reduced, as the material properties of PMMA and the exposure mechanisms of LIGA became more understood and synchrotron sources became more powerful. The subsequent availability of the negative resist SU-8, which is 100 times more sensitive to X-rays than PMMA, has reduced the exposure time from several hours to a few minutes. In all cases, the X-ray mask is manufactured, usually with a gold absorber pattern and a supporting membrane made from beryllium, carbon, silicon, silicon nitride, titanium or glass, or Kapton for a cheaper option. X-ray masks are currently expensive ($7000–10 000) compared with simple optical masks and are only available from R&D institutes.

As synchrotrons cost $100–500 million, conventional commercial depreciation calculations alone are likely to prohibit the industrial use of a synchrotron. In reality, synchrotrons are paid for from public funds, have a predominately science mission and are usually owned by academic institutions. This offers the option for industry to purchase 'spare capacity' at a commercially acceptable hourly rate. Such an arrangement, or similar, will be essential if commercialization of LIGA is to progress. The wafer scanner and cleanroom can be subject to normal capital expenditure and depreciation calculations and even the beamline can be included ($10 million?) if the throughput of wafers is sufficiently high (e.g. 10 000 wafers per year or greater). The commercial rates offered are critical to industry's ability to use X-ray LIGA and rates between $150 and $1000 have been quoted. In the following calculations, $500 per hour will be assumed.

A possible reduction in X-ray source costs, yet to be proven, is by the use of compact synchrotrons, commercially available initially from Oxford Instruments (Helios at the Singapore Synchrotron Light Source[14]) and more recently from Sumitomo [15]. At an estimated capital cost of $20–25 million, including a beamline and operation for 5000 h per year, a cost of $400–500 per hour is possible as a commercial proposition, although the range of applications may be limited by the low-energy output spectrum.

There are 12 synchrotron centers, approximately 15% of those throughout the world, that have contributed to understanding the scientific basis of LIGA and the construction of prototype devices. The required spectrum of energy output, mask making, resist technology and the supporting techniques of metal deposition, injection molding and hot embossing have been investigated so that a basic manufacturing technology can be defined. Only a few of these centers have attempted to engage industry and to promote the commercialization of LIGA [16]. In recent years, the leading synchrotron centers (i.e. owners of synchrotrons) pioneering the industrial use of LIGA have been ANKA at Karlsruhe [17], CAMD at Baton Rouge [18], BESSY at Berlin [19] and ALS at Berkeley[20]. In addition, a number of other centers have started to engage industry, notably SSLS with its Lithography for Micro and Nanotechnology (LiMiNT) process and the Synchrotron Radiation Research Center (SSRC) in Taiwan.

7.3.2
A Cost Model for MEMS Technologies

Understanding and reducing the cost of using LIGA are vital if commercialization is to take place. Little work has been published on MEMS cost models with LIGA or other high aspect ratio microengineering techniques, which are comparable to the robust models that have been developed and verified for the semiconductor industry [21]. However, at least one vendor of semiconductor cost models [22] has turned its attention to MEMS, so that the manufacturing industry may estimate investments and the unit cost of MEMS devices. Unfortunately, the internal assumptions and workings of such models are not openly available. Consequently, a cost model, MEMSCOST [23], has been developed to illustrate some of the important issues concerning the commercialization of LIGA.

MEMSCOST enables the impact of LIGA or any other high aspect ratio step to be compared with the more established thin-film techniques. MEMSCOST has been developed on a conventional fixed-cost basis, plus a demand-dependent variable cost. The model inputs machine-dependent, operational and financial data to determine the output costs calculated as a function of the number of wafer layers and the number of chips per layer. The manufacturing cost of a complete multi-layer wafer or device chip may be calculated from the sum of the layer processing steps and the overall yield of the processes. Each machine is assumed to process one layer of a wafer per sequence, including exposure, alignment, wafer loading and other overheads. Costs for masks and resist processes (consisting of resist spinning, pre-bake, development and post-bake) are included, where appropriate, as part of the process. The algorithms used to calculate the results are described [23].

The capital investment required will influence strongly the cost-of-entry for a manufacturing facility and will depend on the output capacity required of the various high aspect ratio equipment. As production demand increases, capital investment will rise as more machines are required. As expected, the initial equipment cost will be high, whereas the unit costs per wafer and hence per chip will fall with volume. It will be seen that X-ray LIGA offers some investment advantages for very large throughputs (see Figure 7.2).

MEMSCOST also enables the effect of larger wafer sizes and smaller chip areas on manufacturing costs to be estimated and for the overall cost per wafer to

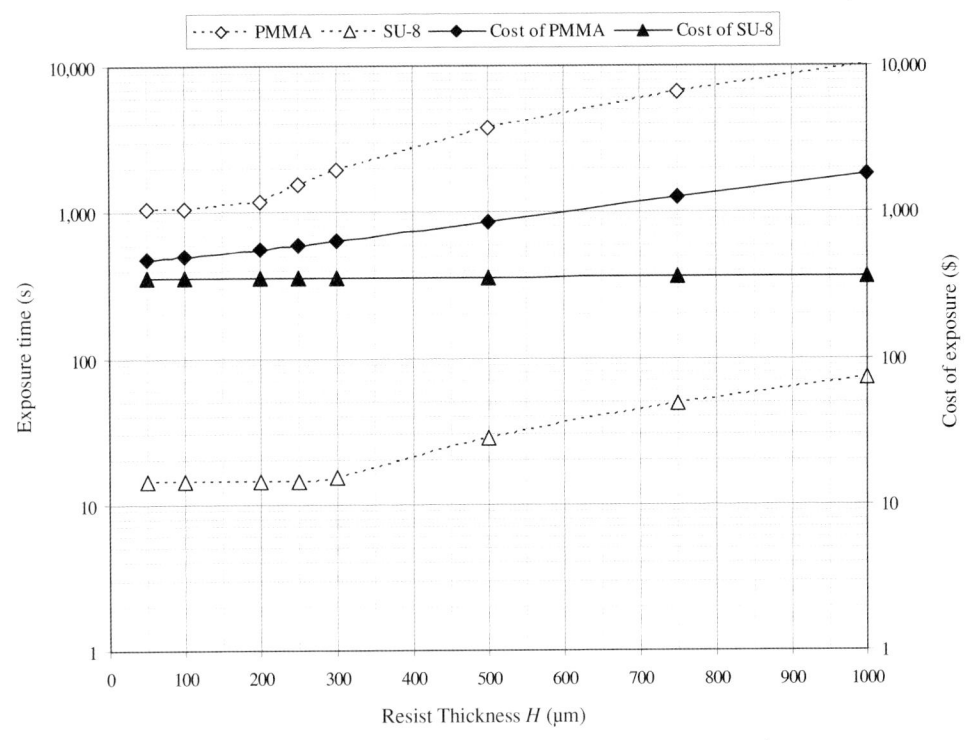

Figure 7.2 Cost of microfabrication, as a function of output ($H = 250\,\mu m$, 200 mm wafers).

include device functional testing and packaging (not addressed here). However, the model shows that testing and packaging can be >50% of the overall cost of a device, even using standardized packages from the semiconductor industry. The final selling price of a device will depend not only on the technology but also the accounting methods used to take account of R&D, marketing and sales, buildings and profit. These costs are not included in the analysis of the full commercialization costs but may be included in the model as fixed-cost overheads.

The time to expose a wafer for a LIGA process will depend on the spectrum of the particular synchrotron used, the beam current available and the preferred exposure dose to the thick resist. The following estimates assume a 2 GeV synchrotron operating with a ring current of 250 mA, delivering a 200 mm wide beam to a 200 mm wafer scanner. Commonly used exposure doses, at the top surface of the resist, are set at $9600\,J\,cm^{-3}$ for PMMA with sufficient aluminum absorber to limit the top-to-bottom ratio to ×3 and $120\,J\,cm^{-3}$ with the top-to-bottom dose ratio limited to ×4.

An estimate of the cost of LIGA is shown in Figure 7.3, where the cost of a LIGA step includes a gold-on-beryllium mask, deposition of the resist to the required depth plus a pre-bake step, X-ray exposure of the resist at the synchrotron followed by development and a post-bake step. A commercial fixed rate of $500 per hour

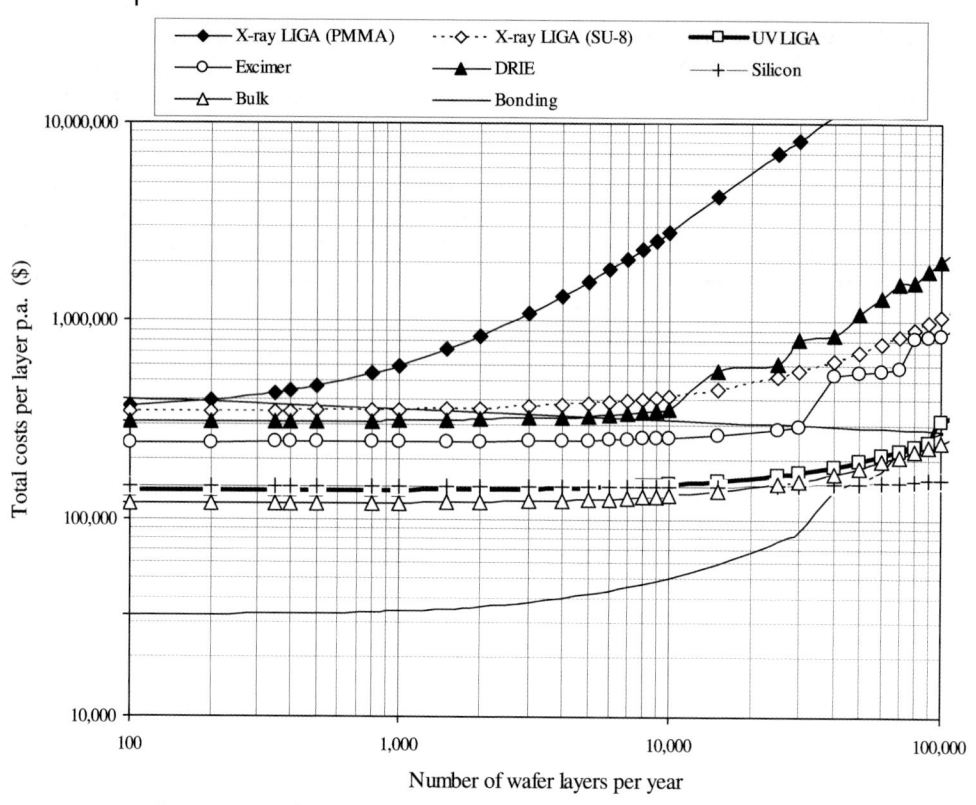

Figure 7.3 Typical LIGA exposure times and estimated costs, as a function of resist thickness.

for synchrotron access is included and a throughput of 1000 wafers per year is assumed for an R&D environment.

It will be seen that the cost advantage of using SU-8 resist instead of PMMA is only a factor of 2.4–5 and substantially less than expected from the 100:1 sensitivity difference. This is partly due to the chosen exposure regimes for the two resists (PMMA:SU-8, 80:1) and partly due to the fixed resist processing costs.

Nevertheless, the difference in costs between the two resists is significant when considering the cost of using LIGA commercially, where even relatively small cost differences can establish a market advantage. For example, a single X-ray exposure for the conditions outlined would cost $864 per wafer with PMMA compared with $243 per wafer with SU-8.

7.3.3
Comparative Costs

Not only are other high aspect ratio techniques competitive in terms of adequate performance for many fabrication tasks, but also the equipment can be owned and placed in the MEMS factory, rather than used remotely and hence perceived as

Table 7.1 Fixed and variable costs ($) and throughput rates for LIGA and competing technologies.

Technology	Equipment		Fixed costs ($ per year)	Variable costs	
	Capital cost ($)	Maximum wafers per year		Wafers per hour	$ per wafer
X-ray LIGA PMMA (scanner only)	1 100 000	11 865	263 600	2.1	245
X-ray LIGA SU-8 (scanner only)	1 100 000	445 974	263 600	78	6.5
UV LIGA SU-8	225 000	98 639	58 400	17	0.6
Excimer laser	900 000	38 717	220 400	6.8	12.4
DRIE	720 000	13 412	177 200	2.4	4.7
Surface micromachining	225 000	456 000	58 400	80	0.2
Bulk micromachining	261 000	55 459	62 000	9.7	1.1
Wafer bonding	162 000	34 545	32 400	3 (pairs)	1.8

the low-cost option. Table 7.1 shows the capital costs and throughput capacity for individual machines, emphasizing the wide range of fixed and variable costs that are, with the exception of X-ray LIGA, well established in industry.

The fixed costs include the capital cost of each equipment type, depreciated over a typical depreciation period (5 years), and fixed maintenance costs. When the annual demand exceeds the capacity of the installed equipment base, additional machines are purchased and added to the fixed costs. The equipment capital cost for X-ray LIGA only includes the scanner and excludes the beamline, which is part of the assumed fixed $500 per charge. The surface microengineering cost is only for the equivalent lithography step, but the bulk micromachining cost includes the selective wet etching process to provide an equivalent microstructure.

The variable costs include manpower and maintenance but exclude materials, which are deemed to be approximately constant for all the techniques. The overall cost per wafer is sensitive to the throughput capability of the equipment.

The cost of microfabricating a single layer of a wafer, for a given technology, is shown in Figure 7.2, assuming 200 mm diameter wafers. As the demand for devices increases, the number of wafer layers that must be processed increases and eventually approaches the maximum number that the installed machines can output. There are two consequences. First, the average time a wafer is queued for machine access increases until it reaches unacceptable levels and manufacturing will suffer from under-capacity. Second, in order to avoid this eventuality, additional machines must be purchased, thus increasing the unit costs in a 'saw tooth' manner (often referred to as 'granularity').

The unit cost of processing a single layer of a wafer, with different microfabrication techniques at various levels of annual throughput, is shown in Figure 7.4. This shows that high aspect ratio techniques are expensive compared with a planar technique, such as surface microengineering. It is also clear that the significant cost of using PMMA urgently requires a much more sensitive positive resist, if complete positive and negative LIGA options are to be available to MEMS designers.

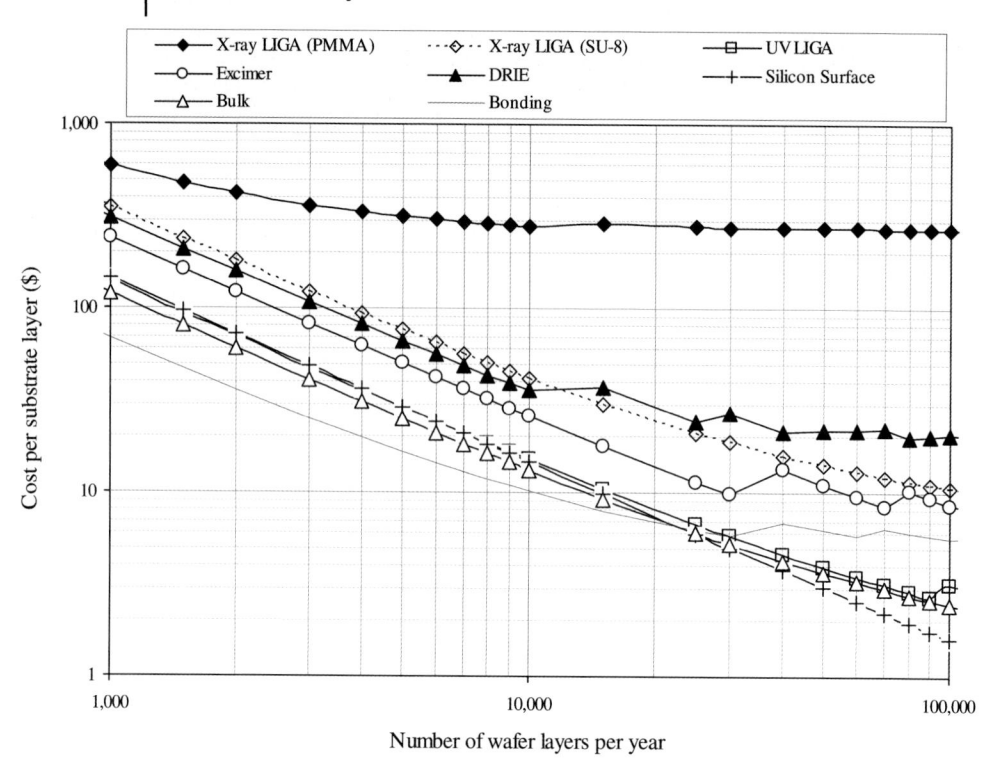

Figure 7.4 Fabrication costs ($ per wafer) for various microfabrication processes as a function of annual wafer production ($H = 500\,\mu m$, 200 mm wafers).

However, using a sensitive resist such as SU-8 at a thickness of $250\,\mu m$ can become cost competitive with DRIE at reasonable levels of throughput, that is, >10 000 wafers per year. This is a significant result for the prospects of exploiting LIGA commercially, as DRIE is already being used for industrial products, particularly those fabricated by integrated CMOS–MEMS technologies. In fact, as the required depth of the MEMS component is increased to gain performance benefits, the cost advantages become even more significant. For example, if the required depth is $1000\,\mu m$ and the throughput requirement exceeds 10 000 wafers per year, then X-ray LIGA also becomes cost competitive with UV LIGA and excimer laser processing. DRIE becomes 2.4 times more expensive.

7.3.4
Device Costs

Figure 7.4 suggests that the various high aspect ratio techniques, particularly X-ray LIGA (even if SU-8 is used), are considerably disadvantaged compared with thin-film techniques. However, for any designs requiring the performance benefits of deep, smooth and near vertical walled structures that are available from LIGA, a cost increase of a small percentage may be acceptable as part of a cost–

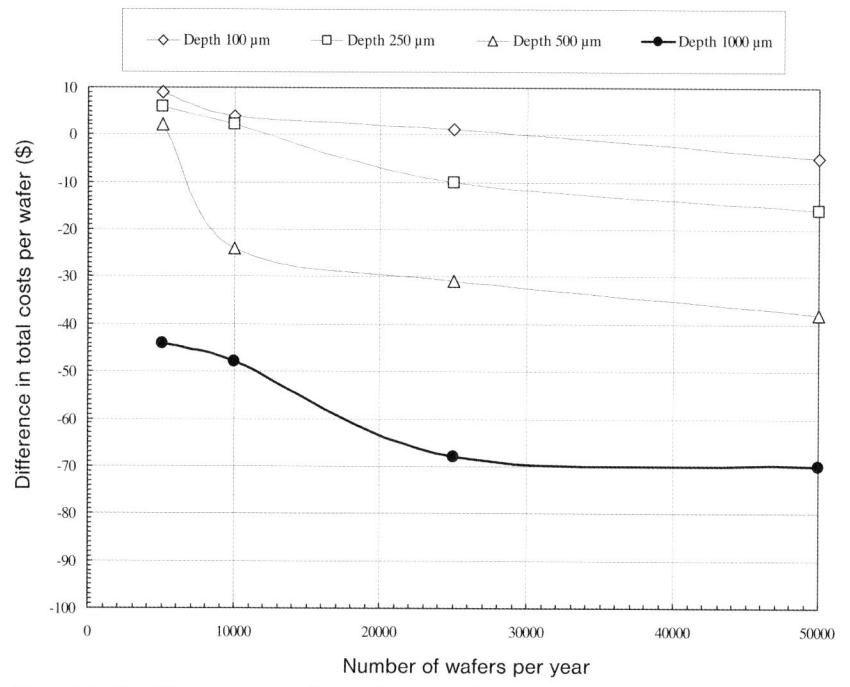

Figure 7.5 The difference in overall costs for an accelerometer chip when replacing a DRIE process with X-ray LIGA as a function of wafer output.

performance maximization process. Thus, a combination of the thin-film and high aspect ratio processes is required to cost a complete device.

For example, the estimated cost of fabricating accelerometers containing CMOS signal processing electronics and a high aspect ratio proof mass, 250 μm deep and fabricated by DRIE on 200 mm wafers, is $1724 per wafer for 5000 wafers per year, $1358 per wafer for 10 000 wafers per year and $872 per wafer for 50 000 wafers per year. This excludes the cost of testing and packaging.

Typically, the DRIE step is only 2–4% of the cost of completely processing a wafer ready for testing, dicing and packaging. Replacing the DRIE step with an X-ray LIGA process, using SU-8 resist, would cost $1730, $1360 and $856, respectively. Figure 7.5 shows the difference in costs if both the required depth and number of wafers per year are increased. It is apparent that there is a regime of design and annual demand where X-ray LIGA can add both performance and cost benefits.

7.4 Foundry Operation

In the long term, commercialization of LIGA may not be determined simply by price but by the production issues concerning the reliability of the synchrotron and the relevant beamline, the stability of the rest processes, the commercial availability of suitable X-ray masks and access to and turnaround time from a synchrotron.

Fortunately, synchrotrons, despite their complexity, have an excellent operational record and reliability (>95%) and hence near 24 h per day availability. A beamline can be dedicated to industrial manufacture at a similar cost to, for example, an optical stepper ($10 million). The installation of a suitable cleanroom in the synchrotron experimental hall is relatively cost effective and can be equipped readily with the special scanners required to utilize the pencil-like beam available from the beamline and the supporting resist processing and metrology equipment. High-quality scanners are available commercially from a well-established equipment vendor [24] which offers a standard machine with suitable high-precision mask and wafer chucks.

Over the last 20 years, researchers have produced prototype devices to demonstrate the technique with little regard to cost, reproducibility of the overall process or the level of skill required to obtain a (single?) result. Commercial exploitation of LIGA will demand that the processes are proven to be stable and the tolerances are clearly defined. These must be embodied in a clear and fixed set of design rules that a foundry operation will guarantee. It should be noted that this has only been done for a few of the competitive high aspect ratio techniques [25], although the semiconductor industry (both in-house facilities and foundries) have tolerances and design rules for thin-film techniques which are well understood and under tight control.

A major concern for potential industrial users of synchrotron facilities is the perception that they are often remote from industrial sites, apparently more concerned with scientific output than industrial production and unfamiliar with the concept of turnaround. This is less of a problem with those (relatively few) establishments that have engaged industry. However, the perception of industry will be much altered if industry takes responsibility for the key aspects of LIGA production and operates a 'one-stop' foundry for manufacturing industry.

Mask making remains both a technical problem and one of availability, exacerbated by the lack of a commercial supply of masks. ANKA, CAMD, BESSY and Sandia have established quality mask making for their internal purposes and there are some early attempts at standardization between them. If a secure commercial supply of masks is to be available to industry then there is an urgent need to involve the optical mask-making industry and for technology transfer to take place.

The semiconductor industry has demonstrated the path for successful commercialization of an apparently expensive process. CMOS has long since ceased to be the reserve of large companies with in-house facilities. Foundries have grown up throughout the world to offer one or more CMOS variants, at an affordable price and close to the cutting edge of technology as it develops. Clear design rules are available and wafer/device yields and performance guaranteed. This offers the customer not only a choice of supplier but also a second source if and when required.

Commercialization of LIGA has to follow the same path. There are a few successes to date, but wider use of LIGA for those applications where its capabilities are highly desirable or essential are dependent on the availability of foundries run by industry, for industry. The foundry will establish the tolerances and design rules, negotiate binding, long-term contracts with the synchrotron operator and deliver to the customer. The customer need not be concerned with, or even aware of, operational details at the synchrotron site.

Axsun Technologies (Billerica, MA, USA) [26] pioneered this route to commercialization of LIGA, first by signing a technology agreement with Sandia Laboratory, thereby gaining access to robust mask-making and resist technologies, and second by making a long-term agreement with the ALS at Berkeley for access to synchrotron radiation using their own dedicated beamline. Mezzo Systems is using the CAMD facilities to manufacture microstructures for various US agencies, including NASA.

ANKA at (Karlsruhe, Germany) has concentrated on providing a cost-effective service to industry. One highly successful commercial application has been the manufacture of gear wheels for the watch industry. In comparison with conventionally manufactured gears, those manufactured by X-ray LIGA were 10 times more accurate with fourfold smoother surfaces [27]. It is expected that prices will be reduced considerably (fourfold?) through the construction and operation of the FELIG fully automated fabrication line [28].

The concept of second sourcing for LIGA is more an organizational than a technical issue. Simulation shows that, provided that beamlines and masks are set up in a similar manner and design rules are sufficiently relaxed to accommodate inevitable differences, mask sets could be moved by the industrial MEMS foundry from one synchrotron to another. A recent study [29] has shown that by adopting the 'standard' processes described for the PMMA and SU-8 resists and manufacturing the masks with dimensional offsets only calculated according to resist thickness, excellent dimensional tolerances can be obtained in the resist image. Synchrotrons with different parameters, ranging from a compact synchrotron (e.g. $0.7\,GeV$, $4.5\,T$ at SSLS) to third-generation synchrotrons (e.g. $3\,GeV$, $1.3\,T$ at Diamond, RAL) now coming on-line, could offer a second source specification on resist images of $\pm 5\%$ tolerance ($\pm 0.05\,\mu m$ tolerance on $10\,\mu m$ and $\pm 105\,\mu m$ tolerance on $20\,\mu m$ lines) in a depth of up to $500\,\mu m$, in PMMA and SU-8 resist, respectively. The costs will vary little with the different beam current operation of the various synchrotrons, due to the relative insensitivity of cost to exposure time.

7.5
Conclusions

There have been a number of barriers to the commercialization of LIGA that have prevented industry from adopting a technology that has several key technical advantages and has been available in various forms of development for over 20 years. The cost of owning or using a synchrotron has been prohibitive, access to a remote, academic center has been seen as difficult and LIGA has been seen as an interesting scientific phenomenon but an immature technology with very little commercial infrastructure.

The situation is now much improved, with synchrotron owners making significant efforts to improve access and, in a few cases, cooperate with industry to construct and dedicate beamlines exclusively for industrial use. Attention is being paid to standardizing aspects of mask making and resist processing, establishing tolerances, improving turnaround time and reducing costs. Although the cost of LIGA remains high compared with alternative high aspect ratio techniques and even higher compared with thin-film techniques, consideration of the overall cost of producing a device reduces the added cost of using LIGA to a small percentage. If the cost of signal processing, digital interfacing, testing and packaging is included, the MEMS function and hence the LIGA component will be an even smaller part of the overall cost of device manufacture.

The exposure times and throughput rates assumed in MEMSCOST illustrate the magnitude of improvements that LIGA must make if commercialization is to be realized. The data used for competitive technologies is typical of currently available equipment and processes and will no doubt be much improved in the future, so LIGA will have to keep pace with or exceed the rate of such improvements. Further commercialization will depend on an increase in demand for the design of more advanced MEMS devices, which need or can benefit from the precision, smoothness and near verticality of the resist walls that LIGA brings.

An industry-based infrastructure is needed, offering commercial sources of masks and a supply of basic materials with quality assurance comparable to that delivered to the semiconductor industry. Although it is often possible to design around the limitations of using only a negative resist, such as SU-8, a similarly sensitive positive resist is needed to replace PMMA.

Potential industrial foundries will need to mimic closely the successful operation of those in the semiconductor industry and offer a complete service to customers, including other high aspect ratio techniques and probably alongside thin-film and even full CMOS/BiCMOS capability.

References

1 Elwenspoek, M. and Jansen, H. (1998) *Silicon Microengineering.* Cambridge Studies in Semiconductor Physics and Microelectronics Engineering, Vol. 7, Cambridge University Press, Cambridge, pp. 10–68.

2 Gower, M.C. (2001) Laser micromachining for MEMS devices. *Proceedings of the SPIE–The International Society for Optical Engineering*, **4559**, 53–9.

3 Rumsby, P., Harvey, E., Thomas, D. and Rizvi, N. (1997) Excimer laser patterning of thick and thin films for high density packaging. *Proceedings of the SPIE–The International Society for Optical Engineering*, **3184**, 176–85.

4 Laermer, F. and Urban, A. (2003) Challenges, developments and applications of silicon deep reactive ion etching. *Microelectronic Engineering*, **67–8**, 349–55.

References

5. Linder, C., Jaecklin, V. and de Rooij, N.F. (1992) Surface micromachining. *Journal of Micromechanics and Microengineering*, **2**, 122–32.
6. Chuang, Y.F., Tseng, F.G. and Lin, W.-K. (2002) Reduction of diffraction effect of UV exposure on SU-8 negative thick photo resist by air-gap elimination. *Microsystems Technologies*, **8**, 308–13.
7. Lawes, R.A. (2005) Manufacturing tolerances for UV LIGA using SU-8 resist. *Journal of Micromechanics and Microengineering*, **15**, 2198–203.
8. Zhang, J., Chan Park, M.B. and Connor, S.R. (2004) Effect of exposure dose on the replication fidelity and profile of very high aspect ratio microchannels in SU-8. *Lab on a Chip*, **4**, 646–53.
9. Griffiths, S.K. (2004) Fundamental limitations of LIGA X-ray lithography: sidewall offset, slope and minimum feature size. *Journal of Micromechanics and Microengineering*, **14**, 999–1011.
10. Schmidt, M.A. (1998) Wafer-to-wafer bonding for microstructure formation. *Proceedings of IEEE*, **86**, 575–1585.
11. Ehrfeld, W. and Schmidt, A. (1998) Recent developments in deep X-ray lithography. *Journal of Vacuum Science and Technology B*, **16**, 3526–34.
12. Meyer, P. *et al.* (2002) An MS-Windows simulation tool for synchrotron X-ray exposure and subsequent development. *Microsystems Technologies*, **9**, 104–8.
13. Lawes, R.A. and Arthur, G.G. (2003) LIGA for Boomerang. *Proceedings of the SPIE–The International Society for Optical Engineering*, **5276**, 307–17.
14. Jian, L.K. *et al.* (2006) Industrial applications of micro/nanofabrication at Singapore synchrotron light source. *Journal of Physics: Conference Series, International MEMS Conference*, **34**, 891–6.
15. (2006) The Sumitomo LIGA foundry using a compact SR ring. Commercialization of Micro and Nano Systems Conference, St Petersburg.
16. Loechal, B., Goettert, J. and Desta, Y.M. (2007) Direct LIGA service for prototyping: status report. *Microsystems Technologies*, **13**, 327–34.
17. Institut fur Microstrukturtechnik (2006) http://www.fzk.de/imt.
18. Centre for Advanced Microstructures and Devices (2006) http://www.camd.lsu/edu.
19. BESSY (2006) http://www.azm.bessy.de.
20. Advanced Light Source (2006) http://www-als.lbl/gov/als/.
21. Goodall, R., Fandel, A.A., Landler, P. and Huff, H. (2002) Long-term productivity mechanisms of the semiconductor industry. Semiconductor Silicon 2002 Proceedings, 9th, Electrochemical Society.
22. IC Knowledge (2007) http://www.icknowledge.com.
23. Lawes, R.A. (2007) Manufacturing costs for microsystems/MEMS using high aspect ratio microfabrication techniques. *Microsystem Technologies*, **13**, 85–95.
24. Jenoptik GmbH (2006) http://www.jo-mikrotechnik.com.
25. Lawes, R.A. (2007) Manufacturing tolerances for UV LIA using SU-8 resist. *Journal of Micromechanics and Microengineering*, **15**, 2198–203.
26. Axsun Technologies-LIGA-based Manufacturing Capabilities http://www.ligafoundry.com.
27. Saile, V. (2004) Mikrosystemtechnik mit Synchrotronstrahlung. www.bessy.de/industrie/IF2004/Programm/Vortraege/if2004saile.pdf.
28. Meyer, P., Klein, O., Arendt, M., Sale, V. and Schulz, J. (2006) Launching into the golden age (3) gears for micromotors made by LIGA process. Commercialization of Micro and Nano Systems Conference, St Petersburg.
29. Lawes, R.A. (2006) A traceable fabrication process for X-ray LIGA–easier access for industry. Commercialization of Micro and Nano Systems Conference, St Petersburg.

8
Polymer Optics and Optical MEMS

Jürgen Mohr

8.1 **Introduction** *205*
8.2 **Optical Components** *208*
8.2.1 Cylindrical Microlenses, Microprisms *208*
8.2.2 Resonant Filters *209*
8.3 **Optical Benches** *211*
8.3.1 Intensity Coupling Element *212*
8.3.2 Multifiber Connector *213*
8.3.3 Bidirectional Transciever- and Receiving Module *215*
8.3.4 Heterodyne Receiver *215*
8.3.5 Achromatic Confocal Distance Sensor *216*
8.4 **Electro-optical Systems** *218*
8.4.1 Triangulation Distance Sensor *219*
8.4.2 Microspectrometer *222*
8.5 **Optical MEMS** *225*
8.5.1 Micro-optical Bypass Switch *225*
8.5.2 Electrostatic Switching Matrix *226*
8.5.3 Chopper for Fiber Applications *227*
8.5.4 FTIR Spectrometer *229*
8.6 **Conclusion** *230*
References *231*

8.1
Introduction

The LIGA process (see Chapter 1) is very well suited to fabricate micro-optical components and systems for the following reasons [1]:

- high precision of the fabricated structures

- exact positioning of different structures on one substrate due to the masking process

Figure 8.1 High aspect ratio microstructure fabricated by X-ray lithography (height, 500 µm; width of the small bar, 5 µm).

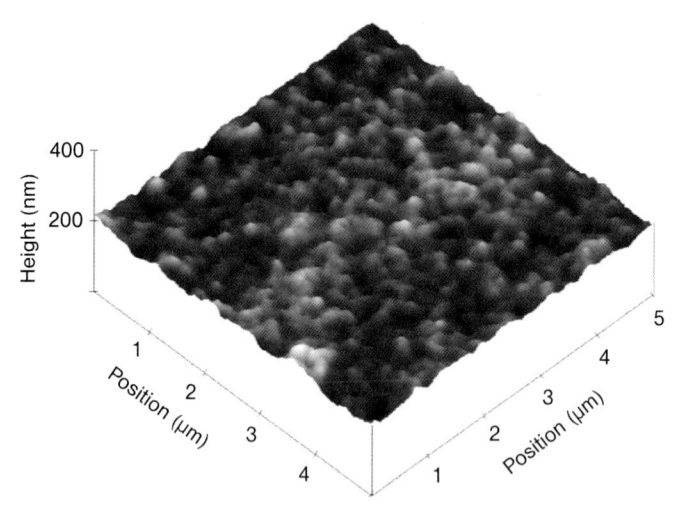

Figure 8.2 AFM image of the side-wall of a microstructure fabricated by X-ray lithography.

- high aspect ratios (Figure 8.1)
- nearly parallel side-walls with an inclination in the range of 0.2 mrad per side-wall for structural heights up to 500 µm
- low roughness of the side-walls in the range 10–20 nm (Figure 8.2)
- good optical transmission in the visible range of the polymers used in lithography [poly(methyl methacrylate) (PMMA), SU-8] and in molding (e.g. PMMA, TOPAS, polycarbonate)
- low shrinkage resulting in low internal stress in the case of hot embossing.

Figure 8.3 Diffraction pattern of a side-wall fabricated by X-ray lithography using an e-beam manufactured mask.

Figure 8.4 Cylindrical lenses with an optimized shape due to high-order corrections of the curvature.

The use of polymers offers the possibility of fabricating simple optical structures such as prisms, cylindrical lenses and grating structures. The low roughness of the structures is a presupposition for a low stray light level of the optical beam. Because of limitations in the mask fabrication by electron beam lithography, periodic structures are produced along the geometry, which will cause interference patterns resulting in light at unwanted places (Figure 8.3) [2]. This is mainly an issue in the case of diffractive structures such as gratings.

The possibility of choosing arbitrarily the lateral geometry of the structures offers the chance to optimize their optical function. Thus, not only circular but also parabolic or elliptical lenses with higher order corrections can be patterned (Figure 8.4). Fabricating the lenses or the prisms on top of a sacrificial layer which allows one to partly under-etch the structures gives the possibility of releasing them with low forces and without any damage from the substrate. This is of great advantage in assembly as the lenses are well positioned and need not be analyzed after gripping as would be necessary in the case of bulky goods [3].

The possibility of fabricating cylindrical optical structures together with fixing structures on one substrate with positioning precision of better than 1–2 μm on an area of several square centimeters allows the fabrication of more complex optical benches [4]. Parallel patterning of micro-optical components and mechanical mounts and guidance structures for passive alignment of glass fibers or other hybridically mounted optical elements (spherical lenses, filters, etc.) allows a passive assembly of the benches and reduces the adjustment and mounting costs considerably.

Using the polymer structures as templates to fabricate metal structures by electroforming offers a new class of optical structures, such as resonant filters, and also the fabrication of optical microelectromechanical systems (MEMS) by adding actuators to the optical benches [5].

In this chapter, examples are given of the different classes of structures, starting with simple optical elements and ending with the most complex micro-optical electromechanical systems (MOEMS) realized by the LIGA process.

8.2
Optical Components

8.2.1
Cylindrical Microlenses, Microprisms

Cylindrical microlenses (see Figure 8.4) and microprisms and also other Manhattan-like structures of optical relevance can easily be fabricated because of the shadow printing characteristics of X-ray lithography. Heights up to 3 mm have already been realized. The high parallelism of synchrotron radiation results in almost no inclination of the side-wall (<0.2 mrad per side-wall for a 400 μm high structure) over more than 80% of the height of the structures. Hence this will have only a marginal influence on light passing through the structure or reflected by the structure.

In the case of X-ray lithography only PMMA and SU-8 can be used as the material to build the optical structures. This results in a restriction of the optical and thermal characteristics of the structures. In the visible and near-infrared (NIR) wavelength range up to 1000 nm the transmission is high, which together with the small dimensions of the structure (maximum several millimeters) results in negligible absorption of the transmitted light. The situation changes for longer wavelengths, which make the structures less attractive if they are used in transmission. To use the polymer structures as reflective elements, they need to be covered by a metal layer. This is done by standard deposition techniques, keeping the temperature of the polymer below the glass transition temperature to avoid size deviations due to a temperature-initiated reflow.

The variety of materials is much broader in replication techniques; in principle, all thermoplastic materials can be used. For optical applications, materials such as polycarbonate and TOPAS have been tested already in injection molding and hot embossing. To follow the demand for functionalized polymers with individually customized optical properties such as refractive index and transmittance, filled polymers have been tested. The use of nanosized ceramic fillers with particle sizes smaller than 40 nm lead to a sum refractive index, which is larger or smaller than that of the pure polymer, depending on the refractive index difference of the filler and polymer matrix. Unfortunately, because of particle agglomeration, pronounced

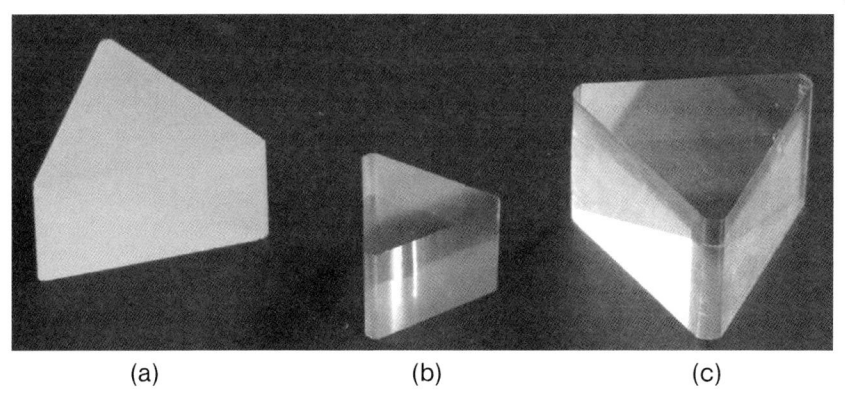

Figure 8.5 Replicated prism structures filled with nanosized ceramic fillers (a) or modified by organic dopants (b and c).

optical damping occurs even at low load. An index change can be achieved also by doping curable reactive resins such as MMA/PMMA- or unsaturated polyester (UP)/styrene-based systems with different organic dopants [6]. Figure 8.5a shows a cast prism using a MMA/PMMA-based reactive resin filled with nanosized ceramic filler with low transparency usable, for example, as an optical diffuser. In Figure 8.5b and c, two prisms made of PMMA modified with organic dopants are shown.

In the case of replication techniques, effects due to stresses incorporated into the structure during the hardening or the demolding process may influence the optical characteristics of the structures. Especially birefringency may occur, which may result in polarization effects when the light transmits the structure.

8.2.2
Resonant Filters

Metal membranes with exactly defined periodically arranged slit apertures, for example cross- or Y-structures, with a large layer thickness, can be used as bandpass filters for the far-infrared region, called resonant gratings [7]. The spectral quality of the filter depends on the shape and dimensions of the slit apertures, but also on the relative thickness of the membrane. The wavelength range in which these filters allow rays to pass through lies in the same order of magnitude as the slit length. The width of the slits and the distance to the neighboring element, which represents the critical dimensions, are about one order of magnitude smaller. Therefore, for a transmission wavelength of about 20 μm, the critical dimensions lie in the order of only a few micrometers, which must be kept highly exact. In order to attain a high separation sharpness, the thickness of the filter should be large compared with the critical dimensions. Because of its capability for fabrication of structures with high aspect ratio and dimensions in the

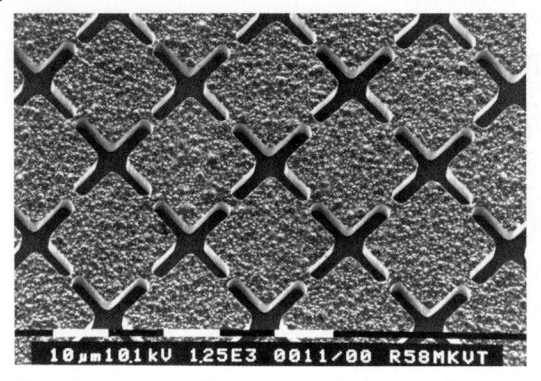

Figure 8.6 Resonant filter structure which acts as bandpass filter in the infrared (slit length, 18.5 μm; width, 3 μm; minimal distance, 2 μm).

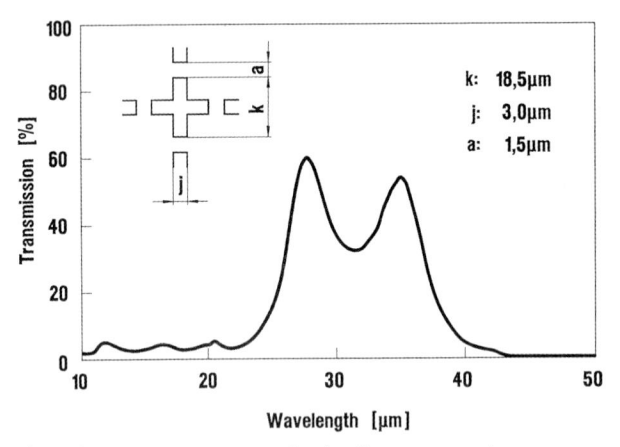

Figure 8.7 Transmission curve for the filter structure shown in Figure 8.6.

micrometer range, the LIGA process is best suited for the production of such filters [8].

As an example of a bandpass filter, a copper membrane with a cross-shaped slit aperture, which has a length of 18.5 μm, a width of 3 μm and a minimal distance to the neighboring elements of 2 μm, is shown in Figure 8.6. The thickness of the membrane is 20 μm. The membrane is produced by X-ray lithography using a PMMA plate on top of a metal plate as resist followed by electroplating. As can be seen from the transmission curve (Figure 8.7), the filter is transparent to wavelengths between about 27 and 35 μm, while the transmission to the high- and low-frequency spectral ranges drops slowly.

High-pass (or low-stop) filters for the far-infrared region are metal membranes with uniform apertures, whose thickness is more than twice the diameter of the

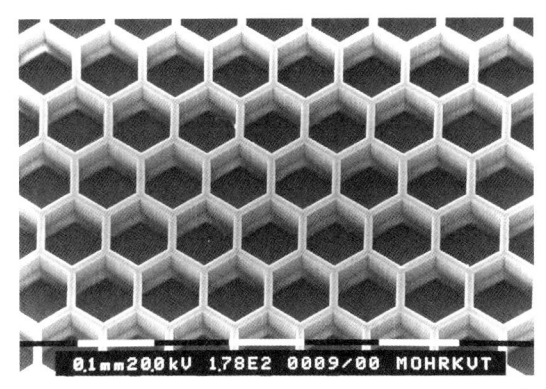

Figure 8.8 Honeycomb structure working as a high-pass filter for the far-infrared region.

hole. Therefore, honeycomb-shaped grids (Figure 8.8) can be used for those filters. A honeycomb grid with a diameter of 80 μm is transparent (up to 95% maximum) to radiation with a wavelength of less than 120 μm. It exhibits a sharp cut-off for higher wavelengths (lower energy).

In view of applications of resonant filter structures also in the NIR and visible wavelength ranges, the X-ray lithography process has been optimized to achieve high aspect ratio structures with submicrometer dimensions. In this case, spin-coated MicroChem 950k PMMA resist, several micrometers thick, was used and exposed to soft X-rays having a peak intensity at 0.4 nm. The exposure dose was set to $4\,kJ\,cm^{-3}$ to achieve an optimized contrast. To reduce diffraction influences, the proximity gap between the mask and resist layer needs to be less than 15 μm. Pre-baking the exposed substrates at 180 °C and adding a fluoro-surfactant to the deionized water rinse in the development step let to a maximum aspect ratio of isolated structures as a function of structure height of 12.5 (height: 10 μm), 8.6 (height: 5 μm) and 6.25 (height: 2 μm) [9]. These structures are transferred to a metal grid by electroplating gold from a sulfitic gold bath. The grid shown in Figure 8.9 is made from a 5 μm high PMMA template out of a column array with a pitch of 2620 nm and 720 ± 30 nm wide trenches which was transferred into a gold mesh of 4.0 ± 0.2 μm height.

8.3
Optical Benches

LIGA-fabricated optical benches with passive optical functionality consist of cylindrical optical structures and fixing structures for mounting optical components such as fibers, lenses and filters. They are fabricated from polymer by either X-ray lithography or molding techniques using mold inserts fabricated by the LIGA process. Also, mold inserts fabricated by combining different precision machining technologies (grinding, laser ablation, e-beam lithography, etc.) have been used to

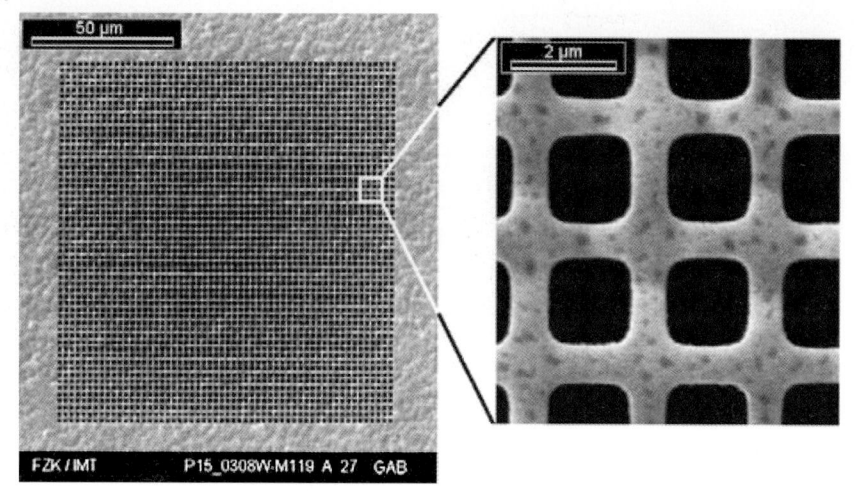

Figure 8.9 SEM image of an infrared filter made of 4.0 μm gold by electroplating into air cavities 5 μm high.

realize more complex 3D polymer structures by molding. To combine refractive and diffractive structures in one optical element, aligned lithography processing or molding is done on substrates with diffractive structures etched into the surface.

To minimize thermal effects in the optical bench due to the high thermal expansion coefficients of polymers, the structures are fabricated on top of a thermostable substrate such as silicon, ceramic or metal. The sizes of the structures are minimized; if possible, they are separated from each others. The approach to fabricate the structures on substrates can be fulfilled easily by hot embossing. This, together with the fact that hot embossing causes low internal stresses in the structures (compared with injection molding) and the possibilities of using opaque molding tools (compared with UV embossing) makes hot embossing most suitable for the fabrication of polymer micro-optical systems.

8.3.1
Intensity Coupling Element

Figure 8.10 shows an intensity coupling element for multi-mode fibers [10]. In this case, a transmission- (fiber 1), a measure-(fiber 2) and a detector fiber (fiber 3) are exactly positioned by fiber fixing grooves to a microprism, placed in the cross connection of the three fibers. This arrangement permits the use of the measure fiber (2) in a bidirectional mode. The light from a light source (fiber 1) is coupled by the coupling element into the fiber and transmitted to a fiber optical sensor. The sensor signal is detected by the same fiber (2) and coupled to the detector fiber (3) by the coupling element. With such an intensity coupling element,

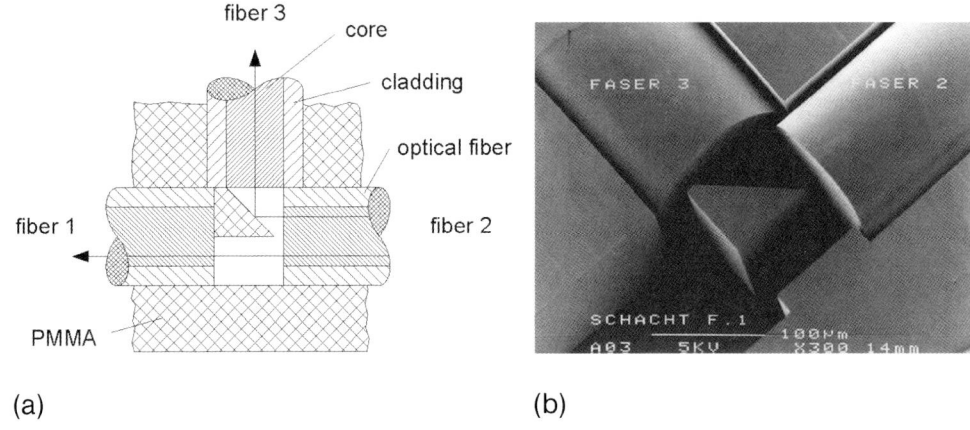

Figure 8.10 (a) Principle of an intensity coupling element for multi-mode fibers. (b) SEM image of a multi-mode fiber coupling element.

every desired intensity ratio can be set by choosing the size and position of the prism.

8.3.2
Multifiber Connector

For use in datacom applications, a 16-multifiber connector has been developed with the aim of mass fabrication by microinjection molding [11]. The mold insert is fabricated using mechanical machining, X-ray lithography and subsequent electroforming.

The connector consists of two micromolded plastic parts: one for the alignment of fibers and guide pins in rows of highly precise alignment structures and the other for fixation and protection (Figure 8.11). The free choice of geometry in the LIGA process allows elastic ripples to be patterned in the side-walls of the alignment structures facilitating the fiber mounting. The decrease in the gap between the alignment structures for the fibers and the pins from 140 µm in the last row to 123 µm in the first row allows very easy assembly and passive alignment of the fibers without the need of active micropositioning. Due to the smaller distance of the alignment structures at the end of the row compared with the fiber diameter, the fiber is exactly positioned in a press fit.

Because of the different diameters of fibers and pins, the cross-section of the connectors features two structural layers. To fabricate the mold insert, a first layer is patterned on a substrate for aligning the fibers and pins vertically in one line by micromilling. The height of this level is 287.5 µm. The rippled alignment structures for the horizontal positioning are fabricated with high lateral precision by X-ray lithography. The resist structure achieved on a two-level substrate is trans-

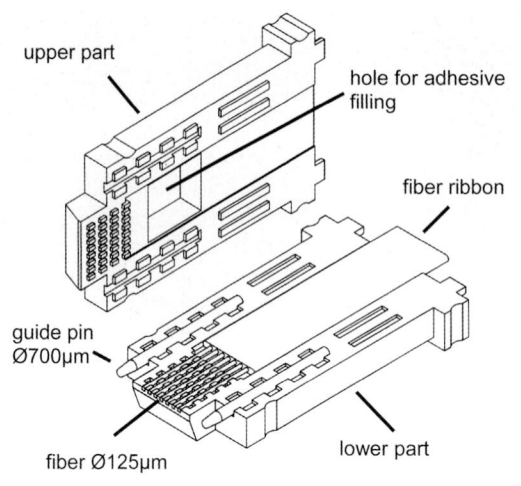

Figure 8.11 Scheme of a multifiber connector based on molded guiding plates.

Figure 8.12 Assembled multifiber connector.

ferred to a metal mold by electroforming of nickel. The two parts of the connector are microinjection molded from poly carbonate and glued together by a simple assembly technique. As the guiding pins are set back compared with the fiber end, the connector end phase with the fibers can be polished easily in the field. The hermaphroditic connectors are coupled together by a coupling element, which has guiding holes for the pins (Figure 8.12).

The loss of the connector depends on the accuracy of the geometric dimensions. Crucial are the level height of the first layer, the pitch of the fiber alignment structures and the flatness of the molded parts in the grooves of the fibers and of the guide pins. The observed deviations from the target value, which have been analyzed statistically, are in the range of the measurement resolution (<1 µm) and confirm the high precision achieved with the technology used. For use with multimode fibers, mean insertion losses of 0.35 dB with a standard deviation of 0.2 dB have been measured.

Figure 8.13 Bidirectional transceiver and receiving element based on a LIGA optical bench.

8.3.3
Bidirectional Transciever- and Receiving Module

A stepped molding tool, fabricated by a combination of mechanical machining and X-ray lithography, is also used to manufacture the micro-optical bench for an optical transmission and receiving module for bidirectional wavelength devision multiplexing (WDM) [12]. Ball lenses, a wavelength filter and glass fibers are positioned next to each others passively by the fixing structures of the optical bench (Figure 8.13). In order to position the fiber on the optical axes of the lenses (diameter 900 µm), the fiber is fixed on a plateau, which lies precisely 387.5 µm above the highest level of the remaining substrate. To avoid misalignment due to temperature changes, the fixing structures are molded on a ceramic substrate by hot embossing. The bench is fixed in a housing where the hermetic sealed active components are adjusted and laser welded relative to the optical setup. The light emitted from the laser diode (on the right side of the housing) is collimated by the first ball lens, runs through the wavelength filter and is focused with the second ball lens to the end phase of the single-mode glass fiber. Light with another wavelength is emitted from the fiber, collimated with the ball lens and reflected by the wavelength filter to the photodiode also welded to the housing (seen the top of Figure 8.13). The optical performance of this device is mainly given by the precision achieved in laser welding of the laser diode to the housing. For well-aligned systems, insertion losses between fiber and photodiode of less than 1 dB could be achieved. The cross talk between the two channels is higher than 40 dB.

8.3.4
Heterodyne Receiver

A heterodyne receiver realized by LIGA technology is another example of an optical bench which includes, in addition to passive elements, also active optical elements [13]. In a heterodyne receiver, an incoming signal will be detected by a coherent

Figure 8.14 Micro-optical bench of a heterodyne receiver.

superposition of the light from a local laser source. This laser operates at a slightly different frequency compared to the signal frequency. From the resulting beats in the superposed signal, the incoming signal can be analyzed. By tuning the local laser, a specific wavelength is filtered out from the incoming signal and can be amplified. The interference can be only realized if the signals have the same polarization, which requires a polarization-sensitive setup. Such a setup has been realized using the micro-optical bench concept (Figure 8.14). The optical bench includes fibers, glass prisms, ball lenses and photodiodes, which are mounted and precisely aligned by the help of fixing structures patterned by X-ray lithography on top of a ceramic substrate. The incoming signal and the signal from the local laser are coupled into the optical bench through fibers. They are mounted on precisely structured fiber mounts to put the fiber on the optical axes of the used ball lenses. The emitted light is collimated by the ball lenses and separated into the two polarization directions by the prisms, whose side-wall is covered by a polarization-sensitive layer. The p-polarized light runs through the prism whereas the s-polarized light is deflected by 90°. On the opposite side of the prism, the light is divided into two equal parts whereas one part is phase shifted by π. Each of the four beams of the incoming signal is superimposed with the signal from the local laser at the position of the photodiode, resulting in the expected beat. The efficiency caused by the superposition was analyzed to be 95% for the operating heterodyne receiver. This is only possible with an overall position tolerance of less than 1 μm for all the elements on the optical bench. Figure 8.15 shows the heterodyne receiver in an electro-optical housing mounted on an SMD circuit board, which includes also the electrical components.

8.3.5
Achromatic Confocal Distance Sensor

To realize a highly precise distance sensor with an outer dimension smaller than 2 mm, a confocal approach has been used. In this case, polychromatic light, passing an aperture, is collimated to the surface to be measured by a highly

Figure 8.15 Housed heterodyne receiver on a SMD circuit board.

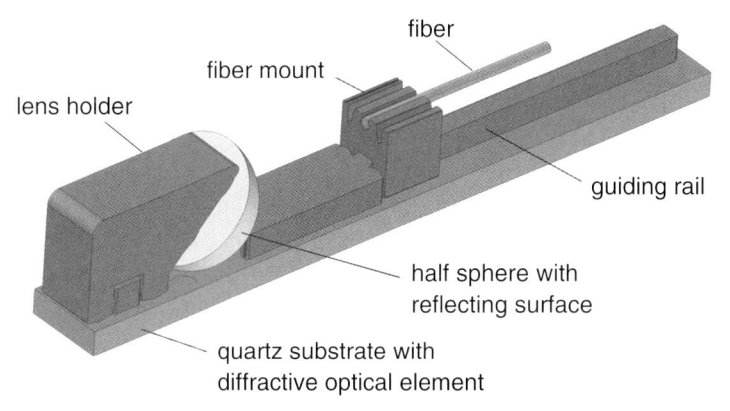

Figure 8.16 Setup of an achromatic confocal sensor based on an optical bench.

achromatic lens. Depending on the distance between the lens and object, a different wavelength will form a sharp focus on the surface. The light is reflected back and more or less only the light with the wavelength which was in focus on the object passes the aperture. Hence by spectral analysis of the transmitted light, the distance can be determined. To achieve a high resolution with respect to the focused wavelength, a monomode fiber with an ~10 μm core diameter is used as aperture. To allow side-wall measurements, the sensor was constructed to measure perpendicular to its optical setup (Figure 8.16).

To achieve sub-micrometer depth resolution, a numerical aperture of ~0.4 was required. The enabling optical elements of this sensor concept are a combination of a refractive and a diffractive optical element (DOE). As a refractive element, a fused-silica half ball lens with a diameter of 1.5 mm is used. The plane surface is coated with a reflective layer. Hence the lens can be used as a redirection mirror and also for collimating and focusing the light beam. The DOE produces the desired chromatic aberration, resulting in different focal planes for different wavelengths. To compensate for aberrations of the half ball lens and to improve the

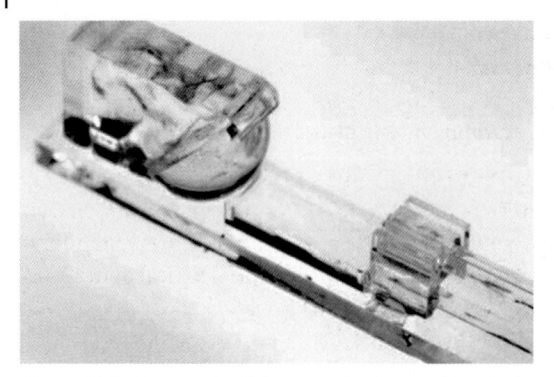

Figure 8.17 Photograph of the assembled optical head of an achromatic confocal distance sensor.

focusing quality, a correction hologram is integrated in the diffractive element. Thereby, a measuring focus diameter of ≤6 µm is achieved [14]. To realize the required high precision of the fiber, the ball lens and the DOE, an optical bench was patterned on top of a glass substrate aligned with the diffractive elements which have already been patterned into the glass.

As optical bench, a rail-rider concept was developed, allowing passive alignment of the fiber and refractive lens. This concept consists of two different kinds of microstructures: 300 µm high alignment structures (rails) on the fused-silica substrate and 750 µm thick fiber and lens mounts (riders) (Figure 8.17). By using the substrate of the DOE also as the carrier substrate for the micro-optical bench, the number of components and the assembly expenditure are reduced and the positioning of the optical elements is relatively independent of the cuts of the substrate when the elements are separated.

8.4
Electro-optical Systems

The way to mount electro-optical components in an optical bench used for the heterodyne receiver is not very efficient. If micro-optical benches have to be combined with laser diodes and detector arrays, a more suitable concept has to be followed which takes into account the fact that the elements are fabricated from different materials by different technologies, which limits the integration possibilities. Therefore, modular concepts become ever more important in photonic applications as they are convenient in terms of price and optical performance.

As the price of the system is still highly dominated by fabrication and assembly costs, a solution has been developed which allows the fabrication of the different modules of various systems by the same and standardized technologies [15]. Therefore, it is most appropriate to separate the system into functional units:

- optical functionality – micro-optical bench
- light generating and detecting functionality – electrical optical base plate
- light modulation – micro actuator platform
- light interfering system – for example, fluidic platform.

If the interfaces between these two components and the qualification methods are well defined, companies which are specialized in the relevant fabrication technique can fabricate both components. In the case of the optical bench, either lithographic methods or replication techniques are used. For the electrical optical base plate, which carries laser diodes, detector arrays and single-element detectors in addition to microelectronics, a mounting technology well known from electronic fabrication can be used. It has to be adapted to the need for higher precision (in the range of 5 µm) and to the handling of unhoused active optical devices. The fabrication of the microactuator platform needs the use of sacrificial layer technology or hybrid mounting; the fluidic platforms are usually fabricated by embossing techniques.

To build the system, the micro-optical bench and the electrical optical base plate are combined to form the micro-optical subsystem, which is characterized by its optical output and its electrical interface. This subsystem is combined with the other modules to form the complex system.

The modular fabrication concept was used at the Forschungszentrum Karlsruhe to build an optical distance sensor based on the triangulation principle and also a microspectrometer.

8.4.1
Triangulation Distance Sensor

There is a strong demand to reduce the size and cost of sensors in industrial automation because the sensors can influence considerably the total size and cost of the industrial machining equipment. For example, small sensors are needed in robot fingers for work position control. Height distribution measurement in printing machines is another example of the use of such sensors. Therefore, a micro-optical distance sensor with a size of a few square centimeters was developed and fabricated using LIGA technology [16]. The sensor works on the triangulation principle. It thereby allows non-contact inspection with high resolution. It consists of a free space micro-optical bench and an electrical optical base plate which are assembled together precisely (Figure 8.18a). The optical bench is used to collimate the beam from a laser diode on to the object to be measured with a spot size of less than 50 µm. Light which is detected through the entrance aperture of the sensor is collimated and directed to the position-sensitive device by the optical elements (imaging mirrors, lenses) on the optical bench. The shape of the mirrors in the detector path is optimized to obtain an aberration-corrected focus on to the PSD in horizontal direction, resulting in a linearized signal. In addition, several light traps are patterned in order to cut out stray light. The bench is fabricated in a height of 750 µm by LIGA technology. The electrical optical base plate carries a laser diode,

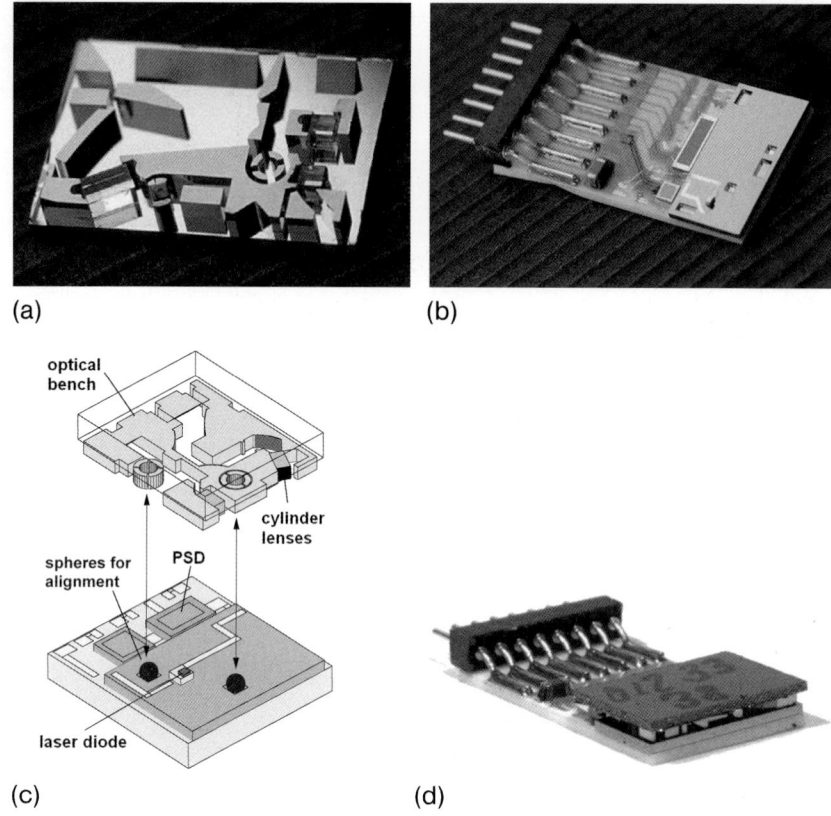

Figure 8.18 (a) Micro-optical bench with micro-optical structures and elements to build a triangulation sensor. (b) Electrical optical base plate with laser diode, PSD and monitor diode to build a triangulation sensor. (c) Passive assembly of the micro-optical bench and electrical optical base plate to build up the electrical optical sub-system. (d) Assembled micro-optical triangulation sensor.

a monitor diode and a position-sensitive device (PSD), which analyzes the change in the light position detected from the target (Figure 8.18b). It can be fabricated with known electronic processing technologies, leading to low-cost fabrication.

To build up the electro-optical subsystem, both base plates are assembled head-over (Figure 8.18c). The alignment is done passively with balls that are first glued into silicon etch grooves whose position is very precise with respect to the laser diode position. The balls fit into corresponding cylindrical tubes fabricated together with the optical structure. By this passive assembly, the laser diode will be positioned in the optical structure with a tolerance of less than 5 µm. Hence light from the laser diode is coupled.

Sensors with a measurement range of 1 mm in a distance of 10 mm have been fabricated and characterized. Figure 8.19 shows a typical signal achieved with

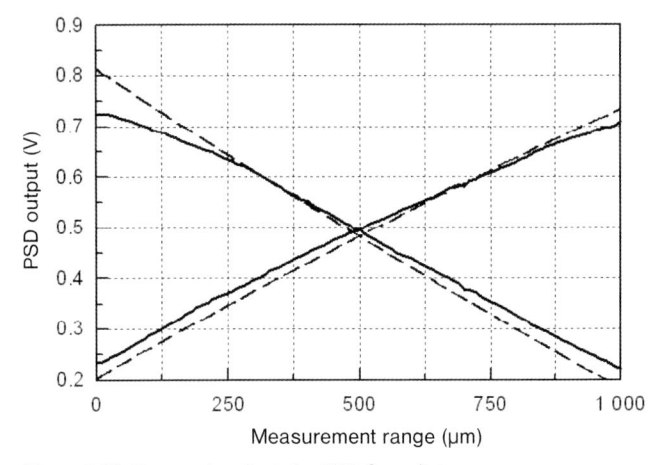

Figure 8.19 Output signals at the PSD for a distance sensor with a measurement range of 1 mm in a distance of 10 mm.

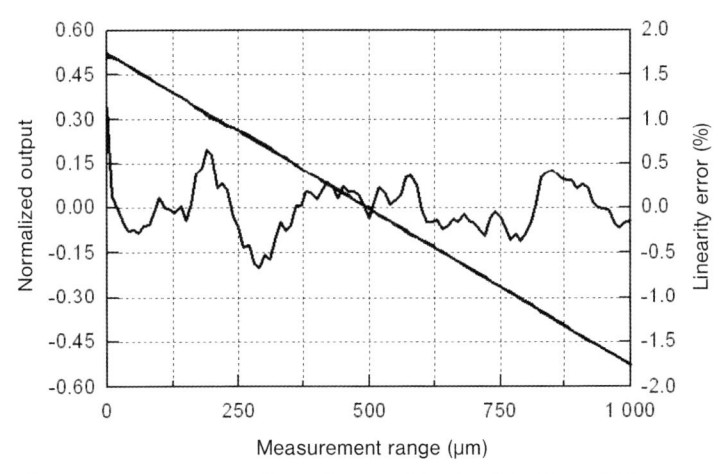

Figure 8.20 Linearity error after evaluation of the signals in Figure 8.19.

white paper as target. The calculated PSD signals versus the measurement range are plotted as dashed lines and the measured signal as solid lines. Both output signals are nearly complementary. Deviations are caused by imaging errors, stray light and adjustment errors of the laser in the optical bench. The normalized output versus the measurement range is given from the ratio of the difference signal and the sum of the signals by $N = (S_1 - S_2)/(S_1 + S_2)$ and should be a slope with values from +0.6 to −0.6 (the active area of the PSD is 60%) and a zero value at the center point of the measurement range (see Figure 8.20, left y-axis). Compared with a linear fit, the linearity error is smaller than 1% over the measurement range (see Figure 8.20, right y-axis). The resolution is about 1/1000 (1 μm) of the measurement range [17].

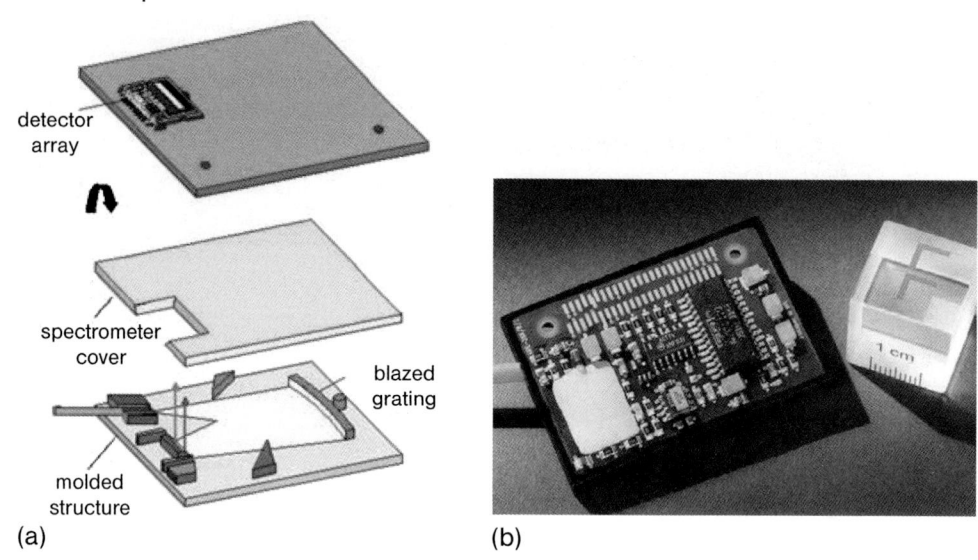

Figure 8.21 (a) Schematic setup of a microspectrometer based on an optical bench and an electrical optical base plate with detector array and electronics. (b) Assembled microspectrometer seen from the back (electronic evaluation board).

A statistically relevant number of sensors have been used to measure deviations and tolerances during automatic assembly. For more than 70% of the sensors the change in the measurement range was less than 10% and the squint angle in the illuminating beam was less than 1.5°.

8.4.2
Microspectrometer

In recent years, microspectrometers have attracted strong interest and a new market has been generated. Although they have lower resolution compared with laboratory spectral analyzers, they offer the possibility of building up hand-held spectrometric systems which are of interest for color detection, chemical analysis, process control and so on. In the 1990s, activities were started to develop a microspectrometer using the LIGA process at the Forschungszentrum Karlsruhe [18]. These developments resulted in a product which was commercialized several years ago by Boehringer Ingelheim microParts.

In Figure 8.21a, the setup of the system is represented schematically for an NIR microspectrometer which uses unhoused InGaAs detectors [19]. It is also constructed with a micro-optical bench and an electrical optical base plate. The optical bench consists of a blazed grating, with perpendicular patterned grating grooves, a fiber fixing groove with entrance slit and a block with a 45° side-wall, which directs the diffracted light to the detector array changing the light path from the

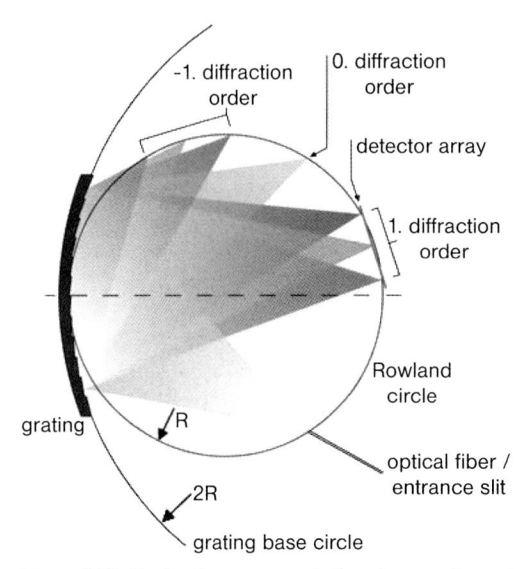

Figure 8.22 Rowland arrangement of a microspectrometer.

horizontal to the vertical direction. The electrical optical base plate carries the detector array and the electronics for signal evaluation (Figure 8.21b). The setup for the microspectrometer for the visible range differs only concerning of the electrical optical base plate. In this case, housed silicon detector arrays are used that are placed directly on top of the cover of the spectrometer. The functionality of the spectrometer is the same in both cases and is given by the grating. All elements of the optical bench are patterned on top of the substrate by means of X-ray lithography or replication techniques.

The design of the microspectrometer is based on a Rowland arrangement, which minimizes changes in the optical performance due to thermal effects (Figure 8.22). In a Rowland arrangement, the entrance slit is imaged on to the detector by a curved grating operating as a concave mirror. This avoids additional optical imaging elements. Due to shadow printing by the LIGA process, the grating does not have a spherical but only a cylindrical shape. As the radius of the grating is twice the radius of the Rowland circle, the entrance slit is imaged on to the Rowland circle independently of the diffraction order and the wavelength. Provided that the entrance slit is next to the normal of the grating, a 1:1 image of the entrance slit on the detector is achieved. Because the detector array cannot be curved like the Rowland circle, it is necessary to linearize the Rowland circle in the position of the detector array. This is done by correcting the shape of the grating by a fourth-order polynomial, which results in a deformation of the Rowland circle at the detector position. As this correction is limited, it limits the size of the detector array depending on the resolution required of the spectrometer. The resolution is also limited by the width of the entrance slit, which should not be too small in order to be able to couple a reasonable amount of light into the spectrometer. In

Figure 8.23 Light distribution in a LIGA microspectrometer. The bright shining area shows the hollow waveguide with the grating at the end. Other light paths are due to reflections or higher orders.

case of the LIGA spectrometer it is in the range 25–50 µm. With an achievable dispersion of 0.1 nm/µm, this will result in a resolution of 2.5–5 nm. To increase the grating efficiency, the grating is blazed to first order. This means that the surface of each single grating tooth is inclined in such a way that the surface operates like a mirror, reflecting the light from the entrance slit towards the position of one wavelength in the first order on the detector called blaze wavelength. The blaze efficiency is limited by the shape of the individual grating tooth, which shows roundings at the edges caused by limits in the (mask) fabrication process. Hence the grating constant of LIGA spectrometers should be larger than 1.3 µm.

The light coupled into the spectrometer illuminates the grating in the horizontal direction according to the numerical aperture of the entrance optics (numerical aperture of the fiber) (Figure 8.23). In the vertical direction, the light is guided by reflections at the bottom side and the cover of the spectrometer (hollow waveguide), which are coated with aluminum (visible range) or gold (NIR range). High reflectivity is needed to optimize the efficiency of the spectrometer. The fact that the light is guided in the hollow waveguide by reflections limits the resolution of the spectrometer because of the so-called mode dispersion. The influence is lower with increasing height of the spectrometer and decreasing numerical aperture in the vertical direction. The light which is coupled into the hollow waveguide illuminates the cylinder–symmetrical blazed reflection grating, where it is dispersed and focused on to the detector, placed above the hollow waveguide. It leaves the spectrometer via the exit window, which is formed by the edge of the deflection mirror and the edge of the cover. The vertical position of the detector is given by the height of the spectrometer and the cover; horizontal alignment is done actively. The covered spectrometer is accommodated in a plastic housing which serves to relieve the fiber and carries the printed circuit board.

Spectrometers with different characteristics for different wavelength ranges have been fabricated in the past. Most prominent is the system for the visible range. It covers the wavelength range from 400 to 780 nm and offers a resolution of 10 nm full width at half-maximum (FWHM). The NIR system covers a wavelength range from 1100 to 1700 nm with a resolution of 16 nm FWHM. Together with the achromatic confocal distance sensor, a spectrometer has been realized which gives a resolution of 3.9 nm FWHM in a spectral range of 750–950 nm.

8.5
Optical MEMS

The possibility of fabricating metallic microstructures with the LIGA process by electroforming offers the chance to build micro-optical benches with hybrid or even monolithic integration of integrated electrostatic or electromagnetic actuators. In the following sections, four examples will be described which have been realized in the past at the Forschungszentrum Karlsruhe.

8.5.1
Micro-optical Bypass Switch

A micro-optical 2 × 2 bypass switch for fiber network application has been realized using an electrostatic linear actuator [20]. The comb drive actuator and fixing and alignment structures are realized by nickel electroforming on top of a silicon wafer (Figure 8.24). The moving part of the comb drive is suspended from thin leaf springs; it is released from the substrate using a sacrificial layer. Four fibers and the associated ball lenses for collimating and focusing the light from and to the fibers are placed in fixing grooves. In the cross point of the light beams emitted from the fibers, the side-wall of the pin attached to the movable comb drive is placed. If the mirror is moved back by applying a voltage to the comb drive, the light can propagate between the fibers without any interaction with the side-wall

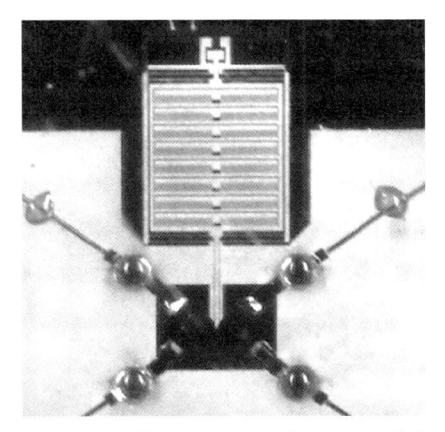

Figure 8.24 Electrostatic 2 × 2 bypass switch for fiber application.

of the pin. Thus, the light from a data communication network (lower right fiber) is transmitted to a consumer via the upper left fiber. From the consumer, the light is transmitted back to the network (lower left fiber) via the upper right fiber. If the actuator is not activated, the side-wall of the pin is placed in the cross point of the light beams. Thus, the light which is emitted from the lower left fiber is reflected directly to the lower right fiber and propagates directly into the network without passing the consumer (bypass situation).

For these elements a loss of 2 dB was measured in the open state. The loss was about 5 dB for the closed state, using the electroformed nickel as reflecting surface without optimizing the reflection coefficient. The cross talk is better than 40 dB. The voltage needed to activate the comb drive was about 70 V [5].

8.5.2
Electrostatic Switching Matrix

The electrostatic linear actuator realized in the bypass switch has a size of 3×4 mm, which is too large to build a switching matrix for rooting applications. In this case, the size has to be reduced to a maximum of approximately 1×1 mm. To fulfill this requirement, a concept based on electrostatic motors has been followed at the Forschungszentrum Karlsruhe [21]. The motors are used to position the mirror element integrated on the circumference of the rotor (Figure 8.25). Two positions are possible and defined exactly by stops. In position 1, which is defined as the beam-deflecting position, the light beam from the entrance fiber is deflected by 90° to the outcoupling beam. If the mirror is positioned in the parking position without any interaction with any light beam, the beam can propagate to the next

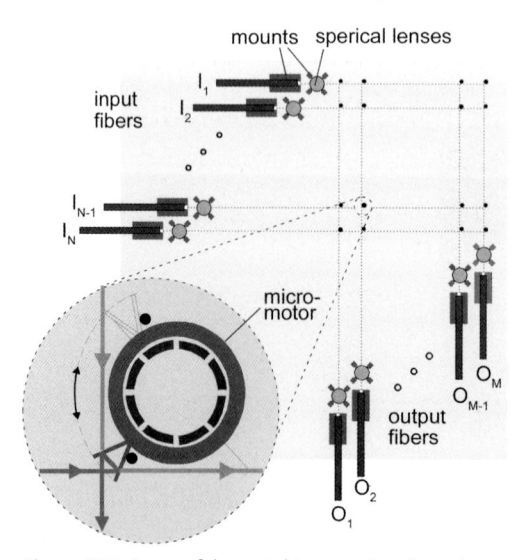

Figure 8.25 Setup of the switching matrix using micromotors.

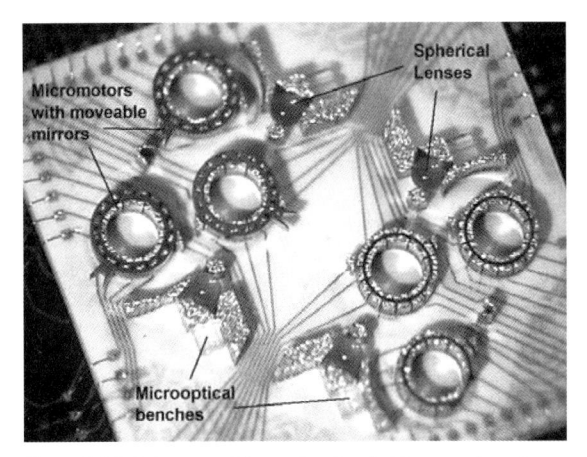

Figure 8.26 Substrate with two 2 × 2 switching matrixes. In the upper left matrix the rotors have been assembled already.

switching element, where it could then be deflected to another outcoupling fiber. The imaging of the light beam between the fibers is achieved by a 4F optic, positioning the fibers at a distance which is the same for all switching possibilities. The fixing structures for fibers and lenses and also the stops are fabricated together with the stators of the micro motor on one substrate in one fabrication sequence by X-ray lithography and subsequent electroforming. This guarantees an exact positioning of the mirrors in the beam. The rotors of the motor are fabricated separately and assembled on the stators.

Figure 8.26 shows two 2 × 2 switching matrixes. In the upper left of the picture the rotors are already assembled in the motors; in the lower part they are still missing, showing the stators. The size of the plate carrying the two systems is 9.3 × 9.3 mm. The size of one switching element is less than 2 × 2 mm for a rotor diameter of 1.7 mm. Insertion losses of 7 dB have been measured using again the electroformed nickel as a reflecting layer without any optimization of the reflection coefficient and without any anti-reflection coating on the lenses and the fiber end phases. The cross talk was better than 90 dB. For an acceleration voltage of 300 V, switching times of 30 ms could be achieved.

8.5.3
Chopper for Fiber Applications

In addition to electrostatic also electromagnetic actuation is used to build up micro-optoelectromechanical systems. As the process to fabricate fully integrated electromagnetic actuators is very complex [22], usually at the least the coil used to generate a magnetic flux is assembled hybridically into the LIGA structure.

As a simple device, a microchopper for fiber use has been fabricated following this concept (Figure 8.27). The device consists of an electromagnetic yoke with an

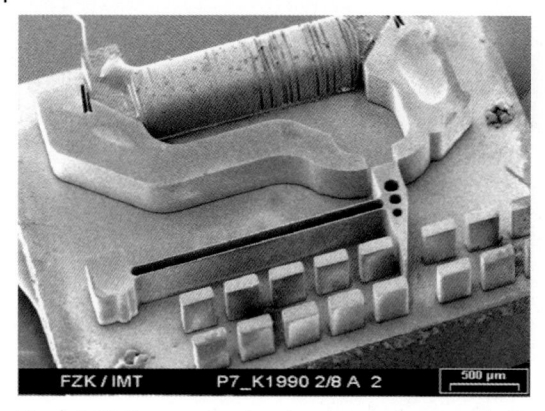

Figure 8.27 Electromagnetic microchopper for use with fibers.

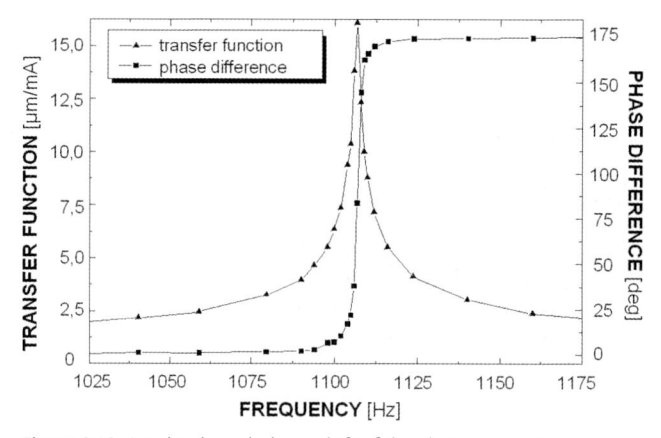

Figure 8.28 Amplitude and phase shift of the electromagnetic micro chopper near resonance frequency.

assembled coil, fixing grooves for fibers and a movable anchor which is fixed to two beams. The anchor is moved into the opening in the yoke when a current is applied to the coil, generating a magnetic flux. The anchor is designed to act as an aperture which is oscillating between the endfaces of two fibers fixed in the fiber fixing grooves formed by walls electroformed together with all the other structures. The anchor is driven in resonance oscillation and interrupts periodically the light propagating between the two fibers. The size of this micro chopper is 3×3.2 mm and the total height is 280 μm [23]. Figure 8.28 shows the amplitude per milliamp and also the phase shift near the resonance frequency, which is about 1100 Hz. At resonance the amplitude per milliamp is about 16 μm; it increases to about 160 μm with increase in current.

8.5.4
FTIR Spectrometer

The Fourier transform infrared (FTIR) spectrometer shown schematically in Figure 8.29a is based on a more complex micro-optical bench. The system consists of a movable and a fixed mirror which, together with a mounted beamsplitter, forms a Michelson interferometer for the light to be analyzed and also for a second light beam from a laser diode with known wavelength, which is used to measure the position of the movable mirror. The light to be analyzed and the light from the laser diode are coupled into the system by fibers and collimated by ball lenses. It is divided in two equal parts by the beamsplitter. One part hits the mirror which is fixed to the substrate and the other part hits the movable mirror. From both mirrors, the light is reflected, superposed after traveling through the beamsplitter

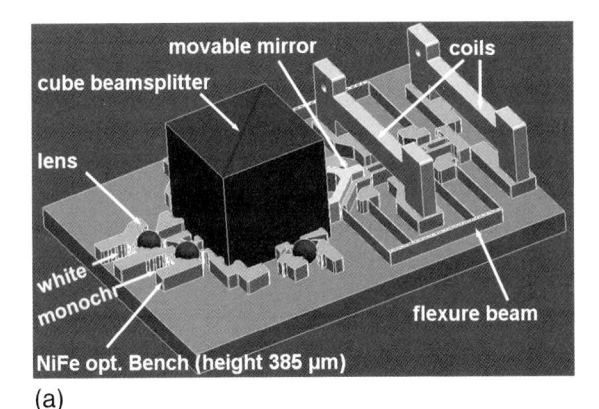

Figure 8.29 (a) Schematic view of the setup of an FTIR system based on a Michelson interferometer with electromagnetic actuators. (b) Assembled FTIR system.

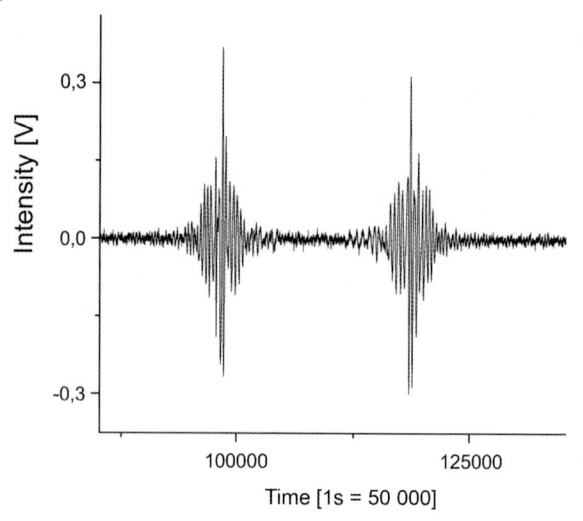

Figure 8.30 Interferogram for a monochromatic peak at 1544 nm achieved with the micro-FTIR system.

and detected by a photodiode which is mounted perpendicular to the substrate and aligned by pins which are fabricated together with all other fixing grooves and the structures for the electromagnetic actuator by electroforming of nickel or nickel–iron in a template fabricated by X-ray lithography. Due to the movement of the mirror, an interferogram is achieved which allows the spectral composition of the light to be analyzed by a Fourier transformation. Figure 8.30 shows the interferogram for a monochromatic peak at 1544 nm which has been achieved with a difference in the optical path length of 108 µm (this relates to a mirror movement of 54 µm). The resolution is 24.4 nm due to the limited pathlength [24]. Optimizing the electromagnetic actuator will result in an optical pathlength of more than 500 µm, resulting in resolutions better than 10 nm.

8.6
Conclusion

The examples described above demonstrate that the LIGA process is well suited to fabricate micro-optical components and systems based on micro-optical benches. Even MOEMS can be fabricated with high quality, making use of electroforming and sacrificial layer technology. Because of the possibility for free shaping, the optical performance of the components and the setups can easily be optimized. This, together with the high precision in shape, is not only used for the fabrication of micro-optical systems in the visible and NIR wavelength ranges: for the first time, it also opens up the fabrication of refractive X-ray lenses from SU-8 by X-ray lithography.

References

1 Menz, W., Mohr, J. and Paul, O. (2001) *Microsystem Technology*, 1st edn, Wiley-VCH Verlag GmbH, Weinheim.

2 Last, A., Hein, H. and Mohr, J. (2004) Shape deviations in masks for optical structures produced by electron beam lithography. *Microsystems Technologies*, **10**, 527–30.

3 Hofmann, A., Gengenbach, U., Scharnowell, R. and Bär, M. (2003) Handling and assembly of micro-optical components – modules and solutions, in Proceeding of the 2nd VDE World Microtechnologies Congress, Munich, 13–15 October 2003, Berlin, pp. 289–92.

4 Brenner, K.H., Kufner, M., Kufner, S., Moisel, J., Müller, A., Sinzinger, S., Testorf, M., Göttert, J. and Mohr, J. (1993) Application of three-dimensional micro-optical components formed by lithography, electroforming and plastic molding. *Applied Optics*, **32**, 6464–69.

5 Mohr, J., Göttert, J., Müller, A., Ruther, P. and Wengeling, K. (1997) Micro-optical and opto-mechanical systems fabricated by the LIGA technique. *Proceedings of the SPIE – The International Society for Optical Engineering*, **3008**, 273–8.

6 Boehm, J., Haussell, J., Henzi, P., Litfin, K. and Hanemann, T. (2004) Tuning the refractive index of polymers for polymer waveguides using Nanoscaled ceramics or organic dyes. *Advanced Engineering Materials*, **6** (1–2), 52–7.

7 Ulrich, R. (1968) Interference filters for the far infrared. *Journal of Applied Optics*, **7**, 1987.

8 Ruprecht, R. and Bacher, W. (1991) Untersuchungen an mikrostrukturierten Bandpassfiltern fuer das Ferne Infrarot und ihre Herstellung durch Roentgentiefenlithographie und Mikrogalvanoformung, Wissenschaftliche Berichte KfK-4825, Karlsruhe.

9 Mappes, T., Achenbach, S. and Mohr, J. (2006) X-ray lithography for devices with high aspect ratio polymer submicron structures. Proceedings of the 32nd International Conference on Micro- and Nano-Engineering (MNE 2006), Barcelona, 17–20 September 2006.

10 Goettert, J., Mohr, J., Mueller, C. and Sauter, H. (1992) Coupling elements for multimode fibers by the LIGA process, in *Micro System Technologies*, 92 (ed. H. Reichl), vde-Verlag, pp. 297–307.

11 Wallrabe, U., Dittrich, H., Friedsam, G., Hanemann, T., Mohr, J., Müller, K., Piotter, V., Ruther, P., Schaller, T. and Zissler, W. (2002) Micromolded easy-assembly multi fiber connector: RibCon. *Microsystem Technologies*, **8**, 83–7.

12 Müller, A., Hehmann, J., Rogner, A., Göttert, J. and Mohr, J. (1995) Hybrid optical transceiver module with a micro-optical LIGA-bench. Proceedings of the European Conference on Optical Communication (ECOC '95), Brussels, 17–21 September 1995.

13 Ziegler, P., Wengelink, J. and Mohr, J. (1999) Passive alignment and hybrid integration of active and passive optical components on a microoptical LIGA–bond. Proc. IEEE Conf. on Micro-Opto-Mechanical Systems CMOEMS 99, Mainz, 30 Oct.–1 Nov. 1999.

14 Lücke, P., Last, A., Mohr, J., Ruprecht, A.K., Pruss, C., Tiziani, H.J., Osten, W., Lehmann, P. and Schönfelder, S. (2005) Confocal micro-optical distance sensor: realization and results. *Proceedings of the SPIE – The International Society for Optical Engineering*, **5856**, 136–42.

15 Mohr, J.A., Last, A., Hollenbach, U., Oka, T. and Wallrabe, U. (2003) A modular fabrication concept for micro-optical systems. *Journal of Lightwave Technology JLWT*, **21**, 643–7.

16 Hollenbach, U., Mohr, J., Oka, T., Wallrabe, U. and Sieber, I. (2004) Novel design and prototyping of a micro-optical distance sensor considering a distributed fabrication. *Proceedings of the SPIE – The International Society for Optical Engineering*, **5459**, 211–18.

17 Hollenbach, U., Mohr, J. and Stautmeister, T. (2005) *Micro optical distance sensor – actual resuslts from prototypes*, in Proceedings of SENSOR 2005 Conference, Nuremberg, 10–12 May 2005, pp. 79–84.

18 Mueller, C. and Mohr, J. (1993) Microspectrometer fabricated by the LIGA process. *Interdisciplinary Science Reviews*, **18**, 273–79.

19 Krippner, P., Kühner, T., Mohr, J. and Saile, V. (2000) Microspectrometer system for the near infrared wavelength range based on the LIGA technology. *Proceedings of the SPIE – The International Society for Optical Engineering*, **3912**, 141–9.

20 Müller, A., Göttert, J. and Mohr, J. (1996) Aufbau hybrider mikrooptischer Funktionsmodule für die optische Nachrichtentechnik mit dem LIGA-Verfahren. Wissenschaftliche Berichte, FZKA5786, Karlsruhe.

21 Ruzzu, A.C.M., Haller, D., Mohr, J.A. and Wallrabe, U. (2003) Optoelectromechanical switch array with passively aligned free-space optical components. *Journal of Lightwave Technology*, **21**, 664–71.

22 Rogge, B., Schulz, J., Mohr, J. and Thommes, A. (1996) Magnetic microactuators fabricated by the LIGA-technique for large displacements or large forces, in Proceeding of 5th International Conferenceon New Actuators, Bremen, 26–28 June 1996, p. 112.

23 Krippner, P. and Mohr, J. (1999) Electromagnetically driven microchopper for integration in microspectrometers based on LIGA technology. *Proceedings of the SPIE – The International Society for Optical Engineering*, **3878**, 144–54.

24 Solf, C., Mohr, J. and Wallrabe, U. (2003) Minaturized LIGA Fourier transformation spectrometer. Proceedings of the 2nd IEEE International Conferenceon Sensors, Toronto, 22–24 October 2003.

9
Refractive X-ray Lenses Produced by X-ray Lithography
Arndt Last

9.1 Introduction *233*
9.2 Microstructured X-ray Lenses *235*
9.3 Materials *237*
9.4 Lens Construction *238*
9.5 Recent Results *239*
9.6 Future Work *241*
   References *242*

9.1
Introduction

Since the discovery of X-rays, there have been many attempts to focus X-rays by means of some kinds of lenses. Wilhelm Conrad Röntgen (1845–1923) tried to focus X-rays like visible light. He did not succeed and concluded that the refractive index of matter was one for X-rays. Over many decades, imaging in the wavelength range of X-rays, for instance, remained restricted to shadow projections, because focusing X-rays with any kind of refractive lens seemed to be impossible.

Since then, many types of focusing elements for X-rays have been developed: zone plates diffracting X-rays to a focus point; capillary lenses, where X-rays are guided in glass microtubes by total internal reflection; crystals bent to focus X-rays; and even refractive lenses. Each system has various limitations with regard to its wavelength range, the minimal focus diameter, the minimally achievable focal length, the price and so on.

X-ray lenses are applicable in many scientific and technical fields. In particular, medical analytical instrumentation, materials science and physics will profit from the potential of refractive X-ray lenses of good quality. For example, these lenses have made X-ray microscopes possible. Due to the short wavelength of X-rays, these microscopes can achieve much higher resolutions than microscopes operating with visible light. Furthermore, it is possible to analyze objects which are opaque for visible light. These microscopes allow for non-destructive imaging and even imaging of live biological samples with sub-micrometer resolution. Today,

Advanced Micro & Nanosystems Vol. 7. LIGA and Its Applications.
Edited by Volker Saile, Ulrike Wallrabe, Osamu Tabata and Jan G. Korvink
Copyright © 2009 WILEY-VCH Verlag GmbH & Co. KGaA, Weinheim
ISBN: 978-3-527-31698-4

scanning electron microscopes even reach nanometer resolution, but the time-consuming sample preparation, the destructive character of the preparation and the fact that scanning electron microscopes reproduce the surface of an object only are limiting disadvantages.

In X-ray spectroscopy, well-collimated X-ray beams are needed to investigate the atomic and chemical composition of samples with high local resolution. X-ray tomography also benefits from focused beams by achieving images of a higher spatial resolution.

Refractive X-ray lenses were patented by Tomie in Japan [1]. These lens designs consisted of aligned holes in a metal block giving a line focus. Subsequently, Kohn and coworkers (Kurchatov Institute, Moscow, Russia) showed mathematically that the intensity gain of refractive X-ray lenses made of light elements can be very high [2]. In the 1990s, Lengeler and Snigirev's group (Physikalisches Institut, RWTH Aachen, Germany) carried out experiments which resulted in functional refractive X-ray lenses [3]. This group still produces metal parabolic lenses. These first refractive X-ray lenses of parabolic geometry are made by embossing 1 mm thick aluminum or beryllium foils from two sides by needles with rotationally parabolic tips. This method allows one to produce radii of curvature of, for example, 200 µm. With a stack of these foils, precisely aligned, focal lengths in the meter range were achieved.

The refractive index for X-rays is slightly below one, depending on the atomic number and the density of the matter under examination. The energy-dependent complex refractive index n may be described by

$$n = 1 - \delta - i\beta \tag{1}$$

where β is the absorption coefficient. The decrement of the refractive index δ from unity was determined to be about 10^{-6} to 10^{-8} for energies between 10 and 50 keV. A decrement of the refractive index δ slightly lower than one means that X-rays entering matter are refracted away from the perpendicular, contrary to the behavior of visible light. This may be understood as follows: the binding energy of electrons in atoms is far below 10 eV. These electrons are forced to oscillate by the electric field of the γ photons of, for example, 10 keV. Consequently, the exciting frequency is much higher than the eigenfrequency of the electrons. The amplitude of the oscillations will therefore be small. The resulting small polarization of the matter gives a refractive index close to one. The electrons will oscillate oppositely in phase with the exciting field. Hence the phase velocity will be larger than the speed of light in vacuum, resulting in a refractive index lower than one.

Taking into account this very small refractive index decrement, the only way to implement focusing refractive lenses with reasonably small focal lengths in the range from a few millimeters to 1 m is to use a large number of lenses with very small radii of curvature. The focal length f_l of such a lens can be approximated as

$$f_l = \frac{R}{2\delta N} \tag{2}$$

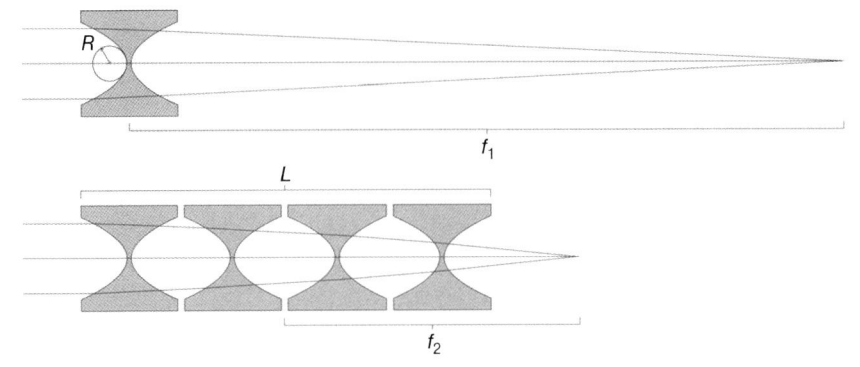

Figure 9.1 Principle of refractive X-ray focusing.

for a focal length that is large compared with the length of the lens, where R is the radius of curvature and N the number of single lens elements. At a focal length comparable to the length of the lens, the length L of the whole lens has to be taken into account. The focal length f_2 of a long lens is

$$f_2 = \frac{R}{2\delta N} + \frac{L}{6} \tag{3}$$

Since the refractive index is lower than one, the lenses have to have the geometry shown in Figure 9.1 for focusing light. For visible light, the refractive index is larger than one, with focusing lenses having a geometry inverse to that shown in Figure 9.1. If the geometry of the lens elements is approximately parabolic, spherical aberrations can be avoided and focal diameters in the sub-micrometer range can be obtained for well-collimated X-rays from a synchrotron.

The absorption of the lens depends strongly on the wavelength, the atomic number of the lens material and its density. The intensity loss caused by the absorption of the lens is compensated by the intensity gain due to the focusing effect of the lens. The intensity gain is the ratio of the intensities at the focal point with and without the lens. It depends on several parameters such as the entrance aperture of the lens, the absorption of the lens material, the size of the X-ray source and the demagnification factor of the optical setup.

Figure 9.2 shows three typical arrangements of the lens elements in refractive X-ray lenses. The incoming beam is marked by an arrow; the focus lines or points are situated at the right side.

9.2
Microstructured X-ray Lenses

To produce refractive X-ray lenses, several requirements have to be fulfilled: a very small radius of curvature of quasi-parabolically shaped lenses and a considerable

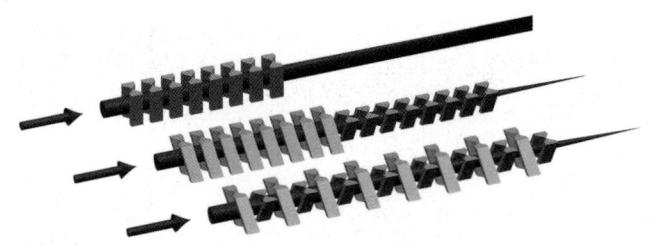

Figure 9.2 Focusing with refractive X-ray lenses: lenses perpendicular to the substrate with a line focus (top) and crossed lenses with a point focus (center and bottom).

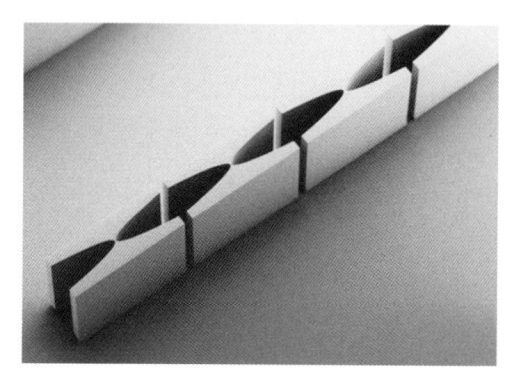

Figure 9.3 Row of parabolic lenses. Lens height: 1 mm.

number of well-positioned and aligned lens elements with very smooth surfaces to avoid scattering. All these requirements can be fulfilled using the X-ray lithographic mask copying step of the LIGA process [4]. The typical advantages of this technique are made use of in the production of lenses consisting of hundreds of single lens elements by a single mask copying step. The resulting single lens elements will be positioned with an accuracy of a few micrometers and the surface smoothness will be in the region of about 10 nm. Radii of curvature of some micrometers can be reached [7]. The process-inherent liberty regarding the choice of the lens geometry makes it possible to manufacture lenses with minimum optical aberration. This includes the compensation of minor changes of the geometry caused by the process. In one production step, a variety of X-ray lenses can be produced on a single substrate. Lenses with a parabolic cylinder shape as shown in Figure 9.3 can be made in good quality. These lenses have a line focus with a focal width of less than 150 nm and an optical intensity gain in the region of several hundred.

For many experiments such as X-ray microtomography and X-ray imaging, a point focus is required. In this case, the process-inherent constraint on 2D sidewall geometries has to be taken into account. When using the LIGA process, the only possible solution to generate a point focus is the combination of two crossed

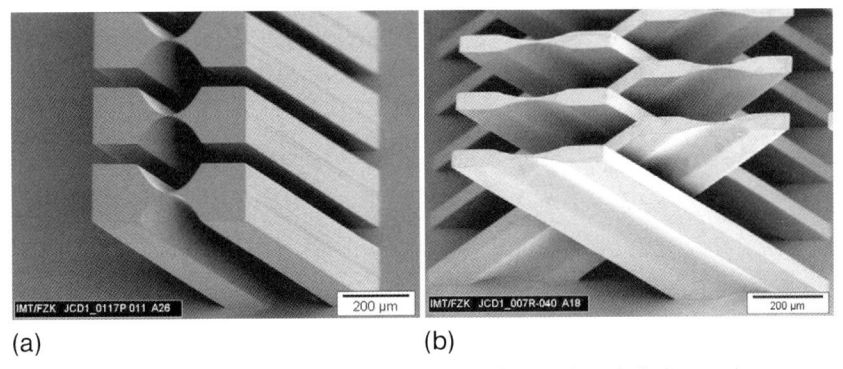

Figure 9.4 Parabolic lenses tilted by 45° (a) and rows of crossed parabolic lenses (b).

cylindrical lenses (Figure 9.4). For this purpose, the mask and substrate are tilted at 45° relative to the exposing X-ray beam for a first exposure and to −45° for a second exposure. Additional suitable apertures protect the areas that are not to be exposed. The structures are about 600 μm high. The intersection area of the −45° and 45° lenses is the usable lens aperture. The resulting point focus reaches dimensions below 400×800 nm and an optical intensity gain of more than 10 000.

9.3
Materials

Materials for manufacturing X-ray lenses have to exhibit low absorption of the radiation and there must be a way to structure them. Depending on the X-ray energy, many materials with a low atomic number are suitable lens materials. For example, Li, Be, B, C, Al, Si and Ni have been investigated. At X-ray energies above 20 keV, lenses made of elements lighter than carbon become transparent and heavier materials must be used.

At the author's institute, the lenses are microfabricated by LIGA technology using the negative resist SU-8 [5]. This resist is fairly sensitive, as a result of which the synchrotron radiation exposure times remain in the region of 1 h. The SU-8 resist shows unique radiation stability. So far, an absorbed dose of several gigagray at an energy of about 16 keV in a linear lens has not been found to reduce the optical performance of the lens noticeably. The properties of the polymer make it suitable for energies between a few keV and 100 keV. Focal lengths depend strongly on the energy. At lower energies, focal lengths of some millimeters can be obtained. At energies above 50 keV, focal lengths of less than 1 m can be achieved. When higher energies are required, electroplated copies of the direct lithographic poly(methyl methacrylate) (PMMA) structures are prepared in nickel [6, 9, 10]. These lenses are suitable for energies up to 1000 keV.

9.4
Lens Construction

When a refractive lens is made for focusing X-rays of a given energy, its ideal geometry is very close to a parabola, as can be seen in Figure 9.3. The optical pathlength in the lens material should be as short as possible to minimize absorption of X-rays in the lens. Unfortunately, the thickness of a parabolic lens increases with increasing lens aperture. For apertures as large as several hundred micrometers, the absorption in the border areas of parabolic lenses rises to such an extent that these border areas scarcely contribute to the total flux. Two lens geometries allow for large apertures and relatively low absorption.

The first type is called a kinoform lens. In this type of lens, the originally parabolic geometry is virtually cut into small parts that are newly arranged like in a Fresnel lens (Figure 9.5). The disadvantage of this geometry is that it has very small details with an enormous aspect ratio, as shown in Figure 9.6a, which are difficult to process. Moreover, the large numbers of side-walls parallel to the optical axis give rise to total reflectance of the X-rays nearly parallel to the optical axis.

The second lens construction is called a mosaic lens. Here, the border areas are fragmented into discrete pillars, as shown in Figure 9.6b. These pillars are chal-

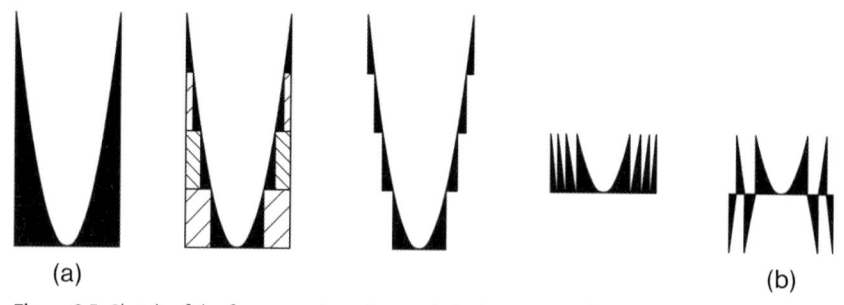

Figure 9.5 Sketch of the fragmentation of a parabolic lens (a) to obtain a kinoform lens (b).

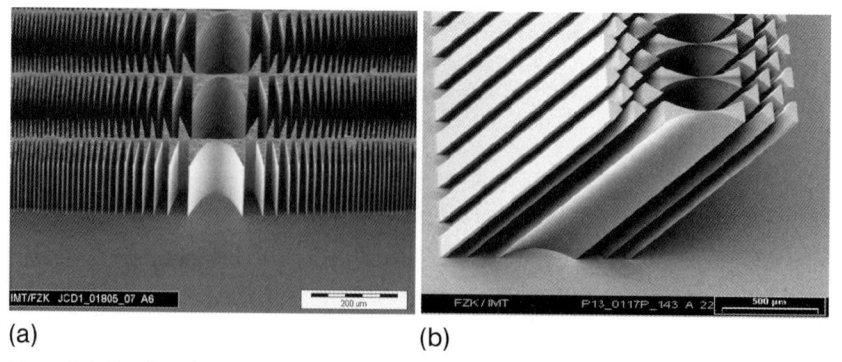

Figure 9.6 Kinoform lenses (s) and mosaic lenses tilted by 45° (b).

lenging to process because of their high aspect ratio: due to capillary forces, they tend to stick together in the drying step after development.

9.5
Recent Results

Refractive LIGA X-ray lenses have been used successfully in many scientific experiments at synchrotrons. Their main characteristics are that they can focus X-rays with energies from above 5 keV up to hundreds of keV, where other X-ray optics tend to fail. In this high-energy range, optics that depend on total internal reflection such as capillary lenses or on Bragg reflection such as crystal mirrors no longer work, because the maximum reflection angles fall to values close to zero for high energies. Lenses using absorbing areas such as zone plates fail at high energies, because absorption decreases strongly at these energies. Refractive X-ray lenses benefit from this reduced absorption, as they become more transparent for high energies.

In Table 9.1, the main parameters of refractive LIGA lenses are listed. The minimum full width at half-maximum (FWHM) values of the focal spot dimensions for lenses with a point focus (250 nm) are larger than those of lenses with a line focus (150 nm). This difference results from a small misalignment in the array of single-lens elements, caused by the necessary double exposure when producing lenses with a point focus.

The minimal working distances achievable for the different X-ray energies by a typical set of LIGA lenses are shown in Figure 9.7. These minimal values are achieved when the lens plate is used unchanged. To adapt the lenses to customers' needs, the focal length of the lenses can be adjusted by removing single lens elements. The resulting working distances will be larger than those of the unchanged lenses. The removal of lens elements is irreversible.

Figure 9.8 shows the intensity distribution in the focal plane of a point-focus lens. The bright peak in the center is the focal spot. As the point-focus lens is

Table 9.1 Main characteristics achieved with refractive X-ray lenses.

Lens parameter	Value
Material	SU-8, Ni or Au
Size of a lens plate	$65 \times 32 \times 1$ mm
Type of focus	Line or point focus
Energy range	5 keV–1 MeV
Focal length	Down to 10 mm
Apertures	33 µm–1.5 mm
Focal spot dimensions	150 nm (line focus) FWHM, 250 nm (point focus) FWHM
Intensity gain	Up to 50 000
Radiation stability	>8.3 GJ kg^{-1} measured at 28 keV

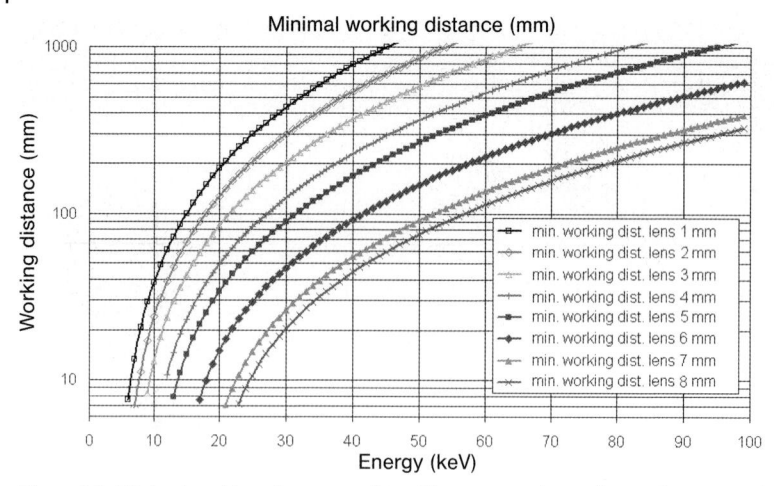

Figure 9.7 Minimal working distances achievable in a typical set of point-focus LIGA lenses.

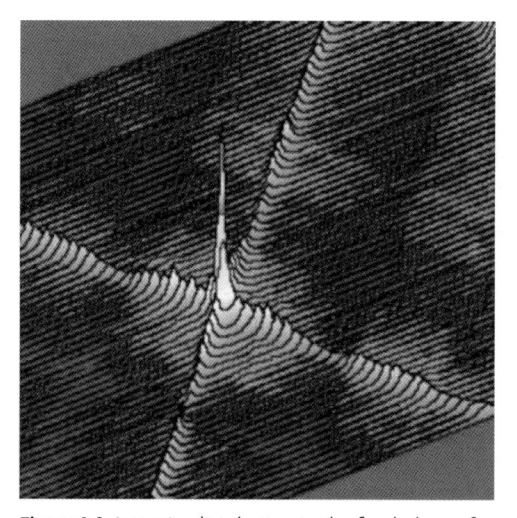

Figure 9.8 Intensity distribution in the focal plane of a point-focus LIGA X-ray lens.

made of two crossed cylindrical lenses, there are two orthogonal lines in the background.

The lenses have been used to construct X-ray microscopes (Figure 9.9), sometimes called 'nanoscopes'. The resolution achieved today with a microscope using LIGA lenses is about 250 nm (Figure 9.10b). Due to the short wavelength of the X-ray light, this resolution may increase to values far below 100 nm [8]. These resolutions are not attainable with microscopes using visible light. The scanning electron microscope (Figure 9.10a) will always have a better resolution, but it cannot show details from inside an object and the samples have to be prepared for the vacuum chamber.

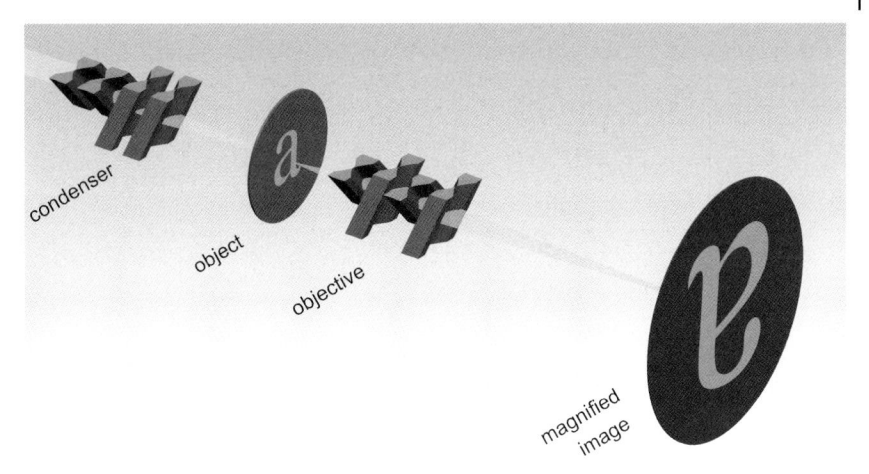

Figure 9.9 Configuration of a 'nanoscope', that is, an X-ray microscope.

(a) (b)

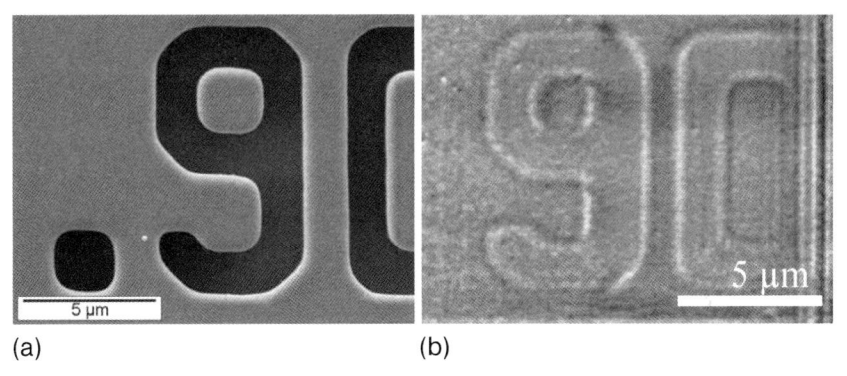

Figure 9.10 SEM image of a silicon nitride membrane with structures (a) and X-ray microscopy of the same structure (b).

9.6
Future Work

The lenses developed so far have small entrance apertures and accept X-rays only when angles under the optical axis are small. Hence their use is mainly limited to focusing the nearly monochromatic light from synchrotron sources. Further efforts will be made to develop lenses with larger apertures and larger angles of acceptance. These lenses would be very useful in focusing light from X-ray tubes. Achromatic lenses will be developed in order to focus white X-ray light on a well-defined focal plane. As soon as these milestones have been achieved, commercial X-ray optical instruments such as X-ray microscopes will be designed in cooperation with industrial partners.

References

1 Tomie, T. (1996) X-ray lens, Japanese Patent 1994000045288, priority 18 February 1994.
2 Snigirev, A., Kohn, V., Snigireva, I. and Lengeler, B. (1996) A compound refractive lens for focusing high-energy X-rays. *Nature*, **384**, 49–51.
3 Lengeler, B., Tümmler, M., Snigirev, A., Snigireva, I. and Raven, C. (1998) Transmission and gain of singly and doubly focusing refractive X-ray lenses. *Journal of Applied Physiology*, **84**, 5855–61.
4 Nazmov, V., Reznikova, E., Boerner, M., Mohr, J., Saile, V., Snigirev, A., Snigireva, I., DiMichiel, M., Drakopoulos, M., Simon, R. and Grigoriev, M. (2004) Refractive lenses fabricated by deep SR lithography and LIGA technology for X-ray energies from 1 keV to 1 MeV. *AIP Conference Proceedings*, **705**, 752–5.
5 Nazmov, V., Reznikova, E., Somogyi, A., Mohr, J. and Saile, V. (2004) Planar sets of crossed X-ray refractive lenses from SU-8 polymer. *Proceedings of the SPIE–The International Society for Optical Engineering*, **5539**, 235–43.
6 Nazmov, V., Reznikova, E., Mohr, J., Börner, M., Mappes, T., Saile, V., Ernst, T. and Simon, R. (2006) Planare refraktive Röntgenlinsen. *Galvanotechnik (Mikrosystemtechnik)*, **4**, 964–9.
7 Nazmov, V., Reznikova, E., Mohr, J., Snigirev, A., Snigireva, I., Achenbach, S. and Saile, V. (2004) Fabrication and preliminary testing of X-ray lenses in thick SU-8 resist layers. *Microsystem Technologies*, **10**, 716.
8 Snigireva, I., Snigirev, A., Drakopoulos, M., Kohn, V., Nazmov, V., Reznikova, E., Mohr, J. and Saile, V. (2004) Near-diffraction limited coherent X-ray focusing using planar refractive lenses made of epoxy SU-8 resist. *Proceedings of the SPIE–The International Society for Optical Engineering*, **5539**, 20–30.
9 Snigirev, A., Snigireva, I., DiMichiel, M., Honkimaki, V., Grigoriev, M., Nazmov, V., Reznikova, E., Mohr, J. and Saile, V. (2004) Sub-micron focusing of high energy X-rays with Ni-refractive lenses. *Proceedings of the SPIE–The International Society for Optical Engineering*, **5539**, 244–50.
10 Nazmov, V., Reznikova, E., Snigirev, A., Snigireva, I., DiMichiel, M., Grigoriev, M., Mohr, J., Matthis, B. and Saile, V. (2005) LIGA fabrication of X-ray nickel lenses. *Microsystem Techologies*, **11**, 292–7.

10
RF Applications

Sven Achenbach and David M. Klymyshyn

10.1 Introduction to RF MEMS *243*
10.2 LIGA-based Vertical Wall RF MEMS Design Approach *248*
10.3 Sample Components *251*
10.3.1 Variable Capacitor (Varactor) 252
10.3.1.1 Varactor Characteristics 252
10.3.1.2 MEMS Varactors 253
10.3.1.3 Vertical Wall Varactors 254
10.3.2 Tall Coplanar Waveguide (CPW) 3 dB Couplers 263
10.3.2.1 Coupler Characteristics and Layout 263
10.3.2.2 Coupler Fabrication 266
10.3.2.3 Coupler Theoretical and RF Test Results 268
10.3.3 Cavity Resonators 270
10.3.3.1 Cavity Resonator Characteristics 270
10.3.3.2 Cavity Resonator Fabrication 272
Acknowledgments 277
References 278

10.1
Introduction to RF MEMS

Traditionally, the term *radiofrequency* (RF) was used to indicate the frequency f of transmission or reception of electromagnetic waves at the antenna of a 'radio' or, more generally, a wireless communication transmitter/receiver (transceiver). This frequency was often in the *microwave* region of the electromagnetic spectrum and thus, today, the terms radiofrequency and microwave are generally regarded as synonymous. RF devices operate with signals at *microwave frequencies*, typically in the range 1–300 GHz with corresponding electrical wavelengths in free space between $\lambda = c/f = 30$ cm and $\lambda = 1$ mm. Signals with wavelengths on the order of millimeters are often also called *millimeter waves*.

RF applications are a comparatively new field in microsystems technologies, remaining a laboratory curiosity in the 1980s and with first devices appearing in the early 1990s. However, this field is now attracting substantial research and development efforts and is experiencing tremendous market growth rates. Market studies, forecasts and roadmaps such as those produced by NEXUS, In-STAT, enabling MNT and WTC predict an almost exponential growth of the market. Growth rates of more than 55% annually (in revenues; 150% in units shipped) are estimated from $125 million in 2004 to more than $1.1 billion in 2009 [1].

The proliferation of commercial wireless devices such as cellular phones and wireless LANs and increasingly radiofrequency identification (RFID) and sensor devices in the early part of the twenty-first century is a strong driver for this. Much of this development has been in the sub-2.5 GHz bands. Such systems are not only placing extreme pressure on an already crowded frequency spectrum in the sub-2.5 GHz region, but are also forcing many bandwidth-intensive applications to higher microwave and millimeter wave frequencies.

Continuing trends towards ever smaller, more power-efficient and cheaper transceivers, while maintaining good performance with increasingly higher frequency broadband systems and in harsher environments with higher interference, are severely challenging traditional technologies. Radiofrequency integrated circuits (RFICs) to a large part have addressed the size and cost issue for the electrically active portions of the transceivers and have made modern communication devices possible. However, passive components such as inductors, capacitors and filters can account for up to 90% of a modern radio transceiver and these cannot easily be implemented in RFICs with acceptable performance. These components must be assembled separately on to the RF electronic circuit.

RF microelectromechanical systems (MEMS) devices can alleviate some of these problems and are expected to become one of the major MEMS product groups. The volume-driven priority application of RF MEMS components to date has been mobile phones. Other important market areas include wireless data networks (WLAN, WPAN and others), global positioning systems (GPS), automotive and other radar systems, satellites/nanosatellites, RFID tags and sensors.

As the name would suggest, microwave signals involve appreciation of *wave* behavior. Electromagnetic waves exhibit similar characteristics to waves in other physical domains (for instance, mechanical and acoustic waves) in that they require consideration of both time and spatial domains. All alternating current (AC) electrical signals exhibit wave behavior. With low-frequency signals, spatial phase variation over the physical circuit dimensions is negligible as the signal wavelength is generally orders of magnitude larger. At microwave frequencies this is not the case, as phase can vary significantly over the physical circuit dimensions, sometimes over several cycles. In this case, the 'electrical length' $\theta = 2\pi l/\lambda$ of the circuit is generally more relevant than the physical length l. Standard electrical circuit theory is not sufficient for describing such 'distributed' microwave circuits. These normally rely on transmission line considerations. A major consideration with microwave circuits is in effectively controlling impedance, which can vary dramatically over the electrical length of a device. In the case of distributed components which are wavelength related, this behavior is desired, whereas in the case of lumped reactive elements, this behavior is detrimental and ultimately limits high-frequency performance. For lumped elements, miniaturization usually means better performance in addition to better integration, a good combination for MEMS applications.

Strictly, 'true' RF MEMS, like other MEMS, would imply a combination of mechanical motion and also electrical control and/or actuation. Many RF MEMS devices such as switches and varactors (variable capacitors) are thus mechanically 'active' devices. This is not to be confused with so-called 'active' RF devices, which generally are electrically active and include transistors with gain. In current terminology, RF MEMS devices are normally considered also to include any passive microstructures that have some mechanical structure, even if these are not mechanically active. A defining feature of all RF MEMS devices is that they are electronic circuits (or parts of circuits) and in addition to possible low-frequency or direct current (DC) control signals, operate with signals at microwave or millimeter wave frequencies in the gigahertz range. The fact that these devices operate simultaneously in multiple coupled physical domains (for instance, passive and active mechanical and electromagnetic) and ultimately as part of a larger electrical circuit, makes them very challenging to analyze.

In addition to electrical size and impedance control, an essential concern in microwave circuits is in signal control and switching. The RF MEMS switch presented by Larson *et al.* in 1991 [2] was possibly the first reported RF MEMS device and, since then, the switch has undoubtedly been the most researched and developed device to date – and with good reason, since in addition to the obvious benefits of miniaturization, RF MEMS switches can operate at frequencies up to 120 GHz and the performance in various categories can be orders of magnitude better than PIN diode and field-effect transistor switches, as demonstrated in Table 10.1 [3]. Various companies have recently developed RF MEMS switches, including Raytheon [4], LG-Korea [5] and Rockwell Scientific [6].

Besides switches [2, 4–6], other emerging RF MEMS devices include high-quality factor Q lumped reactive and tunable reactive elements (capacitors, varactors, inductors) [7–11], couplers and power dividers [12–14], filters and diplexers [15–17], resonant cavity structures [18–20] and micromachined transmission lines [21–23]. Acoustic resonators and filters are also indispensable RF microstructures for wireless applications [24].

Table 10.1 Sample performance comparison of electrostatic RF MEMS, PIN diode and FET switches. After [3]

Parameter	RF MEMS	PIN diode	FET
Cutoff frequency (THz)	20–80	1–4	0.5–2
Power consumption (mW)	0.05–0.1	5–100	0.05–0.1
R_s (series) (Ω)	0.5–2	2–4	4–6
Loss (1–100 GHz) (dB)	0.05–0.2	0.3–1.2	0.4–2.5
Isolation (10–40 GHz)	Very high	Medium	Low

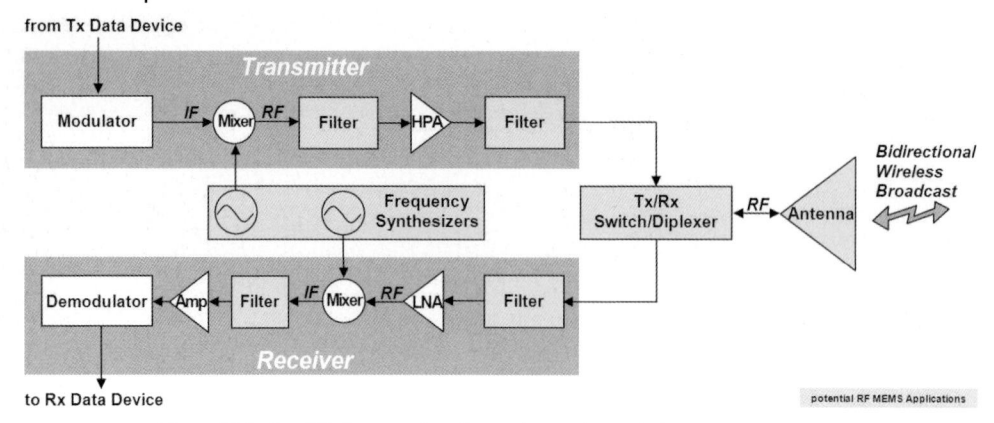

Figure 10.1 Simplified conventional superheterodyne wireless transceiver architecture showing potential RF MEMS applications. With transmitted data Tx, received data Rx, radiofrequency RF, lower intermediate frequency IF, high-power amplifier HPA, low noise amplifier LNA.

These devices find a wide range of potential applications in RF and wireless circuits, for instance as indicated in Figure 10.1, which shows a simplified block diagram of the conventional superheterodyne transceiver architecture that forms the backbone of the RF radio portion of most modern wireless devices for transmitting Tx and receiving Rx data signals. The concept of the superheterodyne receiver today is essentially the same as that patented in 1917 by Edwin Armstrong; however, the components used to implement it and the applications are continuously evolving. The idea is that the high-frequency RF signal at the antenna is converted to/from a common lower intermediate frequency (IF) by non-linear 'mixers' and frequency agile microwave synthesizers. Signal processing such as modulating and demodulating typically is easier with this method, since it can be done at a low frequency where high-gain, high-selectivity components are more readily obtained. Also, the variable frequency synthesizers allow tuning to different RF channels, which can be selected at the receiver IF by a single common high-Q filter, rather than requiring a bank of many high-Q filters.

In such a transceiver, RF MEMS power dividers and switches could form the structures for intricate bi-directional phase shifting and steerable antenna arrays, which could be made tunable to match to different frequencies using RF MEMS = based switches and variable reactance elements. Diplexers with high transmit/receive isolation could be realized with RF MEMS switches and micromachined transmission line or cavity-based structures. MEMS resonators (transmission line, cavity type, lumped reactive, acoustic, mechanical) could be used as frequency-stabilizing elements of frequency oscillators and synthesizers and also as building blocks for low-loss, highly selective filters at both RF and IF frequencies. High-Q RF MEMS lumped reactive components could also be used to replace

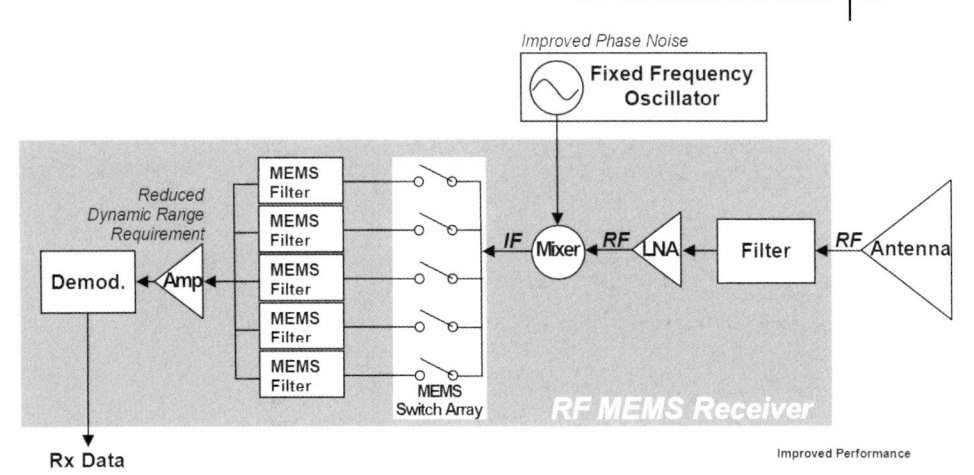

Figure 10.2 Modified wireless receiver architecture with RF MEMS switched filter bank array and possible improved performance or reduced requirements on conventional components.

bulky off-chip components required for electrically active devices such as low-noise amplifiers (LNAs).

Another rapidly advancing trend is in adaptive and reconfigurable 'software' radios, ones that can operate in various frequency bands and automatically adjust to support various telecommunications protocols. Advanced digital signal processing (DSP) techniques and high-speed, low-power processors and field programmable gate arrays (FPGAs) have addressed many of the protocol and adaptive modulation issues, but reconfigurability and adaptability in the RF circuitry portion of the radio are still in their infancy.

RF MEMS devices could produce significant performance improvements and even radically different transceiver architectures. For instance, Figure 10.2 presents an alternative realization of the superheterodyne receiver using an RF MEMS switch array and highly selective filter bank to tune the radio channel a nd a fixed oscillator rather than a frequency synthesizer. With MEMS miniaturization, switched filter banks with possibly hundreds of filters are feasible from a size perspective and could offer lower loss and better frequency selectivity. However, in addition, the use of alternative architectures made possible by RF MEMS could also alleviate the requirements on other circuit components, further simplifying these and improving the performance. For instance, in Figure 10.2, the complicated, usually multiple phase locked loop-based, high power-consuming frequency synthesizer could be replaced with a simple, single-frequency, power-efficient oscillator with much improved phase noise performance. With the high-frequency selectivity of the MEMS filter bank, an amplifier with higher power efficiency and lower dynamic range and distortion requirements could be used.

10.2
LIGA-based Vertical Wall RF MEMS Design Approach

RF MEMS devices are based on various surface micromachining techniques. Most are based on silicon microfabrication technologies [25–27], such as bulk silicon micromachining, silicon surface micromachining and deep dry etching, in an attempt to leverage mainstream fabrication processes. However, low electrical resistivity semiconductor-grade silicon as a substrate material is very lossy at microwave frequencies and can severely degrade RF performance. Mechanical-grade high-resistivity silicon is an improvement, and another approach is to develop the devices on low-resistivity silicon and later transfer them to a low-loss substrate, such as quartz or alumina, but this complicates processing.

RF applications require good, preferably metal, conductors. Semiconducting silicon cannot be applied directly in most cases. Therefore, silicon is either coated with metal thin films or metal membranes and layers of typically 0.5–5 μm gold, copper or aluminum are formed. Alternatively, polysilicon is often applied as the conductive layer with significantly reduced conductivity, but enhanced mechanical properties and a higher degree of standardization. The components are most often laid out as planar, two-dimensional devices with, eventually stacked, layers of functional thin films. Performance is controlled through the lateral geometry in addition to, in most cases, the vertical thickness and separation of the various layers. Figure 10.3 shows an example of such a silicon-based planar RF MEMS device [7]. It is a parallel-plate variable capacitor (295 μm lateral plate size) fabricated using a standard polysilicon surface micromachining multi-user MEMS process (MUMP). A 2 μm thick polysilicon layer is stacked between 1.5 and 0.75 μm thick air gaps. The upper air gap is topped with a second, 1.5 μm thick, polysilicon layer and 0.5 μm of gold.

A fundamentally different approach is to exploit fully the third dimension by flipping the planar membranes 90° upwards. This delivers vertical walls of arbitrary geometry, width, separation and height. These walls are no longer integrated as planar layers on the substrate, but stand up orthogonally. The only limiting

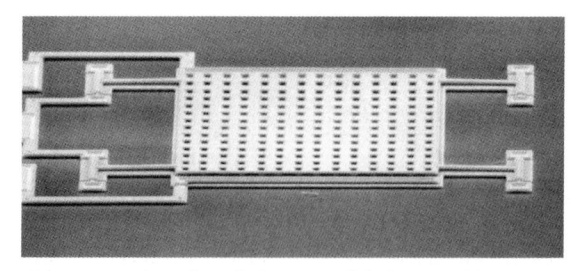

Figure 10.3 Silicon-based planar parallel plate variable capacitor (295 μm lateral plate size) made of two layers of polysilicon (2.0 and 1.5 μm thick), two air gaps (1.5 and 0.75 μm thick) and an upper gold layer (0.5 μm thick). From [7], reproduced with permission of IEEE.

factor is the fabrication technology and the corresponding structure quality achievable. Such an approach has been pursued before, for example to fabricate comblike structures as electrostatic actuators or sensors. A large number of such MEMS devices have been built and even a few RF MEMS devices [28, 29]. However, design rules associated with the fabrication technologies applied have to date limited the performance, especially in the RF context.

Excellent side-wall verticality and high aspect ratios (feature height to minimum lateral feature size), minimal side-wall roughness and structure heights up to millimeters are some of the most important advantages of the LIGA technology compared with competing fabrication processes. Polymer linewidth aberrations as small as 40 nm laterally per 100 μm vertical feature height have been reported for 400 μm high structures [30] and an average of 260 nm laterally per 100 μm feature height for 3 mm high structures [31]. This corresponds to a verticality error of 0.01–0.07°. Aspect ratios as high as 100 have been demonstrated [32] and side-wall mean roughness values (R_a) as low as 5–20 nm have been measured optically and by atomic force profilometry [31]. LIGA technology, especially the X-ray lithography and electroplating steps, therefore appears to be ideally suited to manufacturing vertical wall RF MEMS devices.

The possibility of creating very narrow walls in large structure heights has, in general, four advantages:

- If the final structure is a narrow groove, very good electrostatic performance between the two adjacent vertical side-wall electrodes can be achieved, since most of the electric field is parallel to, and largely unaffected by, the supporting and potentially lossy substrate.

- Higher structures at the same time linearly increase the coupling surface area, which enhances transmittable forces or the sensitivity of a transducer.

- If the final structure is a very narrow metal wall, it can serve as an easily deflectable cantilever that can be actuated with forces or detect low forces.

- If identical forces can be transmitted or sensed at a smaller footprint, the overall lateral dimensions can be reduced with respect to the planar layout approach.

In the case of a lumped device, operational frequencies of the device can theoretically be significantly increased, due to the reduction in the electrical size with reduced footprint.

In the case of a wavelength-dependent, distributed element, the width can also be dramatically reduced by exploiting the vertical dimension. Figure 10.4 illustrates the size impact on changing from the planar to the vertical wall approach, considering for example traditional coplanar waveguide (CPW) and tall CPW 3 dB coupling structures as further described in [33]. In the planar approach (Figure 10.4a), the metal surface area in the gap is prohibitively small due to the thin conductors and the required coupling capacitance can only be met if the two surfaces are located very close together (8 μm separation) and if the surfaces cover a large lateral area (~8000 μm linewidth). In the vertical wall approach, as demon-

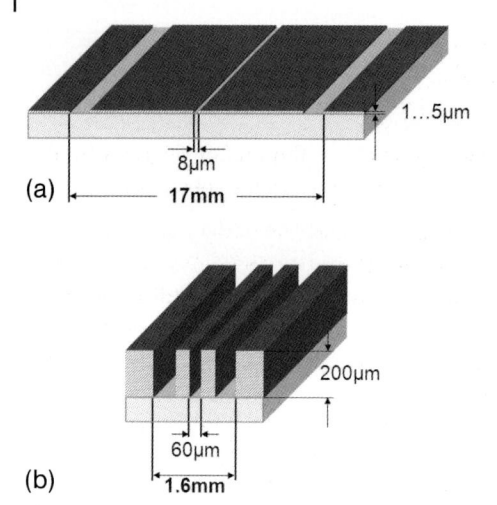

Figure 10.4 Size impact of the vertical wall approach in RF MEMS considering coplanar waveguide (CPW) 3 dB coupling structures, further described in [33]. (a) Planar approach with traditional CPW; (b) vertical wall approach with tall CPW.

strated in Figure 10.4b, the electric field is almost entirely contained in the gap. The 200 μm high metal walls may be as far as 60 μm apart, facilitating fabrication. The metal walls have a much narrower linewidth so that the overall feature size can be reduced by an order of magnitude.

The vertical wall approach has additional benefits for RF MEMS applications in particular:

- Larger fractions of the vertical electrodes are further away from the substrate material. In the case of tall coplanar structures, this implies that more of the electromagnetic field is present above the substrate, resulting in lower effective permittivity and lower dispersion losses from unequal signal propagation velocities in air and substrate.

- The arbitrary shaping and width of metal walls allow adjustment of the conductor cross-section to required values. This limits resistive losses and provides additional freedom to design power handling capability.

Finally, LIGA-based vertical wall RF MEMS structures may additionally benefit from further advantages of the fabrication technology:

- Smooth side-walls limit losses and increase quality factors Q at higher frequency operation. Electromagnetic fields penetrate metal conductor surfaces due to non-infinite conductivity. The penetration depth (skin depth) decreases with increasing frequency and is normally less than 1 μm at most microwave frequencies for high-conductivity metals. As the frequency increases, the skin depth decreases and approaches the roughness dimensions, resulting in more power loss as the fields are concentrated at the rough surface.

- The metal selection is only limited by the plating process and typically includes low-resistivity materials such as gold and copper.

- The substrate material is not limited to silicon with its high dielectric and conductive losses. Lower loss materials, such as alumina or quartz, can also directly be used as substrates, allowing for higher frequency operation.

LIGA-based vertical wall RF MEMS structures therefore imply significant inherent advantages over competing concepts and fabrication technologies.

One of the first mentions of using deep X-ray lithography for RF applications was by Willke and Gearhart at the University of Wisconsin–Madison in 1997 [34], demonstrating the tremendous potential of LIGA for RF. Despite the potential advantages, LIGA-based technologies initially have not experienced widespread application to RF devices since that time. This is likely due to the overwhelming research effort in planar and silicon-based technologies. More recently, several LIGA-based RF MEMS devices have been published by the present authors. Low-loss LIGA-based coplanar waveguides have also been developed at Sandia National Laboratories (Albuquerque, NM, USA) [35] and the application of LIGA processes for the fabrication of 2D and 3D electromagnetic metamaterials for the terahertz range has been proposed [36].

However, the application of LIGA technology also brings about some disadvantages, which are mainly related to fabrication process itself:

- LIGA is less standardized than most silicon-based technologies. Therefore, design and simulation tools are not as readily available and devices cannot be designed as easily.

- Extremely high aspect ratio applications as suggested for some RF MEMS devices may still impose processing challenges and require, among others, high-quality X-ray masks.

- Unlike in silicon-based technologies, electrical integration of LIGA devices requires additional, most often hybrid, fabrication and/or mounting steps.

- Hermetic sealing and housing of the devices usually cannot be accomplished as elegantly as with, for example, bonding of silicon devices.

- Fabrication costs of direct-LIGA devices are comparatively high and may impede some high-volume applications.

10.3
Sample Components

The subsequent sections will highlight selected LIGA-based mechanically active (variable capacitors) and passive (tall coplanar waveguide 3 dB couplers, cavity resonators) RF MEMS components, developed in the microsystems research group in the Department of Electrical and Computer Engineering at the University of Saskatchewan (Saskatoon, Canada). D. Klymyshyn primarily directed RF MEMS device functional development and testing and S. Achenbach primarily directed LIGA and related fabrication development. Fabrication and much of the process development was done at the German national research laboratory Forschungszentrum Karlsruhe, Institut für Mikrostrukturtechnik (FZK/IMT). Collaborating institutes included Telecommunications Research Laboratories (TRLabs) in Canada. Aspects of the design and simulation, device processing and metrology measurements and RF testing and evaluation of the devices will be discussed.

10.3.1
Variable Capacitor (Varactor)

10.3.1.1 Varactor Characteristics
The ability to control reactive impedance effectively is fundamental in most RF applications. Traditionally, distributed elements were required for all but the lowest microwave frequencies, since lumped reactive elements were simply too large and lossy for use at higher microwave frequencies. However, distributed circuits are typically large and their impedance is difficult to tune. With the current trend towards miniaturization and adaptability, RF MEMS varactors are poised to replace conventional devices and permit new capabilities for various tunable impedance and frequency control applications, including impedance matching, phase shifting/modulation and resonators and in lumped element replacement of conventionally distributed circuits such as filters and couplers.

Varactors are characterized by their unbiased nominal capacitance C, the equivalent parasitic series resistance R_S or quality factor Q and associated parasitic inductance L, giving an electrical self-resonance ω_0. Most important tuning characteristics are the tuning ratio and, in the case of electrostatically actuated devices, which have been the most successful to date, the tuning voltage [37].

The capacitance without the application of any tuning signal is called the unbiased nominal capacitance. This value depends on the requirements of the desired application and the frequency of interest. Required capacitance values for typical impedance values can range from several picofarads (pF) in the low0 or sub-gigahertz range to fractions of picofarads for applications above 10 GHz.

In damped oscillations, the *quality factor*, Q, is proportional to the ratio of the stored energy to the energy loss per period:

$$Q = 2\pi \frac{\text{stored enery}}{\text{energy loss per period}} \qquad (1)$$

In the case of a resonator, Q is normally measured at the resonant frequency, where the magnitude of the capacitive and inductive reactances (imaginary part of the complex impedance) and their peak stored energies are equal. This interpretation of Q is not particularly useful for a capacitive reactance element, since the associated inductance is not desired, but causes a parasitic high frequency self-resonance above the useful operating frequency range. The associated parasitic *inductance*, L, and the tunable capacitance determine the *self-resonant frequency*, ω_0:

$$\omega_0 = \sqrt{\frac{1}{LC}} \qquad (2)$$

At frequencies exceeding ω_0, the capacitor starts to behave inductively instead of capacitively. The capacitor may therefore only be operated at frequencies ω considerably lower than ω_0, typically at a maximum of 50% [26]. To increase ω_0, the parasitic inductance must be kept as low as possible.

To determine Q in the capacitive region, the total reactance is normally measured and assumed to be capacitive, although the parasitic inductive reactance effectively reduces the capacitive reactance and lowers the Q. At ω_0, Q drops to zero. The complex *impedance*, Z_{11}, is determined (most likely extracted from the measured scattering parameter, S_{11}) and Q is typically expressed as the ratio of the imaginary portion of the impedance to the real portion:

$$Q = \frac{Z_{11,im}}{Z_{11,re}} \qquad (3)$$

The quality factor increases with decreasing equivalent series resistance and can be expressed for a low-frequency model as

$$Q = \frac{1}{\omega C R_S} \qquad (4)$$

The tuning ratio is the ratio of maximum capacitance to minimum capacitance. Some applications require only a small tuning ratio to fine-tune an impedance value. Many applications, however, require a ratio of at least 2:1 [26].

For a voltage-controlled capacitor, the tuning voltage is the potential required to tune the capacitor through its entire range of values. In most cases, this voltage range is required to be as small as possible, so that it is compatible with system control signals, which has been an early challenge with electrostatically actuated RF MEMS devices.

10.3.1.2 MEMS Varactors

Most existing MEMS capacitor designs feature parallel plates. The capacitance of an air dielectric parallel plate capacitor is defined as $C_{plate} = \varepsilon_0 A/d$ with the *permittivity* $\varepsilon_0 = 8.854 \times 10^{-12}\,\text{F m}^{-1}$, the capacitance electrode *area*, A, and the *air gap thickness*, d. In variable capacitance designs, C is varied by changing the gap d between the plates [7–9]. Traditionally, these plates are limited to planar geometries and lie parallel to the substrate. These devices are constructed from layers that are typically thinner than 5 µm. They are actuated electrostatically or thermally. Many of these devices have been fabricated using silicon-based thin-film processes such as MUMPs. Most planar parallel plate designs have focused on the lower end of the microwave frequency range (1–3 GHz) with capacitance values between approximately 1 and 4 pF.

A lateral comb structure is an alternative configuration [28, 29]. In this geometry, the capacitance is changed by adjusting the overlap of the capacitor fingers. In contrast to planar parallel plate-type capacitors, the direction of actuation is parallel to the substrate surface. Device layers as thick as 80 μm have been reported [38]. These layers were constructed using a highly refined deep silicon etch. Lateral comb capacitors have been targeted for frequencies in the 100 MHz–1 GHz frequency range with capacitance values between approximately 1 and 10 pF.

Smaller capacitors have been realized using shunt-mounted CPW distributed approaches at higher frequencies [38, 39]. These capacitors have demonstrated Q-factors > 100 in the 10–40 GHz frequency range with capacitance values between 0.1 and 0.3 pF.

Few, if any, existing MEMS lumped capacitors have been designed for the 3–10 GHz frequency region with Q-factors > 100. This void can be filled using the LIGA-based vertical wall approach as theoretically demonstrated by simulations for LIGA-MEMS capacitors [40]. Some of these simulations will later be presented in the context of actual RF measurements.

10.3.1.3 Vertical Wall Varactors

10.3.1.3.1 **Concept and Layout** A scheme of a simple two-plate vertical wall capacitor is shown in Figure 10.5a. The capacitor is composed of a comparatively large capacitance electrode and an electrically isolated, thin cantilever beam that is attached to a larger ground structure. The cantilever beam and the capacitance

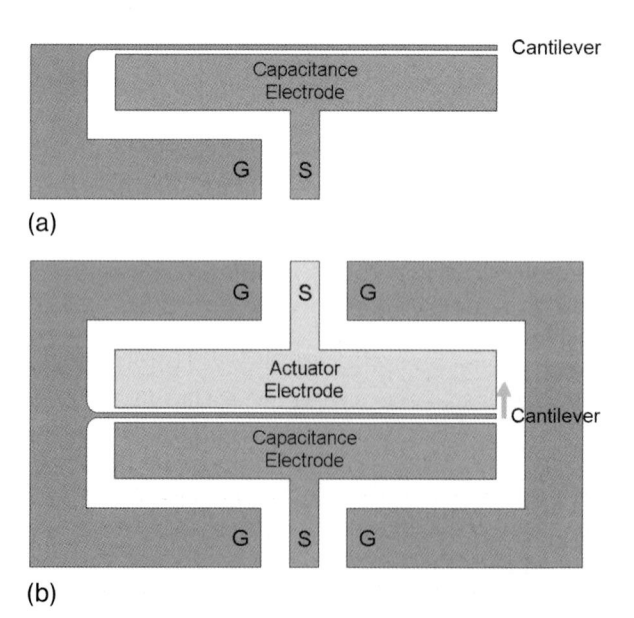

Figure 10.5 (a) Scheme of a high aspect ratio two-plate (ground and signal) capacitor (top view). (b) Top view of a simplified three-plate variable capacitor. The additional ground structure on the right-hand side facilitates testing using standard three-pin wafer probes.

electrode form two adjacent parallel plate electrodes that are separated by a narrow air gap. The capacitance increases with increasing aspect ratio of the air gap, as achieved by a reduced air gap d and/or higher microstructures which proportionally increase the capacitance area A.

For a given device, the capacitance can be varied if the gap d is altered. This is accomplished by introducing an additional actuator electrode (see the three-plate variable capacitor in Figure 10.5b). The thin cantilever is situated between the two electrodes and anchored to the ground piece on the left. As the cantilever is the only structure that is not fixed to the substrate, it can be deflected electrostatically when applying a bias voltage between the actuator electrode and the grounded cantilever. The electrostatic force between the two deflects the cantilever towards the actuator electrode, thus increasing the gap d and inversely proportionally decreasing the capacitance C.

With the given design, the cantilever could be actuated towards either electrode to increase the tuning range further, as was done in [7], but in this scenario RF and control signals must be combined. If the cantilever is actuated towards the actuator electrode only, an increase in tuning range over a two-plate capacitor can be achieved only if the actuator gap is made larger than the capacitance gap [40]. This increase in tuning range comes with an increase in tuning voltage, but RF and control signals can remain separate.

The actual capacitor layouts vary in layout sizes, capacitance values and structural details. Critical dimensions vary around 6 µm cantilever beam width, 6 µm actuator gap width and 3 µm capacitance gap width. Cantilevers are on the order of 1 mm in length and typical overall dimensions are 1450×765 µm. Some devices have a capacitance electrode length that is half the actuator electrode length (500 µm instead of 1 mm), reducing the capacitance for higher frequency applications. This also increases the tuning ratio, since the deflection of the cantilever beam is largest at the end of the beam, leading to a higher average deflection at the electrode.

The capacitor layout has additional ground structures on the right side of the design (see Figure 10.5b). They geometrically facilitate testing using standard three-pin ground–signal–ground (GSG) wafer probes with 150 µm pitch.

One main concern addressed in this layout is ease of electrical isolation of the electrodes and the mechanical detachment of the movable parts of the cantilever.

Usually, devices with moving parts such as the cantilever are fabricated by pre-structuring a thin sacrificial layer before patterning the movable device. In LIGA applications, this sacrificial layer is typically made of titanium. In sacrificial layer processing, the sacrificial layer interfaces to the moving parts of the cantilever and the substrate, while the adhering parts of the cantilever directly attach to the substrate. After selective wet chemical etching of the sacrificial layer, the movable parts of the cantilever will be freely suspended. This is an elegant fabrication sequence. However, it requires an additional lithographic layer and mask in addition to aligned deep X-ray lithography. It considerably increases the potential sources of errors and the cost. Yet more problematic are the sacrificial layer patterning accuracy and alignment requirements, as the sacrificial layer would ideally

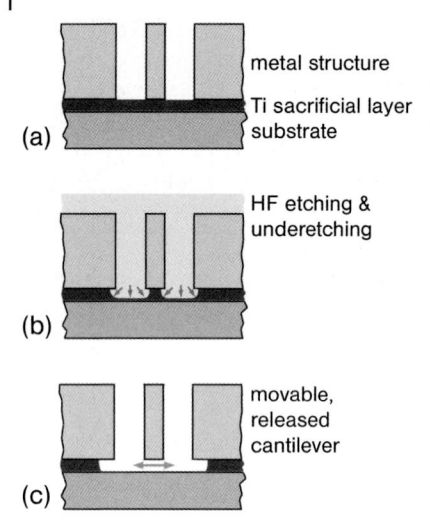

Figure 10.6 Sketch of a time-controlled, layout-dependent sacrificial layer process eliminating the prestructuring of the sacrificial layer.

be aligned exactly with the bottom of the grooves with only a few micrometers of tolerance.

This layout therefore eliminates the prestructured sacrificial layer. Instead, the entire substrate is coated with a metal layer (3 μm Ti/TiO$_x$) which serves as the plating base and also a time-controlled sacrificial layer (see Figure 10.6a). After plating of the cantilever, the plating base is wet chemically etched. All open metal layer areas are dissolved in dilute hydrofluoric acid (HF), electrically isolating the cantilever. The high aspect ratio structures are also slightly underetched (see Figure 10.6b). If this etching process is stopped after an appropriate time, the thin cantilever will be completely underetched and released, while the much wider actuator and capacitance electrodes and the anchoring parts of the cantilever still adhere to the substrate (see Figure 10.6c).

To facilitate this process, the electrodes and anchoring parts of the cantilever are laid out to be at least 20 times wider (150 μm) than the cantilever.

10.3.1.3.2 Fabrication To fabricate the devices, a 150 μm thick PMMA plate (Röhm GS 233) was glued on a titanium-coated 1 mm thick low-dielectric-loss alumina wafer. The sample was irradiated at the electron storage ring ANKA, beamline Litho 2 ($\varepsilon_{\text{cut-off}} = 6.9$ keV) to a bottom dose deposition of 3.5 kJ cm^{-3}. The applied X-ray mask was based on a 2.7 μm thick titanium membrane and manufactured according to standard FZK/IMT processes [41]. Megasonic-supported development was performed for 150 min in GG at room temperature.

A 3 μm Ti/TiO$_x$ coating on the wafer was used as a plating base for 100 μm nickel electroplating. Nickel was chosen as the plating material as most high aspect ratio processing experience in the LIGA process is available for nickel. From an RF perspective, lower electrical resistivity materials, such as gold or copper, would be preferable.

After flood exposure and resist stripping, the titanium was descummed in oxygen plasma and etched for 2 min in 5% HF. This process step dissolves open Ti areas, resulting in electrical isolation of the electrodes. It also leads to under-etching of the nickel structures of approximately 9 μm, which releases the 6 μm wide cantilever but keeps the much larger electrodes still fixed to the substrate. In this way, the plating base was applied as the etch time-controlled sacrificial layer mentioned above, releasing the cantilever beam without an additional lithographic layer [32]. For different samples, the required etch time may vary due to various parameters.

Other sets of samples were electroplated to 70 μm of nickel in 100 μm of PMMA with otherwise comparable process parameters.

Figure 10.7 gives an impression of the high aspect ratio of the fabricated devices. The SEM image is an aerial view (seen from the right side with respect to the sketch in Figure 10.5b), showing the 100 μm deep and 1.6 μm wide capacitance gap with an aspect ratio of 60 in nickel and 100 in PMMA prior to resist stripping. The electrodes are 1 mm long.

The cantilever beam is extremely narrow and long. In order to support the processing of this device, layout variations have been introduced. Some capacitors are equipped with local auxiliary structures such as bulged cantilever supports, adding strength to the actuator without significantly increasing the overall bending stiffness that would resist tuning deflections. These cantilever supports reinforce the capacitance gap resist walls in addition to the cantilever nickel beam, as shown in Figure 10.7. Other auxiliary structures mechanically support the resist walls which

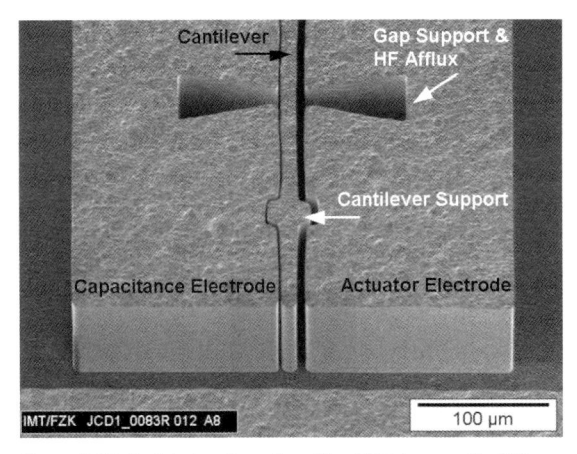

Figure 10.7 Aerial view from the side: SEM image of a 100 μm thick nickel variable capacitor. 1.6 μm capacitance gap, aspect ratio 60 in nickel (100 in PMMA before plating and stripping). 1 mm electrode length. From [32], reproduced with permission of Springer.

form the capacitor gap and facilitate HF liquid afflux during release of the beam, as also depicted in Figure 10.7. Their triangular shape is meant to add maximum strength while least effecting the overall capacitance.

Only the components equipped with support structures were safely processed in the lithography step.

Many of the structures that were well patterned in PMMA, however, were severely deformed in the subsequent electroplating step in an aqueous nickel electrolyte of 52 °C. This is attributed to expansion of the PMMA template with respect to the alumina substrate during the plating process. The relative expansion occurs because of two effects:

- The sample is heated from the ambient temperature (approximately 21 °C) at which it was coated, exposed and developed, to 52 °C during the plating process [42]. PMMA has a thermal expansion coefficient of $\alpha_{PMMA} \approx 70 \times 10^{-6}\,\mathrm{K}^{-1}$, whereas alumina only expands at $\alpha_{alumina} \approx 5.4 \times 10^{-6}\,\mathrm{K}^{-1}$ [31]. The relative change in length scales, $\Delta l_{PMMA-alumina}$, at the differential temperature of 31 K would amount to 120 μm over the entire layout of 60 mm. This distortion is suppressed by firmly attaching the PMMA to the substrate. The built-up stress, however, leads to the detected deformations.

- PMMA takes up water on the order of 2.1% [43, 44]. It therefore swells in the aqueous electrolyte. This effect becomes even more severe at elevated temperatures [43] such as in the electrolyte bath.

The micrographs in Figure 10.8 show such structures that are undeformed as 150 μm thick PMMA structures (a), but become deformed during plating (b; see the arrow indicating deformed capacitance gap).

This particular capacitor features gap support structures, but is not equipped with the cantilever support structures. Figure 10.9 shows the beneficial effect of

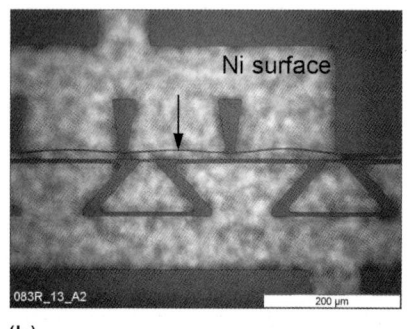

Figure 10.8 (a) Undeformed PMMA walls and grooves. (b) Same feature: deformations of the cantilever beam and the air gaps introduced during the electroplating at 52 °C. 150 μm PMMA, 100 μm nickel. From [32], reproduced with permission of Springer.

Figure 10.9 (a) Undeformed PMMA walls and grooves with additional cantilever support auxiliary structures. (b) Same feature: undeformed cantilever beam and the air gaps after the electroplating step at 52 °C. 150 µm PMMA, 100 µm nickel. From [32], reproduced with permission of Springer.

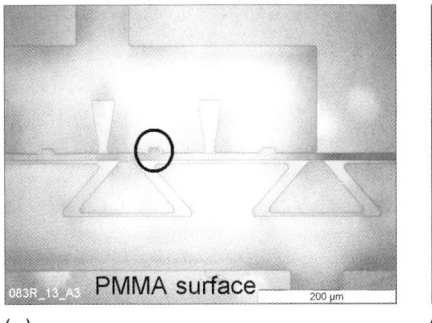

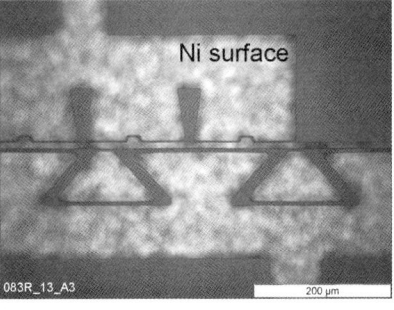

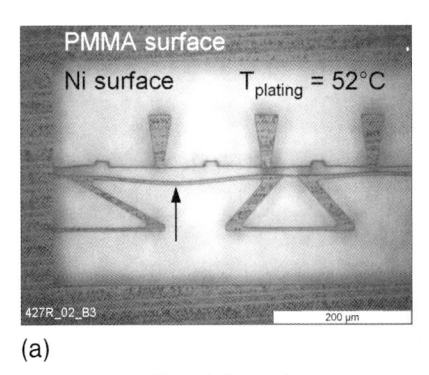

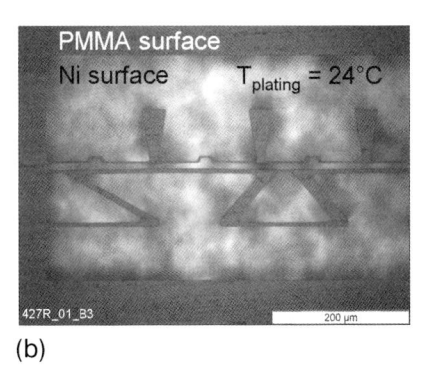

Figure 10.10 Effect of electroplating temperature. (a) Deformations attributed to elevated standard plating temperature of 52 °C. (b) Undeformed structures after ambient temperature plating at 24 °C. 150 µm PMMA, 100 µm nickel. From [32], reproduced with permission of Springer.

the additional cantilever support auxiliary structures [see the circle in (a)]. Even after the plating process, no distortions occur. The process parameters are identical with those used for the structures in Figure 10.8.

The stress during the plating process can be significantly reduced if the plating temperature is reduced to an ambient temperature of about 24 °C. This eliminates thermal distortions stemming from the higher thermal expansion coefficient of PMMA compared with the wafer substrate. It also reduces the resist swelling in the electrolyte [43, 44]. Figure 10.10 shows distorted structures electroplated at 52 °C (a). The same structures are undeformed if plated at ambient temperature (b).

Thinner samples fabricated in 100 μm PMMA and 70 μm nickel are less prone to deformations than the previously mentioned samples; however, the capacitance decreases.

A major challenge is to avoid stiction problems of the cantilever beams. After etching of the sacrificial layer as the final fabrication step, the cantilever could easily stick to the capacitance electrode.

The high surface-to-volume ratio resulting from the scaling laws of miniaturization in MEMS devices makes surface phenomena such as stiction extremely important.

Stiction occurs if two criteria are present: mechanical collapse, and sufficiently strong inter-solid adhesion between the surfaces to hold the surfaces together. Mechanical collapse may occur during fabrication, especially due to capillary forces while drying. Mechanical collapse may also occur during operation of the device due to capillary forces from humidity condensation or due to accidental external forces [45–47].

Stiction in the capacitors that is caused by collapse during fabrication can be relieved by eliminating the mechanical collapse. This is achieved reducing or eliminating the capillary pull during drying by applying modified drying techniques (low surface tension rinsing liquid, dry release, freeze sublimation drying, supercritical drying, etc.) or a modified layout. A simple but not very repeatable alternative, that has been applied in some of the capacitor devices, is to apply an external force to detach the microstructure.

Stiction in the capacitors that is caused by collapse during operation can be relieved by reducing the surface energy so that the inter-solid adhesion can no longer hold the surfaces together. This is accomplished by increasing the surface roughness (which reduces the van der Waals forces) [48] or reducing the contact area in the layout.

Low surface energy hydrophobic coatings, such as self-assembled monolayer (SAM) coatings, are capable of eliminating capillary pull (reducing the contact angle due to hydrophobic properties) [49]. In the case of the capacitors, however, this approach has had only limited success.

10.3.1.3.3 Simulation and RF Test Results

Various varactors between 0.5 and 1.5 pF operating in the 1–5 GHz region were tested using a vector network analyzer connected using coaxial cables terminated with 150 μm pitch tungsten microprobes. DC control voltage was provided to the actuation electrode through one microprobe, and calibrated microwave one-port impedance measurements were obtained using the other microprobe connected to the capacitance electrode. Calibration of the test setup was performed using a commercially available calibration impedance standard.

The measured static capacitance and impedance as a function of frequency for a large varactor with a full-length capacitance electrode fabricated in 100 μm thick nickel and designed for a nominal capacitance of 1.1 pF are shown in the Smith charts in Figure 10.11.

Smith charts visualize impedances and therefore reveal information about Q-factors. They are typically normalized to 50 Ω impedance. The upper half of the

(a)

5 GHz
0.41 - j22.46 Ω
1.42 pF
Q = 55.4

3 GHz
0.41 - j45.55 Ω
1.16 pF
Q = 110.6

1 GHz
0.91 - j148.16 Ω
1.07 pF
Q = 162.4

(b)

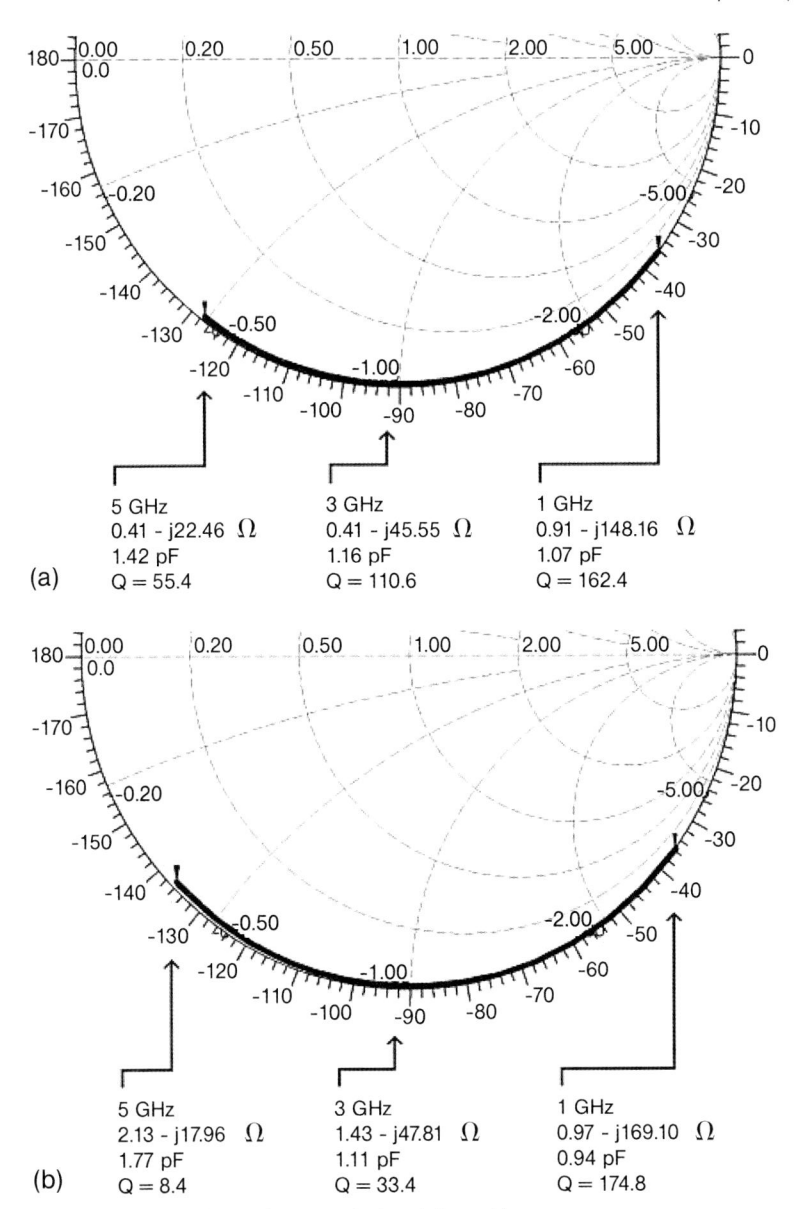

5 GHz
2.13 - j17.96 Ω
1.77 pF
Q = 8.4

3 GHz
1.43 - j47.81 Ω
1.11 pF
Q = 33.4

1 GHz
0.97 - j169.10 Ω
0.94 pF
Q = 174.8

Figure 10.11 Impedance of 100 μm thick nickel variable capacitor. (a) Simulation results; (b) measurements. From [32], reproduced with permission of Springer.

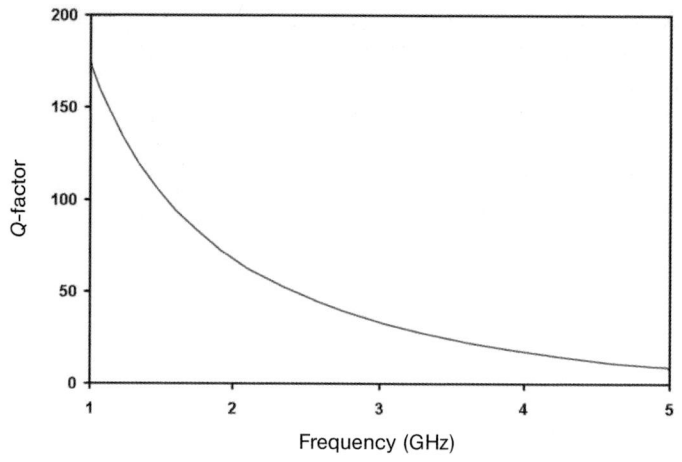

Figure 10.12 Measured averaged Q-factor values as a function of frequency for 100 μm thick nickel variable capacitor.

circle represents inductive behavior and the centerline represents the real part of the impedance (ohmic resistance); the lower half is shown here and represents capacitive behavior. Data points on the outer ring have no ohmic real part and therefore represent purely capacitive devices. An ideal capacitor would have no series resistance and data points would be located on the lower outer ring, moving from right to left with increasing operational frequencies. Their Q-factor would be infinite.

Figure 10.11 displays simulation data in (a) and corresponding measurements are displayed in (b) [32]. Figure 10.12 shows measured averaged Q-factor values as a function of the frequency.

The measurement results show (see Figure 10.11b) a capacitor with a low series resistance, as evidenced by the proximity of the data curve to the outer edge of the Smith chart. The Q-factor of the capacitor was determined using the ratio of the imaginary component of the impedance to the real component. It amounts to 174.8 and a capacitance of 0.94 pF at 1 GHz. At 3 GHz, the measured Q-factor is 33.4 and the capacitance is 1.11 pF. To assess these data better, the capacitor was simulated using the 3-D finite element software Ansoft HFSS (see Figure 10.11a). Simulation results of the same structure show a Q-factor of 162.4 and a capacitance of 1.07 pF at 1 GHz. This result agrees fairly well with the measured results. At higher frequencies, the measured and simulated results begin to differ slightly, possibly due to probing and calibration issues, including nickel top surface roughness and the oxide present on the surface of the nickel that was not included in the simulations.

A slightly smaller varactor also fabricated in 100 μm thick nickel and designed for a nominal capacitance of 0.8 pF was electrostatically actuated through the DC microprobe and tuned over its full capacitance range at 4 GHz. The variable capacitance ratio at 4 GHz is 1.42:1 from 0 to 20 V at a minimum capacitance of 0.59 pF when the cantilever contacts the actuation electrode. The worst case Q-factor, measured at the static capacitance of 0.84 pF, is 36 at 4 GHz.

These are already competitive results when compared with state-of the art silicon-based devices. The series resistance could be further decreased and the Q-factor increased if nickel was replaced by materials with 3–4 times higher conductivity, such as gold and copper as typically used in microwave devices. First samples plated in gold have been fabricated and are awaiting RF measurements.

10.3.2
Tall Coplanar Waveguide (CPW) 3 dB Couplers

10.3.2.1 Coupler Characteristics and Layout

Couplers are extensively used in RF applications for power combining and division. They are used, for example, in balanced mixers and amplifiers, as coupled resonant filter sections and for antenna array feed networks. They are generally of two types, realized either completely with lumped elements [13] or as distributed circuits [14]. Distributed transmission line-based circuits often have a performance advantage over lumped elements, offering, for instance, low loss and good port isolation at high frequencies. However, as demonstrated in Figure 10.4, miniaturization with traditional planar fabrication approaches is often a problem. One such application where compact couplers are critical is the 'smart antenna' electronic beam steering array, where many couplers feed transmit/receive modules (TRMs) [50] performing amplification and phase control. The spacing between TRMs is on the order of $\lambda/2$, presenting a practical integration problem at millimeter wave frequencies where complex networks must fit within in a few millimeters.

The 'coupled line' type of distributed coupler has no physical connection between the input and coupled output port, rather the power is transferred via electromagnetic field interaction through an air gap between transmission line sections in close proximity. The air gap can provide excellent signal isolation between different parts of a circuit, including blocking DC, a non-trivial problem, especially at millimeter wave frequencies. However, thin single-metal layer coplanar edge-coupled line sections generally provide fairly loose coupling levels within the 8–40 dB range [51]. In other words, typically <20% of the input signal power is transferred through the air gap with feasible layout geometries. Many applications require greater levels of coupling, up to 3 dB or 50% of the input power, so technologies providing these tighter coupling levels are desirable.

A 3 dB coupled line coupler operational around the center frequency of 18 GHz was developed. The device is similar in principal to the traditional quarter wavelength-long backward-wave directional coupler (for instance, as described in [52]), but with tall metal coplanar side-coupled conductors rather than thin edge-coupled planar conductors (see Figure 10.4b). Figure 10.13 depicts the rough layout of the device. The six gray areas represent electrically isolated, tall metal structures, four of which are grounded, the remaining two being line traces. Figure 10.13 also incorporates a functional diagram indicating the direction of power flow in the coupler.

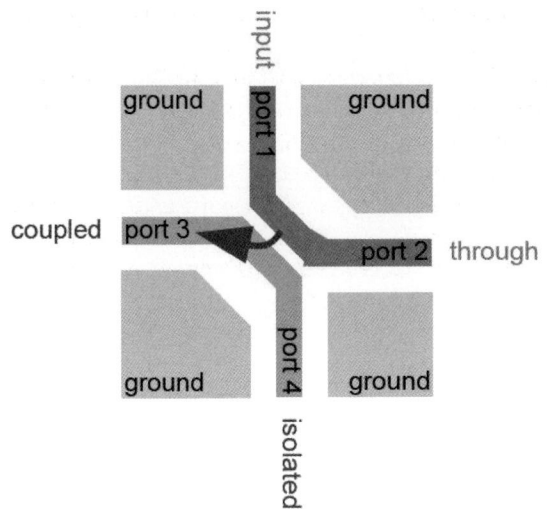

Figure 10.13 Component layout sketch and functional diagram of the backward-wave directional coupler (top view).

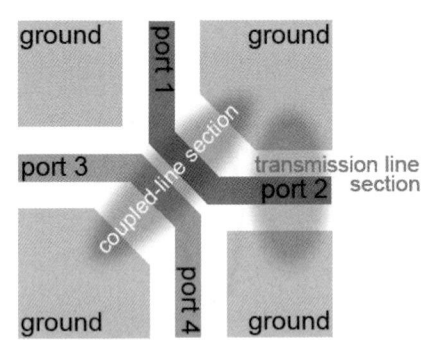

Figure 10.14 Component layout sketch with highlighted coupled-line section and one of four transmission line sections.

The diagram reveals that the coupler actually consists of two separate sections as highlighted in Figure 10.14. The areas in the vicinity of the four ports are controlled *characteristic impedance* (Z_0) CPW *transmission line sections* (comparable to, but smaller in size than, those presented later in [35]), required to transfer power effectively to/from the central area where the two lines meet, which is the actual *coupled-line section*.

At the center frequency where the coupling section is $\lambda/4$, the four-port device can theoretically be made simultaneously lossless, reciprocal and impedance matched to Z_0 at all four ports and the coupling designed so that half of the incident power at port 1 is transferred to port 2 at the end of the same line trace. The other 50% is coupled into the second line trace and fed back to port 3, while port 4 is isolated.

However, for this to be feasible, the characteristics of tall CPW coupled and transmission lines must be well understood. These entirely different waveguiding structures have not been analyzed in great detail to date and analysis is challenging since, for an accurate model, important geometric parameters (such as metal height, substrate thickness and ground width) cannot be assumed to be infinite or zero and most existing analysis techniques for planar CPW-based structures make these types of assumptions (for instance [53–55]). Therefore, the published techniques for analysis of planar CPW structures are generally not appropriate for tall CPW structures. To characterize the important characteristics required for the coupler design, parametric full-wave finite-element method (FEM) electromagnetic simulations on the tall CPW structures were performed.

The important characteristics of the structures are the *characteristic impedance*, Z_0, and *effective permittivity*, $\varepsilon_{r,eff}$, with the coupled-line section being described with a pair of each of these considering bimodal (even and odd mode) excitation ($Z_{0,even}$, $Z_{0,odd}$, $\varepsilon_{r,eff,even}$, $\varepsilon_{r,eff,odd}$). These are generally functions of four variables (trace width W, spacing S between the trace and the ground planes, conductor height H and substrate thickness T) for the transmission line and five variables (gap G between the traces in addition to the four variables listed above) for the coupled lines (see Figure 10.14).

Figure 10.15 presents simulation results revealing dependence of the characteristic impedance of the tall CPW transmission line on its trace width W for various conductor heights H at a fixed spacing S of 300 μm and substrate thickness T of 1 mm. The graph clearly demonstrates the ability of the tall CPW transmission line to attain standard characteristic impedance values at much lower values of the

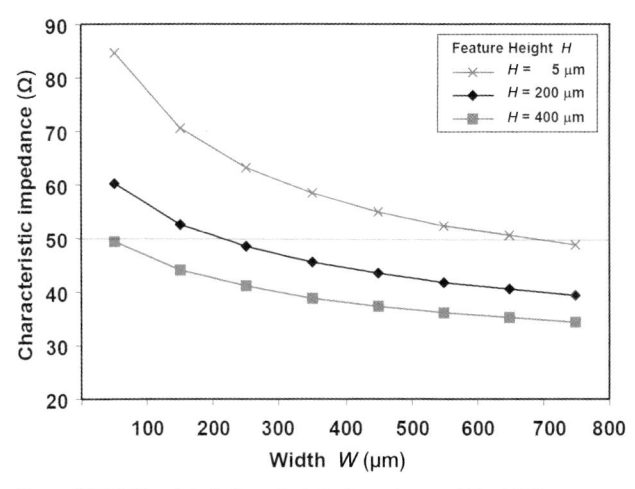

Figure 10.15 Simulated characteristic impedance of the LIGA tall CPW transmission line versus trace width W for various conductor heights H.

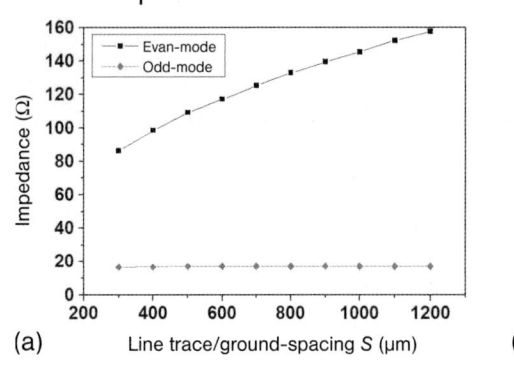

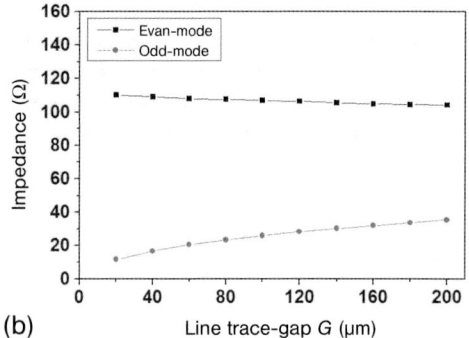

Figure 10.16 Simulated modal characteristic impedances for 200 μm nickel conductor height tall CPW coupled-line sections (top lines, even mode; bottom lines, odd mode). (a) Even mode affected by varying the spacing of line traces and the side grounds; (b) odd mode affected by varying the line traces gap.

trace width, miniaturizing the overall structure widthwise by up to a factor of 10, for instance for a 50 Ω transmission line with 400 μm height. This is understandable, as Z_0 is inversely proportional to transmission line capacitance per unit length and large capacitance is easily obtainable with tall LIGA structures.

Propagation properties in the even mode are mainly affected by the geometric configuration between the line traces and the side grounds. Propagation properties in the odd mode are mainly affected by the geometric configuration between the two line traces [33]. This is shown in Figure 10.16. For a given set of data and with conductor height of 200 μm, variations between the extreme values of spacing between line traces and the side grounds causes 82.2% change in the even-mode impedance, but only 4.5% change in the odd-mode impedance (Figure 10.16a). Alternatively, variations between the extreme values of the gap between the two trace lines causes only a −5.6% change in the even-mode impedance, but a 207% change in the odd-mode impedance (Figure 10.16b).

The geometric data used to design the 18 GHz coupler with nominally 220 μm tall CPW are given in Table 10.2.

10.3.2.2 Coupler Fabrication

The fabrication sequence generally followed that outlined in the capacitors section. A 1 mm thick alumina wafer (ε_r = 9.8 at 10 GHz) was coated with a 3 μm Ti/TiO$_x$ plating base, followed by application of a 340 μm thick PMMA plate (Röhm GS 233). The resist was exposed with a dose deposition of 9.6 kJ cm^{-3} at the top and 3.5 kJ cm^{-3} at the bottom. It was dip-developed for 7.3 h in GG developer. Developed voids were electroplated with nickel to a height of 220 μm. The polymer template was only filled to two-thirds of its height to ensure that the template would not accidentally be overplated, which might electrically short the device. Before being stripped away, the remaining resist underwent flood X-ray exposure to ensure its solubility in GG developer. Finally, the conductors were electrically

Table 10.2 Design data for a 50 Ω tall CPW coupler (transmission line sections and 3 dB coupled-line section).

Parameter	Transmission line sections	3 dB coupled-line section
W (μm)	200	200
S (μm)	250	560
G (μm)	—	60
H (μm)	220	220
T (μm)	1000	1000
Length (μm)	—	2260
Z_0 (Ω)	50	50
$Z_{0,even}$ (Ω)	—	121
$Z_{0,odd}$ (Ω)	—	21
$\varepsilon_{r,eff}$	3.6	—
$\varepsilon_{r,eff,even}$	—	4.3
$\varepsilon_{r,eff,odd}$	—	2.7

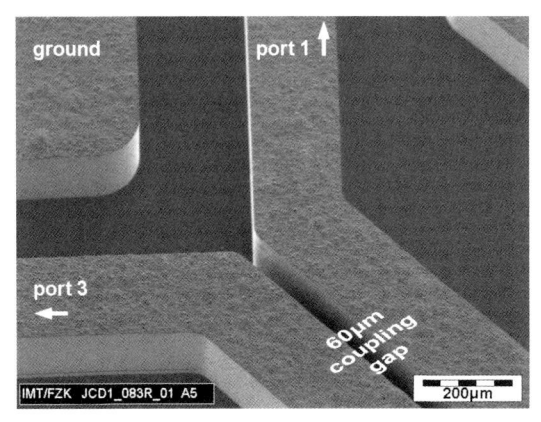

Figure 10.17 SEM image of the 3 dB coupler. 220 μm high Ni conductors on alumina substrate, 60 μm wide gap between conductors.

isolated by removing the plating base with 5% HF for 3 min. To remove any eventual metal residues potentially circumventing electrical isolation, the sample was subsequently exposed to reactive ion etching (RIE) in argon for 10 min.

Figure 10.17 shows an SEM image of such a device. The conductor side-walls have an insignificant roughness typical of the LIGA process. The conductor top surface, however, is rather rough as a result of the plating process. Profiler measurements revealed a *mean roughness* R_a within 0.3 μm and 1.3 μm with typical *peak-to-valley roughness* R_t values between 3 and 6 μm.

This surface turned out to be too hard and rough for standard three-pin GSG beryllium–copper (BeCu) probes, which were easily broken during surface engagement. The sample surface therefore needed to be smoothed. Such a polishing step was not implemented from the beginning as polishing always bears the risk of structure quality deterioration (surface rounding, breaking of structures, etc.). In upcoming samples, such a polishing step would probably be introduced directly after electroplating. With this sample, however, polishing had to be added to the stripped sample. Therefore, low molecular weight PMMA was poured on to the conductors and evacuated during polymerization in order to generate a non-porous lateral support around the conductors. This support was meant to protect the conductors from excessive tangential forces during polishing and also to prevent infiltration of polishing compound into the microstructures. Non-cross-linked PMMA was chosen to facilitate final removal of the resist. The coated and supported conductors were polished for 30 min with acrylic glass polishing compound and rinsed. The PMMA was finally removed by X-ray flood exposure to 4 kJ cm^{-3}, dip development in agitated GG developer and excessive rinse in running deionized water. The surface still had minor scratches, but the roughness had generally decreased to $R_a \approx 0.04\,\mu m$ and $R_t \approx 0.2\,\mu m$. RF measurements could be carried out on this surface.

10.3.2.3 Coupler Theoretical and RF Test Results

For the quarter wavelength long coupled-line section alone without connecting transmission lines at the ports, the theoretical coupling response from ports 1 to 3 (S_{31}) and through response from ports 1 to 2 (S_{21}) are given by [56]

$$S_{21} = \frac{\sqrt{1-k^2}}{\sqrt{1-k^2}\cos\theta + j\sin\theta} \quad (5)$$

$$S_{31} = \frac{jk\sin\theta}{\sqrt{1-k^2}\cos\theta + j\sin\theta} \quad (6)$$

where k is the required coupling coefficient at the center frequency and θ is the electrical length of the coupled-line section. The theoretical performance of the coupled-line section versus frequency for an 18 GHz center frequency with 3 dB coupling is shown in Figure 10.18. It suggests that a broadband coupling response over 12–24 GHz is possible, provided that tight coupling in the coupled-line section can be realized. The required level of coupling, $S_{31} = k$ at the center frequency, can be obtained if the following modal impedances are realized in a coupled-line section:

$$Z_{0,even} = Z_0\sqrt{\frac{1+k}{1-k}} \quad (7)$$

$$Z_{0,odd} = Z_0\sqrt{\frac{1-k}{1+k}} \quad (8)$$

In this example with 3 dB coupling, even- and odd-mode impedances of 121 and 21 Ω, respectively, must be realized in the coupled-line section. The low odd-mode

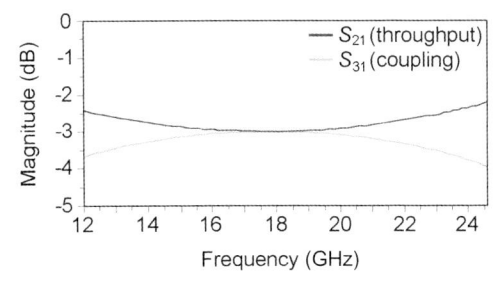

Figure 10.18 Theoretical performance of the coupled-line section without transmission lines connected at the ports.

Figure 10.19 Top view of RF wafer probing measurement setup.

impedance is especially difficult to realize in planar CPW. Figure 10.16 suggests that with tall CPW, metal conductor heights of several hundred micrometers lead to almost independent control over the modal impedances and therefore an extended flexibility in the design.

Additionally, extremely low odd-mode impedance values can readily be obtained with LIGA technology, a prerequisite for tight coupling as explained above.

RF measurements were performed on the fabricated coupler using a setup comparable to that described in the capacitors section. Figure 10.19 is a top view micrograph of a two-port network analyzer wafer probing setup. The three-pin GSG wafer probes are pressed against the top surface of the sample, in this case a tall $50\,\Omega$ LIGA-built CPW calibration standard.

Measurement results are shown in Figure 10.20 over the same bandwidth as for the theoretical results in Figure 10.18. The coupler structure provides tight coupling and close to the theoretical values, which is especially encouraging given that the polishing likely altered the modal characteristic impedances. In addition, the theoretical coupling model does not take into account loss and discontinuity effects from the transmission feedlines. The maximum coupling is demonstrated at 18.2 GHz (slightly off frequency from 18 GHz) and it is −4.07 dB, which is about 1 dB less than ideal. In addition, the device has very good broadband performance.

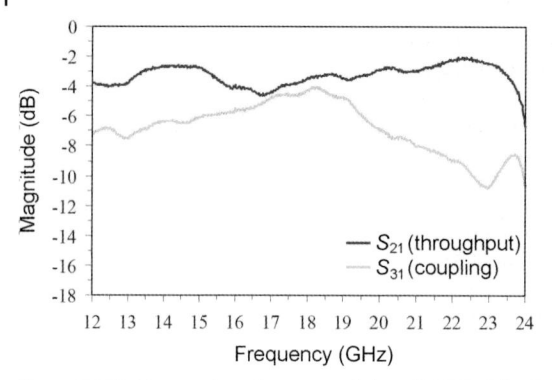

Figure 10.20 Measured performance of the tall LIGA tight CPW coupler.

With respect to the maximum coupling, the 3 dB operating bandwidth is 6.8 GHz and spans from 13.3 to 20.1 GHz.

10.3.3
Cavity Resonators

10.3.3.1 Cavity Resonator Characteristics

One of the most important components for RF applications is the resonator. In transceiver circuits, it determines the RF oscillator frequency and the phase-noise performance [57] and also forms the building blocks for filters and diplexers.

Metallic waveguide cavities have long been among the highest performing resonators, typically having low loss, high Q-factor and high power handling capability. At millimeter wave frequencies, the cavity size becomes small enough that batch fabrication for high-volume commercial applications using microfabrication techniques becomes feasible.

Silicon bulk micromachining has become one popular approach for fabricating RF cavity resonators and filters in the 10–40 GHz range using both anisotropic wet etching [19, 20] and also deep RIE [16] techniques. However, silicon micromachining does not necessarily produce rectangular cavities with vertical and smooth side-walls, potentially leading to increased metal conductor loss and a decrease in Q-factor performance. For instance, anisotropic wet etching can result in side-walls inclined at 54.7° for wafers cut in the (100)-plane. Silicon micromachining furthermore limits the Q-factor as the cavity depth is restricted to the thickness of standard silicon wafers (525–650 μm), unless several wafers are stacked [20].

Polymer-based LIGA fabrication is a promising alternative for ultra-deep microwave cavity structures with excellent side-wall structural and surface quality.

A typical resonator is a box-like, metallic rectangular cavity that can be either air or dielectric filled. Its *resonant frequency*, f_{res}, s related to the physical dimensions of the cavity:

$$f_{res} = \frac{c}{2\pi\sqrt{\mu_r \varepsilon_r}} \cdot \sqrt{\left(\frac{m\pi}{w}\right)^2 + \left(\frac{n\pi}{h}\right)^2 + \left(\frac{p\pi}{l}\right)^2} \qquad (9)$$

10.3 Sample Components | 271

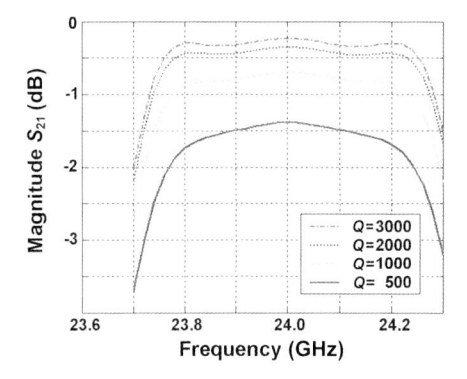

Q-factor	500	1000	2000	3000
Insertion Loss IL [dB]	1.4	0.7	0.3	0.2
Power Transfer P_L/P_{AV} [%]	72	85	93	96

Figure 10.21 Simulated three-pole resonant cavity-based bandpass filter at 24 GHz illustrating the effect of resonator cavity Q-factor on filter response.

with length l, width w and height h. Materials properties of the volume inside the cavity are given by the relative permeability μ_r and the relative permittivity ε_r; c represents the speed of light. The indices for the dominant transverse electric (TE) field TE_{101} mode are $m = 1$, $n = 0$ and $p = 1$.

Smaller dimensions correspond to higher frequencies. An air-filled cavity of 8.83 mm length and width would, for instance, resonate at 24 GHz [58].

An important figure of merit for resonant circuits is the Q-factor, which was defined earlier as the ratio of the average energy stored to the energy lost per cycle. Figure 10.21 illustrates the practical importance of a high resonator Q-factor, with an example of a simulated three-pole coupled resonator bandpass filter at 24 GHz. It is clear that the filter loss (insertion loss IL) decreases significantly, by over 1 dB, with increasing Q values. Expressed in terms of power transfer, the filter with Q-factor resonators of 500 (actually a very respectable resonator Q) transfers only 72% of the available power P_{AV} from the source to the load P_L and 28% of the power is dissipated by the filter. In contrast, a filter with Q-factor resonators of 2000 transfers 93% of the available power from the source to the load and only 7% of the power is dissipated by the filter.

The theoretical unloaded Q-factor at resonance is determined by evaluating the average electric energy stored in the cavity and the loss by the finite conductivity of the cavity walls. The unloaded Q-factor of the cavity represents the resonator itself without any further circuit loading. With ideally smooth but lossy conducting walls and a lossless dielectric, Q_u for TE_{101} mode is

$$Q_u = \frac{(kwl)^3 \cdot h\eta}{2\pi^2 R_{su}} \cdot \frac{1}{2w^3h + 2l^3h + w^3l + wl^3} \tag{10}$$

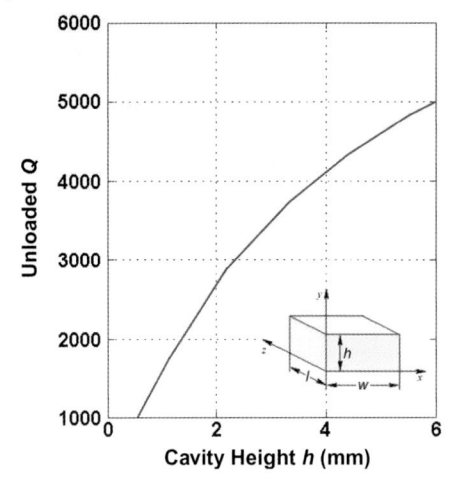

Figure 10.22 Theoretical unloaded quality factor Q_u as a function of the cavity height h. The cavity has a width w and a length l of 8.83 mm, ideally smooth and vertical side-walls and surfaces and a resonant frequency of $f_{res} = 24$ GHz. At $h > 1$ mm, theoretical $Q_u > 1500$.

with wavenumber k, intrinsic impedance η and surface resistivity of the metal walls R_{su}. For air, $k = \omega/c$ and $\eta = 377\,\Omega$.

From this equation, Figure 10.22 depicts the theoretical Q_u, for smooth but lossy conducting walls (assumed in this case to be gold with a conductivity of 4.1×10^7 S m^{-1}) and with a lossless air dielectric, as a function of the cavity height for the above mentioned resonator with 8.83 mm length and width. Clearly, the Q-factor increases dramatically as a function of increasing cavity height and, assuming some margin for conductor loss due to surface roughness and cavity irregularity, Figure 10.22 suggests that obtaining a Q_u of 2000 or more is possible at 24 GHz with cavity depths of at least 1500 μm, well within the fabrication capabilities of LIGA. LIGA capabilities would also allow the development of more complicated waveguide cavity structures, with fine internal features such as the 'split-post' cylindrical post coupled filter proposed in [59].

10.3.3.2 Cavity Resonator Fabrication

Miniaturized cavity resonators could benefit from the specific advantages of LIGA structures. However, the requirements imposed by cavities differ from most MEMS applications. While extremely high metal structures are required (on the order of millimeters), the lateral feature is not as demanding as in other devices (also on the order of millimeters, provided that there are no additional internal features). Required materials properties, such as high electrical conductivity on the inside of the patterned metal walls, lead to high-conductivity metals such as gold instead of nickel. Finally, the metal walls are solely conductors and basically do not need any mechanical strength other than to support themselves.

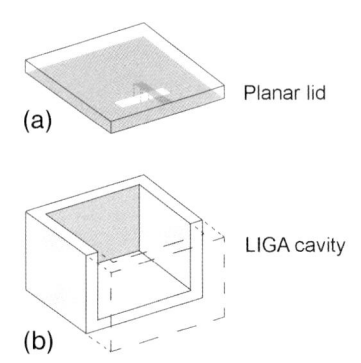

Figure 10.23 Sketch of a fabricated cavity resonator. Part (b) represents the LIGA cavity consisting of four metal walls and a metal base and (a) represents the lid as the sixth cavity wall, incorporating a microstrip transmission line and a ground plane slot line to externally excite the resonator.

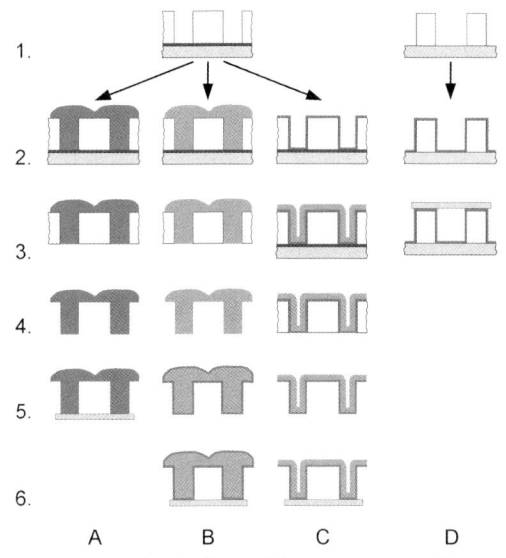

Figure 10.24 Sketch of main fabrication steps of four potential processing sequences (A–D) for the LIGA cavity.

Figure 10.23 visualizes such a cavity resonator. Figure 10.23b represents the high aspect ratio LIGA cavity consisting of four metal walls and a metal base. Fabrication of the cavity (as represented by the white cross-section) will be outlined in Figure 10.24. Inner dimensions for the 24 GHz cavity are 8.83 mm in length and width and heights on the order of 2 mm. This feature height is a compromise between device performance on the one hand and processing reliability, processing time and structure quality on the other [31]. Figure 10.23a represents the lid as the sixth cavity wall. It is fabricated by mechanical micromachining and incorporates a microstrip transmission line and a slot line in the metallized bottom electrical ground plane surface to externally excite the resonator.

Electroplating of the high aspect ratio metal conductors is a key process in this context, partly because massively plated gold walls would drive the device costs, partly because electroplated layers grow comparatively slowly. With typical growth rates on the order of a few micrometers per hour, the plating process could easily take weeks.

Standard LIGA fabrication sequences might be altered to reflect these unusual requirements. Figure 10.24 depicts up to six main steps in four different processing sequences, A, B, C and D.

Sequences A–C are based on 2 mm thick PMMA resist plates (Röhm GS 233) glued to a 3 μm Ti/TiO$_x$ plating base and sacrificial layer which was coated to a 4″ silicon monitor wafer.

X-ray lithography of such thick resist layers (fabrication step 1 in Figure 10.24) requires a special precaution in order to avoid a significant decline in structure quality, such as loss of resist adhesion to the substrate, massive cracking of the resist, foaming of the resist, reduction in structure accuracy and side-wall verticality, excessive resist surface rounding and resist surface attack [31].

This precaution includes a reasonable choice of X-ray mask technology. For the devices presented here, a titanium-based mask was used. Masks based on titanium membranes emit fluorescence radiation that attacks the resist surface. They also tend to heat up in very intense synchrotron beams, leading to thermal distortions of the microstructures and foaming of the resist. A better suited mask membrane material for these types of exposures would be, for instance, beryllium [60]. Unfortunately, a beryllium-based mask was not available. In order to reduce the impact of fluorescence radiation, a layer of 500 μm polyimide was placed between titanium mask membrane and resist surface. The only beamline with a sufficiently hard spectral distribution to expose 2 mm of PMMA at the electron storage ring ANKA is the white light beamline Litho 3. However, its incident beam power of up to 100 W is detrimental when using titanium masks. To reduce the incident power significantly, a metal preabsorber was placed in the beamline. Although this filtering helps to reduce thermal effects in the mask and sample, it also further hardens the spectrum. Consequently, the mask absorber opacity shrinks, reducing the mask contrast (ratio of dose deposition behind transparent to that behind opaque mask areas). Ultimately, this might result in a loss of resist adhesion [61]. In the case of the cavity, however, adhesion of the resist to the substrate does not have to be excellent, because the layout may avoid high aspect ratios and small feature sizes in PMMA mesas and because adhesion is not required for device operation.

The sample was exposed to a dose deposition of 4.6 kJ cm^{-3} at the resist top and 3.2 kJ cm^{-3} at the bottom. Immediately after exposure, dip development in GG developer was applied for 142 h.

After the lithographic resist patterning, a straightforward approach would be as sketched in sequence A, step 2, in Figure 10.24. The polymer voids are directly electroplated with gold. They become overplated until the entire PMMA cavity mesa surface is coated with a structural layer of metal to form later the fifth wall of the cavity. Some of the drawbacks of this approach, such as materials cost and plating time requirements, have already been mentioned. Further than that, little or no experience is available for gold plating of such very thick high aspect ratio microstructures. To allow a more uniform growth rate distribution, unipolar pulse plating was applied in a first test sample fabricated. The pulse on/off ratio was adapted with increasing metal height, depending on the gold concentration in the sulfite-based electrolyte [62].

The third fabrication step is to remove the substrate either by directly dissolving the sacrificial titanium layer in HF or by dissolving the silicon wafer in KOH prior to titanium etching.

The fourth fabrication step removes the remaining PMMA template, for instance by flood exposure from the back side of the sample and subsequent development. This would constitute a five-walled gold cavity with lithographically smooth side-walls.

The sixth side is fabricated separately and added to the cavity (fifth fabrication step).

Sequence B suggests massively plating nickel instead of gold. This basically is an established process and circumvents gold plating issues and high material costs. This process would still take a very long time. The bigger drawback, however, results from fabrication step 5: in order to obtain higher conductivity inner cavity walls, the nickel walls need to be coated with a thin layer of gold. This layer needs to be several skin depths deep, on the order of 2–3 μm. A film of this thickness should not be applied in thin-film techniques, such as sputtering or evaporation. The nickel structure consequently needs to be electroplated. This, however, significantly degrades the side-wall surface roughness, which is critical for the device performance. It would no longer be of lithographic quality; instead, the field distribution during plating to a certain extent determines the growth rate and therefore the surface topology of the sample.

In the third and more adapted sequence C, the second fabrication step consists of applying a thin gold film to the patterned resist surface. This could be achieved, for instance, by multiple sputtering of approximately 50 nm under oblique angles. A drawback of this approach is the marginal adhesion of gold on PMMA. Subsequently, the gold thickness is increased to about 5 μm applying a hard gold alloy by means of thin-film electroplating at 55 °C [62]. In a third fabrication step, the thin gold film becomes supported by electroplating a structural layer of nickel. This is a fast process as the metal does not have to grow all the way from the substrate or laterally overplate the resist mesa surface. Plating starts at all points simultaneously, which also avoids the formation of seams in the middle of the overplated mesa surface. A few hundred micrometers are sufficient. Some blow holes might become trapped inside the nickel cavity walls, which would not affect the performance. Finally, the resist becomes stripped and the lid is added as the sixth side.

In an alternative fabrication sequence D, a negative-tone resist, such as Epon SU-8, is directly coated with a gold film of a few micrometers thickness. Compared with PMMA, gold adheres well on SU-8. This epoxy-based polymer is very resistant to aging and chemical attack and may therefore remain part of the device. Conse-

Figure 10.25 SEM image of a 2 mm tall PMMA cavity template prior to metallization.

quently, this short fabrication sequence is immediately concluded by adding the sixth side. Although this sequence offers the advantages of simplicity and rapid processing, drawbacks include a significant challenge to pattern such thick SU-8 layers. Furthermore, the inner cavity surface is not a lithographic one and will therefore be comparatively rough.

Metal cavities were fabricated according to sequence C in Figure 10.24. Figure 10.25 shows a 2 mm tall PMMA cavity template prior to metallization with gold and nickel. Such templates have been measured for geometric distortions [63]. The PMMA top surface (metal base) can shrink up to 12–16 μm, primarily under thermal impact related to the X-ray mask.

Such a distortion would be detrimental for many applications. In the case of the cavity, however, this translates into a relative shrinkage of only 0.14–0.18%. Simulations revealed that the impact of such distortions on the cavity resonant frequency and cavity quality factor are marginal [58]. Figure 10.26 illustrates the results for isotropic shrinkage (a) and anisotropic shrinkage (b).

Using sequence C in Figure 10.24, a slightly shorter (approximately 1.8 mm tall) cavity with a gold inner surface was fabricated and tested. To facilitate microwave testing of the five-sided metallic LIGA cavity, an available ceramic-PTFE composite substrate with $\varepsilon_r = 10.2$ at 10 GHz was applied. Both sides are coated with comparatively rough rolled copper ($R_{rms} \approx 300$ nm), perhaps an order of magnitude rougher than the other five LIGA surfaces. The copper was mechanically micromachined to fabricate an open-circuit microstrip transmission feed line and ground plane coupling slot (see Figure 10.23a). The ground plane side (with the field excitation slot) was clamped to the LIGA cavity and formed the sixth side, thus completing the cavity.

Excellent resonant performance of the LIGA cavity was measured, showing a very strong resonance at 23.86 GHz, very close to the desired frequency of 24 GHz.

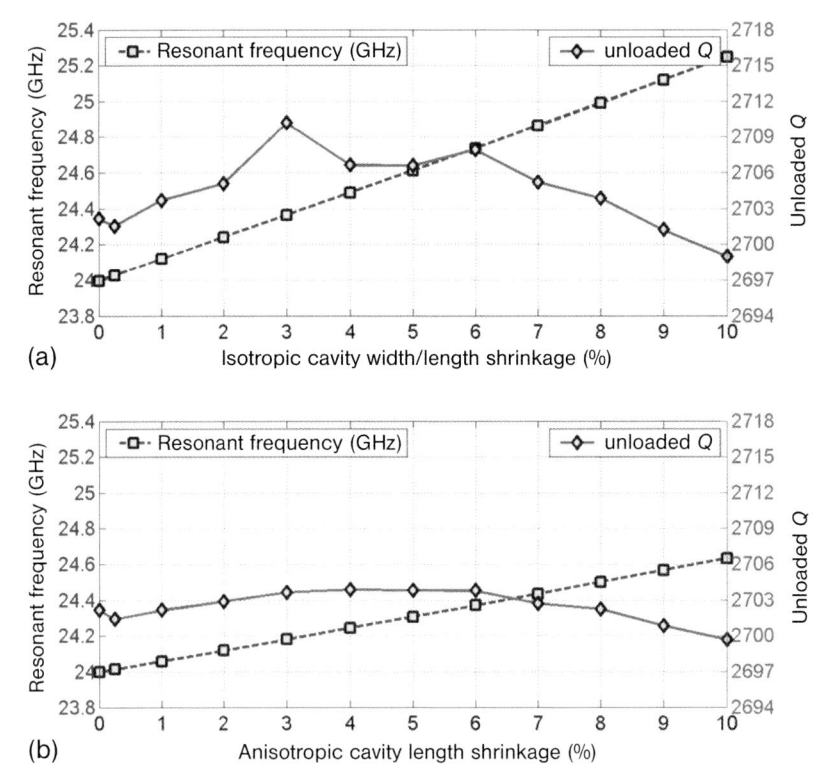

Figure 10.26 Simulated resonant frequency f_{res} and unloaded quality factor Q_u as a function of the cavity structure quality. Cavity nominal dimensions of $8.83 \times 8.83 \times 2\,mm$, nominal resonant frequency of 24 GHz. The cavity is assumed to have a top and bottom substrate surface roughness of $R_{rms} = 40\,nm$ and a side-wall roughness of $R_{rms} = 30\,nm$. Cavity bottom is at nominal size. Side-wall taper leads to shrinkage at the cavity top surface: (a) isotropic shrinkage of square top surface; (b) anisotropic shrinkage in only one dimension of rectangular top surface. From [58], reproduced with permission of Wiley-VCH Verlag GmbH.

Unloaded Q_u was on the order of 1800, which is very respectable, considering the reduced cavity height and relatively rough sixth side test fixture surface.

Acknowledgments

The authors would like to thank their colleagues for the invaluable support without which this work could not have been performed. In particular, Bärbel Rapp, Barbara Matthis, Martin Börner, Nina Dambrowsky and Timo Mappes at FZK/IMT, and Anton Kachayev, Darcy Haluzan, Garth Wells, Venkat Subramanian and Zhen Ma at the University of Saskatchewan and TRLabs in Saskatoon.

References

1 WTC (2005) RF MEMS Market II, 2005–2009. WTC Wicht Technologie Consulting, Munich.
2 Larson, L.E., Hackett, R.H. and Lohr, R.F. (1991) *Micromachined microwave actuator (MIMAC) technology – a new tuning approach for microwave integrated circuits*, in Proceedings of IEEE Microwave Millimeter-wave Monolithic Circuit Symposium, pp. 27–30.
3 Rebeiz, G.M. (2003) *RF MEMS – Theory, Design and Technology*, John Wiley & Sons, Inc., Hoboken, NJ.
4 Goldsmith, C.L., Yao, Z., Eshelman, S. and Denniston, D. (1998) Performance of low-loss RF MEMS capacitive switches. *IEEE Microwave and Guided Wave Letters*, **8**, 269–71.
5 Park, J.Y., Kim, G.H., Chung, K.W. and Bu, J.U. (2000) Fully integrated micromachined capacitive switches for RF applications. IEEE MTT-S International Microwave Symposium Digest, Vol. 1, pp. 283–6.
6 Mihailovich, R., Kim, M., Hacker, J., Sovero, E., Studer, J., Higgins, J. and DeNatale, J. (2001) MEM relay for reconfigurable RF circuits. *IEEE Microwave and Wireless Components Letters*, **11**, 53–5.
7 Dec, A. and Suyama, K. (1998) Micromachined electro-mechanically tunable capacitors and their applications to RF IC's. *IEEE Transactions on Microwave Theory and Techniques*, **46**, 2587–96.
8 Young, D.J. and Boser, B.E. (1996) A micromachined variable capacitor for monolithic low-noise VCOs, in Solid State Sensor and Actuator Workshop, Hilton Head, SC, pp. 86–9.
9 Jou, J., Liu, C. and Schutt-Aine, J. (2001) Development of a wide tuning range two-parallel-plate tunable capacitor for integrated wireless communication systems. *International Journal of RF and Microwave Computer Aided Design*, **11**, 322–9.
10 Ribas, R., Lescot, J., Leclercq, J., Karam, J.M. and Ndagijimana, F. (2000) Micromachined microwave planar spiral inductors and transformers. *IEEE Transactions on Microwave Theory and Techniques*, **48**, 1326–35.
11 Ermolov, V., Lindström, T., Nieminen, H., Olsson, M., Read, M., Ryhänen, T., Silanto, S. and Uhrberg, S. (2004) Microreplicated RF toroidal inductor. *IEEE Transactions on Microwave Theory and Techniques*, **52**, 29–37.
12 Chi, C.Y. and Rebeiz, G.M. (1997) Design of Lange-couplers and single-sideband mixers using micromachining techniques. *IEEE Transactions on Microwave Theory and Techniques*, **45**, 291–4.
13 Lu, L., Bhattacharya, P. and Katehi, L.P.B. (2000) X-band and K-band lumped Wilkinson power dividers with a micromachined technology, in IEEE MTT-S International Microwave Symposium, Boston, MA, pp. 287–90.
14 Ozgur, M., Zaghloul, M.E. and Gaitan, M. (2000) Micromachined 28-GHz power divider in CMOS technology. *IEEE Microwave and Guided Wave Letters*, **10**, 99–101.
15 Brown, A.R. and Rebeiz, G.M. (1999) A high-performance integrated-band diplexer. *IEEE Transactions on Microwave Theory and Techniques*, **47**, 1477–81.
16 Harle, L. and Katehi, L.P.B. (2004) A silicon micromachined four-pole linear phase filter. *IEEE Transactions on Microwave Theory and Techniques*, **52**, 1598–607.
17 Nguyen, C. (1999) Frequency-selective MEMS for miniaturized low-power communication devices. *IEEE Transactions on Microwave Theory and Techniques*, **47**, 1486–503.
18 Kwon, Y., Cheon, C., Kim, N., Kim, C., Song, I. and Song, C. (1999) A Ka-band MMIC oscillator stabilized with a micromachined cavity. *IEEE Microwave and Guided Wave Letters*, **9**, 360–2.
19 Papapolymerou, J., Cheng, J., East, J. and Katehi, L.P.B. (1997) A micromachined high-Q X-band resonator. *IEEE Microwave Guided Wave Letters*, **7**, 168–70.
20 Stickel, M., Eleftheriades, G.V. and Kremer, P. (2001) High-Q bulk micromachined silicon cavity resonator at Ka-band. *Electron Letters*, **37**, 433–5.

21 Herrick, K.J., Schwarz, T.A. and Katehi, L.P.B. (1998) Si-micromachined coplanar waveguides for use in high-frequency circuits. *Transactions on Microwave Theory and Techniques*, **46**, 762–9.

22 Jeong, I., Shin, S.H., Go, J.H., Lee, J.S., Nam, C.M., Kim, D.W. and Kwon, Y.S. (2002) High-performance air-gap transmission lines and Inductors for millimeter-wave applications. *Transactions on Microwave Theory and Techniques*, **50**, 2850–5.

23 Reid, J.R., Marsh, E.D. and Webster, R.T. (2006) Micromachined rectangular-coaxial transmission lines. *Transactions on Microwave Theory and Techniques*, **54**, 3433–42.

24 Ruppel, C.W., Reindl, L. and Weigel, R. (2002) SAW devices and their wireless communications applications. *IEEE Microwave Magazine*, **3**(2), 65–71.

25 De Los Santos, H.J. (1999) *Introduction to Microelectromechanical (MEM) Microwave Systems*, Artech House, Boston, MA.

26 Yao, J.J. (2000) RF MEMS from a device perspective. *Journal of Micromechanics and Microengineering*, **10**, R9–38.

27 Katehi, L.P.B., Harvey, J.F. and Brown, E. (2002) MEMS and Si micromachined circuits for high frequency applications. *IEEE Transactions on Microwave Theory and Techniques*, **50**, 858–66.

28 Yao, J. and Park, S. and DeNatale, J. (1998) High tuning ratio MEMS based tunable capacitors for RF communications applications, in Solid State Sensor and Actuator Workshop, Hilton Head, SC, pp. 124–7.

29 Borwick, R.L., III, Stupar, P.A., DeNatale, J., Anderson, R., Tsai, C., Garrett, K. and Erlandson, R. (2003) A high Q, large tuning range MEMS capacitor for RF filter systems. *Sensors and Actuators A*, **103**, 33–41.

30 Mohr, J., Ehrfeld, W. and Münchmeyer, D. (1988) Analyse der Defektursachen und der Genauigkeit der Strukturübertragung bei der Röntgentiefenlithographie mit Synchrotronstrahlung. KfK 4414, Kernforschungszentrum Karlsruhe, Karlsruhe.

31 Achenbach, S., Pantenburg, F.J. and Mohr, J. (2000) Optimization of the Process Conditions for the Fabrication of Microstructures by Ultra Deep X-Ray Lithography (UDXRL). FZKA 6576, Kernforschungszentrum Karlsruhe, Karlsruhe.

32 Achenbach, S., Klymyshyn, D., Haluzan, D., Mappes, T., Wells, G. and Mohr, J. (2007) Fabrication of RF MEMS variable capacitors by deep X-ray lithography and electroplating. *Microsystems Technologies*, **13**, 343–8.

33 Kachayev, A., Klymyshyn, D., Achenbach, S. and Saile, V. (2003) High vertical aspect ratio LIGA microwave 3-dB coupler, in Proceedings of ICMENS 2003 – International Conference on MEMS, NANO and Smart Systems, Banff, pp. 38–43.

34 Willke, T.L. and Gearhart, S.S. (1997) LIGA micromachined planar transmission lines and filters. *IEEE Transactions on Microwave Theory and Techniques* **45**, 1681–8.

35 Forman, M. (2005) Low-loss LIGA-micromachined conductor-backed coplanar waveguide, in Book of Abstracts, High Aspect Ratio. Micro Structure Technology Workshop, Gyeongju, Korea, pp. 20–1.

36 Casse, B., Moser, H.O., Jian, L.K., Bahou, M., Wilhelmi, O., Saw, B.T. and Gu, P.D. (2006) Fabrication of 2D and 3D electromagnetic metamaterials for the terahertz range. *Journal of Physics*, **34**, 885–90.

37 Haluzan, D. (2004) Microwave LIGA-MEMS variable capacitors. MSc Thesis, University of Saskatchewan, Saskatoon, Canada.

38 Feng, Z., Zhang, H., Gupta, K.C., Zhang, W., Bright, V.M. and Lee, Y.C. (2001) MEMS-based series and shunt variable capacitors for microwave and millimeter-wave frequencies. *Sensors and Actuators A*, **91**, 256–65.

39 Dussopt, L. and Rebeiz, G.M. (2002) High-Q millimeter-wave MEMS varactors: extended tuning range and discrete-position designs, in IEEE MTT-S International Microwave Symposium Digest, 1205–8.

40 Haluzan, D. and Klymyshyn, D. (2004) High-Q LIGA-MEMS vertical cantilever variable capacitors for upper microwave frequencies. *Microwave and Optical Technology Letters*, **42**, 507–11.

41 Schomburg, W., Bley, P., Hein, H. and Mohr, J. (1990) Masken für die Roentgentiefenlithographie. *VDI Berichte*, **870**, 133–54.

42 Pantenburg, F.J., Achenbach, S. and Mohr, J. (1998) Influence of developer temperature and resist material on the structure quality in deep X-ray lithography. *Journal of Vacuum Science & Technology*, **B16**(6), 3547–51.

43 Ruzzu, A. and Matthis, B. (2002) Swelling of PMMA-structures in aqueous solutions and room temperature Ni-electroforming. *Microsystem Technologies*, **8**, 116–19.

44 Aigeldinger, G., Ceremuga, J.T. and Krenz, K.D. (2004) In-situ metrology of swelling effects on critical dimensions in LIGA PMMA molds. Proceedings, ASPE 2004 Annual Meeting, Orlando, FL.

45 Mastrangelo, C.H. (2000) Suppression of stiction in MEMS. *Proceedings of Materials Research Society Symposium*, **605**, 105–16.

46 Kim, B.H., Chung, T.D., Oh, C.H. and Chun, K. (2001) A new organic modifier for anti-stiction. *Journal of Microelectromechanical Systems*, **10**, 33–40.

47 Israelachvili, J.N. (1985) *Intermolecular and Surface Forces*, Academic Press, New York.

48 Tas, N., Sonnenberg, T., Jansen, H., Legtenberg, R. and Elwenspoek, M. (1996) Stiction in surface micromachining. *Journal of Micromechanics and Microengineering*, **6**, 385–97.

49 Zhao, Y. (2003) Stiction and anti-stiction in MEMS and NEMS. *Acta Mechanica Sinica/Lixue Xuebao*, **19**(1), 1–10.

50 Kopp, B.A., Borkowski, M. and Jerinic, G. (2002) Transmit/receive modules. *IEEE Transactions on Microwave Theory and Techniques*, **50**, 827–34.

51 Niehenke, E., Pucel, R. and Bahl, I. (2002) Microwave and millimeter-wave integrated circuits. *IEEE Transactions on Microwave Theory and Techniques*, **50**, 846–57.

52 Mongia, R., Bahl, I. and Bhartia, P. (1999) *RF and Microwave Coupled-line Circuits*, Artech House, Norwood, MA.

53 Wen, C.P. (1970) Coplanar-waveguide directional couplers. *IEEE Transactions on Microwave Theory and Techniques*, **MTT-18**, 318–22.

54 Knorr, J.B. and Kuchler, K.D. (1975) Analysis of coupled slots and coplanar strips on dielectric substrate. *IEEE Transactions on Microwave Theory and Techniques*, **MTT-23**, 541–8.

55 Kitazawa, T. and Mittra, R. (1985) Quasi-static characteristics of asymmetrical and coupled coplanar-type transmission lines. *IEEE Transactions on Microwave Theory and Techniques*, **MTT-33**, 771–8.

56 Kachayev, A. (2003) LIGA-micromachined tight microwave couplers. MSc Thesis, University of Saskatchewan, Saskatoon, Canada.

57 Lee, T.H. and Hajimiri, A. (2000) Oscillator phase noise: a tutorial. *IEEE Journal of Solid-State Circuits*, **35**, 326–36.

58 Ma, Z., Klymyshyn, D., Achenbach, S. and Mohr, J. (2005) Microwave cavity resonators using hard X-ray lithography. *Microwave and Optical Technology Letters*, **47**, 353–7.

59 Ma, Z. and Klymyshyn, D.M. (2005) Cylindrical post coupled bandpass filter suitable for LIGA fabrication, in Proceedings of the 17th IEEE Asia–Pacific Microwave Conference, Suzhou, China, pp. 106–9.

60 Pantenburg, F.J. and Mohr, J. (1995) Influence of secondary effects on the structure quality in deep X-ray lithography. *Nuclear Instruments and Methods in Physics Research B*, **95**, 551–6.

61 Pantenburg, F.J., Chlebek, J., El-Kholi, A., Huber, H.-L., Mohr, J., Oertel, H.K. and Schulz, J. (1994) Adhesion problems in deep-etch X-ray lithography caused by fluorescence radiation from the plating base. *Microelectronic Engineering*, **23**, 223–6.

62 Dambrowsky, N. (2006) Goldgalvanik in der Mikrosystemtechnik – Herausforderungen durch neue Anwendungen. PhD Thesis, University of Karlsruhe.

63 Ma, Z., Klymyshyn, D.M., Achenbach, S. and Mohr, J. (2005) LIGA cavity resonator for K-band applications. in Proceedings of ICMENS 2005 – International Conference on MEMS, NANO and Smart Systems, Banff, Canada, pp. 106–9.

11
Evolution of the Microspectrometer

Reiner Wechsung, Sven Schönfelder, and Andreas Decker

11.1 Introduction *281*
11.2 Product Definition and Strategy *282*
11.3 Product Description and Specification of the Microspectrometer *283*
11.4 Technology *287*
11.4.1 Micromolding *287*
11.4.2 Coating *288*
11.4.3 Microassembly *289*
11.5 Products and Applications *290*
11.5.1 Non-invasive Determination of Bilirubin in Blood *291*
11.5.2 Point of Care Diagnostic of Blood Parameters *292*
11.5.3 Color Measurement of Teeth *292*
11.5.4 Quality Determination of Diamonds *293*
11.5.5 Quality Control in Food and Industrial Production Processes *293*
11.5.6 Other Applications *294*
11.6 Marketing and Sales *295*
11.7 Conclusion and Outlook *296*
References *296*

11.1
Introduction

The development of the microspectrometer as described in this chapter was initiated in order to demonstrate the commercial applications and the potential for the establishment of a new business based on the recently developed LIGA technique [1].

The LIGA technique was developed at the Forschungszentrum Karlsruhe (FZK) in conjunction with STEAG AG under a project entitled 'Micronozzle for Uranium Enrichment'. At the completion of this project, it was decided to spin off the LIGA project group and form a start-up company named microParts GmbH, since 2004 Boehringer Ingelheim microParts GmbH (shortened to microParts in the following). The mission of this new company was to build a profitable, growing business based upon the LIGA technology [2].

Microtechnology offered a number of distinct attributes which could provide attractive benefits to advanced manufacturing industries. The key characteristics of the new products were considered to be:

- high functionality
- reduced size, weight and energy consumption
- improved reliability and robustness
- suitability for low-cost, high-volume production.

The LIGA process allows for the fabrication of structures of any lateral shape and with details in submicrometer dimensions across a total structure height of several hundred micrometers. Thus the application of this technology facilitates the generation of micro-optical modules and systems with the capability to replace complex mechanical fabrication processes that are employed in the production of conventional optical systems with a single microinjection molding process.

In order to commercialize the new LIGA technology, the developers scanned different application fields and searched for products needing the characteristic advantages of microsystems.

In the late 1980s, there was a strong demand to realize higher data rates in information technology. The idea of transferring light signals simultaneously in different wavelengths and the requirements for qualified wavelength multiplexing and demultiplexing systems were the starting point for the FZK to develop the principal design of the LIGA microspectrometer. It was one of a number of product ideas that were investigated and one that had already been patented by the inventors of the LIGA technique in 1987 [3]. The claim of the invention is a monolithic miniaturized grating spectrometer incorporating all the functional elements on a single chip.

This chapter describes the application- and production-oriented product design implementation, the identification of successful applications, their market introduction and the development, optimization and validation of production processes. The principle design aspects of the microspectrometer, the lithography and the mold-making processes are comprehensively described earlier in this book.

11.2
Product Definition and Strategy

In the early and mid-1990s, new trends in the field of molecular spectroscopy became obvious. Driven by the demand especially in medical diagnostics to generate immediate test results, to provide a permanent availability and to save costs per test, application-specific 'dedicated' analyzers were developed and replaced conventional analytical instruments.

microParts anticipated the high potential of the microspectrometer as a cost-effective alternative to expensive laboratory spectrometers for dedicated applications. Particularly miniaturization and the industrial serial production of such a microspectrometer module was seen as the enabler for new high-performance hand-held devices, fully portable instruments and process sensors with an unmet cost–performance ratio.

Optical spectroscopy is applied in

- medical diagnostics
- elemental and gas analysis
- environmental analysis
- color sensors
- industrial, chemical and food process controls
- optical telecommunication.

The wavelength range can be divided into three categories:

- visible   380–780 nm
- near-infrared   780–2500 nm
- infrared   2.5–8.0 µm.

After a thorough analysis of the potential applications and markets, two initially highly attractive proposals had to be excluded because of resolution requirements in the sub-nanometer range and unmet specificity requirements. These were the spectrophotometric infrared gas detection sensor and optical telecommunication wavelength demultiplexer systems.

The product strategy therefore was to develop a product with one basic design verified in two versions for different wavelength ranges:

- visible enhanced: 300–1000 nm
- near-infrared: 1000–1700 nm.

The standard product was designed to incorporate all the necessary functional elements of a classical grating spectrometer on a single chip with no moving parts.

11.3
Product Description and Specification of the Microspectrometer

The LIGA microspectrometer is a Rowland circle spectrograph which integrates fiber groove, entrance slit, the imaging grating, a camera mirror, light traps and supporting structures on a single plastic molded chip (Figure 11.1).

An optical fiber or a microlens images the sample on the entrance slit of the device. From there, the light propagates on to the self-focusing échelle grating. The used order of diffraction is deflected by a camera mirror on to the discrete pixels of a permanently fixed detector array. To access the different wavelength ranges, silicon, InGaAs, PbS or PbSe detector arrays are used.

Typical lateral dimensions of the chips range from 20×13 to 50×20 mm. The optical throughput is defined by the numerical aperture, the entrance slit width of

11 Evolution of the Microspectrometer

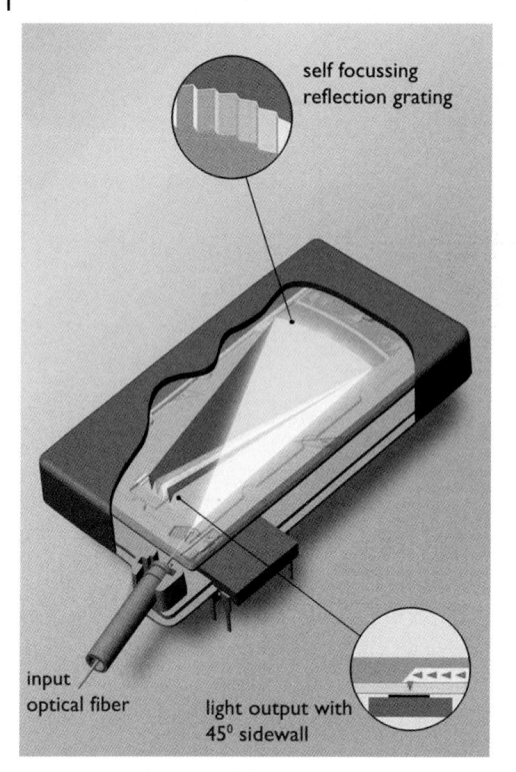

Figure 11.1 Schematic of the LIGA microspectrometer's industrial design.

10–50 μm and the height of the microstructures, which is typical 350 μm and can go up to 750 μm.

Using the Rowland configuration [4] shown in Figure 11.2, the system achieves superior wavelength stability and inter instrument agreement. The only mechanical interface of the polymer waveguide to the diode-array detector is just beside of the slit images in the first diffraction order. Thus, the design provides minimal thermal drift effects and – according to its monolithic nature – no hysteresis. The thermal effect on the wavelength calibration is less than $0.05\,\text{nm}\,\text{K}^{-1}$. Due to this stability and the use of a flat field grating design, the spectrometer achieves a wavelength accuracy of better than 1 nm.

The product design of the microspectrometer and the use of high-grade polymer materials guarantee excellent stability and insensitivity to environmental effects such as mechanical shock and vibration or large temperature shifts of up to 100 K.

In order to match its optical performance to that of expensive and high-quality holographic grating systems, a compensation algorithm for intrinsic stray light

11.3 Product Description and Specification of the Microspectrometer

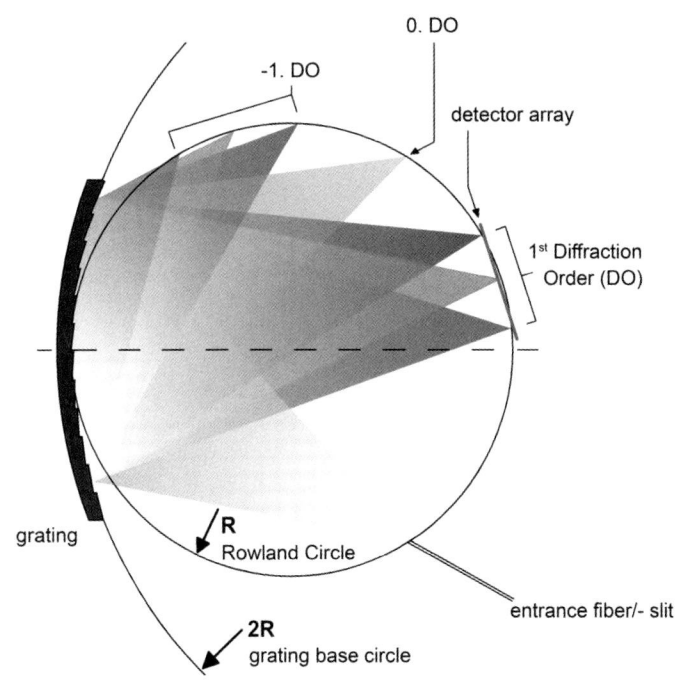

Figure 11.2 Rowland circle arrangement of functional elements of the LIGA microspectrometer.

has been developed. The processed spectra correspond to a stray light attenuation of typically 30–40 dB; the stray light attenuation of the raw data is typically 15–25 dB.

The monolithic setup avoids the need for periodic recalibration of the final device in a majority of applications. The high degree of integration allows for a remarkable reduction in the overall dimensions and weight of the final product. The geometric dimensions of the microspectrometer Original Equipment Manufacturer (OEM) modules are approximately $50 \times 25 \times 10$ mm.

These benefits make monolithic polymer optics the ideal choice for systems addressing a cost-sensitive, high-volume market and systems used under demanding environmental conditions or whenever size matters.

The technical performance data of the microspectrometer as a spectral sensor in specific OEM applications are determined by the design and quality of the spectrophotometer in combination with a specifically tailored light source and optical sampling system. These three functional components need to match the requirements of the calibration algorithm in the specific application. In order to transfer calibration functions from the prototype status to a series of devices on an industrial scale, the three functional elements and the optical signal path

Figure 11.3 Microspectrometer OEM module.

Figure 11.4 Microspectrometer OEM system.

between them need to freeze their spectral transfer function and inter-instrument agreement. This is the reason for the trend towards further integration of the source and the sampling system into the microspectrometer as a 'spectral sensor'.

In serial production quantities, the LIGA process provides the desired combination from targeted optical specification, robustness and reliability with the lowest production cost.

Obviously the optical performance characteristics of the combined light source, sampling system and microspectrometer are defined by the different OEM application requirements (see Figures 11.3 and 11.4).

Because of the variety of different application fields ranging from the UV to the IR spectral range, from 10 down to 0.1 nm bandwidth and from low-cost single-use sensors to complex analytical instrumentation, Table 11.1 presents exemplary data for the microspectrometer product platform in the UV–VIS and near-infrared (NIR) range.

These standard products meet the major performance requirements for a wide range of applications and in a broad spectral range. They provide stable and robust optical performance under demanding environmental conditions and safe transfer of calibration data between instruments.

Table 11.1 Microspectrometer data for the UV–VIS and NIR range.

Parameter	Value
Spectral range	320–1050 nm (1000–1750 nm NIR 1.7)
Resolution	<10 nm FWHM (<15 nm FWHM NIR 1.7)
Entrance fiber	Ø 300/330 µm; NA = 0.22
Spectral accuracy	2 nm (typically 1 nm)
Inter-instrument agreement	0.5 nm
Wavelength stability	0.05 nm K^{-1}
Wavelength repeatability	0.05 nm
Sensitivity with 16-bit ADC	Typically 250×10^{12} counts nm W^{-1} s^{-1}
Signal-to-noise ratio	>5000
Storage temperature	−40 to +60 °C
Operating temperature	0 to +40 °C

11.4 Technology

The LIGA microspectrometer utilizes the X-ray **Li**thography to produce master units, the electroforming processes (**G**alvanoformung) to create the production tools and plastic molding (**A**bformung) processes in order to replicate the highest aspect ratio microstructures in a high-throughput production process.

Comprehensive information about the processes used to generate the mold inserts are provided in earlier chapters of this book. This section will take a closer look at the replication, assembly and test processes of the products.

Efficient industrial production is the base of market success. This task was addressed in cooperation with FZK and a qualified production process for the microspectrometer component was developed in the mid-1990s. The required tooling technology and molding machines were established. Finally, a production line integrating the entire process chain was set up by microParts.

11.4.1 Micromolding

Initially the spectrometer components for the 380–780 nm wavelength range were produced as a three-layer polymer waveguide by hot embossing. With these visible-range products, a number of first OEM applications were generated.

Increasing sales figures and the optimization of materials allowed the enhancement of the production methods. The molding process of the spectrometer component was changed from hot embossing to the more efficient injection molding. The components are now produced as hollow-cavity waveguide spectrometer.

With these design and process changes, both the product quality and the process efficiency advanced significantly.

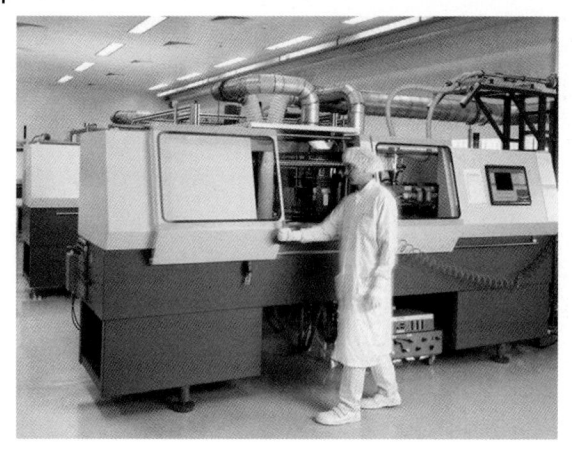

Figure 11.5 Micro injection molding.

Major benefits to the production process performance are:

- A significant reduction in the molding cycle time per unit, which today is just about 15% of the former processing time for hot embossing and bonding.

- Extremely improved endurance of the mold inserts by more than one order of magnitude – the mold insert has now a production life in excess of much more than 10 000 parts per tool.

- The yield of the overall production process, for which the quality and the yield of the molding process are of fundamental relevance, could be improved to meet the industrial standards.

In addition to these outstanding process improvements, the changes to the hollow-cavity waveguide design increased the performance of the microspectrometer significantly. Stray light attenuation could be improved by approximately 40%. Furthermore, the sensitivity could be improved by a factor of three. Most significantly, the inter-instrument agreement between the products and their robustness against harsh environmental conditions was greatly improved by the injection-molded hollow-cavity waveguide product. The hollow cavity also added major functional attributes. It opened up the accessible wavelength range from the visible into the UV and upwards into the NIR and IR spectral ranges. This changed product performance enabled the microspectrometer to be used in various new applications (Figure 11.5).

11.4.2
Coating

The micro injection molding is followed by a PVD process to coat reflective materials for the UV–VIS and the NIR–IR wavelength ranges on the functional elements (Figure 11.6).

Figure 11.6 Aluminum-coated UV–VIS microspectrometer chip.

Crucial to the product performance is to assure a constant reflectivity of the surface and safe contact between the coating material and the plastic substrate. In order to guarantee these properties, highly specified and controlled processes for the UHV processes, the materials and the part handling have been developed.

The performance of the optomechanical performance of the coatings has been verified and validated under extreme environmental conditions and meets the industrial standards according to DIN 58390.

This performance is one of the key requirements for use of the microspectrometer in numerous final applications under harsh environmental conditions.

11.4.3
Microassembly

Once the microspectrometer chip and the lid have achieved their final structure and surface properties, the production moves forward to assembly and test using dedicated cleanroom facilities and mechanized and automated equipment.

The assembly of lid and body parts is eased by precise passive alignment structures on these parts. Alignment to active and passive electro-optical components such as the detector assembly or the integration of light sources is based on active optical processes (Figure 11.7).

Stable and economic production is assured by applying Good Manufacturing Practice (GMP) standards and the utilization of statistical quality management systems such as machine and process stability analysis in combination with comprehensive optical performance tests.

The development, production and business processes at microParts are certified according to ISO 9001 and the medical device guideline ISO 13485 and meet the requirements of a total quality management system.

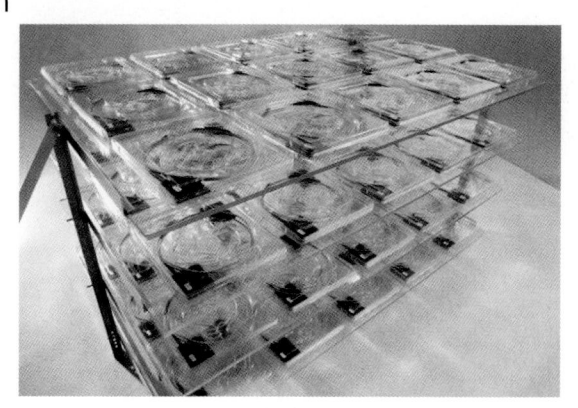

Figure 11.7 OEM microspectrometer products in process.

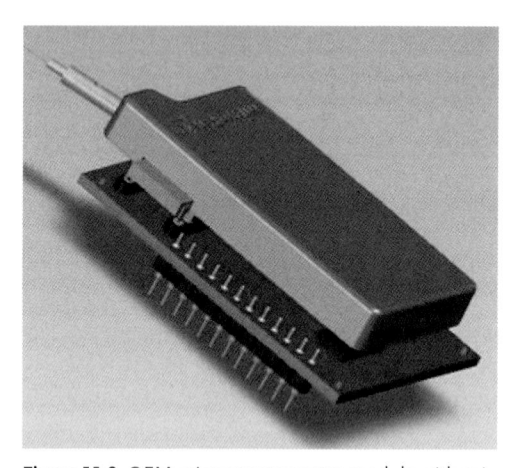

Figure 11.8 OEM microspectrometer module with microprocessor electronics.

11.5
Products and Applications

The standard versions of the microspectrometer are designed to fit into process sensors and hand-held and tabletop instruments with applications in medical diagnostics, analytical and process instrumentation and color analysis (Figure 11.8).

The spectral engine is built into complex microprocessor-controlled devices with specifically developed calibration algorithms. These algorithms and the associated hardware are typically owned by the instrument manufacturer.

There are roughly 150 different applications using the microspectrometer. Today, the most important ones are in the field of point of care diagnostics, biomedical and clinical analyzers, food and pharmaceutical process inspection systems, color sensors and classical analytical instrumentation. Successful new applications often combine the innovative technological approach with innovative analytical methods and processes. Some typical reference applications are described in the following.

11.5.1
Non-invasive Determination of Bilirubin in Blood

A spectacular application is the non-invasive measurement of the blood parameter bilirubin, introduced by the US company SpectRx for jaundice detection in newborn babies. The hand-held instrument BiliChek exposes white light to the baby's skin and analyses the spectrum of the backscattered light to determine the bilirubin concentration in the blood (Figures 11.9 and 11.10). For the first time, it offers a mobile, non-invasive method for the diagnosis of neonatal jaundice that

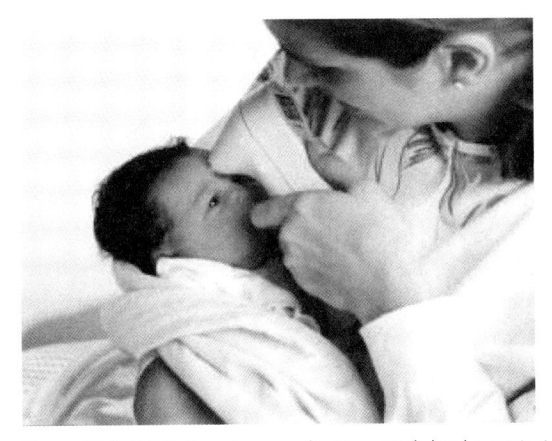

Figure 11.9 BiliChek system: application. With kind permission of Respironics.

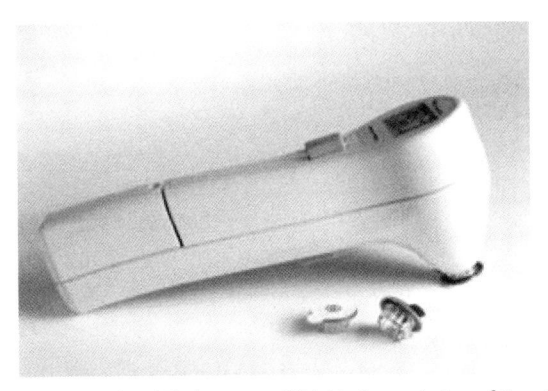

Figure 11.10 BiliChek system. With kind permission of Respironics.

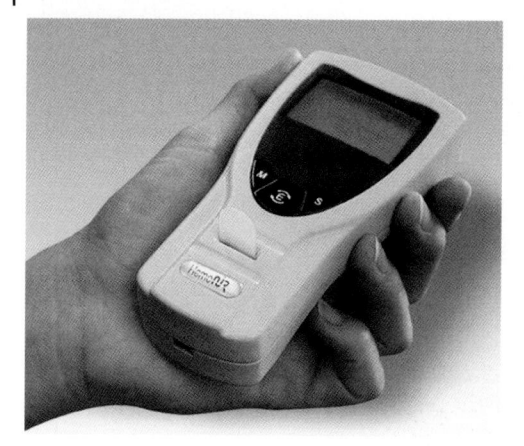

Figure 11.11 HemoNIR system. With kind permission of NIR Diagnostics Inc.

is independent of skin pigmentation or precise age. Results are instantly available while the need for blood samples and the resulting risks of infection and pain trauma are eliminated. The BiliChek instrument has received FDA approval and today is successfully sold throughout the world by the company Respironics.

11.5.2
Point of Care Diagnostic of Blood Parameters

There is an increasing demand for patient-administered measurements of certain blood parameters. A well-known example is the measurement of glucose concentrations by diabetes patients. To minimize pain, very small amounts of blood should be extracted but nevertheless maintain highly accurate results .The blood is sampled on a small test strip or, more frequently, on polymer bio-chips and is then introduced into a small, battery-powered device which measures the specific blood parameter by the evaluation of an electrical or optical signal.

The HemoNIR System of NIR Diagnostics measures total-Hb, met-Hb and carboxy-Hb and other blood parameters from a blood sample of less than 10 µl (Figure 11.11).

The use of the microspectrometer allows a purely optical measurement without the addition of any reagents and therefore a significant reduction in cost per test.

11.5.3
Color Measurement of Teeth

Other typical examples of the use of the microspectrometer are in dental and cosmetic applications.

The IdentaColor II supplied by Identa, Denmark, is a small, portable tooth color measuring instrument (Figure 11.12). It allows dentists to determine the individual tooth shades and to select from a set of color shades the optimum match of a

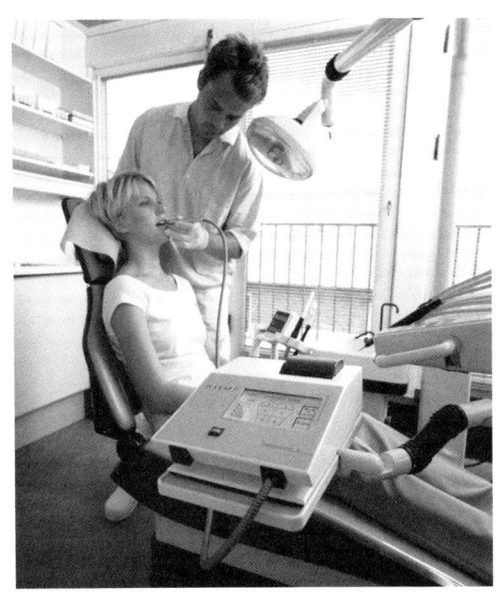

Figure 11.12 IdentaColor II system. With kind permission of Identa A/S.

new ceramic insert. Without the device this is sometimes a difficult task because the color appearance of teeth depends largely on environmental conditions.

11.5.4
Quality Determination of Diamonds

The quality of diamonds and consequently their price depend largely on their color appearance. For an exact and quantitative quality determination, a desktop instrument was developed for use in the jewelers' store. The DC3000 device was developed by Gran Computer Industries and is manufactured and sold by Sarin Technologies (Figure 11.13).

11.5.5
Quality Control in Food and Industrial Production Processes

Quality control of food is routinely performed using NIR spectroscopy. With the availability of an NIR microspectrometer, reliable and economic desktop instruments and in-line sensors have been developed.

The (Meat Compound Analyzer (MCA) desktop instrument developed and produced by FZMB monitors the quality of meat and sausage products (Figure 11.14). Parameters such as water content and protein values are measured in the NIR spectral range.

NIR online devices are used to provide quality determining parameters in on-line measurements directly attached to the production process at high sampling

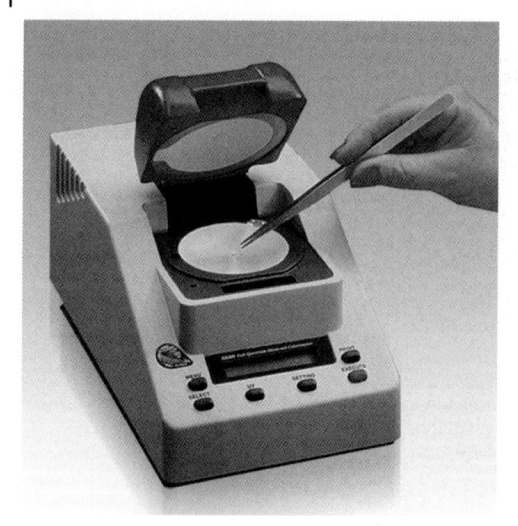

Figure 11.13 DC3000 diamond colorimeter. With kind permission of Sarin Technologies Ltd.

Figure 11.14 MCA meat compound analyzer. With kind permission of FZMB.

rates (Figure 11.15). Among others, parameters such as protein, fat, starch and glucose, residual humidity and water content are monitored.

These in-process sensors typically provide accuracy in the range 1-3% compared with the reference analytics (RA), whereas the stability and reproducibility are typically better than in the RA.

In addition to the food processing industry, this kind of process sensor is frequently used in the chemical, pharmaceutical and other industries.

11.5.6
Other Applications

The major applications of microspectrometers are in quantitative diagnostic and analytical tests. However, solely qualitative tests and identifications are also fre-

Figure 11.15 Online sensor in a food processing environment.

quently done with the product. Among others, these include high-speed material and product sorting processes and the detection of counterfeit products and dangerous substances.

11.6
Marketing and Sales

In general, it is a challenge to introduce an innovative product, such as the microspectrometer, into an established market, particularly when the product has been developed by a new start-up company. Therefore, the right marketing and sales strategy is crucial to the success of the product. microParts addressed this challenge by establishing its own division to develop, market and sell the new microoptical product. The first task of the product management was to conduct an analysis of the existing market in order to identify segments where the microspectrometer had the potential to be incorporated into next-generation product offerings. In discussions with instrument manufacturers and diagnostic companies, innovative new product ideas have been discussed and realized in joint product development programs.

Marketing activities were supported by professionally produced brochures and leaflets, by presentations at international conferences and exhibitions and by the placement of advertising in selected journals.

The economically most important regions of the world for high-tech products are the USA, Europe and Far East. Whereas Europe could be covered directly by microParts' own personnel, distributor agreements were established for the other parts of the world. This strategy has proved to be an excellent choice as the American market in the late 1990s became by far the most successful region for sales of the microspectrometer. In order to address the increasing economic importance of the Far East, a number of distributorship agreements have been established in different countries.

11.7
Conclusion and Outlook

The microspectrometer described in this chapter is a typical example of the benefits of Micro Systems Technology (MST) applications: the high functionality of the product with no moving parts, a miniature size, low cost in high-volume production and high reliability. These features provide the potential for the improvement of existing products by the replacement of components with the microspectrometer. With the steady improvement of the technology, better resolution in the sub-nanometer range will be achieved in the near future; the availability of fast detector arrays not only for the visible range but also for NIR and IR wavelengths will open up new applications in personalized biomedical diagnostic instruments, identification and authenticity tests and also in the industrial process sensor field.

The small size and low cost of this new spectral sensor will allow applications in printers, copying machines and many other systems. In the food and agricultural industries, new hand-held devices will be introduced for online process control and for on-site quality inspection.

The non-invasive determination of bilirubin represents a paradigm shift in point of care diagnostics. In patient self-tests, it is a dream of all diabetics to have a non-invasive, painless glucose measurement. It is expected by the experts in this field that an algorithm for a reliable evaluation of the glucose concentration from an optical spectrum will be available in a few years. This opportunity alone will open up a billion dollar market.

The technological and commercial success of the microspectrometer is an excellent example of the transfer of know-how from research to industry. In addition to patent protection, the complex process know-how is so fundamental that the introduction threshold for competition is very high and the leading position of German research and industry will be extended.

References

1 Becker, E.W. et al. (1986) *Microelectronis Engineering*, **4**, 35.
2 Wechsung, R. (1995) *MST News*, **14**, 13.
3 Procedure of the manufacture of an optical component having one or more echelette-gratings and device thereby obtained. European Patent EP0242574 A2 (1987).
4 Last, A. (2002) Viel Farbe aus wenig Licht. *Physik in Unserer Zeit*, 33.

12
Actuator Manufacture with LIGA Processes
Todd Christenson

12.1 Introduction *297*
12.2 Actuator Scaling *298*
12.3 Materials *303*
12.4 Actuator Types *306*
12.5 DXRL Actuator Applications *314*
12.6 Future Directions *317*
　　References *318*

12.1
Introduction

The use of LIGA-based processing, when in particular practiced with deep X-ray lithography (*DXRL*), has two substantial advantages when applied to the fabrication of microactuators. These advantages include the ability to pattern high aspect ratio geometry and the ability to incorporate a large variety of engineering materials. Together, these capabilities provide the tools to batch fabricate area-efficient actuators with performance as expected from the microscaled physics of their macroscopic counterparts. This process flexibility can also be used to fabricate actuators with a multitude of driving physics methods, including electrothermal, electrostatic, piezoelectric, electromagnetic and hydraulic [1–3]. The application areas for such actuators are broad and include microfluidics, data storage, microwave electronics, micro-optics, fiber optics and microrobotics.

Although complementary to actuators, sensors have to date profited to a much larger extent from microfabrication than actuators. The reasons for this reveal the challenges presented for batch microfabrication of actuators. The use of lithographic-based planar processing has proven a natural approach for the manufacture of miniaturized physical and chemical sensors, which in turn have significantly changed the way in which sensors are manufactured and applied. Processing of this type most commonly takes place on a flat base material or substrate, upon which a sequence of chemical and mechanical processes are conducted to render a number of devices on each substrate simultaneously. This approach therefore carries with it a number of differences relative to serial-based assembly-intensive manufacture. Notwithstanding these manufacturing differences, the impact of micromachining on physical sensing has been tremendous. This result is not surprising when considering that the benefits of scaling many physical sensors accrue directly from diminishing size. As sensor size decreases, the energy of interaction with the environment decreases, which in turn reduces perturbation of the measurand.

An example of how monolithic processing has affected sensor performance is provided by the progress made in integrated force sensing, where sensitivity gains and power requirement reductions have been remarkable. Some of the first micromachined pressure sensors with 25 µm thick membranes and up to 1 cm² area consumed 5×10^{-3} W of power using piezoresistors with force sensitivities near 1×10^{-2} N. By reducing the pressure transducer membrane thickness to 1 µm, a linear size reduction of 100-fold resulted with a corresponding force sensitivity increase to 1×10^{-6} N and power dissipation of a few milliwatts. The advent of vacuum-encapsulated, high quality factor resonant microbeams has provided yet further gains in sensitivity over piezoresistive sensing. By optically sensing resonance of a 1 µm thick × 40 µm wide × 200 µm long beam, force sensitivity of less than 1×10^{-10} N is achieved with an input power of 10^{-13} W. For comparison, atomic force microscopes typically resolve 10^{-9} N, which is the magnitude of van der Waals forces. Magnetic resonance force microscope techniques extend force sensitivity even more with use of 60 nm thick resonant beams that allow sensing to 3×10^{-18} N (a few hundred attograms equivalent force) or the magnetic force due to a single electron spin. The further quest to sense individual nuclear (proton) spins looms large [4].

Actuators have not made similar inroads for implementation in devices of everyday existence or similar advances in performance. Therefore, a summary of the scaling constraints and research addressing microactuator development will be addressed in the following discussion. What is found is that high aspect ratio lithography and micromolding technique provides a powerful tool base for the fabrication of microactuators that offer the ability for deployment in a myriad of uses, consequently presenting an opportunity much like that which has been capitalized on for microsensors.

12.2
Actuator Scaling

One goal for the design of microfabricated actuators is to maintain the scaled forces of their conventionally manufactured and generally larger counterparts. Actuators and corresponding linkage mechanisms are three-dimensional devices that operate via changes in stored energy. One challenge for actuator microfabrication is therefore to be able to maintain geometric flexibility to accommodate intricate electromechanical geometry. Since tolerances play a huge role in manufacturability, a significant problem in this sense is to maintain dimensional tolerances at a small scale. Actuators also typically incorporate a large variety of materials, including good electrical conductors, good electrical insulators and in many cases ferroelectric and ferromagnetic material. Additionally, as actuators are reduced in size, the physical scaling behavior modifies the optimal design of actuators [5]. Finally, Nature, it will be noted, provides phenomenal examples of how to scale mechanical actuation [6].

Perhaps the most important factor for microfabricated actuators that presents itself uniquely in the context of substrate based processing is cost. As in monolithic electronic fabrication, cost scales with the amount of substrate area occupied by the actuator and the accompanying packaging and interfacing needs. Hence an appropriate figure of merit for microfabricated actuators is suggested to be one incorporating chip area as follows. For an actuator with a prismatic working gap defined by a height h and area A, with energy density ρ_E, the force, F_x, with motion normal to the height direction is defined by

$$F_x = \rho_E h \frac{\partial A}{\partial x} \tag{1}$$

Therefore, for force generated lateral to the substrate surface, height is of prime importance. To maximize the energy change which produces this force, however, requires that the energy density be maximized. That is, for a magnetic field, the magnetic saturation of ferromagnetic material is achieved or for electrothermal forces the temperature difference is maximized or for an electrostatic device the electric field desired is nearly the dielectric breakdown field. The ability to maximize energy density is a function of the 'working gap' or the distance through which the field exists. As the working gap increases, the driving force required to maximize the energy density increases to a point to where it typically becomes impractical to maintain. Thus, the driving point impedance of the actuator becomes important, which is largely a function of the ability to pattern the working gap or distributed working gaps with precision.

The previous discussion reveals that an efficient prismatic microactuator, therefore, is one which has substantial thickness out of plane of a substrate with maintained tolerance of associated working gaps. Such structural attributes are well accommodated by DXRL with its ability to define high aspect ratio geometry. A definition of aspect ratio appropriate to a prismatic working gap is the run-out or side-wall angle, which for DXRL is of the order of 1 part in 1000 or less. Lithographic masks as created by laser or e-beam writing can readily resolve submicrometer biases in dimensions which provide a complementary asset to high aspect ratio geometry. In this way, a 100 µm feature may be patterned adjacent to a concentric 100.5 µm feature and when engaged a resulting gap of 0.25 µm may be maintained. This capability has been exploited for electrostatic actuation [7] and for maintaining small inter-component tolerances, as shown in Figure 12.1.

The significance of the ability to maintain small working gap dimensions with even tighter tolerances is revealed for electrostatic and electromagnetic actuators in the graph in Figure 12.2. This graph plots the pressure developed by an electric or magnetic field as a function of working gap distance and impressed voltage for the electric case and impressed magnetomotive force (mmf) for the magnetic case.

Figure 12.1 Assembled nickel gear on nickel shaft and 10× close-up showing 1 μm journal bearing clearance at 120 μm structural thickness.

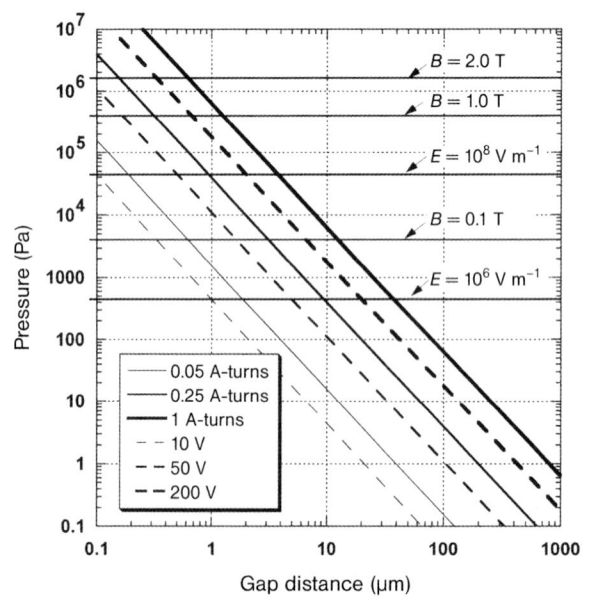

Figure 12.2 Plot of magnetic and electric pressure generated for various electrical drive levels. Solid lines correspond to magnetic pressures for various magnetomotive forces and the dashed lines correspond to pressures due to electric fields for various applied voltages. Maximum pressures for limiting B and E fields are indicated for several values as horizontal lines.

The curves are readily obtained by equations for energy density, where for the electric case,

$$\rho_{E_e} = \frac{\varepsilon_0}{2}\left(\frac{V}{d}\right)^2 \qquad (2)$$

where $\varepsilon_0 = 8.854 \times 10^{-12}\,\text{F m}^{-1}$ is the vacuum permittivity, V is the impressed voltage across the working gap, d, and for the magnetic case,

$$\rho_{E_m} = \frac{\mu_0}{2}\left(\frac{ni}{d}\right)^2 \qquad (3)$$

where $\mu_0 = 4\pi \times 10^{-7}\,\text{Hy m}^{-1}$ is the vacuum permeability and ni is the impressed magnetomotive force (ampere-turns). The assumption made in the magnetic case is that the entire field developed by the impressed magnetomotive force either from a coil or permanent magnet exists across the working gap, which is never the exact case but is approximately true if the relative permeability of soft ferromagnetic material adjacent to the gap is large and the amount of fringing flux outside of the magnetic gap is minimized. Also indicated on the graph are horizontal lines corresponding to the pressures developed for various electric breakdown fields and magnetic saturation flux densities which set energy density maxima for electric and magnetic fields, respectively, and are defined as follows:

$$\rho_{E_e}(\max) = \frac{\varepsilon_0}{2}E_{\text{bkdn}}^2 \qquad (4)$$

$$\rho_{E_m}(\max) = \frac{B_{\text{sat}}^2}{2\mu_0} \qquad (5)$$

The plot indicates that to achieve these maximum pressures with reasonably low impressed electrical drive levels, micron or sub-micron size gaps are desired.

As working gap dimensions decrease, however, additional electromechanical problems arise that are not present in macroscopic devices. As component dimensions decrease, their surface to volume ratio increases. Also, typical surface roughness on DXRL components ranges from 10 to 50 nm. This situation creates the tendency for adjacent moving microstructures to adhere to each other with surface forces greater than can be overcome by those created by inertia. Also, macroscopic actuators that rely on inertia of a rotor or slider for their function often cannot be used on the microscale. One- and two-phase linear and rotary motors, therefore, do not scale well and three phases are required for continuous motion in a particular defined direction.

Some quantities, on the other hand, scale well with size. Hydraulic and pneumatic pressures can be maintained at small scale provided that material strength is maintained and offer very high actuation pressures. Permanent magnets to first order possess magnetizations that are independent of scale. This is not true for electromagnetic coils, which scale detrimentally with miniaturization due to current density limits set by thermal and electromigration constraints. Coils fabricated from superconducting materials also have maximum critical current densities. Halbach stated the issue eloquently: '... when it is necessary that a magnetically significant dimension of a magnet is very small, a permanent magnet will always produce higher fields than an electromagnet ... with permanent magnets one can reach regions of parameter space that are not accessible with any other technology' [8].

Identification of the dimensional scale at which permanent magnets are more capable of generating magnetic fields than coils requires a detailed knowledge of actuator topology, but an approximate evaluation can be made for a simple 'horseshoe' electromagnet geometry that is common to many applications. A ratio of magnetic field produced by a permanent magnet to that produced by an electromagnetic coil of equivalent volume and aspect ratio may be derived as γ_H:

$$\gamma_H = \frac{B_r/\mu_m}{2\alpha\mu_0 t_c J_{max}} \tag{6}$$

where B_r is the remanent magnetic flux density of the permanent magnet material, μ_m is the recoil permeability of the permanent magnet, α is the coil winding efficiency, μ_0 is the permeability of free space, t_c is the coil thickness (dimension normal to the flux path) and J_{max} is the maximum current density of the coil wire. A linear demagnetization response is assumed for the permanent magnet which is accurate for rare earth permanent magnets at moderate temperatures. By plotting this ratio as a function of dimension, an inverse length scale dependence is readily seen, as shown in Figure 12.3.

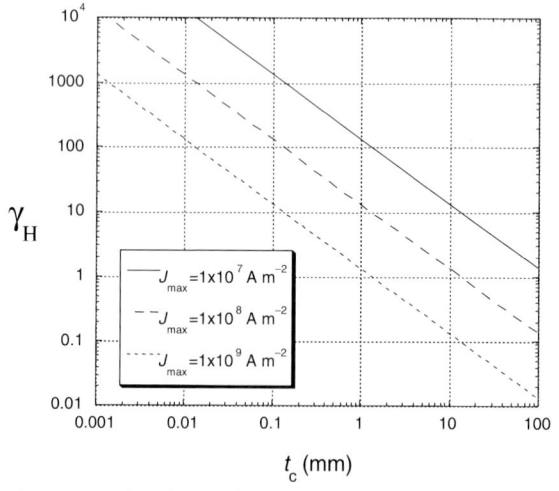

Figure 12.3 Plot of ratio of magnetizing force of permanent magnet to electromagnetic coil from Equation 6 as a function of linear dimension for various maximum wire current densities and with $B_r = 1.2\,T$, $\mu_m = 1.05$ and $\alpha = 0.75$ (75%).

12.3
Materials

Techniques that have been used for mold filling include electroforming, powder filling, flame spray, injection molding and hot embossing. The result of this wide variety of techniques is a material base that includes a number of materials which are not common to planar processing yet are taken for granted in the manufacture of macroscale actuators.

Electroforming can accommodate a number of metals, including many metal alloys, and provides an excellent means to accurately replicate high aspect ratio molds [9, 10, 12]. Soft ferromagnetic materials with high permeability are available with 78/22 Ni/Fe or 78 Permalloy, for example. A B–H response which was measured from an as-electroformed 78/22 Ni/Fe sample is shown in Figure 12.4. One concern in electrodeposition of these types of alloys is a composition variation through the mold thickness. What has been found from microprobe analysis is that composition control of alloy composition can be maintained to a small percentage or less with the use of paddle cells and pulse plating. The 78 Permalloy alloy has a magnetic flux saturation density of approximately 1.0 T and a nearly zero magnetostriction coefficient, which becomes important for microactuators that use ferromagnetic material in flexures. Zero magnetostriction is also important for batch microfabricated magnetic devices, so their function is largely independent of built-in and process-imparted strains. Research on electrodeposition of NiFeCo material has shown the ability to deposit alloys with zero magnetostriction, coercivity less than 2.0 Oe and magnetic saturation flux density greater than 2.0 T [11]. The impact on magnetic actuator performance with this material is dramatic, as the difference between material with B_s over 2 T compared with 1 T is a developed pressure, which is over four times larger and enables direct gap pressures of over 200 psi to be reached.

The ability to form permanent magnets from DXRL patterned molds has been demonstrated with bonded rare earth permanent magnets and fully dense sintered

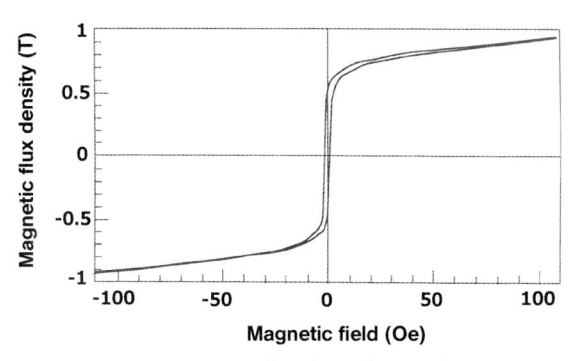

Figure 12.4 B–H response of as-electrodeposited 78/22 Ni/Fe sample with 250 µm plated thickness (maximum relative permeability for this material is approximately 2000).

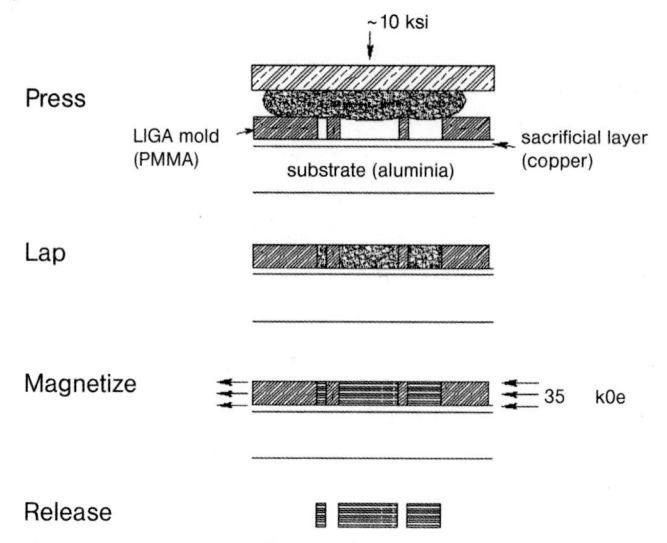

304 | *12 Actuator Manufacture with LIGA Processes*

Figure 12.5 Cross-section of process for batch fabrication of micro miniature precision-bonded rare earth permanent magnets.

Figure 12.6 Scanning electron micrographs of DXRL fabricated bonded Nd2Fe14B permanent micromagnets. Magnetization direction is as indicated and some distortion of the images is present due to the magnetization.

rare-earth material [13]. By using a pressed mold approach as shown in Figure 12.5, the bonded rare earth permanent magnets shown in Figure 12.6 were fabricated. Measurements of an individual NdFeB molded micromagnet are shown in Figures 12.7 and 12.8 and were made with a SQUID magnetometer. Pressing of $Nd_2Fe_{14}B$ powder has also been accomplished with the use of sacrificial plastic molds which volatilize during sintering and have produced micro miniature ceramic magnets with energy products of 19 MG Oe (10^6 gauss oersted). High-rate

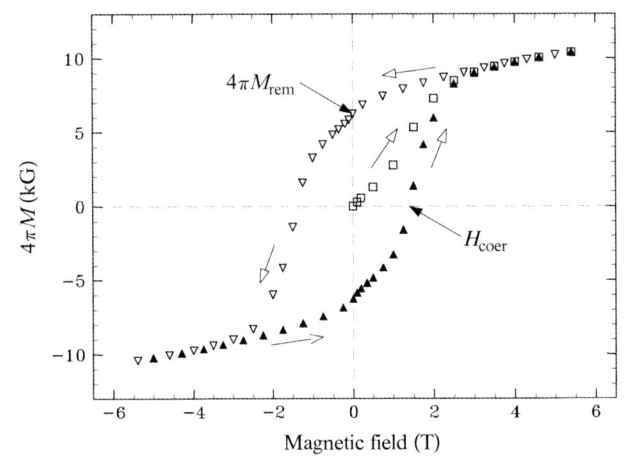

Figure 12.7 M–H characteristic of 20%/vol. bonded $Nd_2Fe_{14}B$ micro-magnet bar as shown in Fig. 12.6 with 0.64 Tesla remanence and 1.4 Tesla coercivity.

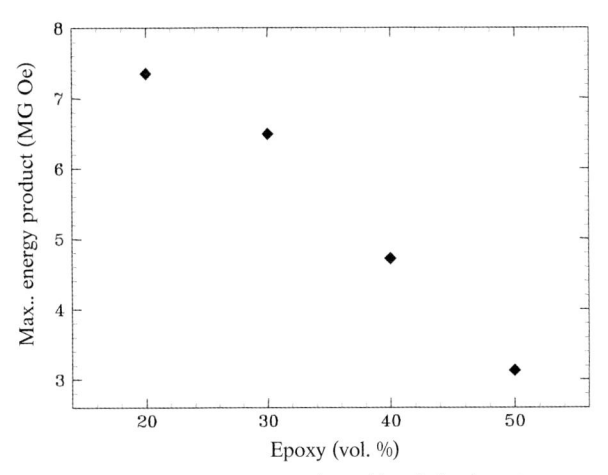

Figure 12.8 Maximum energy product of bonded $Nd_2Fe_{14}B$ bar micromagnets fabricated via DXRL and molding as a function of epoxy composition.

sputtering processes have also been used to deposit thick-film NdFeB magnets up to 300 μm thick [14].

A similar approach to molding permanent magnets can also create precision-molded piezoelectric materials [15–17]. In this process, slurry casting is performed in a plastic mold which may be generated via injection molding from a DXRL defined mold insert. During sintering, the plastic mold is volatilized, leaving a dense ceramic, such as lead zirconate titanate (*PZT*).

12.4
Actuator Types

Examples of microactuators that have benefited from the large material base and high aspect ratio that LIGA processing affords extend to many types and include electrostatic, magnetic–variable reluctance, magnetic–permanent magnet, piezoelectric, thermomechanical, bimetallic and hydraulic [18].

Initial research on LIGA-based actuators focused mainly on electrostatic devices [19, 20]. Electrostatic actuators may be simply fabricated with one DXRL layer and perhaps the addition of a patterned sacrificial layer. Electrostatic comb actuators with a few micrometers air gap and several hundred micrometers height may be readily achieved. Figure 12.9 shows a simple example of a nickel high aspect ratio comb drive fabricated through X-ray lithography and electroforming. Figure 12.10 shows an electrostatic stepping actuator design [21]. As-electroformed metal frequently encounters thickness non-uniformity, particularly when fine geometry such as high aspect ratio lateral springs are desired in a part with larger features. The comb drive device in Figure 12.11 shows this plating rate dependence, where in this case the comb supporting flexure was plated to a thickness less than the comb drive fingers and supports. A method to correct this non-uniformity is to electroform proud of the mold and planarize via lapping, for example, to a uniform thickness [22]. Resulting planarized comb actuator structures are achieved as depicted in Figure 12.13.

Vertical two-dimensional comb arrayed microactuators may also be very efficiently fabricated with DXRL methods, where the advantages of delta CD techniques are maximized [23]. This arrangement is very efficient at generating large

Figure 12.9 Electroformed nickel electrostatic comb drive created from DXRL mold.

Figure 12.10 Electrostatic linear stepping microactuator fabricated from DXRL mold and nickel electroforming.

Figure 12.11 Electrostatic comb drive with flexure showing non-uniformity arising during nickel electroforming.

force per unit substrate area and has unique interfacing capability whereby an entire plate of surface area residing above the actuator finger array is available for mechanical engagement.

For electrostatic comb drives with several hundred micrometers height, forces from tens to hundreds of micronewtons are achievable for actuator and flexure

Figure 12.12 DXRL fabricated bi-directional VR electromagnetic actuator in a 1 × 2 fiber optic switch.

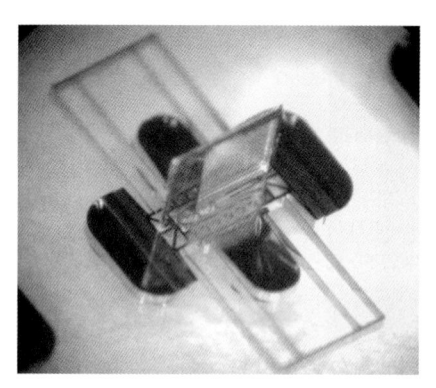

Figure 12.13 Nickel electrostatic microresonator with 300 μm structural height supporting an optical grating.

footprints of several millimeters on a side [24]. Displacements for such actuators are of the order of 100 μm. In air, at atmospheric pressure resonance quality factors of 50–200 are possible with resonant frequencies from tens to hundreds of hertz. For example, the device shown in Figure 12.13 can be tailored for resonant frequencies of 30–100 Hz with resonant amplitude of ±100 μm for an input peak square-wave voltage signal of under 5 V. At the same time, this resonant actuator is readily capable of supporting a few grams of mass or a few tens of millinewtons of force.

Electrostatic rotary actuators may also be fabricated with DXRL-based processing in a relatively simple way. By using a one-layer DXRL pattern with one patterned

Figure 12.14 Nickel electrostatic wobble micromotor.

Figure 12.15 Nickel electrostatic rotary stepping micromotor.

sacrificial layer as with the previously described linear actuator, an electrostatic wobble motor [25] may be constructed as shown in Figure 12.14. Electrostatic rotary stepping micromotors may also be fabricated with DXRL techniques as depicted in Figure 12.15. Such devices may be excited with three-phase signals with voltage amplitude of the order of 100 V. Using a twin stator design, such electrostatic motors are capable of producing approximately 0.1 µN m torque for 2000 µm diameter rotors or 100 µN of force at the rotor periphery [26–28].

Linear electromagnetic actuators have been fabricated with DXRL-based processing mostly using a variable reluctance (VR) approach. In this case, force is generated as

$$F_x = \frac{1}{2} mmf^2 \frac{\partial \Re}{\partial x} \quad (7)$$

where mmf is the magnetomotive force in ampere-turns and $\Re$ is the reluctance in Hy^{-1}, defined as

$$\Re = \sum_i \frac{l_i}{\mu_i \mu_0 A_i} \quad (8)$$

where l_i is the magnetic path length, μ_i the relative permeability and A_i the magnetic cross-sectional area of each respective segment of the actuator magnetic circuit. In practice, the force is limited by the saturation flux density of the soft ferromagnetic material.

Figures 12.16 and 12.17 show examples of linear variable reluctance microactuator construction. Integrated into this device is a stator, which has features to mate with an electromagnetic coil and rotor that is supported by a flexure, typically a folded spring [29, 30]. For a thickness of 250 µm and a total actuator footprint of

Figure 12.16 Linear variable reluctance magnetic microactuator comprised of 250 µm thick Ni–Fe. Coils have been omitted which mate into the two pair of striped magnetic pads.

Figure 12.17 Planar VR magnetic linear microactuator with magnet wire wound coil in flat-pack.

Figure 12.18 Three-phase VR magnetic rotary micromotor with closed-slot stator and wire bonded coils.

a few millimeters on a side, these devices produce hundreds of micronewtons to a few millinewtons of force for strokes of a hundreds to tens of micrometers.

Magnetic VR actuators may also be realized in rotary form, as shown in Figure 12.18. These types of magnetic micromotors spin at rotational rates of 150 000 rpm

Figure 12.19 Three-phase VR stepping magnetic micromotor coupled to variable magnetic brake. The device is fabricated using DXRL defined 250 μm thick 78 Permalloy.

at atmospheric pressure and may have uses as sensors [31]. Stepping motor configurations are also possible so that when coupled to a lossy mechanism, such as the friction occurring in a gear train and an electromagnetic brake, on-chip dynamometry may be performed. An example is shown in Figure 12.19, were it is found that for 52 μW excitation, a maximum speed of 8000 rpm is attained with a maximum output torque of 0.3 μN and with a maximum output mechanical power of 20 μW [32].

Slightly larger rotary VR motors patterned partially with DXRL have been fabricated with DXRL techniques and integrated windings [33]. Such planar minimotors with 6 mm diameter rotors, 17 mm total diameter and six stator core pairs with 108 turns per core, generate output torque of 0.5 μN m at 43 μm core thickness with 20 μm rotor thickness.

Another class of magnetic microactuators has been investigated with the use of DXRL patterned and molded rare earth permanent magnets. By forming precision-bonded rare earth permanent magnets in individual parts of a multipole rotor, brushless DC motors may be constructed. This approach alleviates the difficulty frequently encountered when attempting to magnetize miniature multipole rare earth permanent magnets due to the high magnetizing forces required for rare earth magnets. Similarly, microscale 'magic rings' and Halback arrays may be constructed, as shown in Figure 12.20 [8, 34, 35]. Such magnet array fabrication may be aided by recently devised magnetic and microassembly techniques [36, 37]. Figure 12.21 shows an example of such a motor, which is a low-profile, nine-slot, four-pole, three-phase brushless DC motor with 8 mm total outer diameter, 5 mm

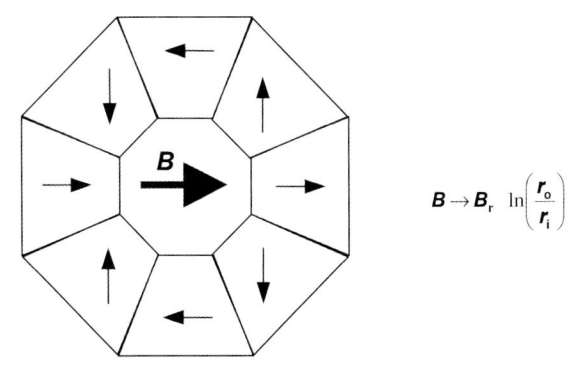

Figure 12.20 Magic ring permanent magnet configuration which generates an internal magnetic flux density greater than the permanent magnet remanence.

$$B \to B_r \ln\left(\frac{r_o}{r_i}\right)$$

Figure 12.21 Planar brushless DC motor with 5 mm diameter four-pole bonded $Nd_5Fe_{14}B$ rotor fabricated via DXRL based molding.

diameter rotor and 1 mm total thickness [38]. The motor incorporates a 1.6 mm OD commercially available ball bearing. The maximum speed is 12 000 rpm with a maximum torque of 25 µN m. Excitation is 0.2 V at 100 mA into 2 Ω at 4 µHy for each phase.

High-speed rotation is of interest for stabilization devices and gyros and other sensors. The previous types of actuators, however, suffer from torque ripple and resulting frequency instability, which may be compensated for by electronic position feedback control. Another type of motor, a hysteresis motor, which derives torque via alternating or rotating hysteresis, possesses a constant torque curve [39]. An appropriate ferromagnetic material for this type of motor is one which is of modest coercivity and thus is neither soft (back-iron) nor hard (permanent magnet).

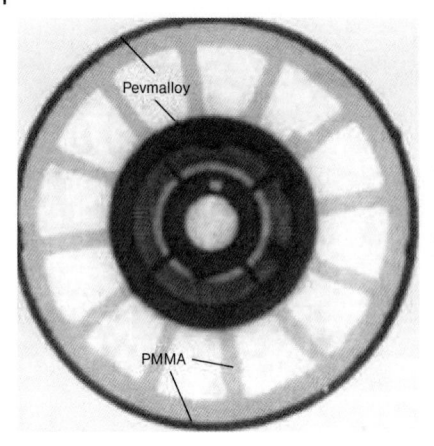

Figure 12.22 Magnetic hysteresis rotor 1.5 mm in diameter and 200 μm thick fabricated via two DXRL exposures

Pure nickel and certain nickel–iron alloys meet this criterion well with a coercivity of 100–200 Oe. Rotors for this type of motor require a relatively thin rim of nickel and an example is shown in Figure 12.22. Few data can be found in the literature for this type of hysteresis micromotor. (What clear from the literature is) the extreme variety of magnetic microactuator types, coupled with their utility at small scale, made clear through careful scaling analyses, has prompted their incorporation in microsystems for many applications [40].

Other microactuator types that have been researched in the context of DXRL processing include piezoelectric [41], thermomechanical [42], ribbon [43] and linear to rotary conversion [44]. Many milli- and microscale actuators are found to be advantaged through mechanical amplifiers or leveraging couplers fabricated via DXRL processing. These include bent beam couplers for piezoelectric actuators [45] and harmonic drives (micromotion) and planetary gear systems (see Figure 12.23) for miniature rotary motors [46]. Hydraulic actuators can attain very high pressures at the microscale if electrically actuated microvalves can be constructed and integrated into hydraulic plumbing. Such devices have been fabricated through LIGA-like processing with the capability of valving 10 MPa (1400 psi) and show great promise for high-force, large-stroke milli and micro linear actuators [47].

12.5
DXRL Actuator Applications

The unique size and performance afforded by DXRL fabricated actuators open entirely new application areas and approaches to electromechanical problems. In particular, actuators with millimeter size and millinewtons of output force are not easily or cost-effectively fabricated with conventional machining techniques and

Figure 12.23 Planetary gear train fabricated from DXRL exposures and nickel alloy electroforming.

are out of reach for many of the planar processing and thin-film microfabrication techniques.

Several electrical components have made use of LIGA-based actuators. The forces created by many of the linear actuators discussed previously provide enough force for electrical relay closure. Considerable opportunity for high aspect ratio microstructures also appears in microwave applications, although there has been limited research in this area. Variable capacitors may be fabricated with a single-layer DXRL formed pattern. Simulation of a vertical parallel plate capacitor fabricated with copper and an integrated electrostatic actuator for movement of the capacitor plate reveals the ability to create high quality factor tunable capacitors at up to 18 GHz [48]. For example, for a capacitor structure with 350 μm thickness, a 0.24 pF capacitor operating at 13 GHz with a quality factor of over 300 and a tuning range of 1.38:1 for a tuning voltage of under 10 V is estimated. This is due in large part to the ability to incorporate a high-conductivity metal such as copper and silver with electroforming processes.

The ability of electrostatic microactuators with high aspect ratio to generate relatively high forces with low voltages and at the same time incorporate tailored flexural stiffness makes their use for micropositioning very attractive. These attributes are well suited for micropositioning in hard disk drives to increase track density via dual servo systems with coarse and fine adjustment [49–52]. Requirements for this type of actuator include motion of ±0.5 μm with an applied voltage of tens of volts, resolution of a few nanometers with a bandwidth of several kilohertz, which can readily be met with HAR (High Aspect Ratio) microactuators.

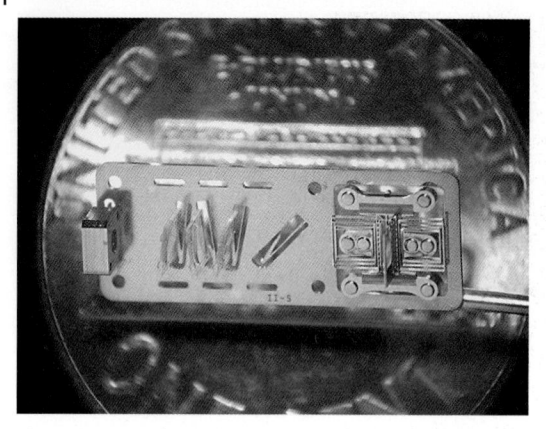

Figure 12.24 Multi-modal miniature microscope with LIGA fabricated optical table, lens mounts and electrostatically scanned amplitude grating.

The ability of LIGA-based microactuators to interface with and accurately position milliscale components makes their use in micro- and milliscale optics natural. Figure 12.12 shows an electromagnetic 1×2 fiber-optic switch fabricated with DXRL technology [53]. This device achieves sub-millisecond switching with 3 V drive and fiber–fiber insertion loss of less than 1 dB.

High aspect ratio metal resonant actuators have been applied to generate structured light sources for use in miniature confocal microscopes. Figure 12.24 shows a micrograph of a miniature microscope developed at the University of Arizona which is being applied to early cancer detection research [54]. The device uses a DXRL-defined 'optical bench' for alignment and support of miniature optics and also a resonant amplitude grating that provides structured light into the microscope when illuminated from a DC source. The amplitude grating, a 1.5×1.5 mm linear array of free-standing ~20 μm wires, is shown supported by the center of the resonant actuator in Figure 12.25. This figure also shows the electrostatic comb drive, which is capable of resonating the grating at ±100 μm amplitude with 10 V with quality factors at atmospheric pressure of over 100.

Demonstration of an entire Fourier transform spectrometer on a chip with integral electromagnetic microactuator has been demonstrated with the use of DXRL-based processing [55, 56]. Within an 11.5×9.4 mm area, an entire optical system with dual-coil electromagnetic actuator with attached conventionally fabricated mirror achieved a resolution of 24.5 nm with a mirror travel of 54 μm at 50 mW of drive power. The design is also amenable to aligned injection molding techniques for mass production.

Several more LIGA-fabricated actuator application examples include devices that combine precision-machined parts with batch-microfabricated components. A miniature mechanism that provides access to triggering a high consequence event only after a unique sequence of binary events has been undertaken has been realized through a LIGA-fabricated counter-meshing gear discriminator coupled

Figure 12.25 Close-up of electrostatically driven resonator with 1.5 × 1.5 mm amplitude grating clamped in perpendicular orientation to the body of the actuator with a bistable clamp.

with permanent magnet motors [57]. The 24 × 18 × 13 mm surety device is capable of discriminating a 24-bit sequence in 15 ms and assuring a probability less than 1 in 10^6 in resisting inadvertent triggering during an abnormal tampering or environmental event. By using magnetic coupling through a fluidic housing, gear pumps with LIGA-fabricated gears with relatively large surface area of interaction that possess embedded permanent magnets have been built to deliver a 2.5 mL min^{-1} flow rate at 3.6 psi when driven by a commercially available 3 mm diameter brushless DC motor [58]. Other fluidic devices have been developed that use LIGA-fabricated fluidic channels and reservoirs coupled with actuated membranes [59]. Research on microgrippers fabricated using LIGA processing also shows the flexibility offered for linkages and flexures with high aspect ratio and fine, nearly arbitrary, geometric features [60, 61].

12.6
Future Directions

Future microactuator development appears to be focused on more complex multilayer topology which makes more efficient use of device volume. Microelectronic and microsensor integration is also important for adaptive actuation and force feedback control. In this regard, LIGA processing typically requires only low temperatures, which makes it suitable for back end of line (*BEOL*) microelectronic processes and thus direct microelectronic integration. Figure 12.26 shows a simple example of microelectronic integration where a magnetic micromotor has been fabricated directly on a silicon substrate with a photodiode array to sense motion. With the ability to drive LIGA-based microactuators directly with silicon microelectronics, a considerable amount of electromechanical integration is possible.

Figure 12.26 Magnetic VR micromotor with 200 μm diameter rotor fabricated on a silicon p–n junction array with single layer metallization. The array is used as a rotary optical encoder.

References

1 Bacher, W., Menz, W. and Mohr, J. (1995) The LIGA technique and its potential for microsystems – a survey. *IEEE Transactions on Industrial Electronics*, **42**, 431–41.
2 Janocha, H. (2000) Microactuators – principles, applications, trends. MICRO. tec 2000, VDE. World Microtechnologies Congress. Expo 2000, 25–27 September 2000, Hannover, pp. 61–7.
3 Wallrabe, U. and Saile, V. (2005) LIGA technology for R&D and industrial applications, in *MEMS: a Practical Guide to Design, Analysis and Applications* (eds J. Korvink and O. Paul), William Andrew and Springer, pp. 853–99.
4 Rugar, D., Budakian, R., Mamin, H.J. and Chui, B.W. (2004) Single spin detection by magnetic resonance force microscopy. *Nature*, **430**, 329–32.
5 Trimmer, W. and Stroud, R.H. (2006) Scaling of micromechanical devices, in the *MEMS Handbook, Introduction and Fundamentals*, 2nd edn (ed. M. Gad-el-Hak), CRC Taylor and Francis, Boca Raton, FL, vol. **1**, Ch. 2, pp. 2-1–2-7.
6 McMahon, T.A., Bonner, J.T. (1983) *On Size and Life*, Freeman, New York.
7 Furuhata, T., Hirano, T., Gabriel, K.J. and Fujita, H. (1991) Sub-micron gaps without sub-micron etching, in Technical Digest, IEEE Micro, Electro Mechanical Systems Workshop, 30 January–2 February 1991, Nara, Japan, pp. 57–62.
8 Halbach, K. (1987) High performance permanent magnet materials. *Materials Research Society Symposium Proceedings*, **96**, 259.
9 Romankiw, L.T. and O'Sullivan, E.J.M. (1997) Plating techniques, in *Handbook of Microlithography, Micromachining and Microfabrication*, vol. 2, *Micromachining and Microfabrication* (ed. P. Rai-Choudhury),

SPIE Optical Engineering Press, Bellingham, WA, Ch. 5, pp. 197–298.

10 Thommes, A., Bacher, W., Leyendecker, K., Stark, W., Liebscher, H. and Jakob, C. (1993) LIGA microstructures form a NiFe alloy: preparation by electroforming and their magnetic properties, in Proceedings of the 1985 Meeting of the American Electrochemical Society, 11–12 October 1993, New Orleans, LA, vol. 94–6, p. 89.

11 Osaka, T., Takai, M., Hayashi, K. and Sogawa, Y. (1998) New soft magnetic CoNiFe plated films with high B_s = 2.0–2.1 T. *IEEE Transactions on Magnetics*, **34**, 1432–4, July.

12 Myung, N.V., Park, D.-Y., Yoo, B.-Y. and Sumodjo, P.T.A. (2003) Development of electroplated magnetic materials for MEMS. *Journal of Magnetism and Magnetic Materials*, **265**, 19–198.

13 Christenson, T., Garino, T.J. and Venturini, E.L. (1999) Deep X-ray lithography based fabrication of rare earth permanent magnets and their applications to microactuators. *Electrochemical Society Proceedings*, **98-20**, 312–23.

14 Dempsey, N.M., Kornilov, N.V. and Cugat, O. (2004) Thick hard magnetic films for mems: some key issues, in Proceedings of the 18th International Workshop on High Performance Magnets and their Applications (HPMA '04), Annecy.

15 Kawanami, S., Kurokawa, M., Taniguchi, M. and Tada, Y. (2001) Development of phased-array ultrasonic testing probe. *Mitsubishi Heavy Industries, Ltd. Technical Review*, **38**, 121–5.

16 Nakamae, K., Hirata, Y., Mizobuchi, H., Hashimoto, A. and Takada, H. (2004) Development of high-resolution and wide-band piezoelectric composite for ultrasonic probe. *SEI Technical Review*, **57**, 26–30

17 Lee, G.S., Jin, Y.Y., Park, S.J., Ajmera, P.K., Malek, C.K., Wang, T.J. and Tang, F. (1999) LIGA-like process for high-aspect ratio PZT microstructures. *Proceedings of the SPIE – The International Society for Optical Engineering*, **3673**, 127–32.

18 Mohr, J., Bley, P., Strohrmann, M. and Wallrabe, U. (1992) Microactuators Fabricated by the LIGA Process. *Journal of Micromechanics and Microengineering*, **2**, 234–41.

19 Burbaum, C., Mohr, J., Bley, P. and Menz, W. (1991) Fabrication of electrostatic microdevices by the LIGA technique. *Sensors and Materials*, **3** (*2*), 75.

20 Wallrabe, U., Bley, P., Krevet, B., Menz, W. and Mohr, J. (1994) Design rules and test of electrostatic micromotors made by the LIGA process. *Journal of Micromechanics and Microengineering*, **4**, 40–5.

21 Suzuki, K. (1990) Single crystal silicon micro-actuators, in Technical Digest – IEEE International Electron Device Meeting, IEDM, San Francisco, CA, pp. 625–8.

22 Christenson, T. and Guckel, H. (1995) Deep X-ray lithography for micromechanics. *Proceedings of the SPIE – The International Society for Optical Engineering*, **2639**, 134–45.

23 Guckel, H., Mangat, P.S., Emmerich, H., Massoud-Ansari, S., Klein, J., Earles, T., Zook, J.D., Ohnstein, T., Johnson, E.D., Sidons, D.P. and Christenson, T.R. (1996) Advances in Photoresist Based Processing Tools for Three-dimensional Precision and Micro Mechanics, Solid-state Sensor and Actuator Workshop, 2–6 June 1996, Hilton Head, SC, pp. 60–3.

24 Kalb, H., Kowanz, B., Bacher, W., Mohr, J. and Ruprecht, R. (1994) Electrostatically driven linear stepping motor in LIGA technique, in Proceedings of Actuator '94, 15–17 June 1994, Bremen, p. 83.

25 Jacobsen, S.C., Price, R.H., Wood, J.E., Rytting, T.H. and Rafaelof, M. (1989) A design overview of an eccentric-motion electrostatic microactuator. *Sensors and Actuators*, **20** (1–2), 1–16.

26 Samper, V.D., Sangster, A.J., Reuben, R.L. and Wallrabe, U. (1998) Multistator LIGA-fabricated electrostatic wobble motors with integrated synchronous control. *Journal of Microelectromechanical Systems*, **7**, 214–23.

27 Samper, V.D., Sangster, A.J., Wallrabe, U., Reuben, R.L. and Grund, J.K. (1998) Advanced LIGA technology for the integration of an electrostatically controlled bearing in a wobble micromotor. *Journal of Microelectromechanical Systems*, **7**, 423–7.

28 Samper, V.D., Sangster, A.J., Reuben, R.L. and Wallrabe, U. (1999) Torque evaluation of a LIGA fabricated electrostatic micromotor. *Journal of Microelectromechanical Systems*, **8**, 115–23.
29 Guckel, H., Christenson, T., Earles, T., Skrobis, K.J. and Klein, J. (1994) Laterally driven electromagnetic actuators, in Solid-state Sensor and Actuator Workshop, 13–16 June 1994, Hilton Head Island, SC, pp. 49–52.
30 Guckel, H., Christenson, T., Earles, T., Skrobis, K., Zook, D. and Ohnstein, T. (1994) Electromagnetic, spring constrained linear actuator with large throw, in Proceedings of the 4th International Conference on New Actuators, Actuator '94, 15–17 June 1994, Bremen, pp. 52–5.
31 Guckel, H., Christenson, T., Skrobis, K.J., Klein, J. and Karnowsky, M. (1993) Design and testing of planar magnetic micromotors fabricated by deep X-ray lithography and electroplating, in Digest of Technical Papers, 7th International Conference on Solid-state Sensors and Actuators–Transducers '93, 7–10 June 1993, Yokohama, pp. 76–9.
32 Christenson, T., Guckel, H. and Klein, J. (1996) A variable reluctance stepping microdynamometer. *Microsystem Technologies*, **2**, 139–43.
33 O'Sullivan, E.J., Cooper, E.I., Romankiw, L.T., Kwietniak, K.T., Trouilloud, P.L., Horkans, J., Jahnes, C.V., Babich, I.V., Krongelb, S., Hegde, S.G., Tornello, J.A., LaBianca, N.C., Cotte, J.M. and Chainer, T.J. (1998) Integrated variable-reluctance magnetic minimotor. *IBM Journal of Research and Development*, **42**, 681–94.
34 Leupold, H.A. (1996) Static applications, in *Rare-earth Iron Permanent Magnets* (ed. J.M.D. Coey), Clarendon Press, Oxford, Ch. 8, pp. 381–429.
35 Abele, M.G. (1993) *Structures of Permanent Magnets*, John Wiley & Sons, Inc., New York.
36 Vikramaditya, B., Nelson, B.J., Yang, G. and Enikov, E.T. (2001) Microassembly of hybrid magnetic MEMS. *Journal of Micromechatronics*, **1**, 99–116.
37 Jones, J.F., Kozlowski, D.M. and Trinkle, J.C. (2004) Micro-scale force-fit insertion. *Journal of Micromechatronics*, **2**, 185–200.

38 Christenson, T., Garino, T.J., Venturini, E.L. and Berry, D.M. (1999) Application of deep X-ray lithography fabricated rare earth permanent magnets to multipole magnetic microactuators, in Proceedings of Transducers '99, June 1999, Sendai, Japan, pp. 98–101.
39 Roters, H.C. (1947) The hysteresis motor–advances which permit economical fractional horsepower ratings. *AIEE Transactions*, **66**, 1419–30.
40 Cugat, O., Delamare, J. and Reyne, G. (2003) Magnetic micro-actuators and systems (MAGMAS). *IEEE Transactions on Magnetics*, **39**, 3607–12.
41 Debéda, H., Mohr, T.v., Freyhold, J., Wallrabe, U. and Wengelink, J. (1999) Development of miniaturized piezoelectric actuators for optical applications realized using LIGA technology. *Journal of Microelectromechanical Systems*, **8**, 258–63.
42 Guckel, H., Klein, J., Christenson, T. and Skrobis, K. (1992) Thermo-magnetic flexure actuators, in Technical Digest of IEEE Solid-state Sensor and Actuator Workshop, June 1992, Hilton Head Isand, SC, pp. 73–5.
43 Numazawa, T., Miura, K., Kawase, K. and Hirata, Y. (2003) Development of a MEMS optical switch composed of ribbon-like actuator. *SEI Technical Review*, **56**, 37–40.
44 Garcia, E.J., Christenson, T.R. and Polosky, M.A. (1997) Design and fabrication of a LIGA milliengine, in Transducers '97, June 1997, Chicago, IL, pp. 16–9.
45 Pokines, B.J. and Garcia, E. (1998) A smart material microamplificiation mechanism fabricated using LIGA. *Smart Materials and Structures*, **7**, 105–12.
46 Michel, F. and Ehrfeld, W. (1999) Mechatronic micro devices, in Proceedings of the 1999 International Symposium on Micromechatronics and Human Science, IEEE, pp. 27–34.
47 Li, B. and Chen, Q. (2005) Design and fabrication of *in situ* UV-LIGA assembled robust nickel micro check valves for compact hydraulic actuators. *Journal of Micromechanics and Microengineering*, **15**, 1864–71.
48 Haluzan, D.T. and Klymshyn, D.M. (2004) High-Q LIGA-MEMS vertical cantilever variable capacitors for upper microwave

frequencies. *Microwave and Optical Technology Letters*, **42**, 507–11.
49 Fan, L.-S., Hirano, T., Hong, J., Webb, P.R., Juan, W.H., Lee, W.Y., Chan, S., Semba, T., Imaino, W., Pan, T.S., Pattanaik, S., Lee, F.C., McFadyen, I., Arya, S. and Wood, R. (1999) Electrostatic microactuator and design considerations for HDD applications. *IEEE Transactions on Magnetics*, **35**, 1000–5.
50 Nakamura, S., Suzuki, K., Ataka, M., Fujita, H., Basrour, S., Soumann, V., Labachelerie, M. and Daniau, W. (1998) An electrostatic microactuator using LIGA process for a magnetic head tracking system of hard disk drives. *Microsystems Technology*, **5**, 69–71.
51 Fujita, H., Suzuki, K., Ataka, M. and Nakamura, S. (1999) A microactuator for head positioning system of hard disk drives. *IEEE Transactions on Magnetics*, **35**, 1006–10.
52 Toshiyoshi, H., Mita, M. and Fujita, H. (2002) A MEMS piggyback actuator for hard-disk drives. *Journal of Micro-Electro-Mechanical Systems*, **11**, 648–54.
53 Norwood, R.A., Holman, J., Shacklette, L.W., Emo, S., Tabatabaie, N. and Guckel, H. (1998) Fast, low insertion-loss optical switch using lithographically defined electromagnetic microactuators and polymeric passive alignment structures. *Applied Physics Letters*, **73**, 3187–9.
54 Lee, J., Rogers, J.D., Descour, M.R., Hsu, E., Aaron, J.S., Sokolov, K. and Richards-Kortum, R.R. (2003) Imaging quality assessment of multi-modal miniatue microscope. *Optics Express*, **11**, 1436–51.
55 Solf, C., Janssen, A., Mohr, J., Ruzzu, A. and Wallrabe, U. (2004) Incorporating design rules into the LIGA technology applied to a fourier transform spectrometer. *Microsystems Technology*, **10**, 706–10.
56 Wallrabe, U., Solf, C., Mohr, J. and Korvink, J.G. (2005) Miniaturized Fourier transform spectrometer for the near infrared wavelength regime incorporating an electromagnetic linear actuator. *Sensors and Actuators A*, **123**–4C, 459–67.
57 Zhang, W., Chen, W., Zhao, X., Li, S. and Jiang, Y. (2005) A novel safety device with metal counter meshing gears discriminator directly driven by axial flux permanent magnet micromotors based on MEMS technology. *Journal of Micromechanics and Microengineering*, **15**, 1601–6.
58 Deng, K., Dewa, A.S., Ritter, D.C., Bonham, C. and Guckel, H. (1998) Characterization of gear pumps fabricated by LIGA. *Microsystems Technology*, **4**, 163–7.
59 Schomburg, W.K., Vollmer, J., Bustgens, B., Fahrenberg, J., Hein, H. and Menz, W. (1994) Microfluidic components in LIGA technique. *Journal of Micromechanics and Microengineering*, **4** (4), 186–91.
60 Ballandras, S., Basrour, S., Robert, L., Megtert, S., Blind, P., Rouillay, M., Bernede, P. and Daniau, W. (1997) Microgrippers fabricated by the LIGA technique. *Sensors and Actuators A*, **58**, 265–72.
61 Carrozza, M.C., Eisinberg, A., Menciassi, A., Campolo, D., Micera, S. and Dario, P. (2000) Towards a force-controlled microgripper for assembling biomedical microdevices. *Journal of Micromechanics and Microengineering*, **10**, 271–6.

13
Development of Microfluidic Devices Created via the LIGA Process

Masaya Kurokawa

13.1 Introduction *323*
13.2 Manufacture of Nickel Tool Part *324*
13.3 Manufacturing of Plastic Microdevices *327*
13.4 Manufacture of Microfluidic Devices *330*
13.5 Conclusion *334*
    Acknowledgment 334
    References 334

13.1
Introduction

Usable plastic chemical microdevices for biological sensors or environmental gas sensors are generally fabricated by casting polydimethylsiloxane (PDMS) on a master plate that had the inverse shapes of micropatterns [1–3]. The inverse shapes on the master plate are able to be produced on a silicon wafer by microelectromechanical systems (MEMS) or on the SU-8 pattern layer formed by conventional photolithography with sub-micrometer accuracy. However, the PDMS devices are difficult to handle because they are too flexible in comparison with glass microdevices. Also, the reproduction of devices with a guarantee of sub-micrometer accuracy is almost impossible. To improve the handling of devices and precise reproducibility in the mass production of microdevices, hard plastic microdevices such as poly(methyl methacrylate) (PMMA) and polystyrene should be developed by manufacturing an injection mold and applying an injection molding technique with the mold [4–6].

In order to produce precisely plastic microdevices having a fine pattern by injection molding, one of the important aspects is to manufacture the tool part with inverse shapes of micropatterns. Usually the pattern making on the tool part is carried out by a machining process such as a cutting process and an electrical discharge machining process. However, the tool part manufactured using this machining process is unsuitable for plastic microdevices, as microstructure patterns ranging from several hundred to several micrometers are demanded, cutting with a tolerance in the sub-micrometer range.

In this chapter, a newly developed manufacturing method for plastic microdevices is described in detail. The method consists of three steps:

1. manufacturing process of a nickel (Ni) tool part with synchrotron radiation (SR) and an electroforming
2. production process of plastic microdevices using an injection molding technique with its Ni tool part
3. making microfluidic devices by bonding microflow devices and cover plates.

It will be possible to achieve reproducibility of nanometer precision in the mass production of the microdevices by an injection molding process if the Ni tool part is manufactured successfully and the injection molding conditions are set appropriately. The new method offers the prospect of obtaining plastic microdevices (microfluidic devices) using the Lithographie Galvanoformung Abformung (LIGA) process.

13.2
Manufacture of Nickel Tool Part

The process to produce the plastic microdevices using injection molding is shown in Figure 13.1. For this process, the Ni tool part having the inverse shape of its micropattern for the microdevice should be made accurately by SR and an elec-

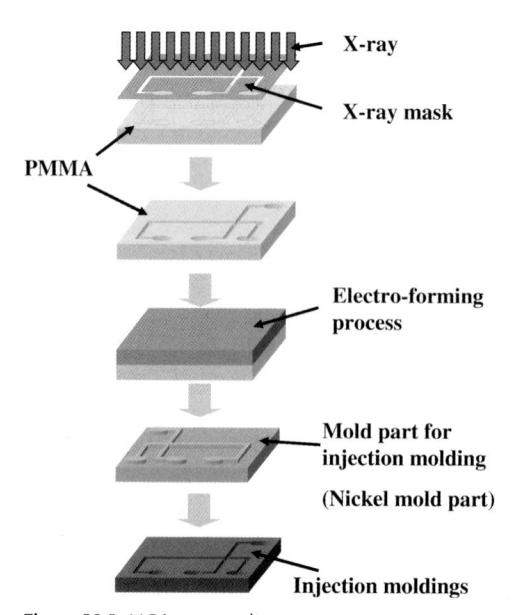

Figure 13.1 LIGA process line.

troforming process. If the Ni tool part can be completed successfully, microdevices with a tolerance of nanometers can be injection molded with its Ni tool part and injection molding technique. Hence the manufacture of the Ni tool part is very important for the production of plastic microdevices.

The X-ray mask shown in Figure 13.1 is made of a frame and a membrane. The membrane supported by the frame has a fine pattern on it, made by photolithography. Generally, gold, copper, tungsten or tantalum is selected as the material for X-ray absorption and silicon nitride, silicon carbide or plastic film such as polyimide is selected as the membrane. By selecting the most suitable membrane with a frame and controlling the photolithography conditions for producing micropatterns on the membrane, the X-ray mask is finished appropriately.

The surface of a PMMA plate (positive resist) was exposed to X-rays until the irradiated amount of the X-ray energy reached an appropriate dose. The relationship between the X-ray dose and pattern depth on the PMMA plate after development of the plate is shown in Figure 13.2. The development, stopper and rinse conditions of the exposed PMMA plate are given in Table 13.1.

By moving the X-ray mask systematically during the exposure process, microstructures having an optional side-wall inclination can be established on the PMMA plate [7–9]. This mechanism is shown in Figure 13.3. The distance between the X-ray mask and PMMA plate should be kept around a few millimeters to move the X-ray mask. However, the longer this distance was, the higher the X-ray diffraction effect became. When the moving mask sequence was set to obtain a 5° side-wall inclination and the distance was set at 5 mm, its inclination was about 8°. For the simulation of X-ray diffraction for exposing micropatterns of 10 μm in diameter through the X-ray mask on the PMMA plate, the distance should be set below 1 mm to avoid the X-ray diffraction effect for a pattern of 10 μm in diameter.

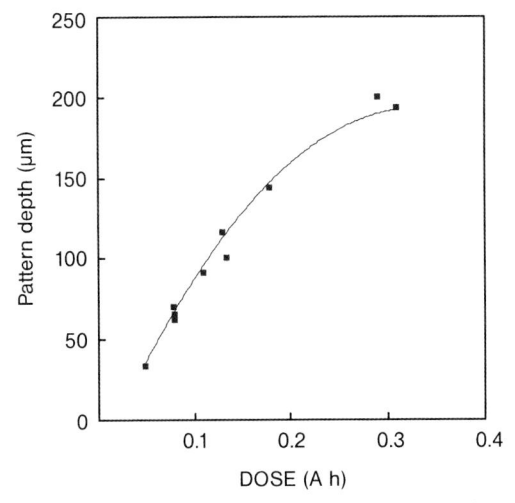

Figure 13.2 Relationship between the X-ray dose and pattern depth on a PMMA plate after development of the plate.

Table 13.1 Development, stopper and rinse conditions of the exposed PMMA plate[a].

Solvent	Concentration (wt%)
Development solution	
2-(2-*n*-Butoxyethoxy)ethanol	60
Morpholine	20
2-Aminoethanol	5
Water	15
Stopper solution	
2-(2-*n*-Butoxyethoxy)ethanol	80
Water	20

a Development conditions: development solution temperature, 40 °C; development time, 120 min. Stopper conditions: stopper solution temperature, 40 °C; treatment time, 10 min. Rinse conditions: rinse solution, water; rinse temperature, room temperature.

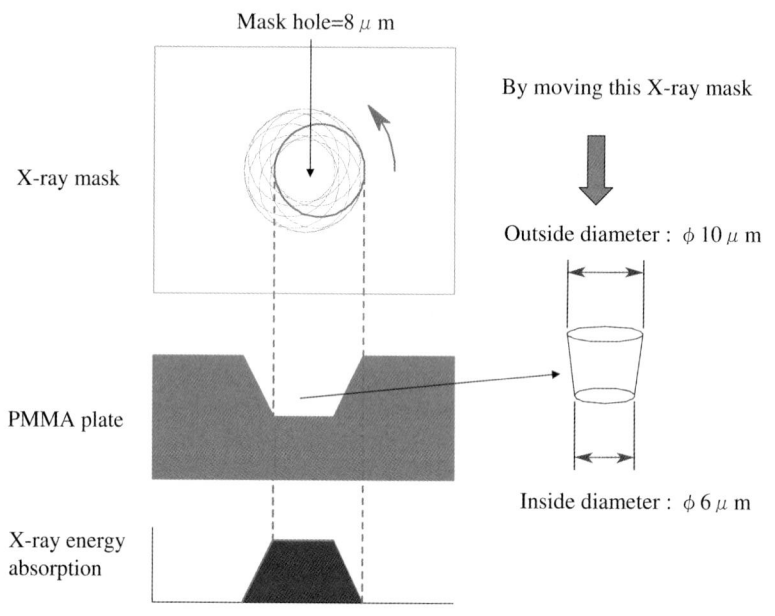

Figure 13.3 Patterning process by SR (moving X-ray mask system).

The simulation results of X-ray diffraction are shown in Figure 13.4. The X-ray diffraction effect shown in Figure 13.4 indicates that the distance between the X-ray mask and PMMA plate is one of the important factors for determining the side-wall inclination.

The Ni tool part is made by an electroforming process for the PMMA plate (master plate) with the formed microstructure and is given electrical conducting

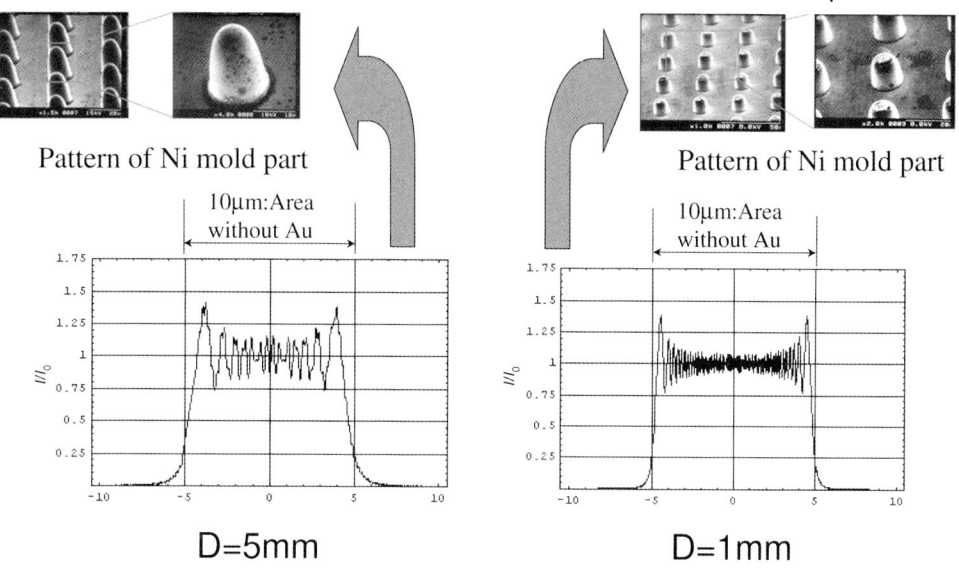

Figure 13.4 Simulation results of X-ray diffraction and SEM images of patterns on the Ni tool part. I/I_0, intensity of X-rays through the X-ray mask; D, distance between the X-ray mask and PMMA plate.

properties on the surface of the PMMA plate. This electroforming process is usually conducted by applying the most suitable conditions such as pH, electric current, solution temperature and so forth, of the nickel sulfamate solution bath, since the electroforming process conditions influence the quality of the Ni tool part (for example warpage and surface properties) [10]. After the Ni tool part has been cut to appropriate dimensions by machining, it is inserted into the injection mold base. The injection mold with the Ni tool part inserted in the base is shown in Figure 13.5.

13.3
Manufacturing of Plastic Microdevices

Reproducibility of nanometer precision can be attained in the mass production of microdevices by the injection molding process if the Ni tool part is manufactured successfully by the electroforming process on the PMMA plate with formed microstructures using the SR process and the injection molding conditions are set appropriately using the injection mold with the Ni tool part inserted in the mold base. Also, the injection molding process can produce molded parts (microdevices) repeatedly with a cycle time of several tens of seconds, as one of the features of this process. Various investigations for forming micrometric or nanometric patterns on molded parts have been conducted [11–13]. A general injection molding machine is shown in Figure 13.6. When plastic microdevices are produced by

13 Development of Microfluidic Devices Created via the LIGA Process

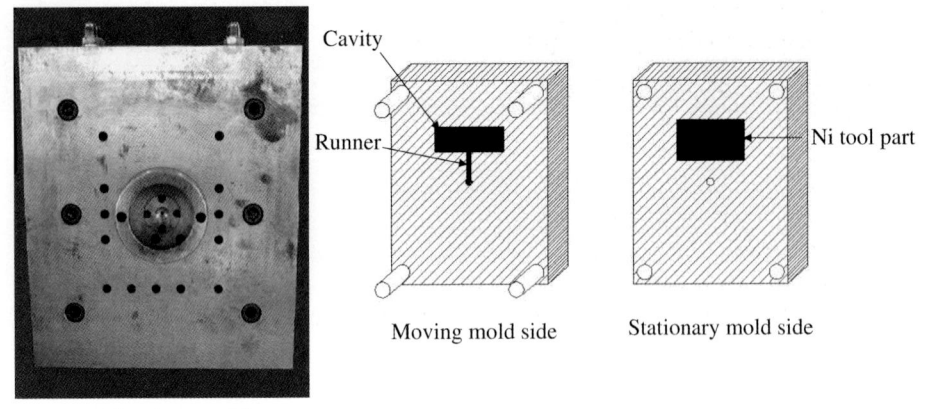

Figure 13.5 Injection mold with the Ni tool part inserted into the mold base.

Figure 13.6 Injection molding machine.

injection molding, high transcription of the micropattern of the Ni tool part to the molded parts is required. Injection molding conditions such as mold temperature, melt temperature and injection pressure for enhancing the transcription of the molded parts are set higher than those for conventional molded parts. However, when the molded parts are obtained with these conditions, the higher the transcription, the larger is the warpage of the molded parts. Therefore, it is very important to produce microdevices having balanced properties of high transcription and low warpage.

To obtain molded parts having the above properties, the following injection molding techniques are usually applied individually or combined, in addition to control of the cylinder setting temperature and the screw revolution to optimize the plasticizing temperature of the polymer:

1. injection compression molding
2. high-speed injection molding
3. injection molding with insulation around the tool part (insulation method)
4. injection molding with heating and cooling around the tool part per cycle (heating and cooling method)
5. injection molding with control of stationary and moving mold temperature individually
6. injection molding with control of mold pressure on packing stage of melt
7. injection molding with gas assistance [14, 15]
8. decision regarding a suitable gate position for the microdevice using computer-aided engineering analysis.

In this work, using an Ni tool part having microstructures as shown in Figure 13.7, an investigation was carried out on the above properties. The result is shown in Figure 13.8. PMMA melt was injected into the mold when the mold temperature had reached a level above the glass transition temperature of PMMA by means of a heating unit. The mold was cooled quickly to the demolding temperature by a cooling unit immediately when the packing stage for injection molding process

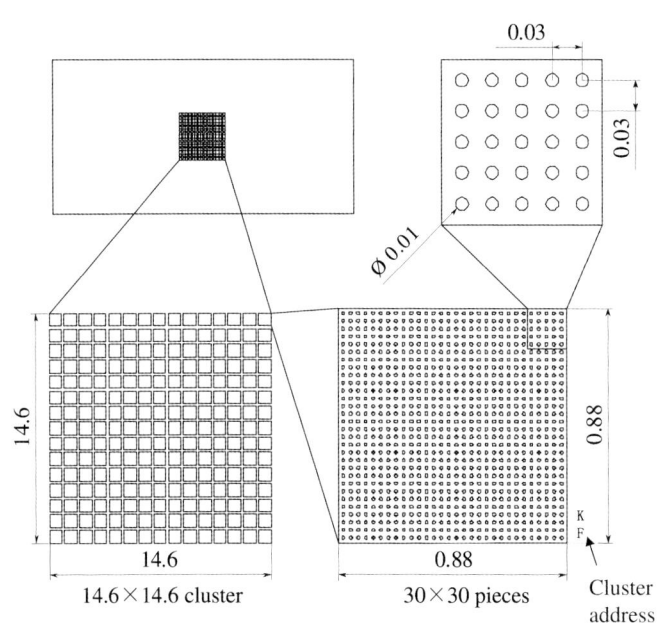

Figure 13.7 Microstructure patterning on a PMMA plate. Microstructure patterning is 15 × 15 clusters with a spacing of 100 μm and each cluster being constituted of 30 × 30 pieces of 10 μm diameter aligned with a pitch of 30 μm.

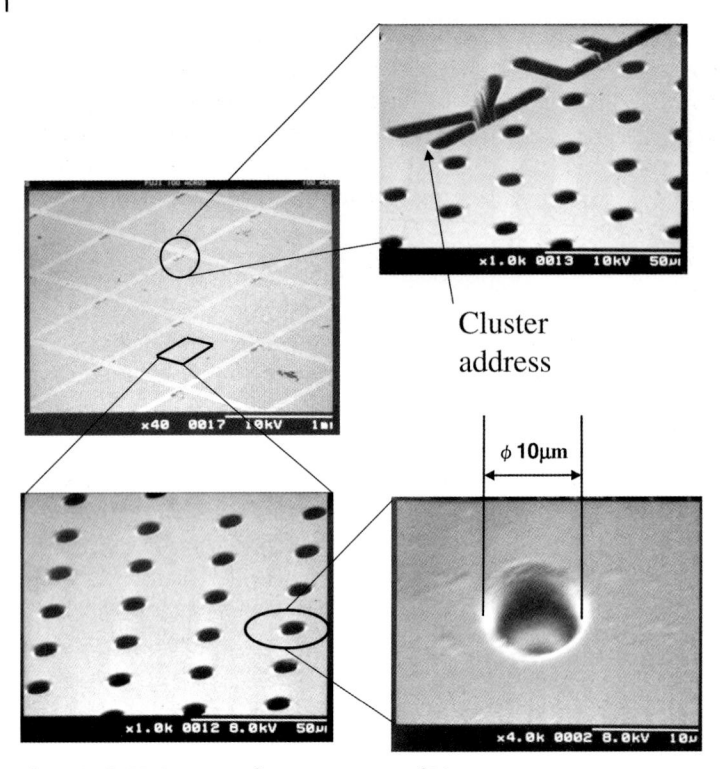

Figure 13.8 SEM images of microstructures of 10 μm diameter formed on the injection molded PMMA device.

had finished. Thereafter the PMMA device was demolded from the mold. It was found that microstructures of 10 μm in diameter and the cluster address from the SEM images in Figure 13.8 had high transcription from the pattern of the Ni tool part. The cycle time was about 90 s. Also, the warpage of the molding became about 10 μm with individual control of the stationary and moving mold temperatures. Molded parts having high transcription and low warpage for microfluidic devices are also obtained by these processes. For other thermoplastics, molded parts having high transcription would be obtained with the same manner of injection molding process. Thus the mold should be heated to above the glass transition temperature when injection starts, and thereafter the mold should be cooled quickly to the demolding temperature as in the case of PMMA.

13.4
Manufacture of Microfluidic Devices

In the case of glass microfluidic devices, techniques of bonding between the glass microdevice and glass cover plate were already established and the anodic bonding process as one of the bonding techniques has been used widely from the reliability point of view. However, a bonding technique for plastic microfluidic devices has not yet been established. Therefore, for the development of bonding techniques for plastic microfluidic devices various bonding processes (for example, heat press process using a heater, laser and vibration and a bonding process using adhesive that hardens on exposure to ultraviolet radiations) have been tried.

In the case of plastic microfluidic devices, the following problems for bonding should be addressed:

1. leakage from bonded positions
2. bubble inclusion between bonded faces
3. deformation of cross-section of the microflow part
4. reduction in bonding time (high throughput).

To solve these problems, a microflow device with ribs along both sides of the microflow path is proposed in this chapter. The microfluidic device can be fabricated by bonding the surface of the ribs and cover plate. In this bonding process, suitable setting of the bonding condition is easily done with a heat press, since the bonding part with the cover plate is limited to only the ribs. Also, bubble inclusion between the microflow device and the cover plate can be eliminated. It is possible to shorten the bonding time considerably in comparison with that for the microflow device without ribs and the cover plate, since the bonding part of the microflow device is limited to only the ribs on this device.

An example of a microdevice with ribs along both sides of the microflow path is shown in Figure 13.9. The rib was set 25 μm wide and 20 μm high for the microflow dimensions (100 μm wide and 100 μm deep). Also, the side-wall inclination was set about 10° for easy demolding of the microflow device with ribs for the injection molding process.

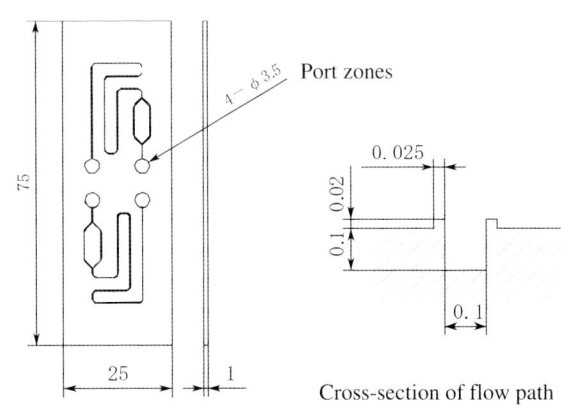

Figure 13.9 Microflow device with ribs along both sides of the microflow path. Microflow path, 100 μm width and 100 μm depth. Rib, 25 μm width and 20 μm height.

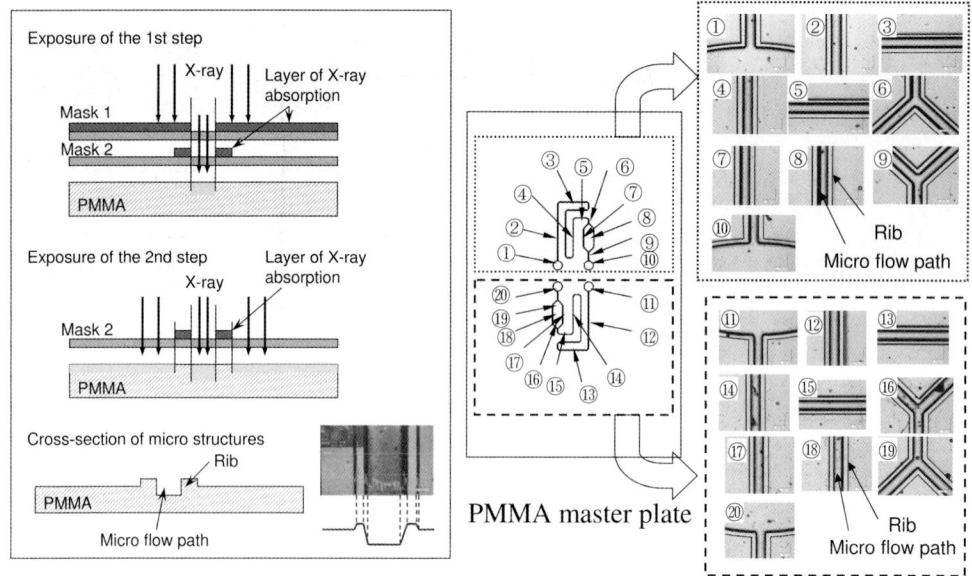

Figure 13.10 Process for manufacturing the master plate of the microflow device with rib and optical micrographs of microstructures formed on the plate.

In the case of manufacturing the microflow device with ribs using the injection molding process, the Ni tool part that accurately transcribes the micropattern of the microflow device must be made by SR and the electroforming process. The master plate for using in the electroforming process was made with the SR process. This manufacturing process is shown in Figure 13.10. As shown, the flow path and the ribs along both of its sides were made by combining two X-ray masks during the SR process. The alignment of micropatterns on the X-ray masks is very important in this process. If there is misalignment of the combination of two X-ray masks, the micropattern will be irregular. After the Ni tool part had been produced, it was inserted into the injection mold base. Then polystyrene (PS) was injection molded to obtain a microflow device with ribs (Figure 13.11). Ribs 25 µm wide and 20 µm high are formed precisely along both sides of the microflow path (100 µm wide and 100 µm deep).

The bonding between the PS microflow device with ribs and PS cover plate was conducted with a heat press. The bonding clearance ranges from 25 µm (corresponding to the height of a rib) to 0 µm (corresponding to plate-to-plate bonding), depending on the conditions of the rib deformation. There is a case where plasma is irradiated on the bonding surfaces of a microflow device and a cover plate in just a bonding process. When plasma of suitable energy is irradiated on the bonding surfaces, there are cases where the bonding strength is higher and the bonding temperature is lower than in the case without plasma energy. A

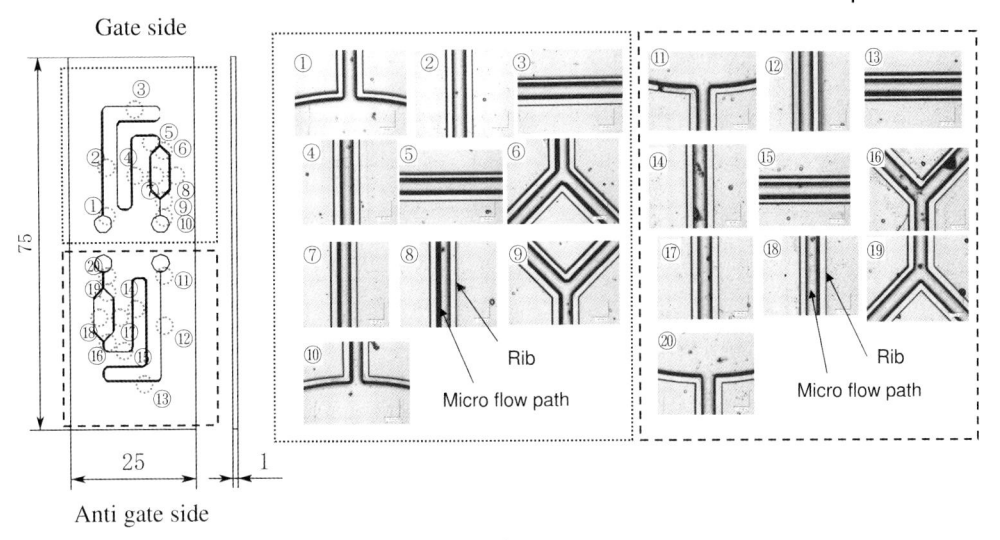

Figure 13.11 PS microflow device with ribs, Specimen dimensions: 75 × 25 × 1 mm.

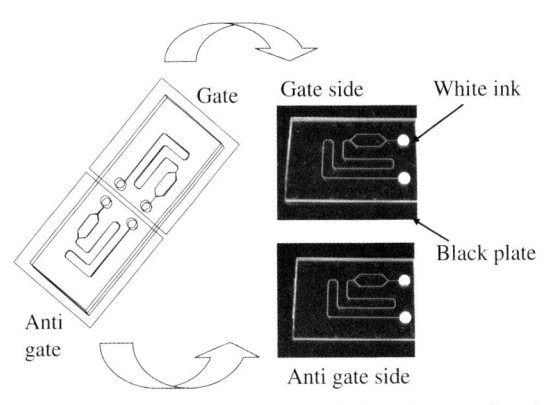

Figure 13.12 Microfluidic device made from the microflow device with ribs and cover plate.

microfluidic device made from the microflow device with ribs and cover plate is shown in Figure 13.12. A leak test was conducted by flowing white ink into the microflow path as shown in Figure 13.12.

The particular investigation of the rib dimensions for the microflow path geometry (for example, the effect of rib dimensions on bonding with the cover plate) needs to have high quality of bonding in the future.

The fluorescence level (which depends on plastic material such as PMMA, PS and polycarbonate) generated from a plastic microdevice usually is higher than that of a glass microdevice when a laser having a wavelength range of 400–600 nm is irradiated on a microdevice. Various investigations of the fluorescence of plastic device are conducted by addition of specific additives to the material or improvement of the surface roughness on a plastic microdevice.

13.5
Conclusion

In this chapter, the manufacturing process for an Ni tool part with SR and an electroforming process as the first step, the production process of plastic microdevices using the injection molding technique with its Ni tool part as the second step and microfluidic devices made by bonding a cover plate and microflow device with ribs along both sides of the microflow path as the third step were described in detail.

Reproducibility of nanometer precision will be able to be attained in the mass production of microdevices by an injection molding process if the Ni tool part is manufactured successfully and injection molding conditions are set appropriately. With regard to the fabrication of plastic microfluidic devices, a microflow device with ribs along both sides of the microflow path was proposed. The bonding time could be shortened considerably in comparison with that for the microflow device without ribs and the cover plate, since the bonding part of the microflow device was limited to only the ribs on this device. Also, the bubble inclusion between the microflow device and the cover plate could be eliminated.

Acknowledgment

The author wishes to express his thanks to Mr T. Minami, Mr N. Oiko, Mr M. Fujihashi and Miss T. Tanaka, Micro Device Development Department, Miniature Precision Products Division, Starlite Co. Ltd, for developing the plastics microdevices using the injection molding technique. Thanks are also due to Prof. Dr O. Tabata, Department of Mechanical Engineering, Kyoto University, for simulating the X-ray diffraction.

The descriptions of the manufacture of plastic microfluidic devices in this chapter are based on data that were provided by research work in the period from 2003 to 2005 in the Nanotechnology Researchers Network Project promoted by the Ministry of Education, Culture, Sports, Science and Technology of Japan.

References

1 Effenhauser, C.S., Bruin, G.J.M., Paulus, A. and Ehrat, M. (1997) Integrated capillary electrophoresis on flexible silicone microdevices: analysis of DNA restriction fragments and detection of single DNA molecules on microchips. *Analytical Chemistry*, **69**, 3451–7.

2 Yokokawa, R., Takeuchi, S., Kon, T., Ohkura, R., Edamatsu, M., Sutoh, K. and Fhjita, H. (2003) Transportation of micromachined structures by biomolecular linear motors. *Micro Electro Mechanical Systems*, 8–11.

3 Leclerc, E., Sakai, Y. and Fujii, T. (2003) A multi-layer PDMS microfluidic device for

tissue engineering applications. *Micro Electro Mechanical Systems*, 415–18.
4 Dang, F., Tabata, O., Kurokawa, M., Ewis, A.A., Zhang, L., Yamaoka, Y., Shinohara, S., Shinohara, Y., Ishikawa, M. and Baba, Y. (2005) High-performance genetic analysis on microfabricated capillary array electrophoresis plastic chips fabricated by injection molding. *Analytical Chemistry*, **77**, 2140–6.
5 Dang, F., Shinohara, S., Tabata, O., Yamaoka, Y., Kurokawa, M., Shinohara, Y., Ishikawa, M. and Baba, Y. (2005) Replica multichannel polymer chips with a network of sacrificial channels sealed by adhesive printing method. *Lab on a Chip*, **5**, 472–8.
6 Yamamura, S., kishi, H., Tokimitsu, Y., Kondo, S., Honda, R., Rao, S.R., Omori, M. and tamiya, E. and Muraguchi, A. (2005) Single-cell microarray for analyzing cellular response. *Analytical Chemistry*, **77**, 8050–6.
7 You, H., Matsuzuka, N., Yamaji, T., Uemura, S., Dama, I. and Tabata, O. (2001) Deep X-ray exposure system with multistage for 3-D microfabrication, *Memoirs of the Sr Center (Ritsumeikan University)*, **3**, 135–46.
8 Tabata, O., Matsuzuka, N., Yamaji, T. and Uemura, S. (2002) 3D fabrication by moving mask deep X-ray lithography (M2DXL) with multiple stages. *Memoirs of the Sr Center (Ritsumeikan University)*, **4**, 75–84.
9 Matsuzuka, N. and Tabata, O. (2003) Algorithm for analyzing optimal mask movement pattern in moving mask deep X-ray lithography (M2DXL). *Memoirs of the Sr Center (Ritsumeikan University)*, **5**, 71–82.
10 Kitadani, T., Utsumi, Y. and Hattori, T. (2002) Fabrication of nickel micro mold inserts by the LIGA process using synchrotoron radiation from 'New SUBARU', in Proceedings of the 19th Sensor Symposium, pp. 327–30.
11 Whiteside, B., Martyn, M.T. and Coates, P.D. (2004) Micromoulding: process evaluation, in Annual Technical Conference Technology Papers 2004, pp. 757–61.
12 Chang, C., Chen, S. and Yang, S. (2004) Precision molding of V-grooved micro parts, in Proceedings of PPS-20, Paper No. 121.
13 Michaeli, W. and Gartner, R. (2004) Injection molding of micro-structured surfaces, in Annual Technical Conference Technology Papers 2004, pp. 752–6.
14 Moore, S. (2001) CO_2 process claimed to enhance part finish. *Modern Plastics International*, **31**, 62.
15 Smith, C. (2004) K2004 first news injection moulding expanding the bubble. *European Plastic News*, **31**, 16–17.

14
Application of Inspection Devices

Yoshihiro Hirata

14.1 Introduction *337*
14.2 Piezoelectric Composite Material for Ultrasonic Diagnosis *338*
14.2.1 Introduction *338*
14.2.2 Production Process *339*
14.2.2.1 SR Lithography *340*
14.2.2.2 Production of Mold Insert *341*
14.2.2.3 Plastic Molding *341*
14.2.2.4 Formation of Superfine Ceramics *341*
14.2.3 Properties of Composites *342*
14.2.4 Evaluation *342*
14.3 Microcontact Probe for IC Testing *344*
14.3.1 Introduction *344*
14.3.2 Production Process *345*
14.3.3 Material for Contact Probe *345*
14.3.4 Design of Tip Shape *346*
14.4 Conclusion *349*
References *349*

14.1
Introduction

In inspection technology, the size reduction trend is well established. Therefore, in this field, the application of the LIGA process is expected. In this chapter, we select two sample applications: piezoelectric composites for ultrasonic diagnosis and contact probes for IC testing.

For ultrasonic diagnosis, the operating frequency is increasing and the wavelength of ultrasound is decreasing. On the other hand, the lead zirconate titanate (PZT) rods in the composites must be finer than one-fifth of the wavelength. Therefore, the size of PZT rods should be finer. In the IC testing field, the size of electrode pads is decreasing, so the contact probe becomes increasingly finer.

As described above, in these two fields, the downsizing is rapid. In addition, the components are mechanical parts and should have a high aspect ratio. The LIGA process is suitable for these fields.

14.2
Piezoelectric Composite Material for Ultrasonic Diagnosis

A ceramic microfabrication process using the LIGA process has been developed for a 1–3 piezoelectric composite (Figure 14.1). This composite material was predicted to be suitable for high-frequency and wide-band ultrasonic transducers used in diagnostic medicine and non-destructive testing. However, no process was available to fabricate micro- and high aspect ratio PZT columnar arrays; therefore, a piezoelectric composite for high frequency was not realized. The LIGA process allowed the production of an array of PZT columns whose cross-section is $25 \times 25\,\mu m$ and whose height is $250\,\mu m$. Using the composite developed for an ultrasonic endoscope, the ultrasonic pulse width was improved from 240 to 180 ns and the bandwidth was expanded from 60 to 150%.

14.2.1
Introduction

In ultrasonic diagnosis used in medicine, ultrasonic waves are transmitted and received via an ultrasonic probe pressed against the patient's body. A tomographic image is obtained using the time between the transmission and the reception of waves reflected back from the surface of an organ to calculate the distance between the surface and the organ.

Ultrasonic diagnosis in medicine is more useful than X-ray computed tomography (CT) or magnetic resonance imaging (MRI), because it causes little radiation injury to patients, permits the measurement of blood flow and yields real-time image data. However, ultrasonic diagnosis cannot replace X-ray CT or MRI because it has low resolution and difficult position recognition. It has been reported that changing the transducer material from a piezoelectric ceramic to a composite material composed of piezoelectric ceramic columnar arrays and resin, as shown in Figure 14.1, is effective in improving resolution, bandwidth and sensitivity [1, 2] for the following reasons:

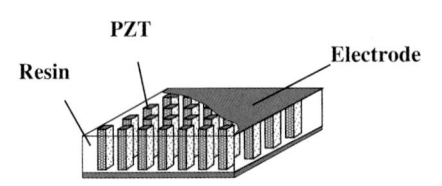

Figure 14.1 Schematic view of 1–3 piezoelectric composites.

- Because the resin functions as a damper, the mechanical quality factor, Q_m, of the material is low. Therefore, attenuation of the oscillation is rapid and the ultrasonic pulse becomes shorter. As a result, the resolution in the depth direction is improved.

- Because the aspect ratio (height/width) of PZT columns in the composites is high, the input energy is effectively converted to oscillations in the depth direction; the electromechanical coefficient in the thickness mode, k_t, is higher than that of bulk PZT. This improves the sensitivity of the measurement and results in expansion of the viewable area.

Piezoelectric ceramic columns must have a sufficiently small cross-section compared with the ultrasonic pulse length. For example, a transducer with a frequency of 10 MHz requires PZT columns having a cross-section less than 40 μm wide (one-quarter of the wavelength), because the wavelength in the human body is about 150 μm. The aspect ratio of PZT is required to be >3 [3]. The conventional fabrication method for micro-PZT columnar arrays is dicing; however, a limitation exists in decreasing the size of the PZT, because PZT is a fragile ceramic material. In this work, a new microfabrication method for PZT columnar arrays was developed and the predicted merits of composites were confirmed.

14.2.2
Production Process [4]

Figure 14.2 shows the process developed. First, synchrotron radiation (SR) lithography was performed on a resist structure to form many fine holes with high aspect ratio on the conductive substrate. The cross-section of the holes was $30 \times 30\,\mu m$ and the depth was 300 μm. Then, nickel electroforming, using the substrate as an electrode, was performed to obtain a nickel mold insert as the negative of the resist

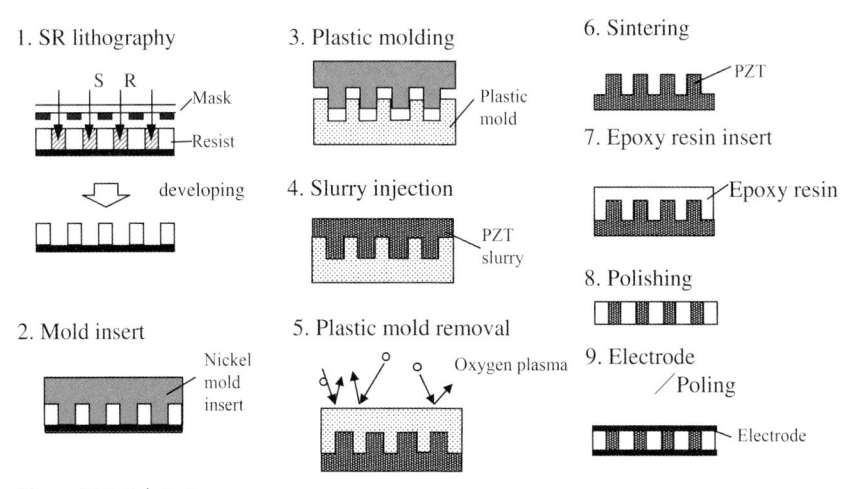

Figure 14.2 Fabrication process.

(a) Resist structure

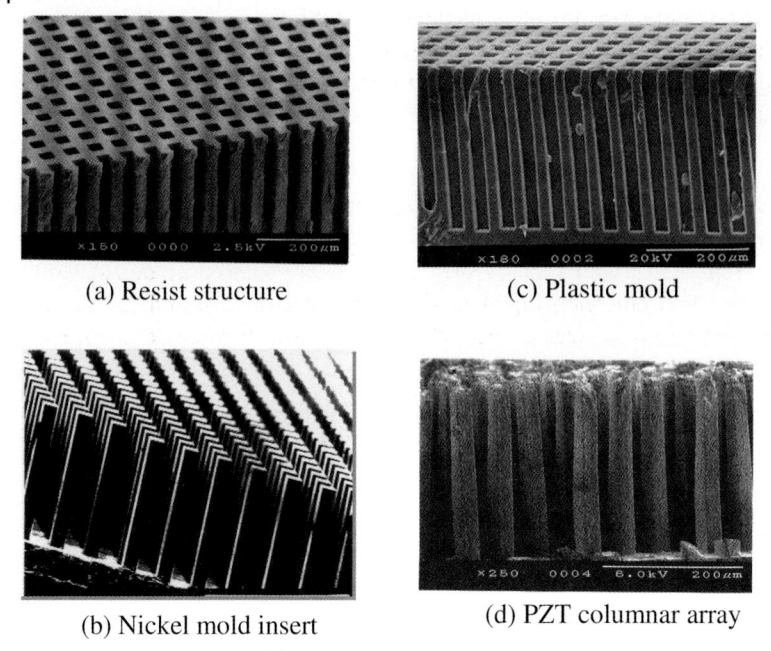

(c) Plastic mold

(b) Nickel mold insert

(d) PZT columnar array

Figure 14.3 SEM images of fabrication process.

structure. Using the mold insert, a plastic mold having the same shape as the resist was manufactured. Then a PZT slurry was injected into the plastic mold. After the slurry had dried and hardened, the plastic mold was removed by plasma etching. Finally, the PZT columnar array was obtained by removing the binder at 1200 °C. The amount of linear shrinkage was 13%; thus, an array of PZT columns whose cross-section was $25 \times 25\,\mu m$ and whose height was $250\,\mu m$ was realized. In the spaces between the PZT columns, epoxy resin was cast in vacuum and any superfluous material on the upper and lower surfaces was removed by polishing to a predetermined thickness. Finally, chromium and gold were sputtered to form electrodes. Figure 14.3 shows scanning electron microscope (SEM) images of the (i) resist structure, (ii) nickel mold insert, (iii) plastic mold and (vi) PZT columnar array.

Using this process, the fabrication of PZT columns with cross-sectional dimensions of $25 \times 25\,\mu m$ and a height of $250\,\mu m$ was realized. In addition, mass production of PZT columnar arrays using molds is possible and the production cost of the process developed is lower than that of the conventional dicing process.

In the following, we describe each process in detail.

14.2.2.1 SR Lithography

To shorten the SR irradiation time, a highly sensitive resist was developed. The resist is a copolymer of methyl methacrylate (MMA) and methacrylic acid (MAA).

It is 10 times more sensitive than poly(methyl methacrylate) (PMMA), which is usually used for the LIGA process [4]. The precision of the shape is comparable to that of PMMA; a perpendicularity of 0.16 µm or less per 100 µm of height can be realized.

14.2.2.2 Production of Mold Insert

By carrying out nickel electroforming using the resist structure made by SR lithography, a fine-structure metal mold insert was formed. For electroforming into superfine holes with high aspect ratio, it was necessary to develop ways to allow the electroforming liquid to permeate, because it is difficult for the liquid to permeate into holes of the resist structure with high aspect ratio. Step-by-step control of the current density and improvement of liquid circulation resulted in an electroforming process that did not induce internal stress or defects.

The Vickers hardness of the mold insert is 300 and the internal stress is less than 30 MPa. The difference between the top and bottom of a column was 0.8 µm when the height of the column was 300 µm. This taper was sufficient to decrease the demolding force and satisfied the size requirement for composites.

14.2.2.3 Plastic Molding

This mold insert has high aspect ratio columnar micropatterns; thus, conventional injection molding is not possible because the high viscosity of the resin destroys the mold insert. To solve this problem, a resin syrup of low viscosity was injected into the mold insert cavity and polymerized thermally. The syrup contained 15% acrylic polymer, whose molecular weight was 300 000 and whose viscosity was 700 mPa s.

For molding with the fine, high aspect ratio mold insert, reducing demolding stress is important. By optimizing the mold release agent and the temperature profile of polymerization, the demolding stress was suppressed to less than 10 MPa and a plastic mold with fine holes was realized.

14.2.2.4 Formation of Superfine Ceramics

After injecting the PZT slurry into the resin molds, the resin molds must be moved before sintering. When fabricating a ceramic structure using the LIGA process, conventional thermal dissolution is normally used to remove the plastic mold. However, micro-PZT columns with high aspect ratio topple in the process. This phenomenon occurs because of the following events. Resist melts and moves around the green PZT columns. The stress applied to the bottom of the columns increases as the columns' diameter and surrounding space decrease. In addition, binder in the columns is dissolved by the heat; hence the PZT columns topple easily. The resin mold must therefore be removed by a non-thermal process; hence O_2–CF_4 plasma etching was developed to remove resin mold [5]. By controlling the radiofrequency (RF) power and other parameters, removing the plastic mold without the collapse of PZT columns was realized.

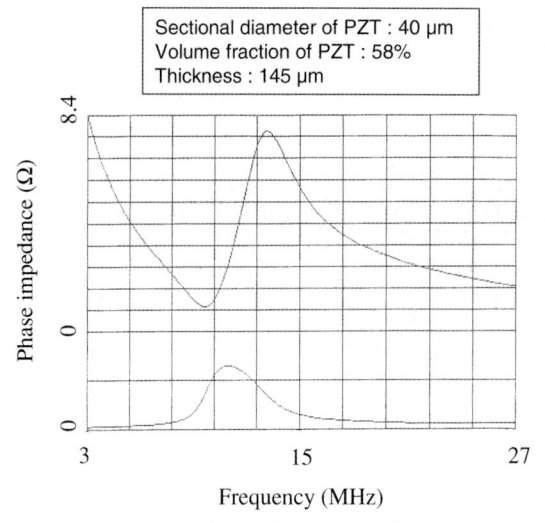

Figure 14.4 Example of impedance curve of piezoelectric composites.

14.2.3
Properties of Composites

An example of the impedance curve of the composites developed is shown in Figure 14.4. There are no small peaks which are usually observed in the impedance curve of bulk PZT. Q_m was 7 and k_t was 69%. As expected from theory, Q_m is lower and k_t is higher than those of conventional materials.

14.2.4
Evaluation

By examining our piezoelectric composite using an ultrasonic endoscope, ENDOECHO, produced by Olympus Corporation, the properties of the probe were compared with those of a conventional probe [6]. Figure 14.5 shows the acoustic pulse shape of each probe. The pulse width of the conventional probe is about 240 ns and that of the probe with our material is about 180 ns. The pulse width of the new probe was about 30% shorter and the resolution of the probe was higher than those of the conventional probe.

Figure 14.6 shows the transfer function of each probe. The relative frequency band [frequency bandwidth (sensitivity −6 dB)/center frequency] of the conventional probe is about 60% and that of the new probe is about 150%. The frequency bandwidth of the new probe is about three times larger than that of the conventional probe. This is useful for doctors because a wide area can be observed at one time.

The image of an actual human stomach, taken using the new probe, may be compared with that taken using the conventional probe. Figure 14.7 (the images

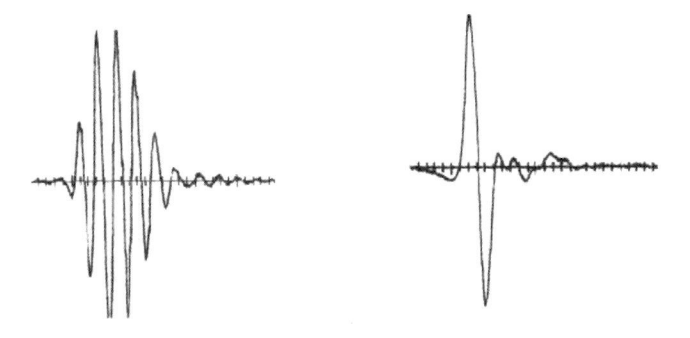

(a) Conventional transducer (b) Piezoelectric composite transducer

Figure 14.5 Acoustic impulse shape [6].

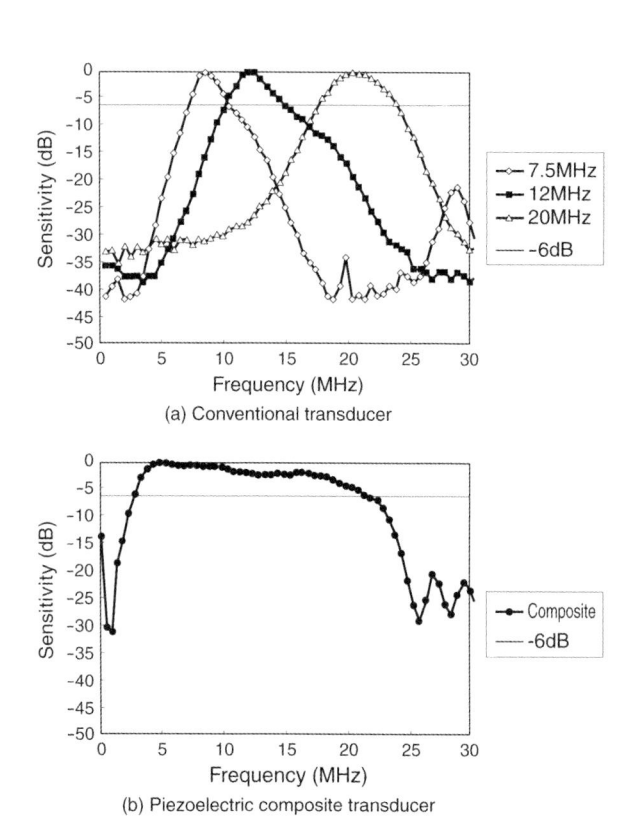

(a) Conventional transducer

(b) Piezoelectric composite transducer

Figure 14.6 Transfer function [6].

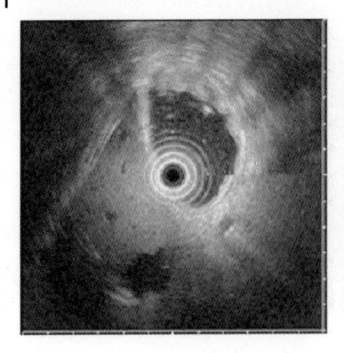

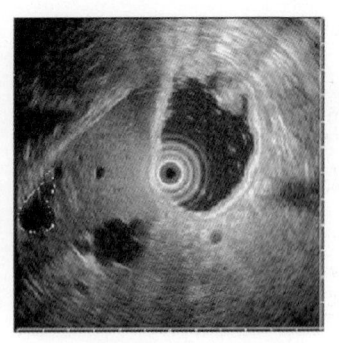

(a) Conventional transducer 7.5 MHz (12 cm range)   (b) Piezoelectric composite transducer C5 mode (12 cm range)

Figure 14.7 Ultrasonic images of actual human stomach [7].

in which were provided by the Internal Medicine Department of Digestive Organs, East Hospital at Kitasato University) shows the maximum ultrasonic depth transfer mode for each probe. In the conventional image, the center part can be clearly discerned, but the other parts of the image cannot be well imaged, as shown in Figure 14.7a. Using the new probe, the image is clear in almost all parts, as shown in Figure 14.7b, particularly the upper right region.

14.3
Microcontact Probe for IC Testing

In recent years, as mobile gadgets such as cellular phones and personal digital assistants (PDA)s have shown markedly higher performance and smaller, semiconductor devices have rapidly become more integrated and faster. Consequently, instead of conventional machining, microelectromechanical systems (MEMS) technology is now being applied to testing and packaging. We developed a microcontact probe for semiconductor devices using the LIGA process and have already supplied units of it to test equipment companies.

14.3.1
Introduction

Mechanically machined cantilevers with tungsten wires are mainly used for contact probes for semiconductor device testing. However, there is a demand for miniature and high-accuracy contact probes due to the reduction in the size and pitch of electrode pads, an increase in test frequencies arising from the increase in the operating frequency of devices and an increase in the volume of the devices under test (DUTs) due to the demand for a reduction in the testing costs of semiconductor devices; accommodation of these demands has become difficult using tungsten probes manufactured by machining. Under these circumstances, we developed a contact probe using the LIGA process.

(1) SR (synchrotron radiation) irradiation

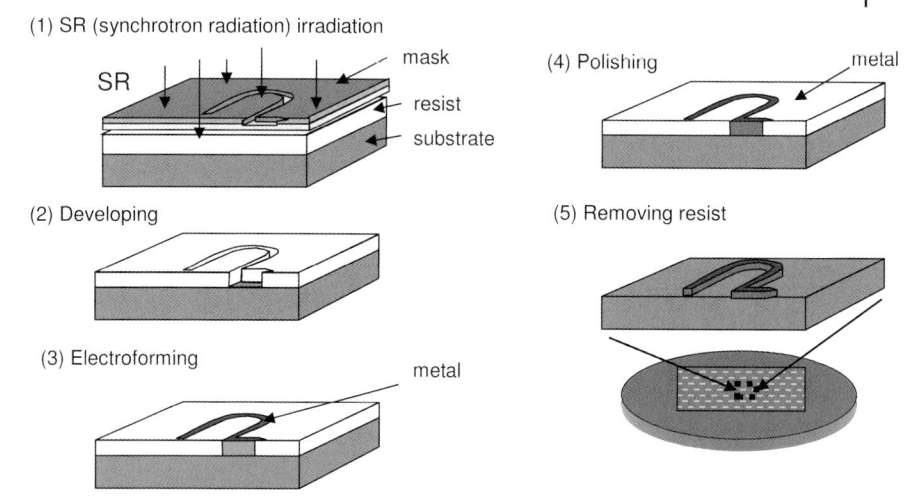

(2) Developing

(3) Electroforming

(4) Polishing

(5) Removing resist

Figure 14.8 Process flow of microcontact probe.

Since the LIGA process permits submicron transcription faithful to the mask pattern, the developed contact probe has a sharp tip shape, which is preferable for scrubbing the surface of an aluminum pad. In addition, its accuracy is ±1 µm; therefore, the variation of the spring constant is suppressed to less than 10%. The developed probe is suitable to fine-pitch testing.

14.3.2
Production Process [7]

Figure 14.8 shows the production process. The dimensional precision of the X-ray mask pattern, which determines the precision of the fabricated structure, was <±0.3 micro;m [5]. After SR irradiation and development, electroforming was carried out using the substrate as a seed layer. The molding process is not applied to fabricate the contact probes; therefore, the efficiency of SR lithography is critical. Sumitomo Electric developed an automatic SR exposure apparatus which has a stepping lithography mechanism (Figure 14.9) and high-sensitivity resist described above. As a result, the throughput was increased and the SR lithography cost was reduced considerably.

Figure 14.10 shows an example of the microcontact probe fabricated by the LIGA process. The tip of the microcontact probe was sharpened by microelectrodischarge machining [8]. On the basis of a touchdown test of 100 000 cycles using the probe, we confirmed that the probe and material had sufficient durability.

14.3.3
Material for Contact Probe

A contact probe is repeatedly used in touchdown hundreds of thousands of times under high-temperature conditions. Both high elastic limit in the spring element

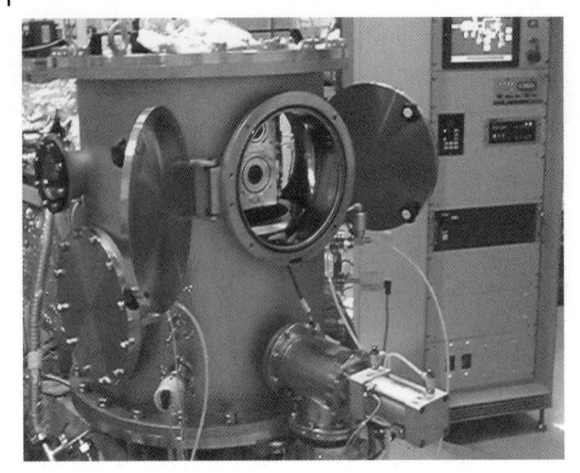

Figure 14.9 Automatic SR exposure apparatus.

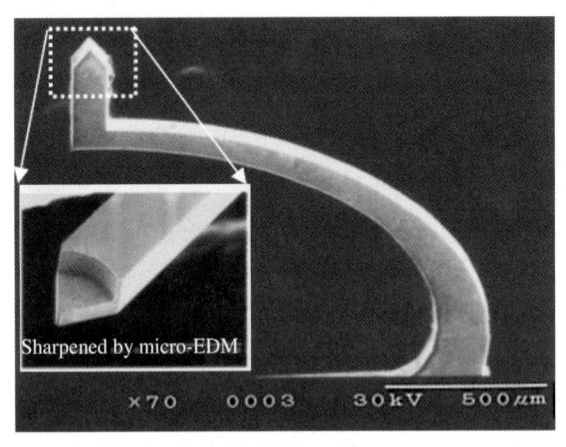

Figure 14.10 Contact probe fabricated by LIGA.

and high hardness at the tip are required for the probe. To increase hardness and toughness, we reduced the grain size of Ni to less than 50 nm. Furthermore, we added Mn in order to suppress the brittleness of the material under high-temperature conditions caused by contamination of sulfur with the additive agent of an Ni electroforming bath. Table 14.1 shows the properties of a new Ni–Mn alloy for a microspring.

14.3.4
Design of Tip Shape

Since the surface of the Al electrode pad was coated with an oxide layer, it was necessary to scrub the pad surface to remove the oxide layer for good electrical

Table 14.1 Properties of the developed Ni–Mn alloy.

Property	Value
Average grain size	<50 nm
Main crystal orientation	(111)
Hardness	Hv650
Young's modulus	190 GPa
Elastic limit	1100 MPa

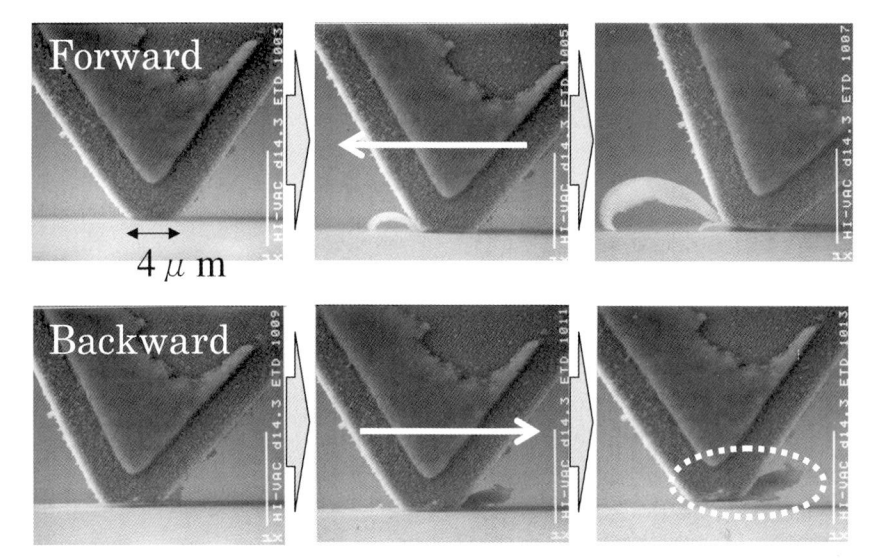

Figure 14.11 SEM images of scrubbing motion on aluminum electrode pad.

conductivity. The technology for observing the contact motion of the probe tip to the electrode pad using SEM was developed. In addition, in order to understand the contact and scrubbing phenomena at the probe tip, contact and wiping simulation technology applying cutting tool simulation using ABAQUS was developed.

First, the motion of the probe tip was observed by dynamic SEM. The probe tip was a truncated pyramid-like MEMS probe tip and was fabricated by the LIGA process. Figure 14.11 shows the observation result. The aluminum debris is generated by two processes: forward scrubbing (increased loading) and return scrubbing (reduced loading). It was found that electrical contact is maintained during forward scrubbing and therefore aluminum debris generation during return scrubbing is not necessary. The aluminum debris generated during the forward motion resembles thin shavings whereas that generated during the return motion resembles thick masses; aluminum debris generated during the return process mainly attaches to the probe tip. This phenomenon was described by ABACUS (see Figure 14.12). Therefore, the new shape which generates no debris was designed by

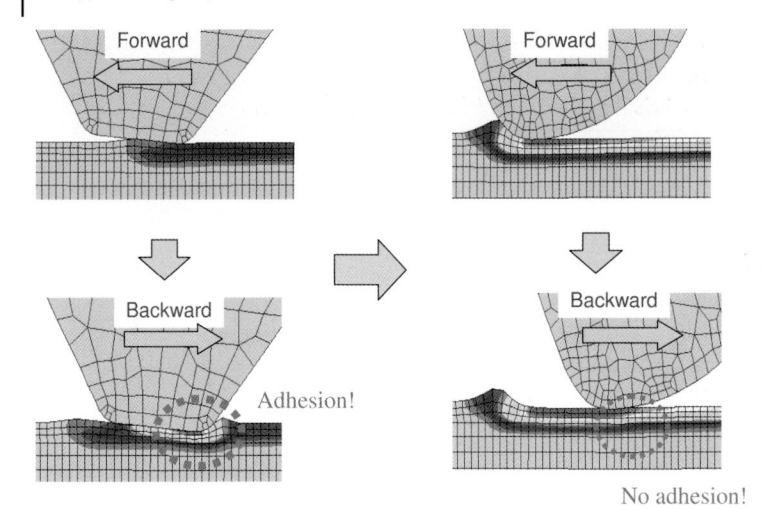

Figure 14.12 Simulation results of scrubbing.

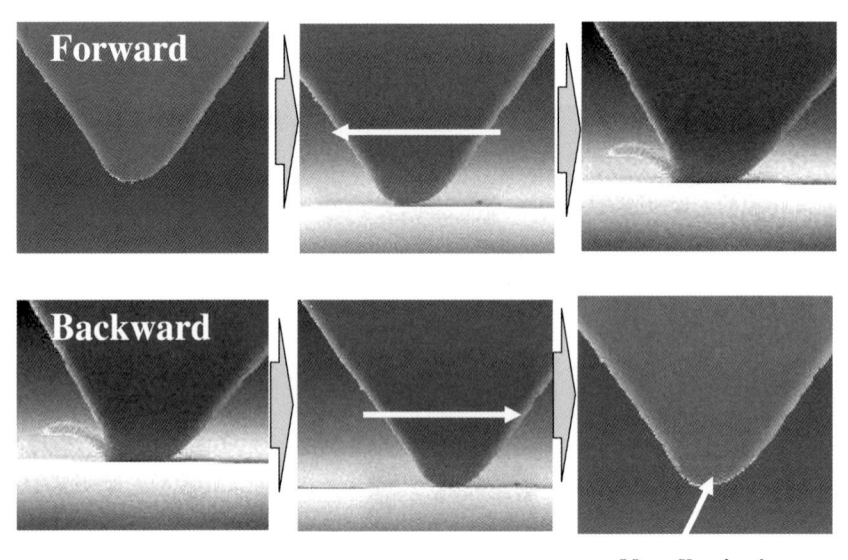

Figure 14.13 SEM images of scrubbing motion of probe with new tip shape.

ABACUS. In a new tip shape, the edge pattern of the forward scrubbing side is maintained, while the shape for return scrubbing is formed so as to avoid needle penetration into the aluminum electrode that may arise by increasing R. This tip shape was fabricated and the effect was confirmed. Figure 14.13 shows the result of SEM observation of the scrubbing motion with the new probe tip shape. During the forward scrubbing, debris similar to that produced by conventional pyramidal tips was generated and excellent electrical conductivity was obtained. During return scrubbing, no debris attached to the tip was observed. After 100 000 scrubbing cycles, no debris was observed. The LIGA process enables us to realize such good performance because of its excellent accuracy.

14.4
Conclusion

Two applications for inspection devices were described. Inspection technology will grow because interest in safety has steadily increased. In these fields, more applications will be realized by the LIGA process.

References

1 Smith, W.A. (1989) In Proceedings of 1989 IEEE Ultrasonics Symposium, p. 755.
2 Gururaja, T.R. (1994) *American Ceramics Society Bulletin*, **73**(5), 50–55.
3 Hirata, Y., Numazawa, T. and Takada, H. (1997) *Japanese Journal of Applied Physics*, **36**, 6062.
4 Hirata, Y., Nakamae, K., Numazawa, T. and Takada, H. (2004) *Sensors and Materials*, **16**, 199.
5 Hirata, Y., Okuyama, H., Ogino, S., Numazawa, T. and Takada, H. (1995) In Proceedings of the 1995 IEEE MEMS Workshop, p. 191.
6 Omura, M. and Takemoto, T. (2003) *Cho-onpa TECHNO*, Jan./Feb., 76.
7 Haga, T., Okada, K., Yorita, J., Nakamae, K., Hirata, Y. and Shimada, S. (2002) In Proceedings of ICEP2002, p. 421.
8 Kawase, K., Hirata, Y., Haga, T., Yorita, J., Mori, C. and Ueno, T. (2004) *SEI Technical Review*, **55**, 12.

15
The Micro Harmonic Drive Gear

Reinhard Degen and Rolf Slatter

15.1	**Introduction** *352*	
15.2	**From Mini- to Micro Harmonic Drive Gears** *355*	
15.3	**Principle of Operation** *358*	
15.3.1	Flexible Gear Wheels *361*	
15.3.1.1	The Flex Spline *361*	
15.3.1.2	Planetary Gear Wheels *362*	
15.3.2	Simulation of the Ultra-flat Gear *363*	
15.4	**Properties of Micro Harmonic Drive Gears** *366*	
15.4.1	Hollow Shaft *366*	
15.4.2	MHD Microgearbox Range *366*	
15.4.3	Experimental Analysis and Data *368*	
15.4.4	Advantages of Micro Harmonic Drive Gears *369*	
15.5	**Direct-LIG: the Production Process of the Micro Harmonic Drive Gear** *371*	
15.6	**Gear Profiles of the Micro Harmonic Drive** *373*	
15.6.1	Micro Harmonic Drive Gear with Involute Profiles *374*	
15.6.2	The IH Profile *375*	
15.6.3	The Novel P Profile *376*	
15.7	**Microdrive Systems** *383*	
15.7.1	The World's Smallest Backlash-Free Servo Actuator *383*	
15.7.2	Applications in Microdrive Systems *384*	
15.7.3	Special Developments for Microassembly Applications *385*	
15.7.4	Nanostage: New Possibilities for Positioning Applications in the Range of a Few Nanometers *388*	
15.7.5	The Microrobot Parvus *390*	
15.8	**Conclusion** *393*	
	References *393*	

Advanced Micro & Nanosystems Vol. 7. LIGA and Its Applications.
Edited by Volker Saile, Ulrike Wallrabe, Osamu Tabata and Jan G. Korvink
Copyright © 2009 WILEY-VCH Verlag GmbH & Co. KGaA, Weinheim
ISBN: 978-3-527-31698-4

15.1
Introduction

The overall trend to miniaturization in many fields has led to the development of several kinds of micromotors as the basis for powerful microactuators [1–3]. In the past few years, a considerable number of different micromotors have been realized to fulfill customer's demands in a variety of different areas of application. The other key constituent element of powerful microactuators is represented by the gear system. Most currently available micromotors are characterized by very high rotational speeds and very low rated torque. Therefore, a gear is often necessary in order to adapt these characteristics to real-world applications, which need higher torques and lower speeds. A wide range of different gear types exist for micromotors, for example, several types of planetary gear systems [4] and spur gear systems [5].

At Micromotion (Mainz, Germany), an affiliated company of Harmonic Drive (Limburg, Germany), the principle of operation of a Harmonic Drive gear system has been applied to a microgear system of only 1 mm axial length and 6 or 8 mm diameter using a modified LIGA technique (Figure 15.1). The Micro Harmonic Drive gear is currently the world's smallest zero-backlash gear system. Originally invented in 2001 by Micromotion, this innovative gear design has successfully been transferred from a research environment into numerous industrial applications. The combination of high reduction ratio, excellent repeatability, high efficiency and high torque capacity offered by this gear principle make it highly suitable for precise positioning applications in semiconductor manufacturing equipment, medical devices, measuring equipment, optical devices, machine tools and even spacecraft.

This microgear, with an outer diameter of down to 6 mm and an axial length of only 1 mm in the smallest currently available size, provides gear ratios between 160 : 1 and 1000 : 1 in a single stage. These high ratios are necessary to convert the very high rotational speeds of up to 100 000 rpm and very low output torques in a

Figure 15.1 Micro Harmonic Drive gearbox and actuator.

range of some μNm provided by current micromotors into lower speeds and higher torques as required by real applications in industrial machines and equipment. The gear is manufactured by a special production process, called Direct-LIG, by which the individual gear components are formed galvanically in a three-dimensional mold produced in a photoresist using X-ray lithography. This production process has been continuously improved to allow the cost-effective manufacture of metallic microgears in small to medium series production.

The gear principle is based on the well-known Harmonic Drive principle, used in the successful 'macro-technologically' manufactured reduction gears of the same name. A major difference with the microgear is the use of a planetary gear to provide an initial speed reduction within the space envelope of the Harmonic Drive gear stage. This planetary stage is provided with radially deformable planet gears, in order to avoid backlash in this reduction stage.

The advantages of miniaturization, for example, low masses, low power consumption and small dimensions open new fields of applications for powerful microactuators in dynamically growing markets. Innovative fields for microgear motors arises for example, in medicine, aerospace, automated assembly and consumer industry. In addition to their miniaturized size and low weight microgear systems must also feature zero backlash and precise angular transmission.

With a repeatability and lost motion of less than 10 arcseconds the Micro Harmonic Drive is ideally suited for applications in high precision micropositioning drive systems. This unique microgear is also the basis for the world's smallest backlash free servo actuator, developed in cooperation with Maxon Motor AG of Switzerland.

Since its market introduction in 2001 the Micro Harmonic Drive has successfully made the transition from laboratory into industrial applications (see Figure 15.2). The microgears and microactuators are already successfully used in microrobots, semiconductor manufacturing equipment and even in spacecraft.

Until now, most machines for precision assembly tasks have been many orders of magnitude larger than the workpieces to be handled or necessary workspace. Micromotion, in cooperation with the Institute of Machine Tools and Production Technology at the Technical University of Braunschweig, has now developed a small-scale SCARA robot featuring a parallel hybrid kinematic structure. This new robot, with a base area of less than 150×150 mm, can position small workpieces with a mass of up to 50 g with a repeatability of better than 10 μm (Figure 15.3).

In the remaining sections of this chapter, the basic principle of operation and components of the Micro Harmonic Drive gear will be described. The unique Direct-LIG manufacturing process will also be described in detail, in addition to practical test results documenting the excellent performance of this microtechnological product. Finally, a number of industrial applications are presented to demonstrate how this novel product has made the transition from laboratory to industrial practice in a very short space of time. The Micro Harmonic Drive gear is now establishing itself as a key enabling technology on the path to the desktop factory.

Figure 15.2 Prototype of the Micro Harmonic Drive gear.

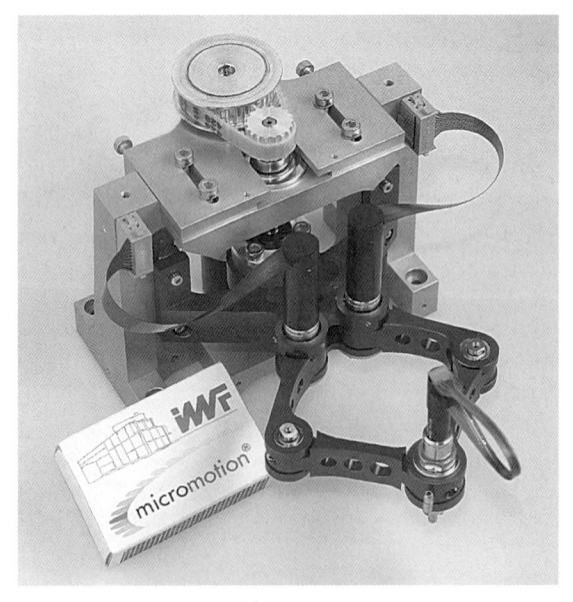

Figure 15.3 Parvus microrobot.

15.2
From Mini- to Micro Harmonic Drive Gears

The trend to miniaturization cannot be overstressed. The use of very small electronic and electro-optical components in a variety of consumer and investment goods is leading to an increasing demand for small-scale servo actuators for microassembly applications in production equipment. As soon as miniaturized systems and hybrid microsystems need to be manufactured in large series, there is a requirement for automated assembly. For small-scale products of this type, the assembly process is often a major cost driver, making up to 80% of total production costs [6]. Manual assembly is either too expensive or does not achieve the required process stability. Automated microassembly requires, in turn, specialized production equipment for handling miniature components. The assembly process typically requires movements in several degrees of freedom, which are enabled by power transmission components, such as motors, gears, ballscrews and so on.

Until recently, the physical size of these drive components was much larger than that of both the components to be handled and the necessary workspace, with the result that many machines and robots for microassembly have dimensions far in excess of the necessary working area. There is now a clear trend to equip physically smaller machines with microdrive systems. These machines have a smaller footprint and often higher assembly accuracy than the previous generation of machine.

Microgear systems represent a key element in microdrive systems. Microgears are not a particularly recent development and micro-spur gears or micro-planetary gears have been available on the market for a number of years. However, these products suffer from poor positioning accuracy and are therefore rarely used for positioning applications in machines. These previous solutions either have backlash or permit only very light loads.

What is needed are microgears that are not only very small in size, but also feature high repeatability, zero backlash, high reduction ratios and a low parts count. These requirements inspired the development of a new microgear, the Micro Harmonic Drive gear [7] (Figure 15.4). Until now, there have been no microgears suitable for precise positioning applications. Microgear systems must not only be extremely small, but also exhibit the following features:

- high repeatability
- zero backlash
- high reduction ratio
- a low parts count.

The solution is the principle of the Harmonic Drive gear system. This kind of gear system stands out compared with other gear principles, for example spur gears (Figure 15.5a) and planetary gear systems (Figure 15.5b), because of its high precision and zero backlash transmission properties. Its exceptional properties have been proven for many years in the fields of industrial robots, machine tools, measuring machines, aerospace and medical equipment [8, 9]. Harmonic Drive

Figure 15.4 Comparison between the smallest conventionally manufactured Harmonic Drive gear and a Micro Harmonic Drive gear component set.

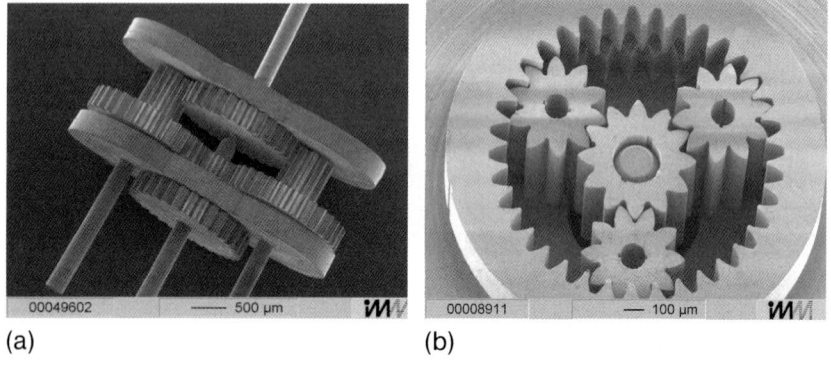

Figure 15.5 (a) Micro-spur gear with two trains; (b) micro-planetary gear system.

gear systems can be classified into the flat type and the cup type. The flat-type gear system offers the following advantages, which are particularly important with reference to microgear systems:

- small number of components
- a compact design
- the high reduction ratio necessary for micromotors reachable in a single gear stage.

Only by using suitable microgear systems is it possible to apply existing micromotors in a wide range of different applications. To access new innovative fields of application in the range of microdrive systems, Micromotion has developed a new generation of high-precision and zero backlash microgear system: the Micro Harmonic Drive. The Micro Harmonic Drive gear is currently the world's smallest zero backlash gear and in combination with a specially developed motor from

Figure 15.6 World's smallest zero backlash actuator (diameter 8 mm, length 31.3 mm).

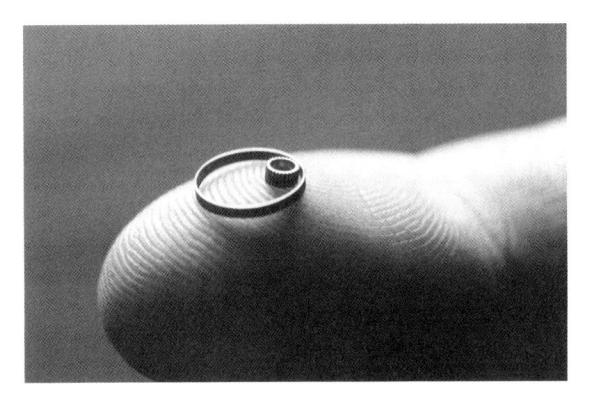

Figure 15.7 Microgear components.

Maxon Motors, Switzerland, forms part of the world's smallest zero backlash positioning actuator (Figure 15.6).

The Micro Harmonic Drive gear was introduced into the market in 2001 as the world's smallest backlash-free microgear. It is manufactured using a modified LIGA process, called Direct-LIG (Figure 15.7). This allows the cost-effective production of extremely precise metallic gear components. Subsequently this gear has been implemented in a range of miniaturized servo actuators, which provide zero backlash, excellent repeatability and long operating life.

The Micro Harmonic Drive gear component set has an outer diameter of just 6 or 8 mm and an axial length of 1 mm. It can provide reduction ratios between

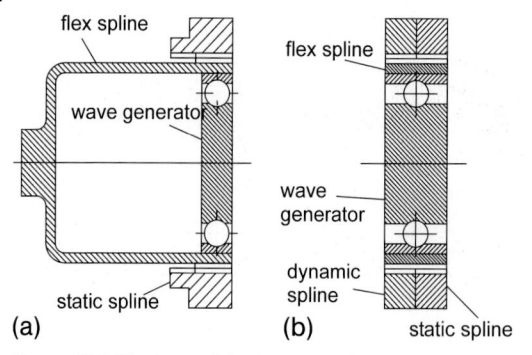

Figure 15.8 Versions of the Harmonic Drive gear: (a) cup type; (b) flat type.

160:1 and 1000:1 in one stage. In order to allow easy integration in a wide range of different applications, the component set is typically mounted inside a microgearbox of the MHD series with shafts mounted in preloaded ball bearings. The MHD gearboxes are available in two sizes with 8 or 10 mm outer diameter, either with an input shaft or for direct coupling to commonly available micromotors from Arsape, Escap, Faulhaber, Maxon, Mymotors, Myonics, Phytron and others.

15.3
Principle of Operation

To construct a gear with the operating principle of a Harmonic Drive, there are two design alternatives (Figure 15.8): the cup-type design and the flat-type design. The flat-type design is ideal to realize a ultra-flat microgear system with a high reduction ratio.

The basic elements of the flat-type harmonic drive gear system are the elliptical wave generator and the three gear wheels

- flex spline
- circular spline
- dynamic spline.

There are three basic configurations for the wave generator: an elliptical ball bearing, a planetary gear arrangement and a two-roller link (see Figure 15.9).

With respect to the miniaturization of the Micro Harmonic Drive, the planetary gear configuration for the wave generator possesses the following advantages:

- All gear components can be manufactured using the high-precision LIGA technique.

- The assembly effort can be minimized, because the wave generator consists of only three components.

15.3 Principle of Operation

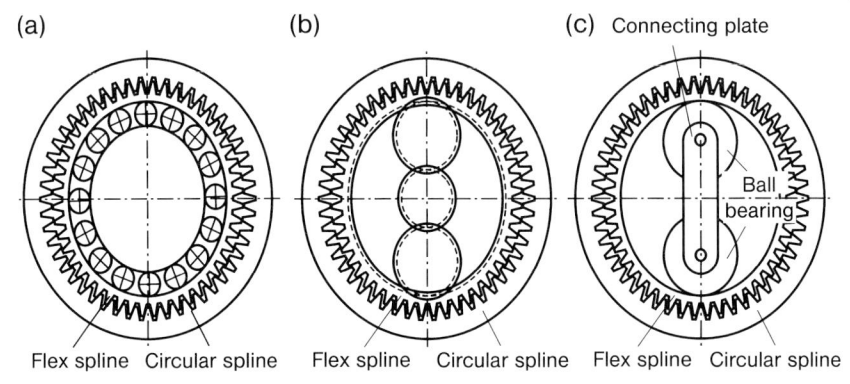

Figure 15.9 Wave generator: (a) elliptical ball bearing; (b) planetary gear; (c) two roller link.

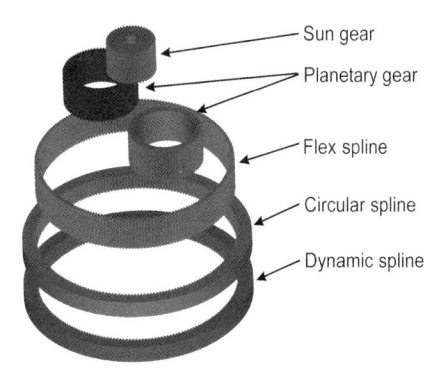

Figure 15.10 Components of Micro Harmonic Drive gear.

- The total reduction ratio of the gear increases due to the planetary gear. This design can therefore flexibly adapt the very high rotational speed of micromotors in only one stage to the specific requirements of a given application.

- This variant of the wave generator possesses only a low moment of inertia and therefore permits a highly dynamic positioning performance

By using a planetary gear for the wave generator, it is possible to vary the total ratio of the Micro Harmonic Drive over a wide range.

The basic elements of the Micro Harmonic Drive gear system are the wave generator consisting of two planetary wheels and a sun gear wheel and the three gear wheels flex spline, circular spline and dynamic spline (Figure 15.10). The wave generator deflects the elastically deformable flex spline elliptically across the major axis. Due to that, the teeth of the flex spline engage simultaneously with the two ring gears – circular spline and dynamic spline – in two zones at either end of the major elliptical axis (Figure 15.11). Across the minor axis of the elliptically deflected flex spline there is no tooth engagement.

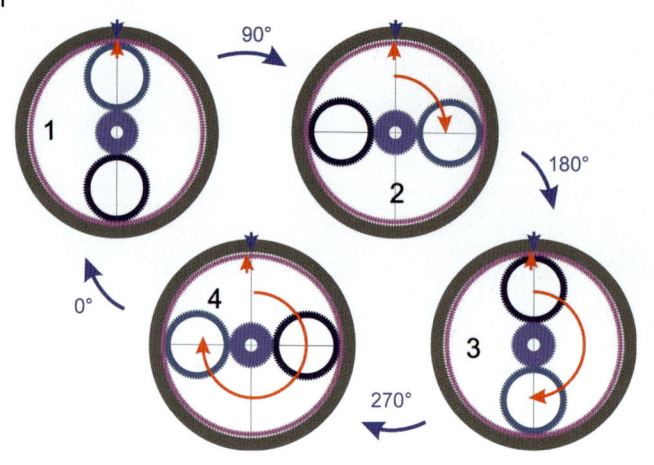

Figure 15.11 Operating principle of the Micro Harmonic Drive gear.

When the sun wheel of the wave generator rotates, the zones of tooth engagement of the flex spline travel with the angular position of the planet wheels of the wave generator. A small difference in the number of teeth between the flex spline and the circular spline (the latter has two teeth more) results in a relative movement between these gear wheels. After a complete rotation of the planet wheels of the wave generator, the flex spline moves relative to the circular spline by an angle equivalent to two teeth. The dynamic spline is used in the flat-type gear system as the output element and has the same number of teeth as the flex spline and therefore the same rotational speed and direction of rotation.

A gear module down to 34 μm must be used to realize the necessary high reduction ratio and the small dimensions simultaneously. The single gear wheels of the Micro Harmonic Drive are manufactured by electroplating and consist of a nickel iron alloy. Due to the high yield point of 1500 N mm^{-2}, the low elastic modulus of 165 000 N mm^{-2} and its good fatigue endurance [10], this electroplated alloy possesses the necessary properties for perfect functioning of the flexible gear wheels of this microgear system.

The Micro Harmonic Drive (Figure 15.12) offers the following advantages, which are particularly important with reference to microgear systems:

- zero backlash yet miniaturized dimensions
- excellent repeatability
- high torque capacity
- only six components and therefore high reliability
- high efficiency
- extremely flat design
- low weight

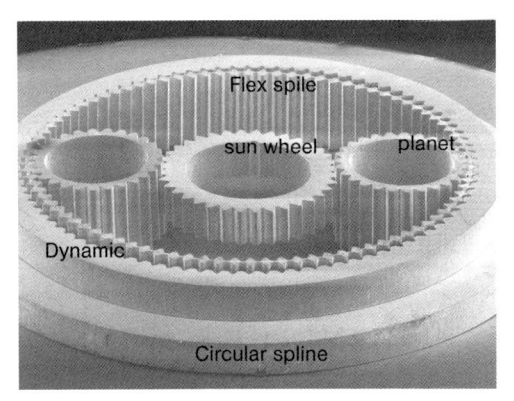

Figure 15.12 Components of the Micro Harmonic Drive gear.

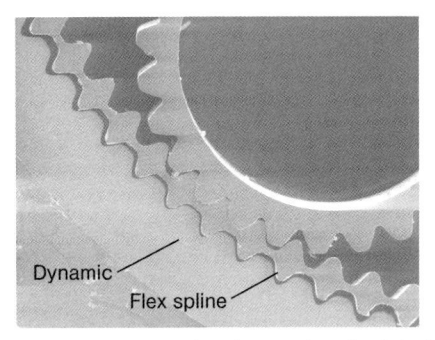

Figure 15.13 Internal and external teeth of the flex spline.

- compact dimensions
- the high reduction ratio necessary for micromotors that can be reached in a single gear stage.

By using a planetary gear for the wave generator, it is possible to vary the total ratio of the Micro Harmonic Drive over a large range. For the gear size shown, reduction ratios from 160 to 1000 can be realized in a single stage.

15.3.1
Flexible Gear Wheels

15.3.1.1 The Flex Spline

The flex spline represents the most challenging component of the Micro Harmonic Drive (Figure 15.13).

In contrast to conventional gear systems based on the Harmonic Drive operating principle, the flex spline of the Micro Harmonic Drive needs, in addition to its very thin ring thickness, both internal and external teeth. This duplex toothing is necessary due to the planetary gear configuration of the wave generator. To achieve trouble-free operation, the flex spline must exhibit uniform deflection behavior. This is realized by using the same number of teeth for the external and internal toothing. The production of the duplex toothing and the thin ring thickness necessary for low bending stresses when the flex spline is deflected is made possible by using the LIGA technique [11]. Because of this technique, it is possible to realize a ring thickness in the tooth root down to only 26 μm for a tooth width of 1000 μm.

15.3.1.2 Planetary Gear Wheels

Another component contributing essentially to the zero backlash and precise operating behavior of the Micro Harmonic Drive is the flexible planet wheel of the wave generator. Both planet wheels have the primary task of realizing the exact deflection of the flex spline. Additionally, the planet wheels have to compensate for errors of fabrication and wear of the gear system while still providing an exact deflection of the flex spline. This error-compensating property of the planet wheels is made possible by their design as a spring element.

Therefore, the flexible properties of a tube with a thin ring thickness acting in a radial direction can be used. The planet wheel is designed as a thin ring providing simultaneously enough flexibility to compensate for errors yet rigid torsional stiffness. The flex spline is pressed by the planet wheels simultaneously into engagement with the circular spline and the dynamic spline. Consequently, errors in both zones of tooth engagement are compensated for by their spring travel (Figure 15.14).

As a result, both external and internal teeth of the flex spline are brought into contact with the leading and return faces of the teeth of the meshing gear wheels. The preload of the gear system provided by the flexible planet wheels is the basis for the zero backlash transmission behavior and high positioning precision of the Micro Harmonic Drive.

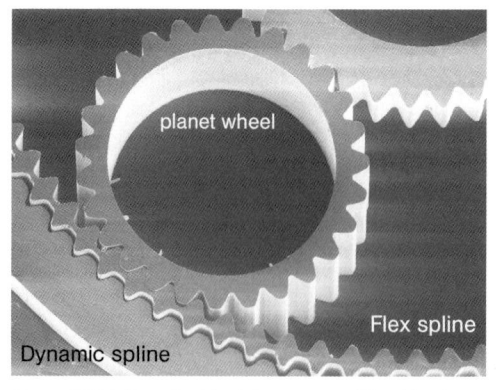

Figure 15.14 Zero backlash by means of flexible planet wheels.

15.3.2
Simulation of the Ultra-flat Gear

The design of the geometry of the tooth flanks of the gear wheels flex spline, Circular Spline and Dynamic Spline represents a key stage in the dimensioning of the Micro Harmonic Drive gear. In contrast to conventional gear systems the relative meshing of the flex spline and circular spline teeth involves not a rolling but a primarily radial movement. To be able to execute an exact dimensioning of the teeth, in particular of the geometry of the flanks, it is necessary to compute the exact curves of movement of the flex spline teeth in relation to the circular and dynamic spline (Figure 15.15).

These curves of movement serve as basis for an exact dimensioning of the teeth of all the gear wheels of the Micro Harmonic Drive. Thereby, trouble-free functioning of the gear system and the precision of movements needed for positioning drive systems can be realized.

The main goal of the finite element method (FEM) analysis was to dimension the flexible wheels of the gear system: the flex spline and the planet wheels. The deformed gearwheels were examined by FEM analysis to determine the occurring stress exactly. The results of the maximum stress in flex spline and the planet wheel are shown in Figure 15.16a and b. The data for the nickel–iron (NiFe) alloy and fatigue strength serve as the basis of the analysis. As a result of the geometric optimization of the gearwheels, a maximum stress of about 100 and 120 N mm^{-2}, respectively, could be reached, which is well below the fatigue strength of NiFe (600 N mm^{-2}).

The optimization of the geometric dimensions of the flexible gear wheels of the Micro Harmonic Drive was executed by FEM analysis.

The mechanical stress in the flex spline resulting from the deflection through the wave generator and an external torque load can be exactly determined as a function of the geometric dimensioning (Figure 15.17a). Additionally, the FEM simulation allows the exact calculation of the deflection of the loaded flex spline (Figure 15.17b).

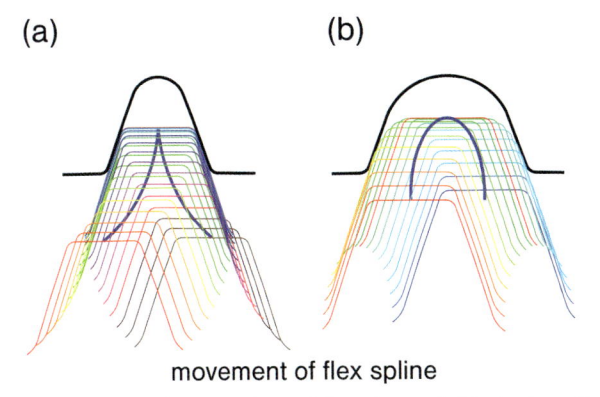

Figure 15.15 Relative meshing of flex spline with (a) circular spline and (b) dynamic spline.

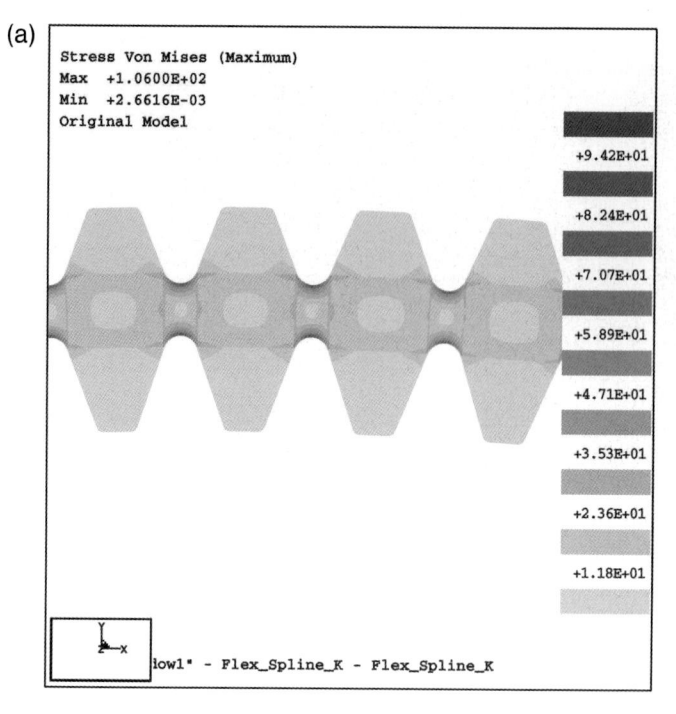

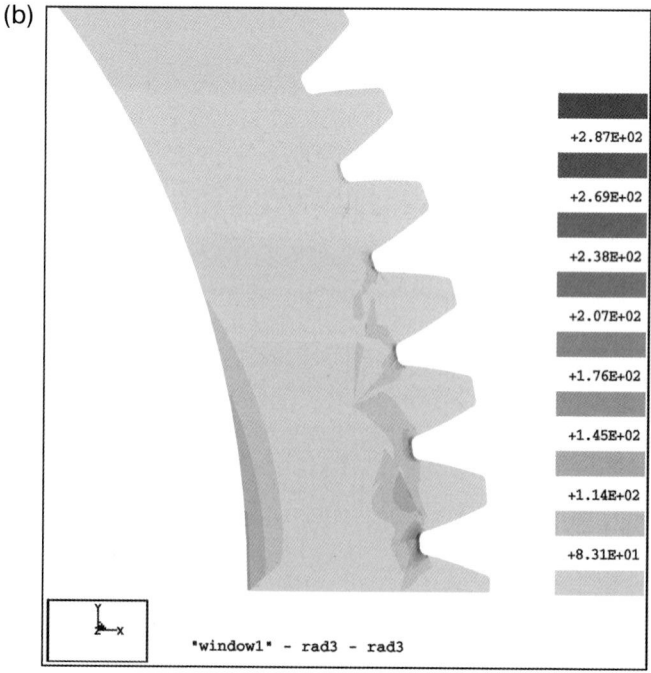

Figure 15.16 FEM analysis: (a) stress in flex spline; (b) stress in planet gear.

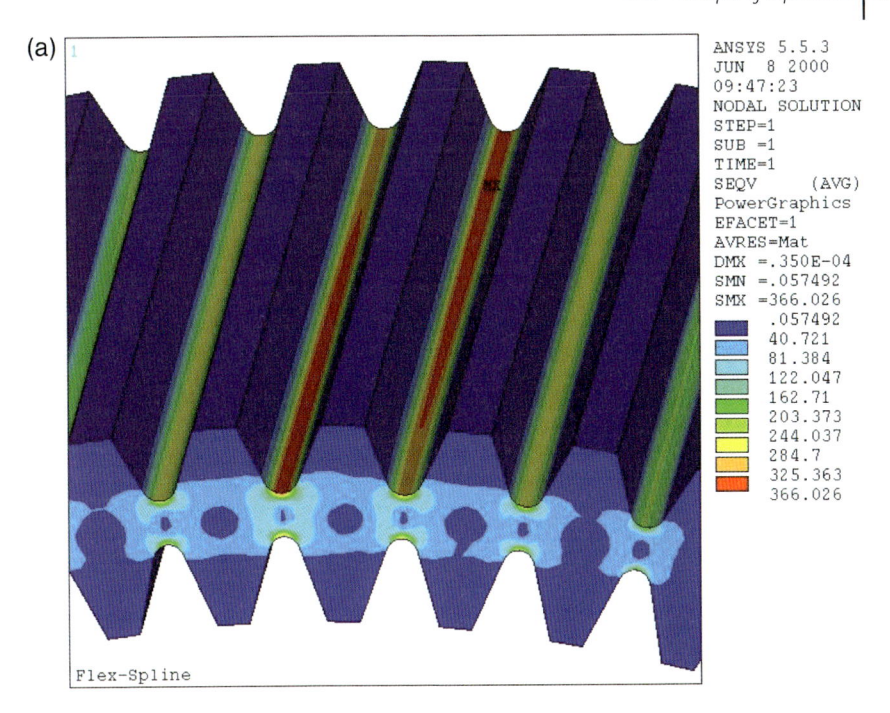

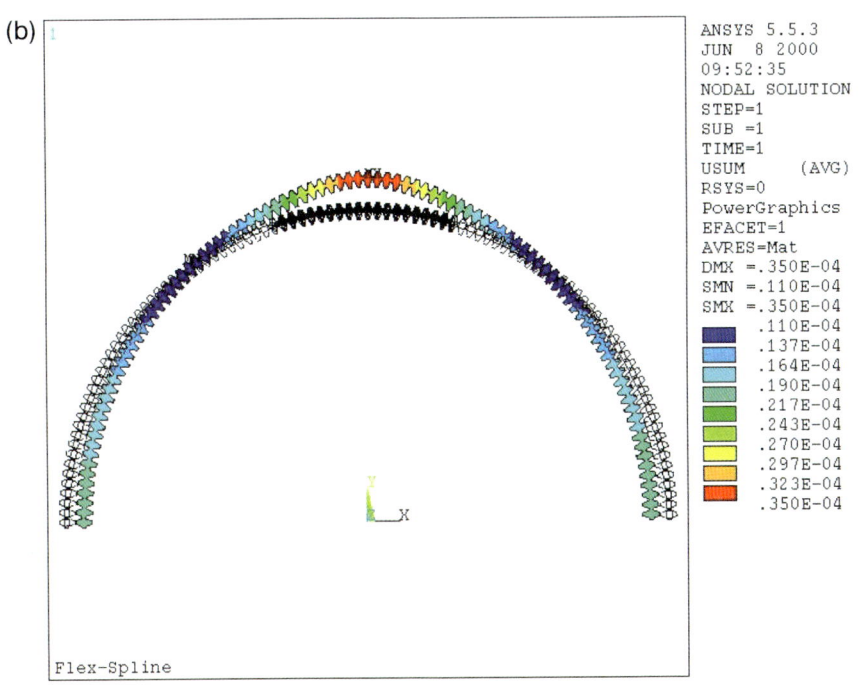

Figure 15.17 (a) Stress under load; (b) deflection of the flex spline.

15.4
Properties of Micro Harmonic Drive Gears

15.4.1
Hollow Shaft

Due to its special design, the Micro Harmonic Drive offers not only the possibility of realizing a very high transmission ratio without backlash. Another important property for systems which are optimized with respect to their outer dimensions is the possibility of realizing a hollow shaft. This means that a hollow shaft can be passed straight through the gear system along the central rotational axis. This hollow shaft, which passes through the sun gear (Figure 15.18), is important for a wide range of different functions.

Especially in the field of devices optimized with respect to outer dimensions, the possibility of a hollow shaft offers significant opportunities to reduce the outer dimensions of the whole system. Different kinds of signals or media needed for the application can be transported through the hollow shaft. The hollow shaft can be used, for example, for sensors or for optical fibers. Additionally, vacuum or air can be transmitted through the hollow shaft to the output side of the gear.

15.4.2
MHD Microgearbox Range

The gear component set is typically mounted inside a gearbox (Figure 15.19) with an output shaft mounted in preloaded ball bearings. The gearbox can either be directly coupled to a micromotor or can be provided with an input shaft, so that the motor can be mounted off-axis.

Two sizes of the Micro Harmonic Drive (MHD) microgearbox have been developed:

- MHD-8
- MHD-10.

Figure 15.18 The central bore in the sun gear permits a hollow shaft.

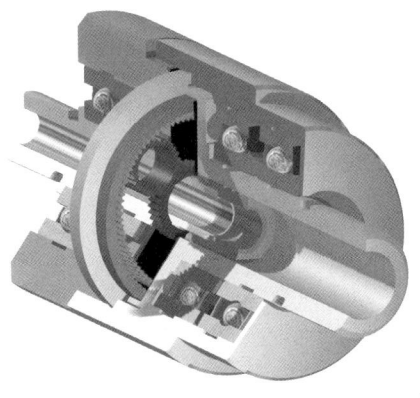

Figure 15.19 Micro Harmonic Drive MHD gearbox.

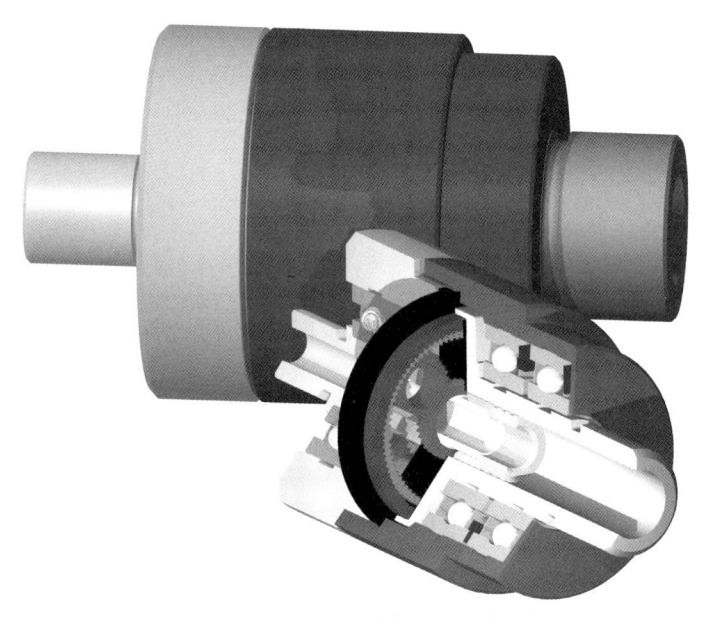

Figure 15.20 The new microgearbox models, MHD-10 and MHD-8, shown as a unit with hollow shaft.

The MHD-10 model uses the gear component set with an outer diameter of 8 mm. The MHD-10 model has a housing diameter of 10 mm and a centering shoulder of 9 mm serving simultaneously for axial location. The MHD-10 model is provided with the reduction ratios 160, 500 and 1000 and can be built up both with and without a hollow shaft (Figure 15.20).

The MHD-8 model uses a newly developed and further miniaturized gear component set with an outer diameter of only 6 mm. The MHD-8 model features a housing diameter of 8 mm with a centering shoulder of 7 mm. Reduction ratios of 160 and 500 are available for this model. A version with a hollow shaft is also available for this gearbox size. Both models are available as a unit with an input and output shaft or directly coupled with several types of currently available micromotors.

15.4.3
Experimental Analysis and Data

The very low friction torque of this zero backlash microgear system is based on the exact dimensioning of the gear wheels and the high precision reached by using the LIGA technique for manufacture. In spite of the preload of the wave generator, which is necessary to realize a zero backlash gear system, the maximum measured friction torque is only 16 µN m (Figure 15.21a).

The measured maximum value of the efficiency of the Micro Harmonic Drive gear amounts to 40% for a transmission ratio of 500 (Figure 15.21b). The measuring results illustrated in Figure 15.21b show the steady increase in the efficiency with increasing output torque. Due to the monotonically increasing trend of the measured points and the progress of the theoretical curve, a further increase in efficiency may be expected towards still higher torques.

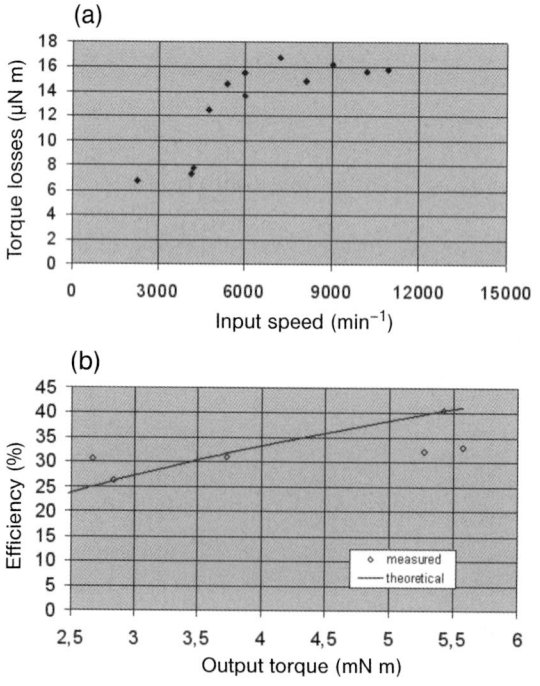

Figure 15.21 Correlation between (a) input speed and friction torque and (b) output torque and efficiency.

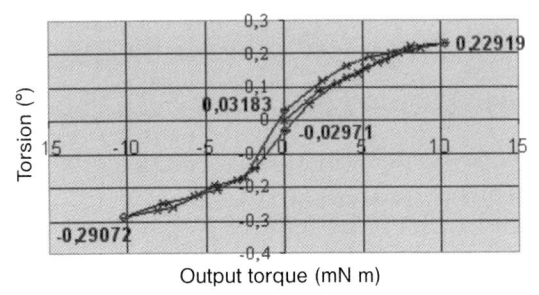

Figure 15.22 Hysteresis of the Micro Harmonic Drive.

Table 15.1 Key performance data for MHD gearboxes.

Gearbox parameter	MHD-8		MHD-10		
Reduction ratio	160	500	160	500	1000
Peak torque (mN m)	14	20	24	36	48
Rated torque (mN m)	7	10	12	18	24
Repeatability (arcsec)	10	10	10	10	10
Outer diameter (mm)	8	8	10	10	10
Weight (with input shaft) (g)	3.5	3.5	5.7	5.7	5.7

Additionally to its low friction torque and high efficiency, the Micro Harmonic Drive is distinguished especially by its excellent transmission qualities in comparison with other gear systems. The repeatability, lost motion and the hysteresis are suitable criteria to describe the quality of the transmission of a zero backlash gear system operating in positioning drive systems. The hysteresis describes the effects of a changing output load of the angular position of the output shaft of the gear and simultaneously its torsional stiffness. The value of the lost motion describes the angular error, which results by positioning movement from opposite directions.

The high efficiency and precise transmission behavior of the zero backlash Micro Harmonic Drive is shown clearly by the narrow hysteresis curve and the resultant low hysteresis losses of less than 0.1° (Figure 15.22). The most important data and measured values of the realized Micro Harmonic Drive gears are listed in Table 15.1.

Due to its properties, especially its high repeatability and its lost motion of less than 10 min, the Micro Harmonic Drive is ideally suited for applications in high-precision micropositioning drive systems.

15.4.4
Advantages of Micro Harmonic Drive Gears

The use of the Micro Harmonic Drive gear provides the machine designer with numerous advantages:

1. Miniature dimensions yet zero backlash.

   The Harmonic Drive gear stage is backlash-free by nature and the elastically deformable planet wheels eliminate backlash in the planetary stage.

2. Excellent repeatability for precise positioning.

   The zero backlash of the Micro Harmonic Drive gear provides a repeatability in the range of a few seconds of arc. This allows positioning tasks to be carried out with sub-micrometer accuracy.

3. High dynamic performance for fast indexing applications.

   The high torque capacity and low moment of inertia permit extremely fast accelerations of up to $550\,000\,\text{rad}\,\text{s}^{-2}$ at the input shaft. This corresponds to an acceleration of the motor shaft from 0 to 100 000 rpm in 25 ms. This, in turn, allows extremely fast angular movements, for example, a rotation of 180° in less than 80 ms.

4. Very long operating life.

   The MHD microgearboxes have an operating life of 2500 h at rated operating conditions, that is, at rated input speed and rated output torque. This corresponds to many million operating cycles in practical applications and the operating life of the microgearbox is typically equivalent to or longer than the expected operating life of the machine in which it is used. The 'life-cycle costs' are therefore considerably lower than for other solutions with a lower initial cost.

5. Very high reliability.

   The MHD gearbox has a significantly higher mean time between failure (MTBF) rating than other microgears. This is mainly the result of the far lower number of parts compared with other gears. A planetary microgear with a reduction ratio of 1000 : 1 typically has 25 individual gear wheels, whereas the comparable Micro Harmonic Drive gear has just six.

6. High efficiency to avoid power losses.

   The Micro Harmonic Drive gear has an efficiency of up to 82% at rated operating conditions. This is also significantly higher than for other microgears. The reason lies in the small number of tooth engagement areas. A planetary gear with ratio 1000 : 1 has 30 regions of tooth engagement, whereas the comparable Micro Harmonic Drive has just eight.

7. Extremely flat design for compact gearbox dimensions.

   The axial length of the MHD microgearbox is independent of the reduction ratio and is less than half the length of other microgearboxes for the same output torque and reduction ratio.

8. Low mass for applications in portable devices or in moving structures.

   As can be seen from Table 15.1, the gearboxes weigh just a few grams. In practical applications, this means that the moving masses in the machine can be minimized. This, in turn, can contribute to greater thermal stability and lower temperature rise, both of which are essential in high-precision machines. Furthermore, this permits higher accelerations and/or smaller feed drives.

9. High reduction ratios for low-loss torque conversion and easy control.

    The high reduction ratios greatly reduce the load moment of inertia reflected at the motor shaft. The result is that in most practical applications the motor is hardly influenced by the load inertia. In combination with the low input-side moment of inertia of the gear, this has the effect that the control of the motor is almost independent of the load inertia over a very wide range of load inertias. This makes the control of the motor and setting-up of the control system very easy.

10. Hollow shaft capability.

    The optional hollow shaft can be used to pass laser beams, air/vacuum supply or optical fibers through the center of the gear or actuator along the central axis of rotation. This can greatly simplify the design of machines where otherwise the laser beam or fiber would need to be diverted around the actuator.

11. Robust, accurate output bearing arrangement.

    The high load capacity of the output bearings (preloaded ball bearings in an O-configuration – see Figure 15.19) mean that no additional support bearings are needed for the load in most applications. Furthermore, the accurate geometric tolerances (axial and radial run-out less than 5 µm) allow the attachment of load components, for example mirrors, filters or lenses, directly to the output shaft.

12. Applicable under extreme environmental conditions.

The use of high-quality materials, such as stainless or high-alloy steels, for the gearbox housing, input/output shafts and bearings provides a high level of corrosion resistance, even for standard MHD microgearboxes. The Micro Harmonic Drive gear, which is manufactured in a high-strength nickel–iron alloy, can be sterilized and can be used over a very wide temperature range (−160 to +150 °C). It can also be applied in a vacuum [12], using grease, oil or dry lubrication, depending on the specific requirements of the application.

This combination of features makes the Micro Harmonic Drive gearbox very attractive for precise assembly applications. The high repeatability means that components can be oriented with very high accuracy, and the high dynamic performance means that assembly speed must not be sacrificed.

15.5
Direct-LIG: the Production Process of the Micro Harmonic Drive Gear

The manufacturing of the tiny structural dimensions of the gear wheels of the Micro Harmonic Drive gear is carried out by means of photolithographic processes. In order to be able to keep tolerances in the sub-micrometer range and also to exploit the properties of metallic gear wheels, the Direct-LIG process is used. The Direct-LIG process is based on the LIGA process [13] and includes the two steps lithography and electroplating (Figure 15.23).

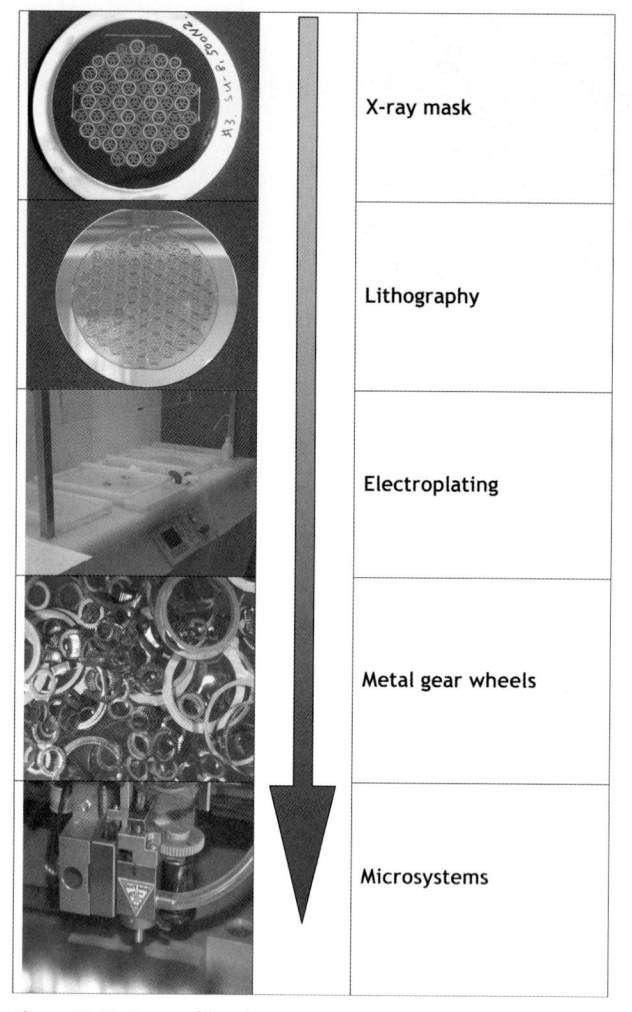

Figure 15.23 Steps of the direct LIG-process.

The structures of the gear wheels are situated as a gold absorber layer on an X-ray suitable mask. The mask pattern is copied through a projection step with high precision into a thick photoresist. To be able to manufacture structures with a height up to 1 mm and simultaneously to keep tolerances less then 1 µm, synchrotron radiation is necessary [14]. After the irradiation, the unexposed areas can be developed with a particular solvent. The negative mold of the gear wheels inside the photoresist is galvanically deposited using a nickel–iron electrolyte. Due to the high yield point of 1800 N mm^{-2}, the low elastic modulus of 135 000 N mm^{-2} and its good fatigue endurance, this electroplated alloy possesses the necessary properties for perfect functioning of the flexible gear wheels of this microgear system (see Figure 15.24).

Figure 15.24 Microgear wheel fabrication on a silicon wafer.

In addition to the extremely high accuracy and the possibility of having a resolution in the range of a few nanometers, the Direct-LIG process offers additional new possibilities to shape the profile of a tooth not available when using classical cutting production methods. Using photolithographic production methods, there is no need during the design of the profile of gears to consider the kinematics of tools, the fixture during machining or the behavior of tool wear. Because the tooth shape is copied through a projection step, the freedom of design in the lateral direction is much higher. Additionally, there is no conjunction between the lateral complexities of the components to be manufactured and the production costs, that is, the production costs are independent of the number of teeth or whether an inside or outside toothing is created. However, in the direction of the third dimension, it is only possible to vary the width of the teeth. In this direction, it is not possible to vary the shape of the tooth profile.

15.6
Gear Profiles of the Micro Harmonic Drive

Until recently, conventional gear profiles were applied to the Micro Harmonic Drive gear – either involute profiles or the IH profile, as invented for the 'macro' Harmonic Drive gear in the late 1980s. These profiles have been optimized for production using conventional gear manufacturing techniques, such as hobbing, shaping and grinding. These techniques necessarily place limitations on the gear profiles that can be economically manufactured. The 'Direct-LIG' process, on the other hand, allows more freedom for a true optimization of the gear profile used. The use of X-ray lithography means that almost any two-dimensional profile can be used to produce three-dimensional gear teeth.

This section will describe the development history of a unique new tooth profile, which is leading to a dramatic improvement in the performance characteristics of the Micro Harmonic Drive gear. This new profile, which was first shown to the public in early 2005, leads to a 200% increase in torque capacity and a 200% increase in torsional stiffness. This has the result that this zero backlash gear

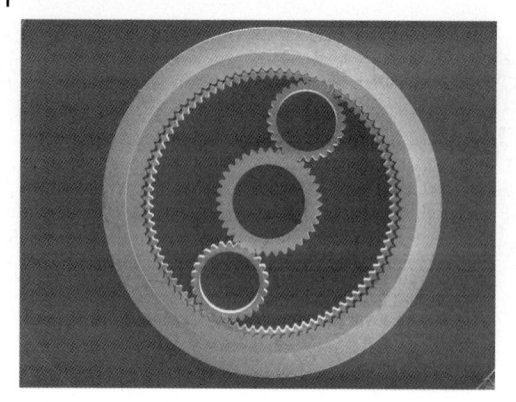

Figure 15.25 Basic components of a Micro Harmonic Drive gear component set.

system is now not only significantly more accurate than other microgears, but also offers a substantially higher power density and torsional stiffness. The section will also show how this improved performance is extending the application area of this microgear even further.

Due to the similar tooth numbers between flex spline and the two internal gear wheels, circular spline and dynamic spline, the Harmonic Drive gear systems feature a huge tooth engagement region and a high power density in comparison with other functional principles (Figure 15.25).

However, the use of a flexible mechanical transmission element, the so-called flex spline, results in unusual kinematic properties in a Harmonic Drive gear system. In contrast to 'conventional' gear systems, the proportion of sliding dominates, compared with the proportion of rolling between the teeth of the flex spline and the two internal gear wheels.

15.6.1
Micro Harmonic Drive Gear with Involute Profiles

In spite of the different kinematic properties featured by Harmonic Drive gear systems, this type of gear system also uses tooth profiles based on the involute curve. The advantages of using the involute for the tooth profile results from the simple manufacturing and care of the cutting tools and the possibility to revert to commonly known calculation schemes for dimensioning. The main disadvantage of using the involute profile is that the particular kinematic properties of the functional principle of a Harmonic Drive gear system are not considered in the profile and therefore:

- Only a smaller tooth engagement region can be realized.
- The torque capacity is lower.
- The efficiency is lower.

In Figure 15.26a, a Micro Harmonic Drive gear component set is illustrated using an involute tooth profile with a modulus of 34 µm.

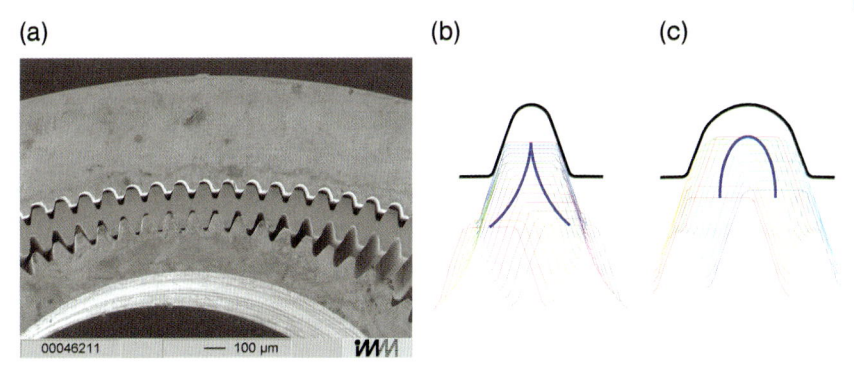

Figure 15.26 (a) Gear teeth of a Micro Harmonic Drive with an involute profile; (b) simulated motion sequence between flex spline and circular spline; (c) flex spline and dynamic spline.

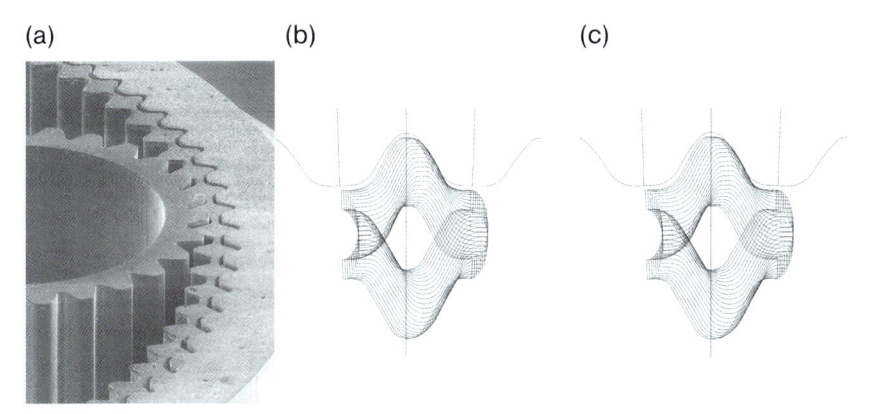

Figure 15.27 (a) Micro Harmonic Drive with IH-tooth profile; (b) simulated motion sequence between flex spline and dynamic spline; (c) flex spline and circular spline.

15.6.2
The IH Profile

To compensate for the disadvantages of an involute tooth profile, a novel tooth profile has been developed by Harmonic Drive Systems in Japan. The so-called IH-tooth profile is derived from the special kinematic properties of a Harmonic Drive gear system (Figure 15.27).

An important boundary condition for the development of the IH-tooth profile was the need to manufacture this profile by hobbing, shaping and grinding. As this tooth profile considers the particular kinematic properties of a Harmonic Drive gear system, the region of tooth engagement can be doubled, compared with the involute profile, therefore providing higher torsional stiffness, an increased torque capacity and a significantly increased life expectancy. For Micro Harmonic

Drive gear system purposes, the disadvantages of the IH-tooth profile are that the profile is optimized to the tooth engagement properties of cup-type Harmonic Drive gear systems, which uses an elliptical ball bearing as wave generator, that is, the tooth engagement between flex spline and dynamic spline and also the boundary condition of using a planetary pre-stage as wave generator are not considered by the dimensioning of the tooth profile.

15.6.3
The Novel P Profile

The requirements for the novel P profile for the Micro Harmonic Drive gear system are based on the use of a planetary stage as wave generator and the construction as a so-called flat-type Harmonic Drive using a second internal gear wheel, the dynamic spline. For the novel P profile, the shape of the tooth is simul-

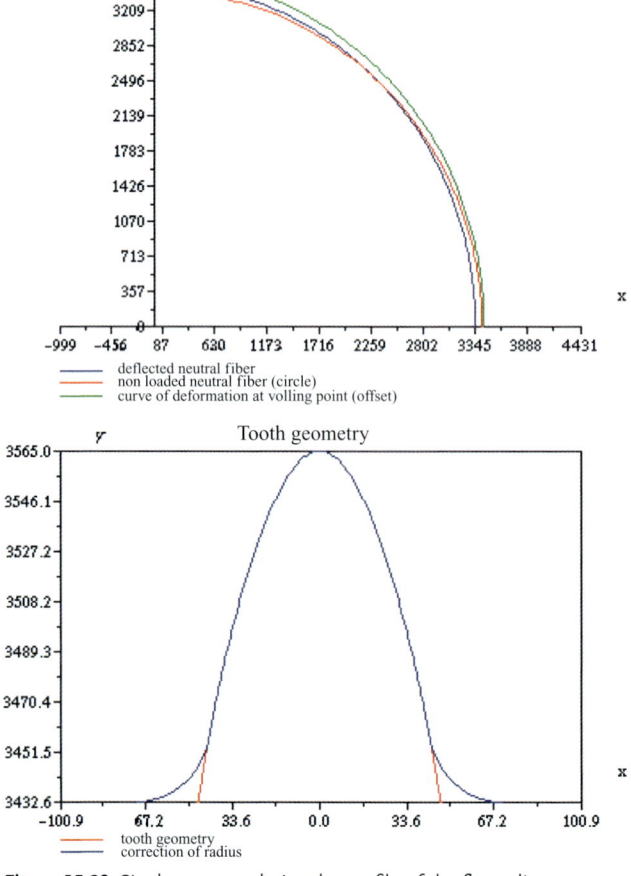

Figure 15.28 Single steps to derive the profile of the flex spline.

taneously optimized to the kinematic conditions of both the tooth engagement between flex spline and circular spline and the tooth engagement between flex spline and dynamic spline. Another important boundary condition is given by the production method applied. Because manufacturing is not carried out using conventional gear cutting processes, the freedoms offered by the Direct-LIG process become available. The aims of the development of the new profile are increases in the torque capacity, the torsional stiffness and the transmission accuracy.

The derivation of the tooth profiles of the single gear wheels is carried out based on a calculation of the geometry of deflection of the elastically deformed flex spline. The motion sequence between flex spline and circular spline and also flex spline and dynamic spline is used to calculate the geometry of deflection of the flex spline. Based on these two motion sequences, it is possible to derive the profile of the teeth of the flex spline. Subsequently, this profile, which is derived directly from the kinematics of the gear system, is provided with corrections in the region of the bottom and the top of the teeth (Figure 15.28).

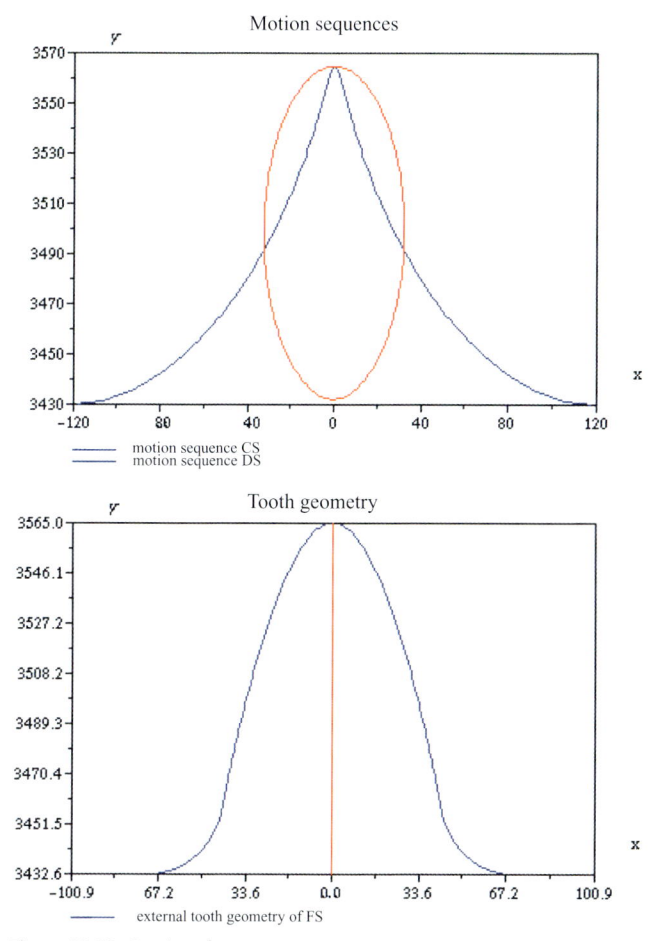

Figure 15.28 *Continued*

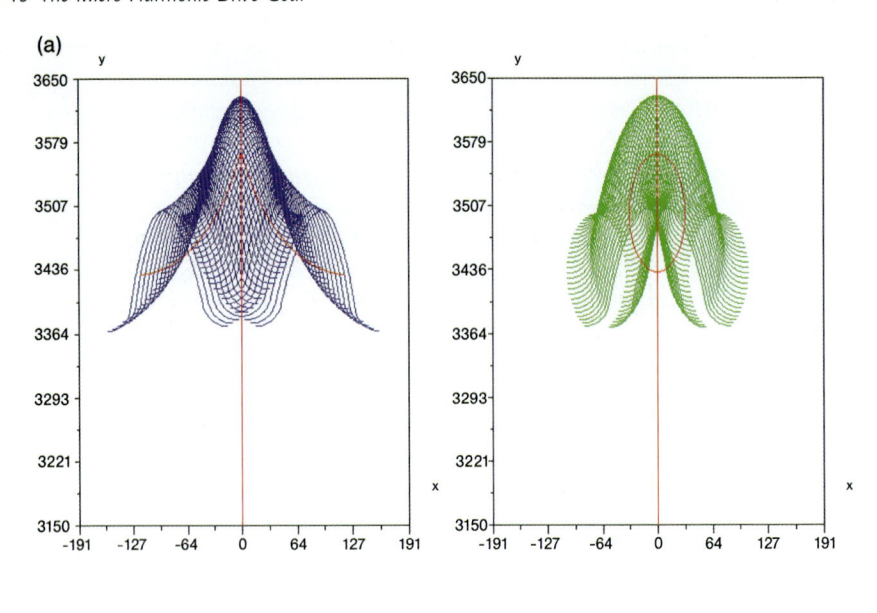

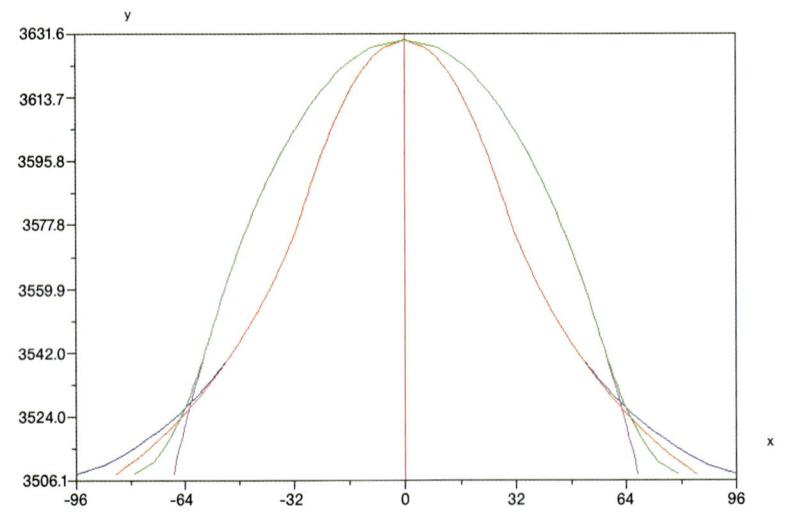

Figure 15.29 Derivation of the tooth profiles of the circular spline and the dynamic spline.

The derivation of the profile of the dynamic spline and circular spline is based on the tooth profile of the flex spline and the respective motion sequences. The envelopes representing the tooth profile of the dynamic spline and circular spline can be calculated from both the tooth profile of the flex spline and its motion sequences. Subsequently the tooth profile of circular spline and dynamic spline is provided with corrections in the regions of the bottom and the top of the teeth (Figure 15.29).

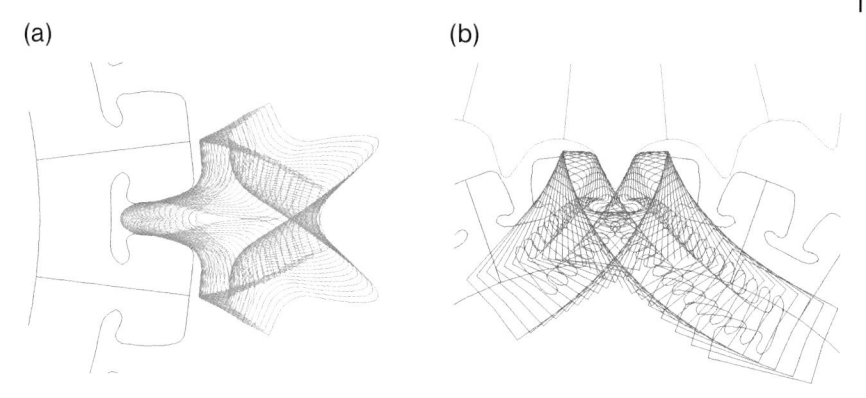

Figure 15.30 Derivation of the tooth profiles of the wave generator planet gear.

Table 15.2 Geometric comparison between IH and P profiles.

Profile	Dynamic spline	Circular spline	Flex spline	Planet wheel	Sun wheel
IH profile					
P profile					

The method for deriving the tooth profile of the gear wheels of the planetary prestage is carried out in an analogous manner and is based on both the motion sequence and the profile of the internal teeth of the flex spline. Thereby the profiles of the planetary gear wheel and the sun gear wheel are provided with corrections in the regions of the bottom and the top of the teeth (Figure 15.30).

The shapes of the single gear wheels of a Micro Harmonic Drive gear system with an IH-tooth profile are compared with the P tooth profile in Table 15.2.

The aims of the optimization of the profile of the flex spline are the following:

- an increase in the tooth engagement region additionally between flex spline and dynamic spline

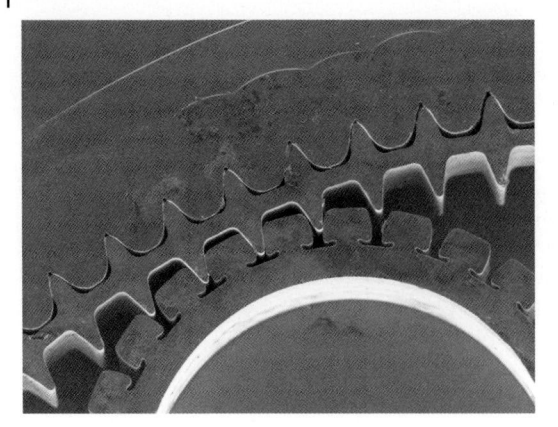

Figure 15.31 Optimized tooth profile of the flex spline.

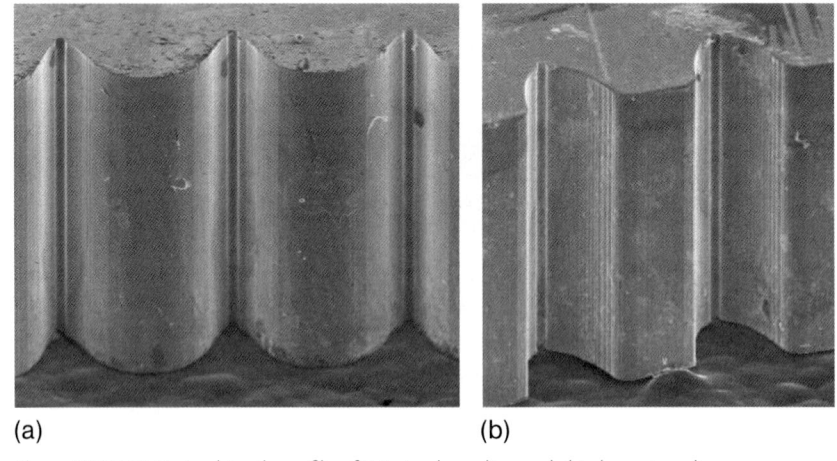

Figure 15.32 Optimized tooth profile of (a) circular spline and (b) dynamic spline.

- optimization of tension inside the flex spline
- an increase in flexural stiffness of the flex spline.

The optimized P profile for the teeth of the flex spline of a Micro Harmonic Drive gear system is illustrated in Figure 15.31. The increase in the addendum results in a larger tooth engagement region. The decrease in tooth thickness reduces the mechanical tensions inside the ring of the flex spline and allows an increase in the ring thickness and therefore the flexural stiffness of the whole flex spline.

Due to the optimizations in the tooth profile of the dynamic spline and the circular spline, the following properties could be realized (Figure 15.32):

15.6 Gear Profiles of the Micro Harmonic Drive

- increased tooth engagement region
- well-defined regions for running in and running out
- well-defined regions of contact and guiding the flex spline
- well-defined shaping of the clearance.

Due to the small thickness of the flex spline teeth, it is necessary to realize a large tooth thickness for the planetary gear wheel. However, this required tooth thickness conflicts with the desired radial compliancy of the planetary gear wheels. To be able to set up the radial spring properties of the planetary gear wheel independently from the tooth profile, there are undercuts integrated in the region of the tooth root (Figure 15.33).

The complete component set of a Micro Harmonic Drive gear with realized P profile is illustrated in Figure 15.34.

In first experimental investigations with the P profile, measurements of torque capacity, friction torque, torsional stiffness and efficiency could be made.

(a)   (b)

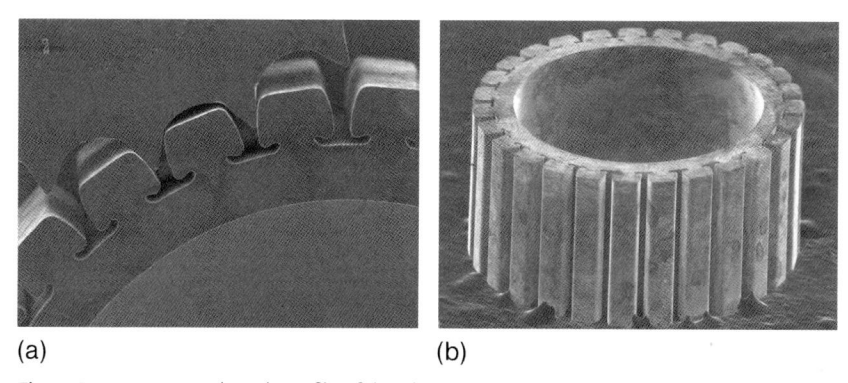

Figure 15.33 Optimized tooth profile of the planetary gear wheel.

Figure 15.34 Component set of a Micro Harmonic Drive gear with transmission ratio 160:1 featured with the optimized P profile.

The measurements were carried out with a gearbox of the size MHD-10 with a transmission ratio of 160:1: MHD-10-160-PH. The ratcheting torque, that is, the torque that deforms the gear wheels sufficiently that they come out of engagement, achieves with the P profile values over 95 mN m. This corresponds to an increase in torque capacity of approximately 200%. Furthermore, the friction torque of the input shaft remains constant at approximately 20 µN m. The trend of the curve of efficiency in relation to the output torque is illustrated in Figure 15.35.

Figure 15.36 illustrates the result of the measurement of the hysteresis curve. The hysteresis curve is acquired by applying a torque to the output shaft with the input rotationally locked. The hysteresis curve describes the relationship between the torque and the angle of torsion.

The value of the hysteresis loss accounts for the measured gearbox about 1.5 arcmin. The measured gearbox has a torsional stiffness of 10 Nm rad^{-1}.

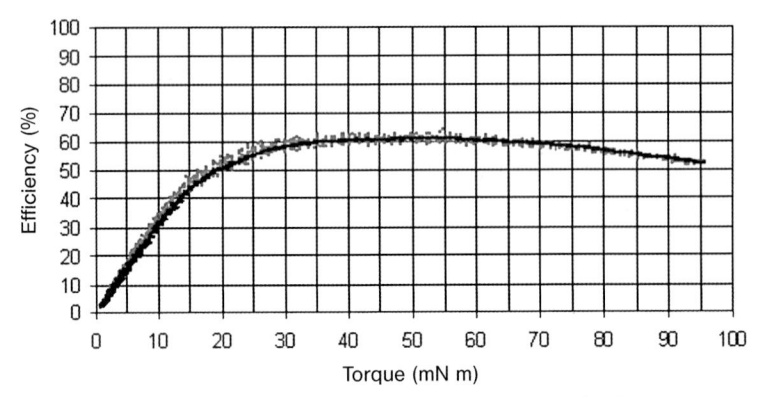

Figure 15.35 Curve of efficiency in relation to the output torque for the MHD-10-160-PH gear.

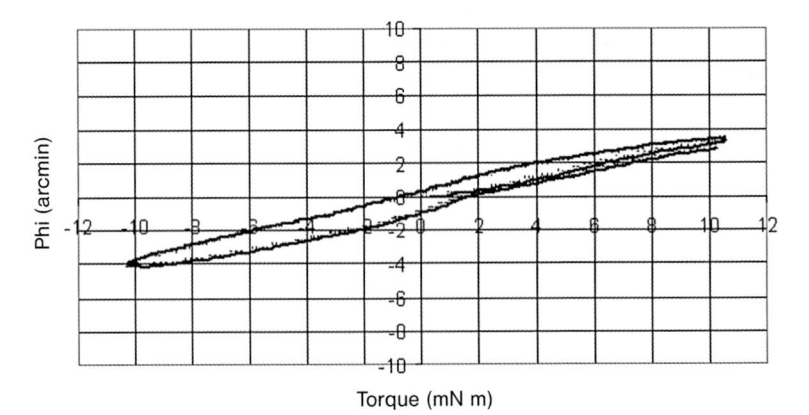

Figure 15.36 Hysteresis curve of the MHD-10-160-PH gear.

Figure 15.37 Ultra-flat microdrive system

15.7
Microdrive Systems

The Micro Harmonic Drive can be combined with all currently available micromotors, for example, stepping motors, AC or DC motors and pancake motors. The combination of the Micro Harmonic Drive with a pancake motor represents a powerful and geometrically matching microdrive system (Figure 15.37).

The pancake motor is distinguished by its small diameter of 12.8 mm and especially by its extremely flat height of 1.4 mm. By combining the ultra-flat Micro Harmonic Drive and the pancake-shaped penny motor [15], it is possible to realize a microactuator with only 4.3 mm axial length and 13.4 mm diameter. This microactuator provides an output torque of up to 10 mN m and rotational speeds up to 100 rpm with an operating weight of only 4.3 g.

15.7.1
The World's Smallest Backlash-Free Servo Actuator

A further innovation is represented by the world's smallest backlash-free servo actuator. The main objective of this joint development is to provide a reliable and robust actuator for precise industrial positioning applications. The servo actuator comprises a Maxon EC6 motor, an MR Encoder and the MHD-8-160-SP microgearbox with hollow shaft. It has the following properties (Figure 15.38):

- hollow shaft with Ø 0.65 mm
- transmission ratio of 160:1.

The Maxon EC6 is an EC motor with only 6 mm outer diameter and with integrated Hall sensors for commutation. The motor has a maximum output speed of 100 000 rpm. The maximum continuous current is 500 mA and the maximum continuous torque is 0.260 mN m.

The digital MR encoder from Maxon supplies 100 pulses per revolution. It provides the TTL signals needed for servo-drives: channels A and B. This gives the

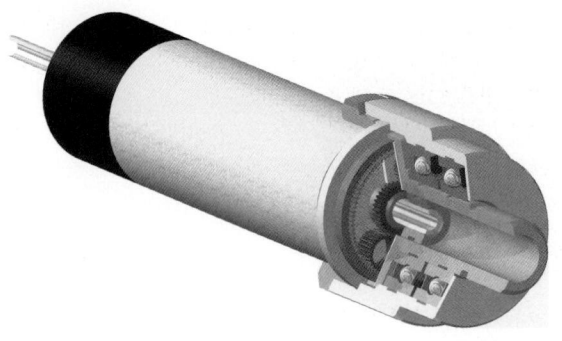

Figure 15.38 World's smallest backlash-free servo actuator.

Figure 15.39 Comparison between the servo actuator and an optical fiber connector.

encoder signals that suit the user's individual requirements. Figure 15.39 shows a size comparison of the complete servo actuator with a standard fiber-optical connector for single mode fibers.

15.7.2
Applications in Microdrive Systems

Additionally to their small dimensions, microactuators incorporating the Micro Harmonic Drive gear offer new advantages due to their low mass, low inertia and low power consumption despite their excellent positioning accuracy and highly dynamic performance. The precision microgears and microactuators from Micromotion are a key enabling technology for a new generation of miniaturized devices in a wide range of application areas. The Micro Harmonic Drive is ideally suited to precise positioning applications in the following fields:

- optics, for example, to adjust lenses and mirrors
- medical equipment, for example, to dose drugs or to drive surgical instruments
- optical communication, for example, to switch or adjust fibers
- semiconductors, for example, to assemble, handle and adjust semiconductor components
- robotics, for example, to drive axes of microrobots with high accuracy
- laser technology, for example, to adjust the beam by means of mirrors or lenses

- biotechnology, for example, to dose expensive materials and to adjust pipette probes
- measuring machines, for example, to adjust non-contacting sensors
- aircraft and spacecraft, for example, to control nozzles or valves in nanosatellites.

15.7.3
Special Developments for Microassembly Applications

In this section, some practical examples will be described where Micro Harmonic Drive gears are being used successfully in industrial microassembly applications. One of the main application areas for microassembly equipment is in the electronics industry. The production process can be divided into a 'front-end' process comprising the lithographic structuring of the silicon wafer and a 'back-end' process, starting with the dicing of the wafer into individual chips and ending with the packaging of the electronic components, ready for subsequent final assembly.

So-called 'die attach' machines are used in the assembly phase of the 'back-end' process. Alphasem is one of the world's leading manufacturers of 'die attach' machines, which are used to assemble the chips in a protective package and connect the chip to the outside world. To do this, the chips, which are today no larger than a speck of dust with dimensions of just 0.25×0.25 mm, must be orientated and positioned highly accurately. The new Easyline 8032 machine from Alphasem incorporates a new 'rotary bond tool', including a Micro Harmonic Drive gearbox, to realize simultaneously high accuracy in the range of a few millirad, a short positioning time in the range of a few milliseconds and an extremely low weight of about 30 g. This space and mass optimized unit is used to position the chips with high accuracy at any desired angle of rotation.

With this costumer-specific bond tool, the motor is mounted off-axis and so permits a hollow shaft to be passed through the center of the reduction gear. This hollow shaft is used for a vacuum feedthrough, which is used to hold the chip in place during the positioning and assembly cycle. The hollow shaft also allows the use of an optical sensor, which looks through the center of the reduction gear and output shaft to check that the chip has been correctly gripped. Figure 15.40 illustrates a low-weight rotary bond tool for high-speed applications.

At the heart of the rotary bond tool is a Micro Harmonic Drive gearbox in a custom-made design (Figure 15.41). The output shaft is mounted in preloaded ball bearings, which ensure that the radial and axial run-out of the output shaft is minimized. The output shaft of the gearbox serves simultaneously as a fit for the costumer and process-specific tool. Due to the high transmission ratio and the zero backlash of the Micro Harmonic Drive gearbox, is it possible to drive these units with simple and robust stepper motors. A section of such a costumer-specific rotary bond tool made by Micromotion is illustrated in Figure 15.42. Due to the integration of several functionalities into the gearbox, it is possible to realize a very low weight and high dynamic system for high-accuracy assembly applications.

Figure 15.40 Rotary bond tool with hollow shaft and optical sensor. Photograph courtesy of Alphasem AG.

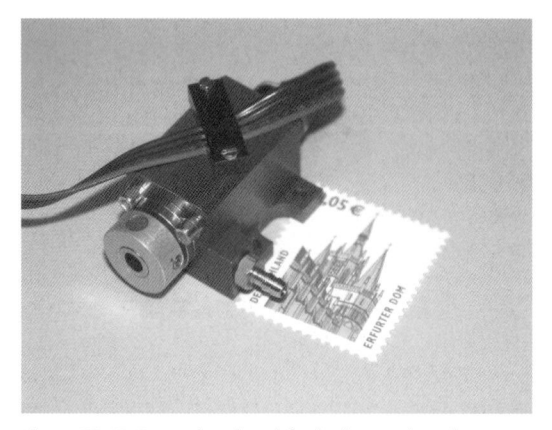

Figure 15.41 Rotary bond tool for high-speed applications. with a total weight of 22 g and a repeatability of 0.005.

Figure 15.42 Rotary bond tool section.

Figure 15.43 Three-axis micromanipulator.

The complete electromechanical sub-assembly is assembled and tested by Micromotion and allows chips to be placed with a repeatability of less than 1 μm and high speed. In the field, rotary bond tools of this design have achieved more than 30 million cycles without any loss of accuracy.

The rotary bond tools described above allow highly accurate rotational positioning of a work piece, but there are many microassembly tasks requiring movements with three degrees of freedom. For this type of application, Micromotion has developed a three-axis micromanipulator (Figure 15.43).

This compact device, with a diameter of only 36.2 mm and an axial length of less than 50 mm, features two linear and one rotational axis. The linear axes are driven using a cam arrangement, which move a small table in the X- and Y-directions. The table carries the eight-axis actuator, which drives the tool directly (Table 15.3).

Table 15.3 Key performance data for three-axis micromanipulator.

Parameter	X-axis	Y-axis	Theta-axis
Stroke/angle	1 mm	1 mm	>360°
Max. speed	2 mm s^{-1}	2 mm s^{-1}	100 rpm
Resolution:			
Full steps	<0.3 µm	<0.3 µm	0.0860
Microsteps	<0.02 µm	<0.02 µm	0.005
Repeatability	<0.3 µm	<0.3 µm	0.010
Forces/torque	10 N	10 N	5 mN m

This design offers following advantages:

- sub-micrometer accuracy
- easy controllability (stepping motors are used for all axes)
- low mass (<50 g)
- highly dynamic performance.

Importantly, the long strokes for the linear axes, easy controllability and high stability under production conditions are superior to solutions based on piezo actuators.

Typically, this device is used for fine positioning and is mounted 'piggy-back' on high-speed coarse positioning axes. Here the low weight is of particular importance. The trend to shorter assembly cycle times is leading to more dynamic primary positioning axes, typically featuring linear direct drive motors. If the mass of the 'piggy-back' micromanipulator can be minimized, then the temperature increase of the linear motors is less for the same duty cycle. This can, in turn, avoid problems due to thermal instability of the machine, which can dramatically affect the positioning accuracy of the machine.

15.7.4
Nanostage: New Possibilities for Positioning Applications in the Range of a Few Nanometers

Current solutions for positioning systems for applications needing a resolution in the range of single nanometers involve some significant problems. The idea of the invention of the nanostage is to use a conventional servomotor technique in combination with high-precision compliant mechanisms. Hence the behavior of the actuator is well known and easy to control.

Most positioning systems with a resolution in the range of a few nanometers are based on piezoelectric effects. When using such a technique, there are several drawbacks:

- loss of position if there is an interruption in the power supply
- local abrasion, particularly for inch-worm type drives

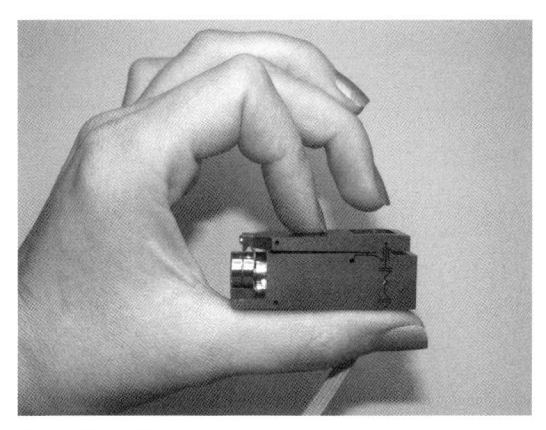

Figure 15.44 The nanostage.

- need for a direct measurement system
- need for a cost-intensive controller
- overshooting during positioning
- short travel range in relation to the size.

The particular feature of the new nano positioning system, called 'nanostage', is the combination of a high-precision and high-resolution eccentric mechanism with the kinematic structure of a monolithic flexure hinge. The eccentric mechanism is realized with a stepper motor combined with a Micro Harmonic Drive gear system. This drive unit provides a very simple controlling technique due to the stepper motor and simultaneously a very high resolution due to the transmission ratio of 1000:1 of the Micro Harmonic Drive gear (Figure 15.44).

The flexure hinge fulfils the following functions:

- The guidance system is realized with two parallel flexure hinges suitable for movements in the range of single-digit nanometers.

- An additional reduction system transforming the large movement of the eccentric mechanism is used to generate a resolution in the range of single-digit nanometers.

The build-up of the flexure hinges is done in a monolithic manner. As a result, no assembly is necessary and therefore influences due to errors of assembly can be avoided, such as asymmetric mechanical tension in the hinges. The kinematic chain consists of the following components:

- a microstepper motor with 20 full steps (40 half steps) per revolution
- a high ratio Micro Harmonic Drive gearbox with a transmission ratio of 1000:1
- an eccentric mechanism with 1 mm eccentricity built up with preloaded ball bearings
- a monolithic lever as a compliant mechanism with a further reduction ratio of 50:1.

Table 15.4 Key performance data for nanostage.

Parameter	Value
Dimensions	$20 \times 20 \times 50$ mm
Travel range	$40\,\mu m$
Resolution half step	<3 nm
Linear speed	$20\,\mu m\,s^{-1}$
Forces	10 N
Actuator	Stepper AM1020
Gear system	MHD-10-1000-M

Key performance data for nanostage are listed in Table 15.4.

Typical applications of such a new device are in particular in the following fields:

- photonics
- fiber technology
- laser systems
- integrated optics
- microscopes
- semiconductors
- medical techniques
- space applications.

15.7.5
The Microrobot Parvus

The simplest classification of robots is into serial, parallel and hybrid structures. The first category covers Cartesian robots. These are typically very large in comparison with the components to be handled and are often very expensive. However, they can provide a repeatability between 1 and $3\,\mu m$. The second category covers SCARA robots, which have a large workspace in relation to their physical size, but only achieve a repeatability of $10\,\mu m$, even in the case of the most accurate designs. In the field of parallel robots, there are few examples of industrial use, achieving a repeatability down to $5\,\mu m$.

Nowadays, the existing high-precision robots are relatively large and expensive. However, a growing market demand for smaller and cheaper robotic devices for precision positioning and assembly is apparent. The minimization of conventional industrial robots is in progress. Due to the possibilities of Micro Harmonic Drive gears, further miniaturization of such industrial robots is possible, using proven control technology and avoiding the complexity of alternative actuator technologies such as piezo actuators.

The typical structure of the micro-parallel SCARA robot Parvus is shown in Figure 15.45. Both active joints A_1 and A_2 are equipped with Micro Harmonic

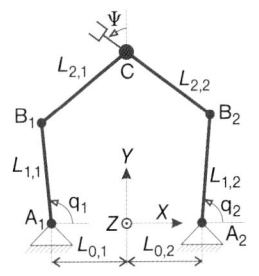

Figure 15.45 Parallel structure of the robot.

Drive gears combined with Maxon DC motors (q_1, q_2). They drive the plane structure in the X–Y direction, whereas joints B_1, B_2 and C are passive in this case.

The plane parallel structure offers two translational degrees of freedom in the X–Y plane. The Z-axis is integrated as a serial axis in the base frame of the robot. The easy handling of the whole plane parallel structure driven in the Z-direction is possible because of its minimized drive components and light aluminum alloy structure. The Z-axis is driven by a Harmonic Drive servomotor combined with a conventional ballscrew. Additionally, the rotational hand axis Ψ was executed as a hollow rotational axis integrated in the passive joint C as the Tool Center Point (TCP) of the parallel structure. This allows media such as vacuum to be passed along the hand axis. A Micro Harmonic Drive gear combined with a Maxon EC motor drives the Ψ-axis. All joints of the robot are provided with preloaded angular contact ball bearings that are strained by springs. The resulting joints are nearly free of backlash and have low friction.

All servomotors support a high-resolution encoder feedback signal. Additionally, the Parvus is equipped with magnetoresistive position sensors for initialization of the robot and an emergency stop function. The control of the first prototypes of the Parvus robot was developed on a real-time system from dSPACE. The system features a PowerPC750 digital signal processor (DSP) running at 480 MHz, a digital I/O board, high-resolution analog I/O boards, an encoder board and a serial I/O board. The dSPACE system was chosen because of the powerful hardware solution that is available from one supplier and the good operability of creating a graphical user interface. The first two functional prototypes of Parvus, shown in Figure 15.46, are being used for further research at present. The specifications of the first two prototypes of the Parvus are shown in Table 15.5.

A Matlab analysis based on the kinematic transfer functions led to a linear speed diagram shown in Figure 15.47.

Furthermore, performance specifications regarding accuracy were measured. The test method applied conforms to the ISO standard EN ISO 9283. The repeatability was measured at five definite points within a rectangular workspace shown in Figure 15.47. Not including point P5, the repeatability of this first prototype was between 5 and 10 μm. Figure 15.48 shows the plots of the best and worst results. With further optimization of the robot, it is very likely that the theoretical resolution can be achieved.

15 The Micro Harmonic Drive Gear

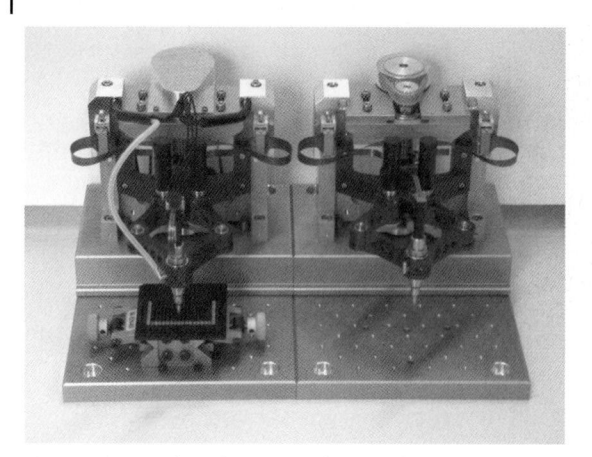

Figure 15.46 Desktop factory: the first two functional models of the Parvus.

Table 15.5 Technical specifications of Parvus.

Parameter	Value
Dimensions (rectangular)	$60 \times 45 \times 20$ mm
Footprint (area of robot base)	130×170 mm
Theoretical resolution	<3 μm
Linear speed	>100 mm s^{-1}
Rotational speed (Y-axis)	187 rpm[a] or 60 rpm[b]
Angular resolution (Y-axis)	0.022[a] or 0.007[b]
Payload	50 g

a Ratio 160:1.
b Ratio 500:1.

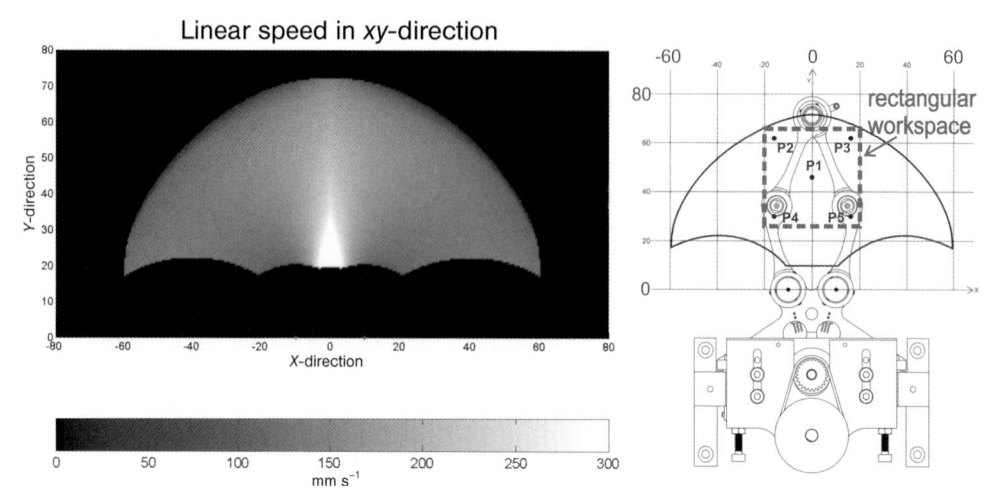

Figure 15.47 Linear speed diagram and measured workspace of Parvus.

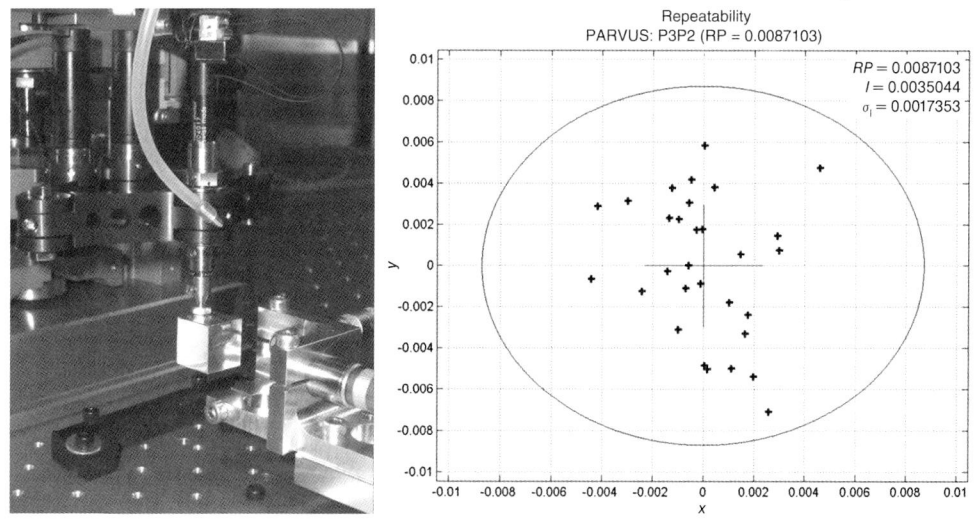

Figure 15.48 Results of repeatability tests.

15.8
Conclusion

New positioning applications in medical equipment, optics, microrobotics and semiconductors need new drives and gears with extremely small dimensions. Additionally to the small size, these new applications need high positioning accuracy and precise control. These requirements cannot be achieved using existing solutions of microgear systems. The preferred functional principle of the existing solutions is represented by the planetary gear system. This gear operating principle needs several stages and therefore a lot of parts. The main disadvantage of the available products is their backlash of several degrees. Therefore, a high positioning accuracy is not possible with existing solutions. The Micro Harmonic Drive sets new standards. This gear system combines the advantages of a compact build-up, a high power density and excellent properties of positioning. All this is realized using only six components. The consequences are that the Micro Harmonic Drive is more precise, smaller, simpler and therefore more reliable than existing solutions.

References

1 Michel, F. and Ehrfeld, W. (1999) International Symposium on Micromechatronics and Human Science MHS'99, Nagoya, Japan.
2 Berg, U., Begemann, M., Hagemann, B., Kämper, K.-P., Michel, F., Thürigen, C., Weber, L. and Wittig, TH. (1998) Series poduction and testing of a micromotor, in Proceedings of Actuator'98, Bremen, pp. 552–5.
3 Nienhaus, M., Ehrfeld, W., Stölting, H.-D., Michel, F., Kleen, S., Hardt, S., Schmitz, F. and Stange, T. (1999) Design and realization of a penny-shaped micromotor.

Proceedings of Design, Test and Microfabrication of Mems and Moems, **360** (Part 2), 592–600.

4 Thürigen, C., Ehrfeld, W., Hagemann, B., Lehr, H. and Michel, F. (1998) Development, fabrication and testing of a multi-stage microgear system, in Proceedings of Tribology Issues and Opportunities in MEMS, November 1997, Columbus, OH, Kluwer, Dordrecht, pp. 397–402.

5 Degen, R., Berg, U., Broerkens, G., Ehrfeld, W., Michel, F. and Neumeier, M. (1999) Ultra small micro positioning system for high force, high velocity and long travel range with a built-in linear encoder, in Proceedings of MOEMS 1999, Mainz, pp. 113–19.

6 Hesselbach, J. and Raatz, A. (2002) *mikroPRO – Untersuchung zum internationalen Stand der Mikroproduktionstechnik*, Vulkan Verlag, Essen.

7 Degen, R. and Slatter, R. (2002) Hollow shaft micro servo actuators realized with the Micro Harmonic Drive, in Proceedings of Actuator 2002, pp. 205–12, Bremen.

8 Slatter, R. (2000) Weiterentwicklung eines Präzisionsgetriebes für die Robotik, *Antriebstechnik*.

9 Slatter, R. and Koenen, H. (2004) Lightweight harmonic drive gears for service robots, in Proceedings of Mechatronics and Robotics, Aachen.

10 Abel, S. (1996) Charakterisierung von Materialien zur Fertigung elektromagnetischer Mikroaktoren in LIGA Technik. Dissertation, Universität Kaiserslautern.

11 Ehrfeld, W. and Lehr, H. (1995) Deep X-ray lithography for the production of three-dimensional microstructures from metals, polymers and ceramics. *Radiation Physics and Chemistry*, **45**, 349–65.

12 Slatter, R. and Degen, R. (2004) Micro actuators for precise positioning applications in vacuum, in Proceedings of Actuators, Bremen.

13 Kirsch, U., Degen, R. and Slatter, R. (2005) Advantages and possibilities of the direct LIGA-process and its applications, in Proceedings of COMS, Baden.

14 Menz, W. and Mohr, J. (1997) *Mikrosystemtechnik für Ingenieure*, 2nd edn, Wiley-VCH Verlag GmbH, Weinheim.

15 Kleen, S., Ehrfeld, W., Michel, F., Nienhaus, M. and Stölting, H.-D. (2000) Ultraflache Motoren im Pfennigformat, *F&M, Jahrg. 108, Heft 4*, Carl Hanser Verlag, Munich.

16
Microinjection Molding Machines

Christian Gornik

16.1 Brief History of Injection Molding *395*
16.2 Components and Design Concepts of Injection Molding Machines *396*
16.2.1 Clamping Unit *396*
16.2.2 Injection Unit *398*
16.2.3 Screw Plastification *399*
16.2.4 Non-return Valves *402*
16.2.5 Drive Systems *403*
16.2.6 Additional Features *404*
16.3 High-precision Moldings of Microparts *407*

16.1
Brief History of Injection Molding

The history of injection molding machines starts with the invention of a substitute material for ivory as billiard balls in 1872. Due to a lack of materials which could be processed with the first injection molding machines, the concept fell into oblivion. In 1926, the first series-produced injection molding machine was introduced by the company Eckert & Ziegler. With a combination of a toggle and a spring-loaded mechanism, the closing of the mold and the injection were realized. Later, compressed air was used to inject the melt into the mold. In 1932, the company Franz Braun built the well-known Isomaten. These machines used an electric motor and a spindle for injection. The maximum shot weight was in the range 30–35 g. The injection molding machines of those days had just a plunger for injection. The raw material was put into a hopper and from there it trickled into the heated barrel. Melting of the material occurs only because of heat conduction. Therefore, the shot weight was limited to some tens of grams and the homogeneity of the melt was bad. A major step forward towards an injection molding machine which produces a better melt quality was the introduction of the reciprocation screw in 1956 by Hans Beck and the company Ankerwerk. Beck filed a patent for that system in 1943, but it took more than a decade until this concept

Advanced Micro & Nanosystems Vol. 7. LIGA and Its Applications.
Edited by Volker Saile, Ulrike Wallrabe, Osamu Tabata and Jan G. Korvink
Copyright © 2009 WILEY-VCH Verlag GmbH & Co. KGaA, Weinheim
ISBN: 978-3-527-31698-4

was realized. Nowadays, with just a few exceptions, most injection molding machines have a reciprocating screw for plastification of the material and injection. Further important steps are the introduction of microcomputers for the control of the injection molding machine in the 1970s and the development of special processes such as multicomponent injection molding, gas- and water-assisted injection molding and foaming with chemical or physical blowing agents.

16.2
Components and Design Concepts of Injection Molding Machines

Figure 16.1 depicts the most important components of and injection molding machine. These are:

- the clamping unit
- the injection unit
- the plastification unit
- the control unit
- the frame of the machine
- the hydraulic system or electric drive motors.

16.2.1
Clamping Unit

The mold is fixed on the clamping unit either mechanically or by magnetism. It is important that the mold is centered by proper means on both the movable and the fixed platen. Especially for the replication of fragile structures and micropatterns, one has to take care that there is no movement in the vertical direction between the two mold halves on the first few millimeters of the opening stroke. Hence the platen parallelism of the machine platens in both the open and the

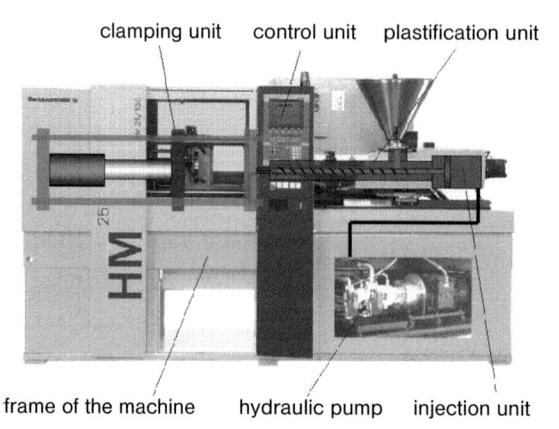

Figure 16.1 Injection molding machine.

closed positions is necessary to avoid destruction of the micropattern. The clamping unit builds up the clamping force which is necessary to avoid opening of the mold during injection because of the pressure created inside the mold cavity. This pressure, the so-called cavity pressure, multiplied by the projected area of the part is the minimum clamping force needed on the injection molding machine. The cavity pressure depends on the flow behavior of the material, the wall thickness and the ratio of the flow length and the wall thickness of the part. For high-precision parts, high stiffness of the mold and the platens of the machine is essential to avoid deformation and therefore dimensional changes of the part.

There are different executions of the clamping unit. A toggle system has the advantage of high dynamics so this system is used mainly for machines in the packaging industry where short cycle times are of high economic importance. Most all-electric machines use a toggle which is moved by a servo motor-driven spindle. Another clamping unit system is the full-hydraulic system. The movement of the platens and the build-up of clamping force are realized by hydraulic cylinders. Clamping forces up to 80 000 kN can be achieved by such systems. For microinjection molding, the clamping forces required are in the range of some tens of tons. Hence a servo motor-driven spindle can be used as a direct drive. This results in high accuracy in the velocity and force control. Although molds with up to 128 cavities and in some cases even more are in use for mass production, for high-precision parts it is advantageous if the mold has only a few cavities. This gives less variation of the part dimensions. With multi-cavity molds it is not only a challenge to produce cavities with exactly the same dimensions, but also the balancing of the runner system is crucial. In order to achieve higher productivity also for the production of parts with tolerances in the range of a few micrometers, a turntable on the machine is used. There are two equal mold halves mounted on the movable platen. In one station the injection molding is done and in the other station the parts can be removed from the mold by a robot system. After mold opening, the turntable rotates 180° and the mold is closed for the next shot and the removal of the new parts. This system, in combination with a spindle-driven clamping unit, is shown in Figure 16.2.

An important task of the clamping unit is the so-called mold protection function. When the mold closes and there is a part in between the two mold halves, the clamping unit should stop in order to avoid destruction of the mold cavities. This is realized by a reduction in speed and force within the last few millimeters before the mold halves touch each other. If the clamping movement stops because of the low force in the mold safety stroke, there will be an alarm.

The ejector is placed on the movable platen. In some cases the function of the ejector for the production of micropatterned surfaces is underestimated. The movement for ejection of such parts should be smooth and the movement should start with slow speeds until the micropattern is demolded. Hence in the case of a hydraulic machine, a proportional valve should be used for the ejection movement instead of the standard valves usually used. A rough ejection can cause serious destruction of both the micropattern of the LiGA cavity insert and the molded part.

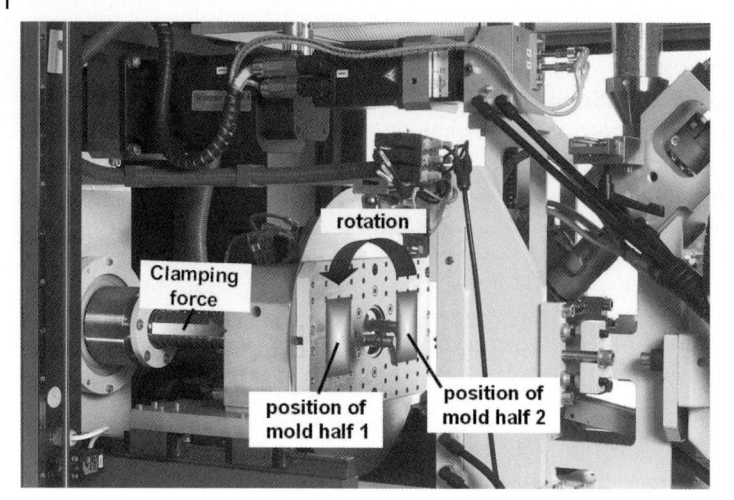

Figure 16.2 Clamping unit and turntable of a microinjection molding machine.

16.2.2
Injection Unit

The injection unit is the unit which moves either the injection piston or the reciprocating screw. On hydraulic injection molding machines there is a hydraulic piston behind the reciprocating screw which is pressurized during injection. On an all-electric machine there is usually a servo motor-driven spindle which moves the reciprocating screw. The created pressure in front of the screw is called specific injection pressure and differs from the hydraulic pressure by a factor in the range of 10; for example, a hydraulic pressure of 210 bar creates a specific injection pressure of 2100 bar. The reciprocating screw pushes the melt through the orifice of the nozzle into the mold cavity. During injection, the velocity of the injection system is controlled and the pressure needed to achieve this velocity is provided by the hydraulic system. The injection molding switches from the injection stage to the holding stage just before the mold cavity is volumetrically filled. During the holding stage, the pressure is controlled and therefore the shrinkage of the polymer is compensated in order to avoid sink marks or voids inside the part. When the metering takes place, the screw rotates and creates a certain pressure in front of the screw tip, the so-called back-pressure. The screw speed and the back-pressure are controlled by the injection molding machine.

The selection of the injection unit depends on the shot volume and on the pressure which is needed to fill the part. The shot volume is calculated by the part weight and the melt density of the polymer. There are P–V–T data (pressure, specific volume and temperature) provided by the material supplier where the melt density is given as a function of the processing temperature and the metering pressure. The metering stroke used should be in the range of 1–3 times the screw diameter; for example, for a screw with a diameter of 25 mm the metering stroke should be between 25 and 75 mm, which is 12.3 and 36.8 cm³ in terms of volume. If the metering stroke is less than one screw diameter, the consistency of the closing behavior of the non-return valve could influence the part quality. In addition, the residence time could be too long, which causes thermal degradation of the polymer. If the metering stroke is more than three screw diameters, the temperature homogeneity of the melt is usually bad so that precision molding is not possible. The injection unit is designated by the international size number, which is calculated from the maximum shot volume and the maximum pressure of the machine:

$$\text{Internation size} = \frac{\text{maximum shot volume} \times \text{maximum injection pressure}}{1000}$$

It is a number which describes the work capacity of the injection unit but not the performance in terms of power. For example, an injection molding machine with an injection unit 130 has a maximum shot volume of 61.4 cm³ and a maximum pressure of 2218 bar. For the selection of the proper injection unit, further data have to be considered: the maximum injection speed, the plasticizing rate, the maximum screw torque and the maximum screw rpm. The maximum injection speed is important if thin-walled parts are produced. It is important to fill the part before significant portions of the part freeze. The plasticizing rate is given in grams of plasticized polymer per second. The expected metering time can be calculated from this number. One has to be aware that the plasticizing rate depends not only on the screw speed but also on the metering stroke, the back-pressure and the processed polymer.

16.2.3
Screw Plastification

The plastification unit has to prepare a homogeneous melt without entrapped air. Degradation of the material inside the barrel has to be avoided, so the temperature setting and control of the barrel temperature have to be considered carefully. Table 16.1 shows the processing temperatures of polymers used in applications with micropatterned surfaces. In some cases a melt temperature sensor in the nozzle is used to measure the real melt temperature. It has to be considered that the higher the melt temperature, the lower should be the residence time in the plastification unit in order to avoid degradation of the polymer. Thermal degradation leads to discoloration of the plastic and to a reduction in the mechanical properties.

The execution of the plastification unit has to be chosen depending on the materials which it is intended to process on the injection molding machine. In addition to the screw geometry, the materials used for the plastification unit are important. Fillers, for example glass fibers, and additives, for example flame retardants, can cause excessive wear and corrosion on the components. A wear-resistant execution usually consists of a bimetallic barrel and a through-hardened

Table 16.1 Processing temperatures of engineering and high-performance polymers.

Polymer	Processing temperature (°C)
Polypropylene (PP)	200–260
Polycarbonate (PC)	260–320
Poly(methyl methacrylate) (PMMA)	210–280
Polyacetal (POM)	180–220
Poly(ether ether ketone) (PEEK)	350–380
Liquid crystal polymer (LCP)	270–310

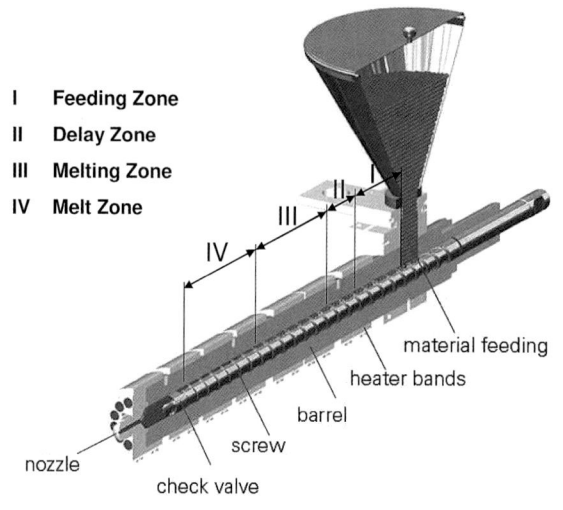

I Feeding Zone
II Delay Zone
III Melting Zone
IV Melt Zone

Figure 16.3 Functional zones in the plastification screw.

screw. The bimetallic barrel has an inner wear-resistant layer with a thickness of about 2 mm. The outer material is a conventional steel grade which has to withstand the injection pressure.

Some polymers tend to adhere on metal surfaces. In such cases, a physical vapor deposition (PVD) coating of the screw, for example, titanium nitride (TiN), can help to avoid black specks in transparent parts due to degradation of polymer on the screw surface. TiN is also used as a coating material of cavities to improve the wear resistance. For some other applications, multi-layer PVD coatings have been developed.

As today in most cases a screw is used for plastification of the material, the basic principles of conveying and melting in a plastification screw are described in the following. Figure 16.3 depicts the functional zones inside the plastification unit. In the feeding zone there is only solid material. Under the throat there is a loose agglomeration of pellets. However, after a few turns there is a plug flow, which means that there is no relative movement between the pellets. This is due to the increase in pressure inside the plastification unit. The frictional forces on the screw and on the barrel can be deduced from the relative velocities of the plug. The conveying in the feeding zone works like a screw and a nut. If the nut is fixed and the screw rotates, the nut will move in the axial direction. If the nut is not fixed and the screw rotates, the nut will rotate and there will be no axial movement of the nut, which means no conveying of material in the screw. For a proper feeding behavior, the screw surface should be smooth and the surface roughness of the barrel in the feeding section should be higher.

In the so-called delay zone, a thin melt film is created on the barrel surface due to the heat conduction from the heated barrel. As the melt film is thinner than the clearance between the screw and the barrel and also the temperature of the screw in that area is lower than the melting temperature of the polymer, there is no melt inside the screw channel. The thickness of that melt film increases rapidly because of the dissipation created in this thin melt film. As soon as the thickness of the melt film is greater than the clearance between the screw flights and the barrel, the melt is scraped by the pushing flight of the screw. In the melting zone, the portion of the melt in the screw channel increases and the portion of the solid material decreases. As the melt film adheres to the barrel wall and the solid bed moves through the screw channel, shear is introduced into the melt film. This shear causes a dissipation, which means the creation of additional heat and therefore an increase in thickness of the melt film. The melt is transported in front of the pushing flight of the screw as soon as the thickness of the melt film is greater than the clearance between the screw and the barrel. A melt pool is formed. In order to visualize the formation of the melt pool in experiments, some colorant pellets have been added to polypropylene (PP). The PP can be mixed with the colorant pellets only if both components are melted. Figure 16.4 depicts the cross-section of the material in the screw channel after cooling the barrel in order to freeze the mixing situation. The complete material should be melted before the end of the compression zone of the screw. In the melt zone, the melt is mixed in a helical flow, which leads to a homogeneous melt. As soon as there is only melt inside the screw channel, the conveying mechanism is based on the fact that the melt adheres to both the screw and the barrel surface. This leads to a velocity profile along the height of the screw channel. The velocity component v_z in the direction of the screw channel leads to the volume flow through the screw. However, there is also a cross-channel flow caused by the velocity component v_x. The volume flow depends not only on the screw rpm but also on the pressure conditions inside the plastification unit.

In many cases a plastification screw with three geometric zones is used. It consists of the feeding zone, the compression zone and the metering zone. The compression ratio of the screw, which is the ratio of the channel depth in the feeding zone to the channel depth in the metering zone, is usually between 2.0:1 and 3.0:1. For special materials, for example feedstocks for powder injection molding, it is about 1.5:1. In addition to the three-zone screws, mixing screws and barrier screws are available for plastification. A mixing screw has a mixing element after

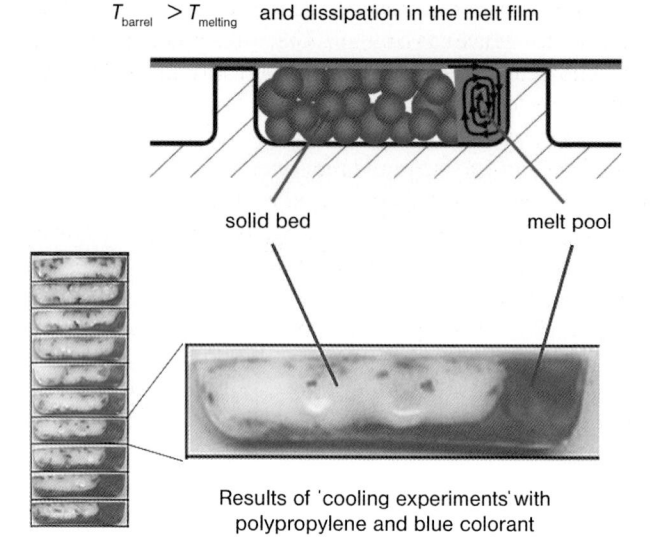

$T_{barrel} > T_{melting}$ and dissipation in the melt film

solid bed

melt pool

Results of 'cooling experiments' with polypropylene and blue colorant

Figure 16.4 Melting in a plastification screw.

Screw tip   Sealing ring   Stop ring

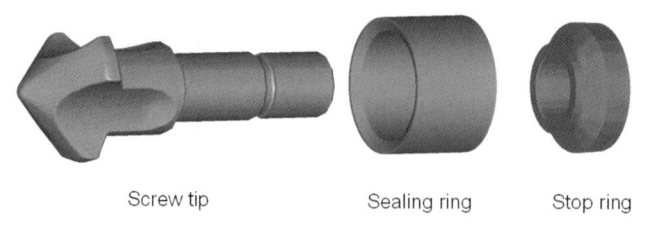

Figure 16.5 Ring non-return valve.

the metering section of the screw, which gives a better distribution of colorants. In a barrier screw, the mid-section of the screw is double-flighted with a barrier flight in between the two screw channels. Just completely melted material can pass the barrier flight and flow in the following metering section. Due to this separation of melted and unmelted material, higher plastification rates can be achieved.

16.2.4
Non-return Valves

In a reciprocating screw, the non-return valve prevents the backflow of melt into the screw channel during injection. The most common type is the ring non-return valve (Figure 16.5). It consists of the screw tip, the sealing or check ring and the stop ring. During metering, the pressure at the end of the screw channel is higher than that in front of the screw tip. Therefore, the sealing ring is pressed against the flights of the screw tip and melt can flow in front of the screw. Due to the tight clearance between the outer diameter of the sealing ring and the barrel, the sealing ring does not rotate. Hence the relative movement between the rotating screw tip and the sealing ring in combination with the force which presses the sealing ring against the flights of the screw tip can lead to enormous wear. Therefore, the selection of the materials for the screw tip and the sealing ring is of great importance. When the injection starts, the pressure in front of the screw increases and this causes the sealing ring to be pressed against the stop ring. Hence the space in front of the screw is sealed against the screw channel. The closing behavior of the non-return valve is very important when high-precision parts are molded. The geometry of the screw tip and the sealing ring should allow a quick and consistent shut-off and there should be no sharp edges where stagnation of the melt flow could occur.

There are many other special designs available, for example, the ball-type and spring-loaded check valves, which show advantages for some applications.

16.2.5
Drive Systems

The most common type of injection molding machine is the all-hydraulic machine. An electric motor drives a hydraulic pump system, which creates a certain volume flow of oil and a certain pressure in the hydraulic system. Each movement of the injection molding machine is controlled by a hydraulic valve. In case of the movement of the injection piston, closed-loop servo valves are used in order obtain high precision and repeatability of the injection process. At the beginning of the 1990s, another type of injection molding machine was developed in Europe: the all-electric machine. Due to the high investment costs of all-electric machines, nowadays this machine type is used only for products with exceptional demands on precision. On an all-electric machine, all axes are driven by servo motors. The clamping unit of all-electric machines is usually a toggle mechanism which is moved by a spindle. The toggle clamping system has the advantage that less power is needed for building up the clamping force compared with a direct clamping mechanism. In addition, the toggle can be opened smoothly, which is an advantage if micropatterned surfaces are produced. The all-electric injection molding machine consumes about 50% less energy than an all-hydraulic machine of the same size. Another advantage of all-electric machines is that all movements can be done in parallel, so the ejection can take place in parallel to the opening of the mold and the metering can still continue during mold opening. This can save cycle time. Regarding the suitability of running the injection molding machine under cleanroom conditions, the all-electric machine also has a slight advantage over the all-hydraulic machine because the fumes of hydraulic oil are a serious source of contamination. Another advantage of the all-electric machine is the lower emission of heat to the ambient surroundings. Therefore, the investment costs of the air conditioning system are low. A disadvantage of the all-electric machine is that additional investment for a hydraulic powder unit is necessary if molds with hydraulic core pullers are used. Servo motors have the restriction that they cannot create a high torque without or with only very low speeds over a long period of time. Therefore, long holding pressure times of a couple of minutes reduce the lifetime of the servo motor tremendously. The same applies for the movement of the carriage of the injection unit. Due to this circumstance, some all-electric machines use a hydro-pneumatic system for that movement.

The so-called hybrid machines combine hydraulic and electric technology. If a high plastification rate is needed, the drive of the screw is done by an electric motor and all other movements are hydraulically driven. For high-precision molding, the injection unit is executed as an all-electric unit and the complete clamping unit is hydraulic.

16.2.6
Additional Features

In order to achieve a consistent injection molding process, the conditions during transfer of the polymer from the hopper to the cavity should be as constant as possible over the production time. Therefore, it makes sense to use a drier for the material. This ensures a constant temperature of the plastics pellets before they reach the throat. Hygroscopic polymers, for example polycarbonate, have to dried for a couple of hours before processing to avoid moisture in the plastic pellets, which can cause streaks in the final part.

A good mold concept includes a venting system which ensures that there is no exceptionally high resistance for the melt to fill the cavity due to the compression of air inside the cavity. The worst case would be if the so-called Diesel effect occurs. This means that the temperature at the flow front increases substantially and causes degradation of the polymer. Unacceptable surface quality and a reduction in the mechanical properties are the result. For conventional molds, venting is usually done through small channels with a depth in the range of a few micrometers which extend from the expected end of the flow front to the outer side of the mold. In some cases, the clearance between the ejector pins and the corresponding holes is designed to allow the air to flow through during the mold filling stage. As most micropatterned surfaces have the structure of blind holes, the air cannot escape from those areas. This would result in a bad replication quality of the cavity insert as the melt cannot flow to the bottom of the microstructure due to the entrapped air which is in between. Therefore, the evacuation of the mold prior to injection is usually used to overcome this problem. A vacuum pump on the injection molding machine and a vacuum valve which is switched on and off by the control unit of the machine are recommended in order to achieve good replication.

However, it is not only the entrapped air in the blind hole-like surface structure that can reduce the quality of replication, it is also the tendency of the plastic melt to freeze if it comes in contact with the mold surface. Therefore, the mold temperature and the mold temperature distribution on the cavity surface are very important. The mold temperature is usually controlled by one or more mold temperature control units which heat water or oil to a set temperature. The heated fluid circulates through the heating channel inside the mold. As a convenient feature, an interface between the temperature control units and the control unit of the injection molding machine is available. Hence the setting of the mold temperature can also be done on the molding machine. For conventional injection molding, the temperature of the mold is far below the crystallization or so-called 'no flow' temperature. This causes the flow front to stop before penetrating the microstructure in the surface. Even high holding pressures could not fill these structures completely. Therefore, in some cases, especially when high aspect ratios have to be replicated, a so-called variotherm process is necessary. The aspect ratio is the depth divided by the lateral width of the microstructure. With a variotherm process, the mold is not operated at a constant temperature but it is heated before injection and cooled before ejection. The first temperature is near the melting temperature of the particular polymer. Hence the freezing of the melt is avoided and the viscosity during mold filling stays in a range where the melt can penetrate the microstructured surface. Two phenomena have to be considered: the creation of a frozen layer section and the viscosity for the relevant flow length to thickness ratio. Due to the high shear rates that usually occur in such microstructures, in most cases a prediction of the needed mold temperature by simulation is not possible. After injection, the injected part has to be cooled to acquire a certain strength for ejection. Therefore, the heating of the mold is switched off and the cooling is switched on. These steps are integrated in the control unit of the machine, which has certain interfaces to the mold temperature controls. The heating could be done with oil, by electrical heating elements, by induction heating, by flow of hot gases through the cavity or by novel heating sources which are applied directly on the surface of the cavity. The cooling is usually done by water or by oil eventually in combination with Peltier elements.

On conventional hydraulic machines, the pressure is controlled by a hydraulic pressure transducer which is located close to the injection cylinder. Better control of the injection and holding stage is achieved by using a melt pressure transducer to control the process. The melt pressure transducer is located in the nozzle of the barrel. Its pressure-sensitive membrane is in direct contact with the melt. All frictional forces which occur in the hydraulic cylinder, between the plastic material in the screw and the barrel and between the sealing ring and the barrel are of no importance if a melt pressure transducer is used to control the process. The stated frictional forces vary slightly from shot to shot. The melt pressure transducer is very close to the mold, so high-precision molding is possible, similar to using a cavity pressure transducer. A cavity pressure transducer is located in the cavity, but for small parts usually there is not enough space for such a transducer.

In some cases, a melt temperature sensor is used to obtain the real melt temperature in front of the screw. The melt temperature is displayed on the control unit of the injection molding machine, but this signal is usually not used for closed-loop control because there are many factors which influence the melt temperature, for example the barrel temperature profile and the shear inside the plastification screw. The melt temperature sensor can be either a thermocouple or an infrared temperature sensor. Infrared temperature sensors measure the infrared radiation of the plastic melt and they have a response time of about $20\,\mu s$.

The disadvantage of infrared temperature sensors is that they need to be calibrated to each processed polymer due to the influence of the abortion coefficient on the temperature signal. Especially fillers such as colorants have a significant influence on the absorption coefficient of infrared radiation. In addition, the radiation of layers with different thicknesses contributes to the total measured infrared radiation. Hence the temperature signal is an average volumetric temperature in the region of the sensor head. Infrared temperature sensors are relatively expensive and therefore they are not used very often.

For thin-walled parts, it is necessary to achieve a high-volume flow. The injection of the melt has to be finished before the material freezes in the region of the gate. Therefore, hydraulic machines can be equipped with a hydraulic accumulator. In the accumulator, a certain amount of hydraulic oil is stored under a high pressure. If the valve on the accumulator is opened, high energy is immediately available for injection. The acceleration of the injection piston is much higher if an accumulator is used compared with a conventional hydraulic system. The accumulator is pressurized during the other process steps. Another possibility for reaching high injection speeds is to use a hydraulic system with a double pump. This means that the electric motor drives two hydraulic pumps. Hence for fast injection, the volume flow of oil of both pumps can be used.

For high-precision molding of parts with a microstructured surface, it is important to consider the ambience of the injection molding machine. The ambient temperature should be constant and there should be no air stream directed directly on the machine. Such an air stream, which can originate from an open window near the machine, causes cooling of the outer side of the barrel. This influences the temperature control of the barrel, which can result in slight variations in melt temperature. However, not only is the ambient temperature of the production cell important, but also the purity of the air in the region of the mold and the demolded parts. As the features on the surface of the mold are sometimes in the range of several micrometers, there should be no dust particles in the mold area. Therefore, a laminar flow box (LFB) can be mounted on top of the clamping unit. The laminar air stream created by the LFB is filtered, which ensures dust-free surroundings. Another possibility is to place the machine in a cleanroom.

For the production of parts with dimensions in the sub-millimeter range, reliable quality inspection is necessary. As the microstructures could hardly be seen by the naked eye, special vision systems are used for quality assurance. These systems consist of lamps for the proper lighting of the relevant dimensions of the part, lenses for magnification and cameras. The cameras take a picture of each part and send it to a computer, which compares relevant details of the measured part with programmed values. If a microstructure is missing, it is recognized by the software, which sends a signal to the control unit of the injection molding machine. This signal can be used to separate the good parts from reject parts or it causes an alarm because there is the risk that some material is still in the mold cavity, which can cause damage to fragile regions of the mold on closing the mold. The latter is only possible if the inspection of the parts and the evaluation of the picture are finished before the start of the next cycle. This has to be considered when choosing a vision system. In addition to the visual inspection system, the so-called quality table function of the injection molding machine can be used to visualize the trends of important processing parameters, for example injection time, maximum injection pressure and switch-over pressure. These parameters are recorded by the control unit of the machine and can be stored for documentation purposes. Especially if microparts for the medical industry, for example implants, are produced it is necessary to store the files with the relevant quality parameters.

16.3
High-precision Moldings of Microparts

The steadily increasing interest in parts with dimensions in the sub-millimeter range and parts with microstructures on their surface, so called m2M parts (micro-to-Macro), led to the development of special process equipment for the reliable production of such parts (Figure 16.6). Conventional injection molding machines with reciprocating screw have a lower limit in shot volume of approximately 1 cm^3. This is due to the limitation of the screw diameter of the reciprocating screw of 14 mm. Screw diameters smaller than 14 mm have a high risk of fracture due to the small remaining cross-section of the core. Furthermore, it is not possible to obtain trouble-free feeding with pellets of normal size. For plastics in the shape of powder and for small pellets (microgranulate), diameters of 12 and 10 mm have been successfully used. The combination of plastification and injection with the same element, the screw, leads to a compromise between residence time of the material in the screw channel and reproducible injection stroke. That means that for the production of parts with small shot volumes on injection molding machines

Figure 16.6 Microinjection molding machine (Microsystem 50).

with large screw diameters, there is a risk of degradation of the polymer in the plastification unit and the closing behavior of the non-return valve has a significant influence on the shot weight consistency. Furthermore, the resolution of conventional linear transducers for measuring the injection stroke in combination with large screw diameters is not sufficient to achieve reproduction of shot weights in the range of some cubic millimeters.

In the past, the production of small parts was a compromise between high quality and high output. Large sprues and runner systems have been used to increase the total shot weight and therefore the injection stroke and throughput of material in the screw. However, this leads to bad control of the situation inside the cavity because there is a relatively large melt cushion between the part and the pressure-generating system, the reciprocating screw. Therefore, if there is a slight change in the process parameters, for example, mold temperature or melt viscosity, the part is either a short shot or it is overpacked. Another possibility for producing microparts is by injection through the cavity of the part. The material portion that is too much for the part is collected in a neighboring cavity. Hence even if there is a variation in the volume injected the part itself is always completely filled. A process which is also in use for the production of microparts is to place the part on a plate, which means that the part itself is just a structure on the surface of the produced plate. Therefore, the shot weight is large enough to produce such parts with conventional injection molding machines. After injection molding, the parts are separated from the plate by grinding off the plate or similar process steps. The mentioned techniques for the production of microparts have their drawbacks and therefore special micromolding machines have been developed.

The injection of the melt on such microinjection molding machines is usually done by a piston or so called plunger with a small diameter. It was mentioned previously that reciprocating screws can only be used down to a diameter of 14 mm. Injection pistons can have smaller diameters, which leads to a much better resolution of the injected volume. By comparing a screw with a diameter of 14 mm with a piston with diameter of 5 mm, a resolution of the injection stroke of 27.3 and 9.8 µm is calculated, respectively. The stroke for a part weight of 1 mg is 5.6 µm for a diameter of 14 mm and 44.6 µm for a diameter of 5 mm. This means that the shot weight consistency of the injection piston is much better than for the 14 mm screw. In the early days of injection molding, piston injection was used. However, with these systems the melting of the material was done only by heat conduction from a heated barrel. Due to the low thermal conductivity of polymers, the homogeneity of the melt temperature was bad. This would lead to problems in the production of microparts. Therefore, in microinjection molding machines a screw plasticizes the material. The combination of thermal conduction and shear leads to a high melt quality. In the injection unit, shown in Figure 16.7, the plastification and homogenization of the material are done by an extruder screw with a diameter of 14 mm. As there is no retraction of the screw during metering as with conventional reciprocating screws, the screw length can be shorter. This leads to a reduction in volume of material inside the screw channel and therefore to a reduction in residence time for the polymer during production. There is also a reduction in

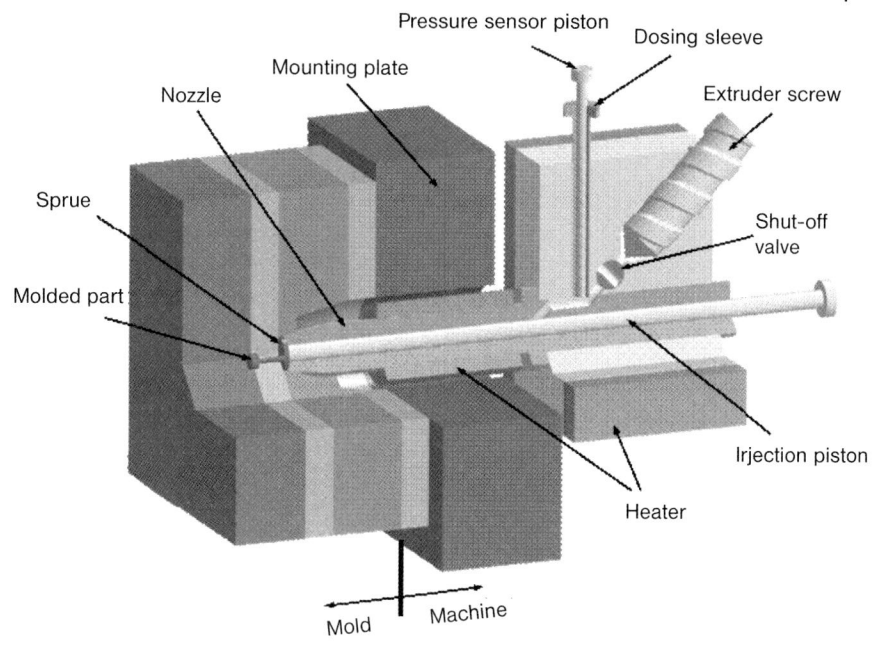

Figure 16.7 Injection unit of a microinjection molding machine.

screw channel width compared with conventional screws because the plastification rate is not crucial but the residence time is. A screw diameter of 14 mm ensures reliable feeding of the plastic pellets. For processing of special materials, special screw designs are available, for example, for powder injection molding and liquid silicon processing. The screw is usually a three-zone screw and it has to provide a homogeneous melt. As the shot volume of microparts is sometimes less than the weight of one plastic pellet, the homogeneity of the raw material has a large influence on the part quality. In some cases, microgranules are used which are homogenized in a compounding process and which are delivered in smaller pellet sizes to ensure proper feeding.

The plasticized melt is conveyed into a vertical dosing cylinder. There is a pressure sensor behind the dosing piston which controls the back-pressure of the conveyed melt. The linear transducer which measures the stroke of the piston ensures a variation of the dosing volume of less than 1 mm^3. After the set dosing volume has been reached, the screw stops rotating and therefore no more melt flows into the dosing cylinder. The shut-off valve closes and separates the plastification region from the dosing region. The clearance of the shut-off valve is crucial and it has to be chosen according to the melt viscosity of the processed material. Just before the injection takes place the dosing piston pushes the melt into the horizontal injection barrel. At that moment the injection piston is the rear position. Then the injection piston is accelerated to a set speed and pushes the melt towards the mold. This piston is driven by a cam or a similar apparatus, which is driven by an electric servo motor. To avoid excessive back-flow of the melt between the piston and the barrel, the proper clearance between the two components is important. This clearance depends on the processing temperature and it has to be checked before production. When processing a low-viscosity material, there has to be a tight clearance. The melt fills the cavity and if the switch-over criterion is reached the injection system switches to the holding pressure. The deceleration of the injection piston has to be done very quickly to avoid over-packing the mold. Therefore, there is a special mechanism which ensures that there is no movement of the injection piston during the deceleration of the electric servo motor. Although the deceleration of an electric servo motor takes just a few milliseconds, the mentioned mechanism is necessary for better process control. For good reproducibility of the shot weight, the course of the cavity pressure has to be the same for each shot. Hence in most cases the cavity pressure is used as a switch-over criterion for the production of microparts. The pressure sensor usually leaves a visible mark on the parts and therefore it should be placed in an area where it does not interfere with the functionality of the part. In most cases the pressure sensor is placed in the area of the gating system.

The front region of the plastification barrel, the dosing barrel and the rear region of the injection piston have the same temperature, which is controlled by the control unit of the machine. To ensure a homogeneous temperature distribution, it is encapsulated with a material which has a high thermal conductivity. The injection piston injects the melt directly into the split line of the mold. During the injection stroke, the piston leaves the region of high temperatures towards the nozzle tip, which has a temperature that is almost the same as the mold temperature. Owing to that special injection system, no conventional molds with sprues can be used. The advantages of the injection system are depicted in Figure 16.8. Not only can the material that is needed for the sprue be reduced, but there is also no requirement for a melt cushion. In a conventional reciprocating screw, the material in the melt cushion is used to balance the variation in metered melt volume. This material of the melt cushion is compressed inside the plastification unit when the injection takes place. Until the next shot, this melt portion is injected. In the microinjection molding machine, the volume which is needed for one shot is plasticized and completely injected. Hence there is no melt portion which undergoes a previous compression. The dosing piston and the injection piston have a diameter of 5 mm. Therefore, the resolution of the dosing volume and the injection stroke is much better than those of a conventional reciprocating screw with screw diameter 14 mm. The comparison of the volume which is passed through during deceleration from an injection speed of 250 mm s^{-1} gives a value of 2 mm^3 for microinjection molding machines with electric servo motors and 770 mm^3 for hydraulic machines with a screw diameter of 14 mm

Although the weight of microparts is in the range of several milligrams or even less, it is necessary to obtain a relatively high throughput of material to ensure both an acceptable residence time of the material at the processing temperature and highly economic production. This is achieved by using parallel sequences. There is a rotary table on the movable clamping plate on which two mold halves

16.3 High-precision Moldings of Microparts

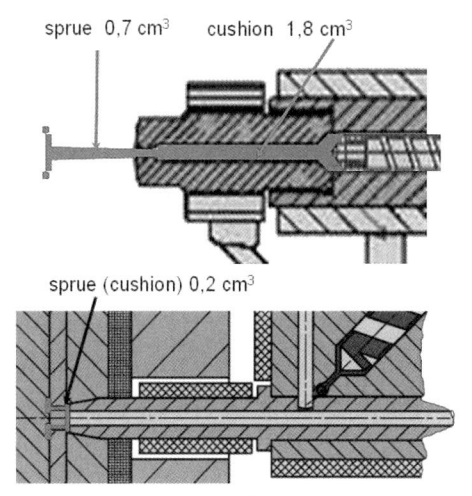

sprue 0,7 cm³   cushion 1,8 cm³

sprue (cushion) 0,2 cm³

Figure 16.8 Comparison of a conventional injection molding machine and a microinjection molding machine concerning the gating system.

with the same geometry of the cavity are mounted. After the mold has been closed, the melt is injected in one position and in the other position the previous injected parts can be removed by a robot system at the same time. The parts can be removed by a mechanical gripper which is mounted on the linear robot system or by a suction apparatus. In both cases, centering elements that ensure alignment of gripper and mold are used. These centering elements of the gripper need to have the same accuracy as the centering system of the mold halves. During the holding and cooling stage, the handling system puts the part in front of a camera system, which checks the dimensions of the part and it checks also if some microstructures are missing. Depending on the result of this evaluation, the parts are placed in their positions for packaging or in the reject box, respectively. For further utilization of the parts, they are put in blister packaging or on trays. In the case of powder injection molding, the green parts are placed on sintering trays. The part which comes out of the injection molding machine is named the green part. It consists of a mixture of powder with a grain size in the range of a few micrometers and a polymeric binder system. There are up to 1000 parts on one tray. These trays are piled up and then put directly into the debinding furnace and, after debinding, into the sintering furnace. Hence there is no need for further handling steps of the small green parts. The cycle times which can be achieved by microinjection molding machines are about 4 s or sometimes even less. The cooling time of such small parts with thin walls is usually negligible. However, the cycle time depends not only on the cooling time of the injected parts but also on the mold opening and closing times. When parts with fragile structures and high aspect ratios are produced, the opening and take-out speed has to be low to avoid destruction of the part.

Multi-cavity molds are used to increase the economy of production. For simple parts, up to 128 parts or even more can be produced in one shot on conventional injection molding machines. In such cases, usually hot runner systems are used to distribute the melt in each cavity. The rheological balancing of the mold, which means the simultaneous filling of all the cavities, is difficult. In some cases mold filling simulations are used for the technical design of the runner system. For parts with dimensions in the range of millimeters or sub-millimeters and tolerances of a few micrometers, the use of multi-cavity molds is crucial. The production of two or more mold cavities with identical dimensions and the balancing of such small cavities are major challenges. If the cavities are not filled exactly at the same time, there are different pressure conditions for each part, which leads to differences in the shrinkage and therefore in the dimensions. In addition, the larger the mold the more difficult is it to achieve an even temperature distribution on the mold surface. However, if the temperature in each mold cavity is different, the cooling behavior for each part is different. In the case of semicrystalline polymers, the shrinkage and the mechanical properties are highly influenced by the cooling rate. Due to the mentioned problems, molds with not more than four cavities are usually used for high-precision parts. The lower the melt volume between the pressure generating system and the part cavity, the lower is the pressure drop and the better the control of the process. Therefore, piston injection directly into the split line of the mold makes sense as it is realized on microinjection molding machines.

The basics of the replication of microparts and microstructures on the surface of parts are more complex than for conventional injection molding. Due to the sections with small cross-sections for the melt, the description of the viscosity curve should also include shear rates up to 10^{50}s^{-1} or even 10^6s^{-1}. For some polymers, it is reported that at such high shear rates there is a second Newtonian plateau. Due to this second Newtonian plateau, the approximations usually used for the viscosity curve lead to wrong results in simulation. The first Newtonian plateau is at low shear rates. This is followed by a shear rate range with distinct shear thinning. which can be described by the power law model or similar. Conventional injection molding can be described by an extended power law model which incorporates the temperature dependence of the viscosity. However, it is not only the viscosity which has to be considered more carefully when microparts are produced by injection molding: the thermal properties of the melt are also of major importance. During mold filling, a frozen layer is created due to the mold surface temperature, which is lower than the melting temperature of the polymer. Therefore, the cross-section for the melt to flow through becomes smaller and smaller. For thin-walled parts, the mold filling has to be completed before the part sections close to the gating system freeze completely. High injection speeds are necessary to ensure complete part filling. For the selection of the proper machine, not only the nominal maximum injection is of importance but also the maximum acceleration. Nowadays, with conventional injection molding machines maximum injection speeds of 1000mm s^{-1} can be achieved by using hydraulic accumulators. With microinjection molding machines, the acceleration of the injection piston is

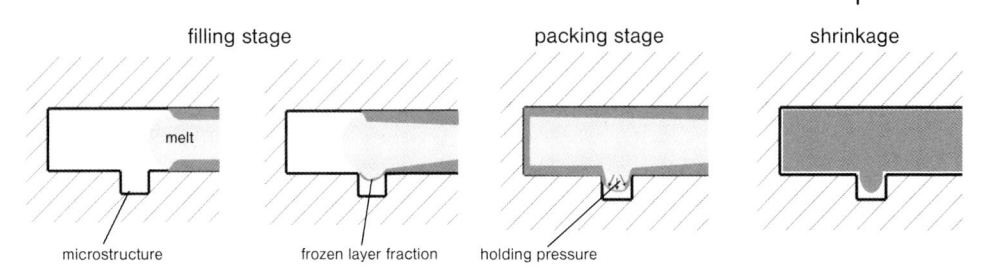

Figure 16.9 Filling, packing and shrinkage stage during the injection molding of parts with microstructured surfaces (m2M parts).

not crucial because there is a sufficient length of the barrel which is used for acceleration of the melt. The thickness of the frozen layer depends on the melting temperature, the mold temperature, the no-flow temperature, the contact time and the thermal diffusivity. The thermal diffusivity is the thermal conductivity divided by the density and by the specific heat capacity. As all three factors are a function of temperature, the calculations are usually done with a so-called effective thermal diffusivity that is considered to be constant over a certain temperature range. The frozen layer and the formation of a highly viscous skin at the flow front lead to a filling sequence of microstructures as shown in Figure 16.9. There is an initial filling phase of the microstructures. Hence there is incomplete filling of the microstructure during mold filling because there is not enough pressure in that region. After the mold has been filled completely, there is the packing phase and holding pressure is applied to the melt. At that moment, it depends on the thickness of the frozen layer whether there is complete replication of the microstructure or not. If the frozen layer is already too thick, even with very high holding pressures it is not possible to move the melt front into the microstructure. The mold temperature and the melt temperature influence the thickness of the frozen layer at the end of the filling stage and the efficiency of the pressure during the packing stage. It was shown experimentally that high cavity pressures lead to better replication qualities. However, is not only the filling and packing stage which influences the final shape of the microstructure, but also the shrinkage of the polymer. Due to the different situation inside the microstructure, the shrinkage factor in those regions can be different from the shrinkage of the macroscopic dimensions. Also, the trapped air inside the blind holes of the microstructures has to be considered, as already mentioned above. Hence an evacuation of the mold prior to injection is necessary to ensure good replication quality.

Another process for replicating microstructured surfaces is hot embossing. The greatest disadvantage of this process is that one cycle takes several minutes. A combination of injection molding and hot embossing, the so-called injection compression process, combines the advantages of both processes. The mold prior to injection is not closed completely but there is still a small gap of about 1 mm between the two mold halves. After injection of a portion of the total shot volume,

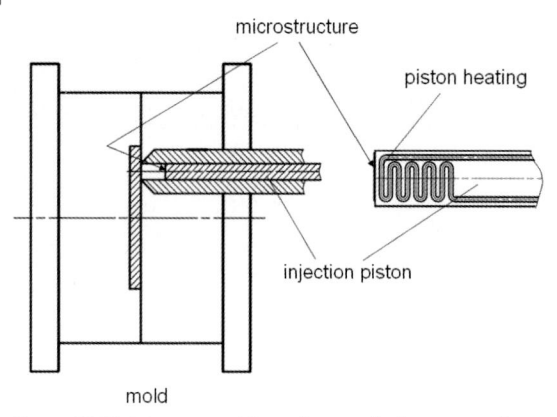

Figure 16.10 Injection molding of parts which are partially microstructured with piston injection.

the clamping unit closes the mold and builds up the clamping force. In parallel with that, the injection continues. Special software in the control unit of the machine and a mold with a vertical flash face are necessary. The injection compression leads to lower internal stresses and usually to better surface replication. With conventional injection molding, there is a pressure gradient from the gating to the end of the flow front, which results in an uneven pressure distribution over the whole part. With injection compression, the applied pressure over the complete surface of the part is more or less the same. Injection compression is limited to flat parts and it is not possible to replicate undercuts. Another possibility for replicating microstructured surfaces by injection molding is to place the LiGA insert not in the mold but on the injection piston, as shown in Figure 16.10. Also in this case there is an even pressure distribution over the complete microstructured area. The melt comes in contact with the microstructured piston tip before injection. Hence there is time to heat up this area by the melt, which leads to a reduction in melt viscosity in the microstructures during injection. After injection, the melt cools because of the mold temperature, which is lower than the melting temperature of the polymer. The last area to cool below the melting temperature is the microstructured surface on the tip of the piston. If the heat from the melt is not sufficient to keep a low viscosity in the microstructures, the tip of the piston can be heated to ensure good replication of the microstructure. This can be a constant heating or cyclical as in a variotherm process.

17
Filled Resist Systems

Thomas Hanemann, Claas Müller, and Michael Schulz

17.1	**Introduction**	*416*
17.2	**Variants of Lithography**	*417*
17.2.1	UV Lithography	*417*
17.2.2	Deep X-ray Lithography	*418*
17.2.3	e-Beam Lithography	*419*
17.2.4	Microstereolithography	*419*
17.3	**Properties of Filler–Polymer Matrix Composites**	*420*
17.3.1	Flow Properties	*421*
17.3.2	Optical Properties	*423*
17.4	**Application of Filled Resist Systems**	*424*
17.4.1	Deep X-ray Lithography on Filled Preceramic Polymers	*424*
17.4.1.1	Description of Preceramic Polymers	*424*
17.4.1.2	Determination of the Crosslinking Mechanism	*424*
17.4.1.3	Powder-filled Resist Material	*425*
17.4.1.4	Exposure	*427*
17.4.1.5	Influence of Suspension Composition and Treatment	*428*
17.4.2	Filled SU-8 Resists	*428*
17.4.2.1	Electrically Conductive SU-8 Composites	*429*
17.4.2.2	Magnetic SU-8 Composites	*429*
17.4.2.3	Silica-filled SU-8 for Microelectronic Applications	*430*
17.4.2.4	SU-8–Alumina Composites	*430*
17.4.3	Nanoparticle-filled Resists for e-Beam Lithography	*432*
17.4.4	Filled Resists for Microstereolithography	*433*
17.5	**Miscellaneous**	*437*
17.5.1	Photosensitive Polyimide–Silica Composites	*437*
17.5.2	Water-based Filled Photoresists	*437*
17.5.3	Three-dimensional Lithography Using Sol–Gel-based Compositions	*438*
17.6	**Conclusion**	*438*
	Acknowledgments	*439*
	References	*439*

17.1
Introduction

All modern lithographic methods in microelectronics and microsystem technologies are based on the interaction of photosensitive monomers, oligomers and polymers with electromagnetic radiation. The kind of photoreaction, which depends strongly on the chemical and physical properties of the resist material used and on the applied radiation type and energy, determines the application possibilities. Different photoreactions such as photopolymerization, photocrosslinking, photodegradation and photothermal reactions have been described in the literature [1]. With respect to lithography, photopolymerization and photodegradation are the most important ones. If a monomer or oligomer polymerizes under irradiation conditions independently of the applied wavelength, the material is defined as a negative resist, that is, a negative image of the pattern on the used mask will be created. If a polymer is destroyed by light via chain scission, the material behaves as a positive resist (Figure 17.1). Both mechanisms are exploited in the realization of electronic and printing materials, device fabrication in microelectronics and microsystem technology and packaging and in different surface modification techniques for post-processing of films, textiles or fibers [1].

With respect to further miniaturization of micro- and nanostructures with variable aspect ratios in combination with further progress in the different established lithographic methods and new developments such as laser lithography in the deep-UV range, new resist systems have been developed in recent years [2–4].

The addition of ceramic or metal fillers to photoresists is motivated by the following aspects:

- improved sensitivity to electromagnetic radiation
- improved resolution
- improved chemical and mechanical resist stability

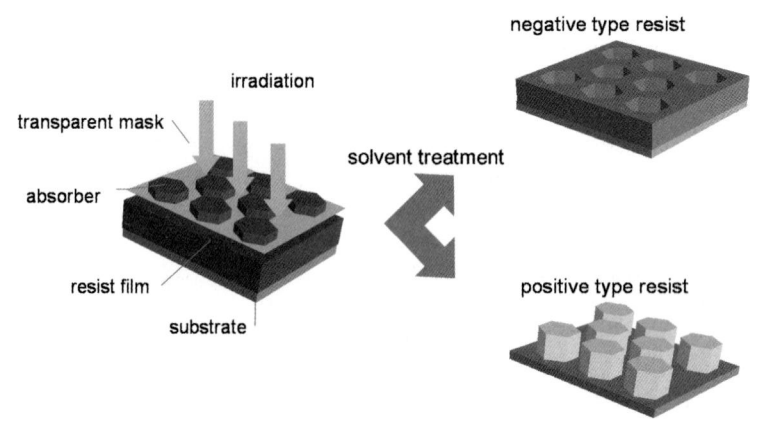

Figure 17.1 Principal resist behavior during irradiation.

- introduction of new functionalities (electrical conductivity, magnetic properties)
- direct fabrication of ceramic or metal microcomponents using lithographic microstructuring in combination with thermal post-processing (debinding and sintering)
- rapid prototyping of ceramic and metal parts.

17.2 Variants of Lithography

In microelectronics and microsystem technology, three fundamental lithographic techniques have been established. For a long period, mercury vapor arc lamps were used for photolithography in the visible or near-UV range; today, powerful laser sources in combination with special chemical amplified resists permit UV lithography in the deep-UV range down to the realization of 100 nm structures [2, 3]. High aspect ratio microstructures can be realized via deep X-ray lithography. Geometric details in the nanometer range can be generated via e-beam lithography, which is essential for the optical mask fabrication. Due to shadow exposure, the above-mentioned lithography variants allow only the patterning of 2.5D structures.

In addition, real 3D microstructures can be fabricated via microstereolithography due to the layer-by-layer irradiation of a negative resist using a dynamic mask generator. In the following, the fundamental principles and the requirements for the applied resist materials are described.

17.2.1 UV Lithography

UV lithography is an optical lithography technology for wavelengths of 365 nm or shorter and is already widely used for the fabrication of microsystems and advanced CMOS components. The lithography process transfers a pattern from a mask to a photoresist. First the substrate is coated with a photoresist being sensitive to UV radiation. This is normally done by dropping the resist on top of the substrate, which is then spun at a high speed to obtain a uniformly thick resist layer. The substrate with the resist is then exposed to UV radiation through a mask that protects parts of the resist. Finally, the substrate is put in a developer. SU-8 is a typical example of a negative resist and is widely used for the fabrication of high aspect ratio structures. SU-8 components are useful for MEMS, fluidic and packaging applications. There are two different ways to expose the resist to UV light, shadow exposure and projection exposure. In shadow exposure, the sample is put in contact or close contact (called proximity exposure) with the resist and the pattern will be directly transferred to the resist with a scale of 1:1. In projection exposure, the mask and pattern are set wider apart and the optical system has a magnification up to 10×. Magnification of the optical system opens up the possibility of having a mask demagnification ratio, thus making mask fabrication easier [3, 4].

17.2.2
Deep X-ray Lithography

Deep X-ray lithography is normally performed using synchrotron radiation. To prevent the synchrotron radiation from being absorbed on its way from the bending magnet to the sample, the lithography beamline is highly evacuated. Separated by a beryllium window and a shutter, the concluding scanner chamber is evacuated and flooded with 100 mbar of helium after the sample is mounted. In the scanning chamber, the sample holder, carrying the mask and the resist-coated substrate, is moved up and down in front of the beam (Figure 17.2) [3].

The X-ray scanners used allow the exposure of sample areas of up to 10 cm^2. Since large-angle deflection of the synchrotron radiation is not possible, the substrates must be fixed vertically in the scanner. These boundary conditions influence the working properties of the resist material. After the coating of the substrate, a drying step is required. No volatile fractions are allowed, in order to protect the chamber from damage, and the resist must not be liquid because of the vertical mounting.

A specific range can be extracted from the synchrotron spectra by using appropriate filters in the beamline, whereby the gradient between top and bottom dose can also be adjusted. The calculation of the exposure dose necessary is based on the radiation source parameters, the filters and windows in the beamline and also on the atomic composition and resist density.

The absorption of X-ray in matter is caused by interactions between photons and electrons [4]. Emerging scattered radiation and pair formation effects are not

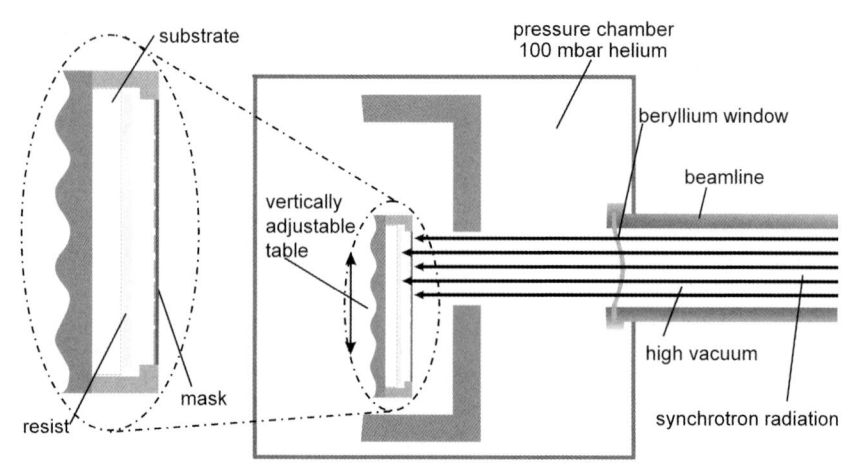

Figure 17.2 Scheme of an X-ray scanner according to [3].

taken into account for the resist change. Based on the values of the mass absorption coefficient, the absorption of the resist material can be calculated following the empirical Equation 1 with the mass absorption coefficient σ, a wavelength dependent constant $c(\lambda)$, the material density ρ, the atomic number Z and the wavelength used λ [5]:

$$\text{Mass absorption coefficient} \quad \sigma = c(\lambda)\rho Z^3 \lambda^3 \quad (1)$$

17.2.3
e-Beam Lithography

Nanoscale structures can be generated using e-beam lithography (EBL). EBL is used for either mask generation or direct writing of patterns on the wafer. Most direct writing systems use a small electron beam spot that is moved relative to the wafer to expose the pattern one pixel at a time. Therefore, the systems can be classified as raster scan or vector scan with fixed or variable beam geometries. After generation from a cathode, the electrons are accelerated and focused by electrostatic lenses and a variety of apertures and knife edges [3, 4].

When electrons penetrate the resist layers, they are scattered and therefore experience a deviation from their original direction. Depending on the electron energy and the atomic mass, the angle of deviation varies. The large atomic mass of the substrates causes so-called back scattering, leading to exposure of the resist from underneath in areas which are not supposed to be irradiated. Forward scattering of the electrons on the atoms of the resist leads to strong broadening of the beam. These proximity effects are a major challenge in EBL [3, 4].

17.2.4
Microstereolithography

Microstereolithography was the first rapid prototyping process in microsystem technology. Rapid prototyping has been established in the macroworld and also for some years now in the microworld for rapid product development with respect to a significant reduction in the time-to-market factor and the product development costs. A comprehensive overview of rapid prototyping in micro- and nano-technology can be found in [6]. Microstereolithography was developed in the early 1990s using a photocurable resin as a negative resist material and a xenon discharge lamp or an argon ion-laser as a light source fabricating a simple microfluidic system or a microclamping tool [7, 8]. The smallest geometric features were in the micrometer range and aspect ratios around 10 and more could be achieved. The original stereolithography technique was invented in 1984 by 3D Systems [9]. Unlike shadow mask-based lithographic techniques, the 3D microstructure is built by the in-layer-solidification of low-viscosity photocurable resists such as acrylates, urethanes or epoxies, doped with suitable photoinitiators adjusted to the light sources used (Figure 17.3).

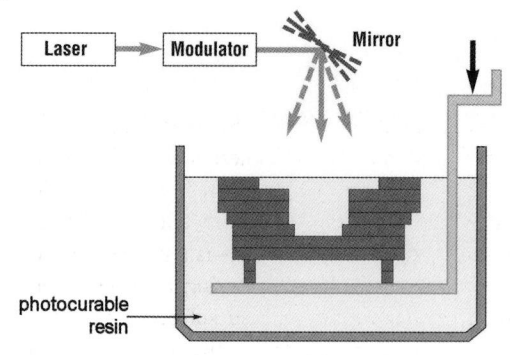

Figure 17.3 Principal microstereolithography setup.

At the beginning of the microstereolithography process, the platform is located just below the resist's surface. An optical imaging system including a spatial light modulator such as a transmission liquid crystal display or a digital micromirror device acting as a dynamic pattern generator allows the transfer of the 2D CAD information (slice) into the polymer via solidification. The first slice data are written into the photosensitive resist via polymerization during irradiation. After finishing, the platform is moved a small distance towards the bottom of the reaction container and another layer can be written. The sequence can be repeated until the physical model is finished. After removing the part from the container, a post-exposure curing via thermal treatment has to be performed for complete polymer solidification. The microstereolithography technique developed by Bertsch and coworkers [10, 11] allows the fabrication of microsized parts with smallest dimensions of 5 µm. More recently, the so-called two-photon polymerization process was developed, allowing further miniaturization and smaller objects with structural details in the nanometer range [12].

17.3
Properties of Filler–Polymer Matrix Composites

The addition of inorganic or organic fillers or dopants to resist materials has a strong impact on almost all physical properties depending on the filler concentration, average particle size, particle size distribution, specific surface area, surface chemistry and molecular and electronic structure. If an organic dopant is soluble in the resist, the resulting composite properties are mainly affected by the dopant's molecular structure changing, for example, the light absorption, refractive index and important thermomechanical properties such as glass transition temperature depression due to plasticizing or embrittlement and others. Inorganic materials are insoluble in resist systems, hence the fillers have to be dispersed in the liquid matrix, the resulting composite shows a typical multi-component, multi-phase behavior such as light scattering and significantly changed flow behavior.

17.3.1
Flow Properties

In most of the lithography variants, a resist film has to be generated via spin coating or casting prior to the patterning step. The thickness and quality of the resist film depend on, among other process parameters, the resist viscosity. The addition of solid inorganic or organic fillers affects the viscosity and the principal flow behavior of polymer- or resin-based fluids. The viscosity is a measure of the internal friction within a fluid on a molecular level covering all interactions between individual polymer chains, solvent molecules and solved or dispersed particles. Following the simple approach for ideal liquids, the viscosity can be described as the ratio between shear stress τ and shear rate γ (Equation 2), hence the viscosity η is constant at different shear rates (Newtonian flow behavior). In a first approximation, Newton's viscosity law can be used for almost all inorganic and organic liquids. In the case of non-ideal liquids, the viscosity varies with the applied shear rate and two additional fundamental flow behaviors occur. If the viscosity rises with increasing shear rate, a dilatant flow can be observed. The opposite phenomenon, a viscosity drop with increasing shear rate, is named pseudoplastic flow, which is typical of filled polymer resins or melts. The addition of small filler amounts has only a small impact on the resulting composite viscosity and can be described using Einstein's viscosity law originally defined for solutions and later extended to dispersions (Equation 3) [13, 14]. The introduction of the relative viscosity η_{rel} (Equation 3) as the ratio of the measured composite viscosity η_{comp} to the viscosity of the pure liquid matrix η_M allows a comparison of different fluid–filler composites. The relative viscosity changes with the volume filler content Φ, introducing a constant factor k_E, which can be estimated to a numerical value of 2.5 for small solid loads below 15 vol.%, neglecting particle–particle interactions and considering monosized, non-deforming particle spheres [15]. A further power series extension of the Einstein approach introduces further parameters for geometric particle–particle interaction (k_H) (Equation 4). Finally, Thomas summarized a large number of experimental results in a semi-empirical approach (Equation 5) for the change in the relative viscosity with filler load [16].

$$\text{Newton's viscosity law} \quad \eta = \frac{\tau}{\gamma} \tag{2}$$

$$\text{Einstein} \quad \eta_{rel} = \frac{\eta_{comp}}{\eta_M} = 1 + k_E \Phi \tag{3}$$

$$\text{Extended Einstein model} \quad \eta_{rel} = 1 + k_E \Phi + k_H \Phi^2 + \cdots \tag{4}$$

$$\text{Thomas} \quad \eta_{rel} = 1 + 2.5\Phi + 10.05\Phi^2 + 0.00273 e^{16.6\Phi} \tag{5}$$

At low filler loads, the viscosity is mainly determined by the matrix viscosity. In the case of polymer-based resist materials, it is dominated by the intermolecular polymer chain friction. At higher filler concentrations, filler–filler interaction occurs due to the reduced impact of the polymer fluid lubricant, which results in a pronounced increase in the composite viscosity. The maximum filler load is achieved if the solid particles are in direct contact. The resulting viscosity increases to an infinite value. The observed experimental results can be described in different semi-empirical approaches introducing the maximum filler load Φ_{max} in an exponential expression. The most important descriptions were developed by Krieger and Dougherty [17] (Equation 6) and Quemada [18] (Equation 7).

$$\text{Krieger–Dougherty} \quad \eta_{rel} = \left(1 - \frac{\Phi}{\Phi_{max}}\right)^{-k_E \Phi_{max}} \quad (6)$$

$$\text{Quemada} \quad \eta_{rel} = \left(1 - \frac{\Phi}{\Phi_{max}}\right)^{-2} \quad (7)$$

The different models for the description of the relative viscosity change with filler load have been applied successfully in the case of photocurable resist–ceramic composites [19], thermoplastic filler feedstocks for injection molding [15, 20] and micro- and nanofillers dispersed in UV-curable composites [21, 22].

In addition to the filler content, the composite's viscosity depends on the particle size distribution, the specific surface area, particle size, deformability and packing density. Table 17.1 shows exemplarily the properties of different micro- and nanosized aluminas. Micro-sized alumina powder such as CT3000SG and RCHP are well-established fillers for feedstock preparation used in microinjection molding. It can be seen from Table 17.1 that the measured average particle sizes (d_{50}) of the micro-sized aluminas differ by a factor of only 2 from the nanosized aluminas. The aluminas obtained from Degussa and Nanophase were described as nanosized materials with primary particle sizes of 13 and 38 nm, respectively, which agglomerate due to surface energy minimization. The latter aluminas possess a significantly larger specific surface area. The impact of the different micro- and nanosized aluminas and different filler contents on the relative viscosity of curable unsaturated polyester-based composites is demonstrated in Figure 17.4.

The increase in the specific surface area results in a depression of the maximum achievable filler load; using the micro-sized aluminas CT3000SG and RCHP, ceramic moieties of more than 40 vol.% can be achieved under the given experi-

Table 17.1 Comparison of micro- and nanosized alumina [23].

Type	Vendor	Specific surface area ($m^2 g^{-1}$)	d_{10} (μm)	d_{50} (μm)	d_{90} (μm)
CT3000SG	Alcoa	7	0.071	0.266	1.002
RCHP	Baikowsky	8	0.242 (d_{20})	0.335	0.575
TM DAR	Taimei	14.5	0.104	0.165	0.297
Nanotek	Nanophase	34	0.125	0.155	0.192
C	Degussa	107	0.125	0.155	0.205

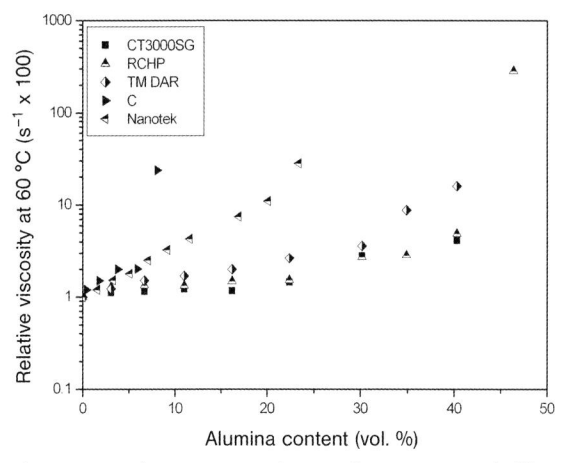

Figure 17.4 Relative viscosity change of composites with filler content.

mental conditions (cone–plate rheometer). The use of the nanosized alumina allows only small filler loads due to the large surface area, which fixes huge amounts of the liquid phase. As a conclusion, the larger the filler's specific surface area, the higher is the resulting composite viscosity. Thin-film processing such as spin coating or casting of filled resists is influenced significantly by the filler, as described later in the investigated SU-8–alumina composites.

17.3.2
Optical Properties

The addition of inorganic or organic fillers to polymer resist materials affects directly the optical properties. Particles with geometric extensions larger than the wavelength of light show absorption. If the average particle size is in the same range as the applied wavelength, Mie scattering occurs, causing the Tyndall effect. In the case of smaller particles, Raleigh scattering can be observed; the intensity is proportional to $1/\lambda^4$ of the wavelength used and to the square of the volume of the particles used [24]. Therefore, as a rule of thumb, only the particles with a diameter smaller than one-tenth of the light wavelength used permit acceptable transmission.

For the realization of microstructured ceramic parts using photolithographic techniques and photocurable ceramic composites, an estimation of the curing depth is essential. Following the modified Lambert–Beer law (Equation 8), the curing depth C_d is proportional to the average particle size d and the logarithm of the ratio of the incident light E_0 to the minimum intensity for photopolymerization E_c and indirectly proportional to the volume fraction Φ of the filler [25]. For highly filled composites, the scattering ability can be described in terms of the scattering efficiency Q, which is a function of the refractive index difference Δn between the filler and the surrounding polymer matrix, the interparticle spacing S, a form factor F depending on the particles shape and the wavelength used λ (Equation 9) [25, 26].

$$\text{Curing depth} \quad C_d = Fd \frac{1}{\Phi} \frac{1}{Q} \ln\left(\frac{E_0}{E_c}\right) \tag{8}$$

$$\text{Scattering efficiency} \quad Q = (\Delta n)^2 \frac{S}{\lambda} \tag{9}$$

17.4
Application of Filled Resist Systems

17.4.1
Deep X-ray Lithography on Filled Preceramic Polymers

17.4.1.1 Description of Preceramic Polymers

Preceramic polymers are polymers which contain in the molecular structure in addition to carbon, hydrogen, nitrogen, oxygen, silicon and sometimes boron and titanium atoms. Unlike hydrocarbon-based materials, the combustion delivers ceramic residues depending on the gas atmosphere used. The pyrolysis of siloxanes (containing silicon and oxygen) under air or oxygen results in silicon dioxide; the firing of silazanes (contain silicon and nitrogen) under air or oxygen also results in silicon dioxide. Using inert gas and temperatures around 1200 °C, amorphous silicon carbonitride (SiCN) can be obtained. These materials show a homogeneous elemental distribution, are free of sinter additives and exhibit good high-temperature stability and oxidation resistance [27]. The thermoplastic silazanes can be structured using X-ray lithography and subsequently converted into ceramics [28, 29]. Several silicon-containing polymers that serve as precursors for SiCN ceramic were developed in the past, such as the ABSE shown in Figure 17.5 [27, 30].

The selection of suitable precursors was based on the criteria listed in Table 17.2. From Table 17.3, it can be seen that none of the polymers examined fits all requirements. Several other polymers such as SiBCN precursors were examined, but owing to their improved sensitivity to moisture, their use as resist material is impossible. The light sensitivity can be improved by adding suitable photoactive compounds as photoinitiator as in other resist materials.

17.4.1.2 Determination of the Crosslinking Mechanism

The mechanism of photoinduced crosslinking of a selection of these preceramic polymers was investigated using spectroscopic methods (FT-IR, NMR). It is notice-

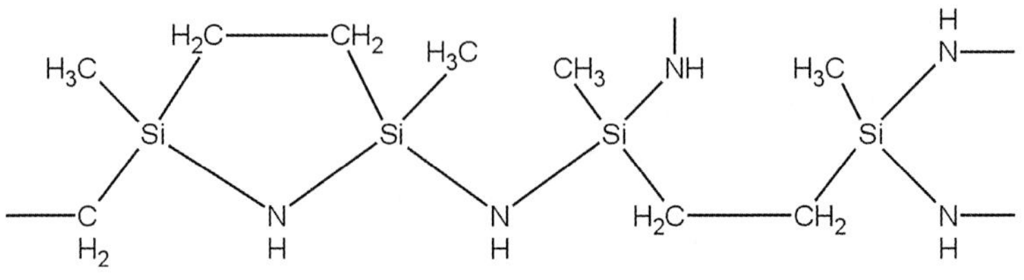

Figure 17.5 Chemical structure of the preceramic polymer ABSE [30].

Table 17.2 Required physical and chemical properties with respect to lithography [28].

(a) Solid aggregate state under ambient conditions; in the case of UV-lithography, an adjustable highly viscous state was acceptable

(b) Soluble in standard aprotic organic solvents

(c) Formation of stable films on substrate surfaces

(d) Processability under ambient conditions for several hours, that is, improved chemical stability against oxygen and moisture

(e) Ready and selective solubility in developing solvents

(f) High sensitivity to photochemically induced crosslinking due to the presence of chromophoric side-groups

(g) Availability and small batch variation

Table 17.3 Investigated ceramic precursors[a] [28].

Type	Vendor	(a)	(b)	(c)	(d)	(e)	(f)	(g)
NCP-200	Nichimen	+	+	+	+	+	+	−
SLM465012VP	Wacker	+	+	+	+	+	+	−
Ceraset	KiON	−	+	−	+	+	+	+
VL20	KiON	−	+	−	+	+	+	+
ABSE	CME Bayreuth	+	+	+	++	−	−	0

a Evaluation: ++ = very well suited; + = well suited; 0 = suited; − = not suited.

able that the exposure dose needed to measure an effect in IR and NMR spectroscopy was more than 20 times higher than the dose needed for the microstructuring of the polymer using X-ray lithography. Obviously a considerably lower dose is necessary to crosslink the material to an insoluble state, that is, few linked atoms are sufficient. Absorption of the incident X-rays in the top layer causes a decrease in the deposited dose from the top to the bottom of the sample. For dose investigation, the specimens were turned around after the first half of irradiation time to minimize the dose gradient and a middle dose was deposited. Figure 17.6 shows the dose distribution within the specimen thickness calculated concerning the absorption coefficient of the ABSE material [28].

17.4.1.3 Powder-filled Resist Material

Due to the evaporation of volatile combustion products during pyrolysis, substantial shrinkage of the samples of up to 50% occurs. This often leads to the destruction of microstructures adhering to the substrate. The idea was to minimize crack formation by mixing the polymer with 30–70 wt% of an Si_3N_4 powder (UBESNE10)

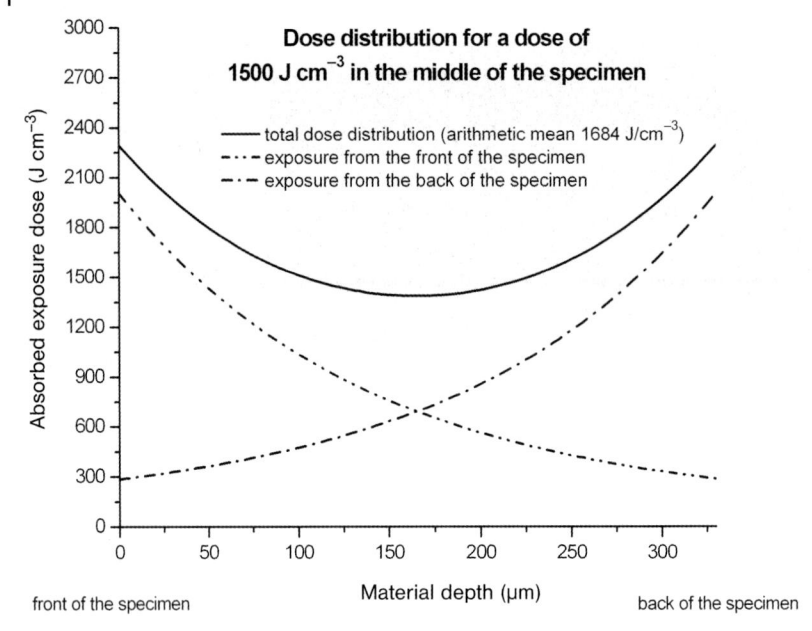

Figure 17.6 Determination of the dose distribution in X-irradiated ABSE samples calculated concerning the absorption behavior [28].

related to the polymer content to increase the ceramic yield. Two different preceramic polymers were used in the experiments, the liquid Ceraset and VL20, and the solid ABSE precursor in a 50 wt% solution in *n*-octane [28, 29].

Delamination of the structures occurred during the concluding pyrolysis step because of the mismatch of the thermal expansion coefficient of the alumina substrate and the preceramic polymer film. To handle this effect, native substrates were produced from composites containing Si_3N_4 powder in the preceramic polymeric matrix. A tape casting process, concluded by a thermal crosslinking step, was established that allows the production of substrates with a thickness around 500 μm. Solutions of preceramic polymer and suspensions with different silicon nitride filler contents were coated on the native substrates. After evaporation of the solvent, the samples were exposed using X-ray lithography with a microstructured test mask. The primary particles of the ceramic powder (d_{50} = 0.5 μm, manufacturer's data) are smaller than one-tenth of the dimensions of the replicated structures. To achieve high resolution in the lithographic process, it is necessary to deagglomerate the powder. A suitable deagglomeration of the α-Si_3N_4 powder was obtained using both an ultrasound disintegrator and a dissolving stirrer.

The silicon nitride powder absorbs the synchrotron radiation to a minor extent. The mass absorption coefficient of the elements contained in the ceramic is small, which means that the filler used is almost transparent in the X-ray range [31]. The exposure dose required for the different polymer–ceramic composites was calculated before the exposure concerning the content of Si_3N_4. Although the mass

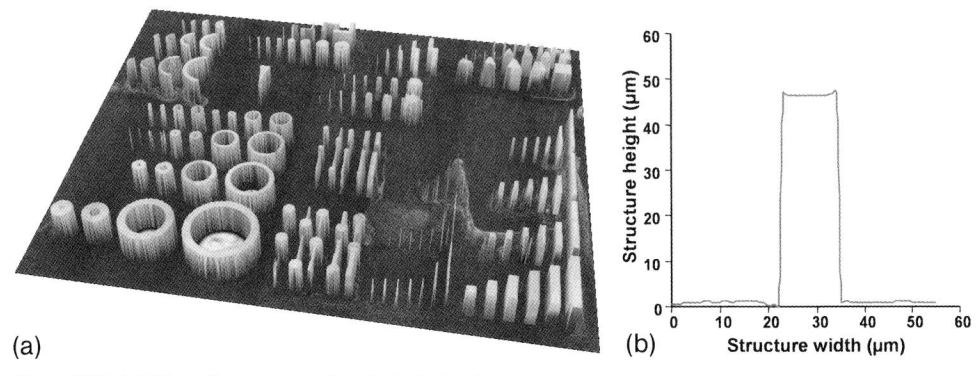

Figure 17.7 (a) 3D surface topography of a test structure mode of ABSE after deep X-ray lithography; (b) height profile of a structural detail [32, 33].

absorption coefficient of the Si and N atoms is low, for samples filled with powder the desired dose rises steadily with filler content [5, 28, 31].

17.4.1.4 Exposure

Samples were exposed at the ANKA synchrotron facility at the Forschungszentrum Karlsruhe (Karlsruhe, Germany) and the CAMD facility of Louisiana State University (Baton Rouge, LA, USA). The microstructures were converted into ceramics after the patterning and development steps at temperatures up to 1500 °C. Using a bottom minimal dose of 1.5 kJ cm^{-3}, aspect ratios of up to 20 were realized. The test mask layout and the resulting 3D surface topography of a sector of the exposed and developed structure based on ABSE polymer, recorded using multilayer optical microscopy, is presented in Figure 17.7a [32].

Structured layers with a thickness of up to 200 µm have been realized [33]. The edge precision of unfilled polymer is excellent; side-walls are perpendicular to the surface of the substrate and the pattern. The edge precision is <1, as can be seen on the profile of a bar structure (Figure 17.7b) from the middle of the mapped sample. In Figure 17.8a, a test structure is shown before and after pyrolysis. Whereas the unfilled resist material, except for the smallest features, cracked upon pyrolysis, the powder-filled structures remained stable. The surface roughness of the samples filled with powder is low compared with the side-wall roughness, as can be seen from the SEM images in Figure 17.8b. When the dimensions of the microstructure meet the powder particle size, the quality of reproduction decreases.

Using deep X-ray lithography, a value for the contrast of the unfilled ABSE material was calculated. The contrast describes the sensitivity of a resist material in lithography. Following Equation 10, a contrast value for the ABSE material of around 0.8 was determined. D_0 is the dose without any effect on the resist and D_{100} the related value for a complete solidification. With a sensitivity to crosslinking by synchrotron radiation of 500–1000 J cm^{-3}, the preceramic polymer ABSE,

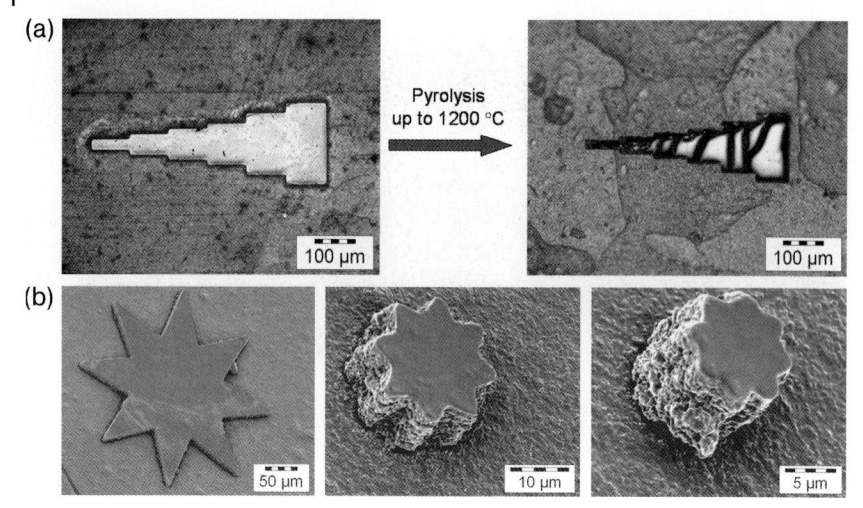

Figure 17.8 Test structures made by X-ray lithography on native substrates. (a) Wedge structure before and after pyrolysis of the unfilled ABSE polymer; (b) powder-filled ABSE precursor after pyrolysis.

which acts as a negative resist, ranges between the values for the standard resist materials SU-8 (15 J cm^{-3}, negative resist) and poly(methyl methacrylate) (PMMA) (4000 J cm^{-3}, positive resist).

$$\text{Contrast} \quad \gamma = \frac{1}{\log\left(\dfrac{D_{100}}{D_0}\right)} \tag{10}$$

17.4.1.5 Influence of Suspension Composition and Treatment

The treatment of the suspensions has a strong influence on the porosity, density and hardness of the resulting fired ceramic. The extended deagglomeration by ultrasonic treatment caused a significant decrease in the porosity (Figure 17.9). The influence of the aggregation on the substrate quality can also be seen from the Vickers hardness. Maximum hardness of 1068 ± 36 HV was achieved at 50 wt% (24.5 vol.%) for ultrasound-treated samples (for comparison: sintered silicon nitride $HV_{10} = 1400$).

Admittedly, for the pyrolytic conversion into the amorphous SiCN ceramic, film thickness should be limited as crack formation becomes worse with increasing dimensions, especially in the case of unfilled preceramic resist materials [27].

17.4.2
Filled SU-8 Resists

The main purpose of adding fillers to SU-8 is the exploitation of the outstanding sensitivity to UV and X-radiation. A number of different attempts to introduce electrical or magnetic properties into the resist for the realization of composite

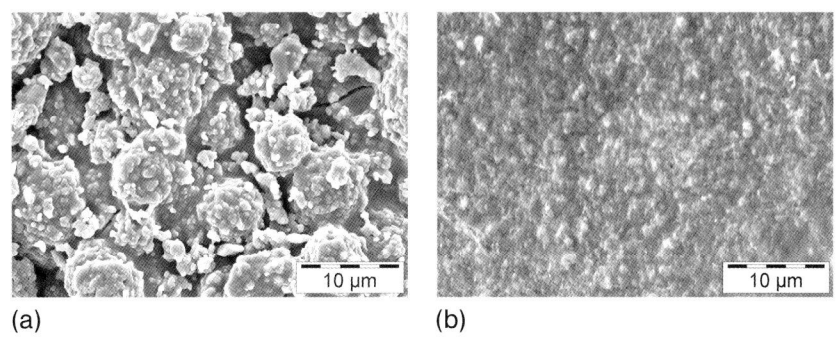

Figure 17.9 SEM images of the pyrolyzed substrates. Tape casting suspensions treated (a) without and (b) with ultrasound.

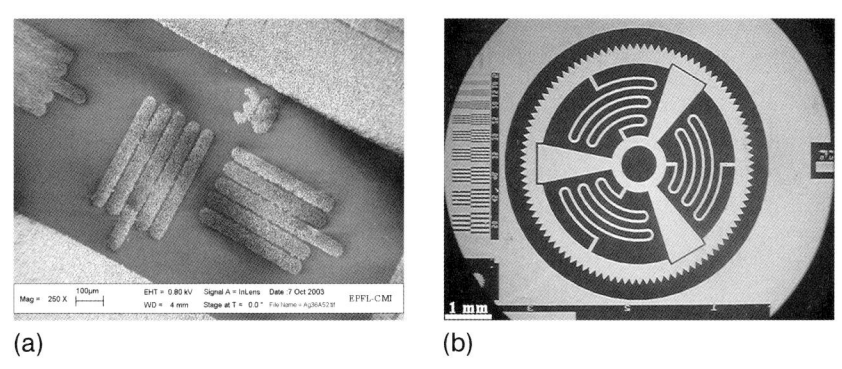

Figure 17.10 Images of SU-8–silver test structure (a) and wheel (b) [34].

microcomponents have been described in literature. In a different approach, SU-8–alumina composites can be microstructured via deep X-ray lithography. After thermal post-processing, ceramic microcomponents can be obtained.

17.4.2.1 Electrically Conductive SU-8 Composites

SU-8-based composites containing up to 40 vol.% micro-sized silver particles (average diameter around 1.5 μm) show electrical conductivity [34]. As in other described resist–filler systems (see Section 17.4.4), the curing depth decreases with increasing silver content. At filler contents larger than the electrical percolation threshold of 6 vol.%, the polymerized layers have a thickness of less than 20 μm. Using the UV-LIGA technique, simple test structures (thickness 8 μm) for resolution tests (Figure 17.10a) or even wheel structures (Figure 17.10b) can be realized [34, 35].

17.4.2.2 Magnetic SU-8 Composites

Ferromagnetic resists can be realized by incorporating micro- or nanosized ferromagnetic materials such as nickel or BaFe [36, 37]. The addition of nanosized nickel powder (average particle size 80–150 nm) decreased the optical transmission properties even at very low loads of 4 wt%. Due to scattering, reflection and diffraction of light, patterning via UV lithography was successful using wavelengths of 365 and 405 nm at thin films with a thickness around 70 μm. The magnetic resist was furthermore used for the fabrication of cantilevers as micro-mirror substrates.

17.4.2.3 Silica-filled SU-8 for Microelectronic Applications

The addition of transparent microsized silica particles to SU-8 improves the thermal stability up to 360 °C and adjusts the coefficient of thermal expansion (CTE) to common solder material [38]. The filled resist can be used as a solder mask during circuit fabrication. The filler content can be varied between 30 and 70 wt% with respect to the targeted CTE. Depending on the filler content, the filled SU-8 can be photocured with a dose around 0.05–0.4 J cm^{-2} in the UV region.

17.4.2.4 SU-8–Alumina Composites

For the realization of ceramic microcomponents fabricated using deep X-ray lithography, a similar process sequence to that described earlier was applied but now using SU-8 as photoresist filled with two different sub-micrometer aluminas (Alcoa CT3000SG and Reynolds RC-HP withouto Mg; for properties, see Table 17.1). The process comprises the individual process steps of composite formation, film processing, deep X-ray lithography and thermal post-treatment such as debinding and sintering to the final alumina part [31].

The addition of ceramic fillers to reactive resins or resists results in a moderate viscosity increase at low filler load as described earlier. Entering the critical region around 40 vol.% solid content, the viscosity increases dramatically over several decades. For the realization of ceramic microcomponents, a ceramic filler content of the composite should be around 45–50 vol.% in order to avoid enhanced part warpage or anisotropic shrinkage effects. To evaluate the flow behavior of SU-8–alumina composites, the change in the relative viscosity with alumina load was investigated (Figure 17.11). The dependence of the relative viscosity on the filler load is similar to that described in Section 17.3 and shown in Figure 17.4. At low filler loads up to 30 vol.%, a moderate viscosity increase can be observed; further addition of alumina up to 42 vol.% yields a relative viscosity increase by a factor of 20.

The increase in the relative viscosity with the alumina load follows the Quemada approximation as described in Section 17.3.1. Due to the high composite viscosity, the use of spin coating as a film formation method is excluded, therefore a modified casting technique was developed as described in [31]. Different layer thicknesses up to 500 μm could be realized; the subsequent soft-baking step had to be adjusted to the individual layer thicknesses.

In deep X-ray lithography, beryllium mask membranes deliver the best contrast due to the low atomic number of the material. As the aluminum oxide powder increases the absorption of the X-rays in the resist, the minimum dose was expected to be considerably higher than the bottom dose used for a pure SU-8

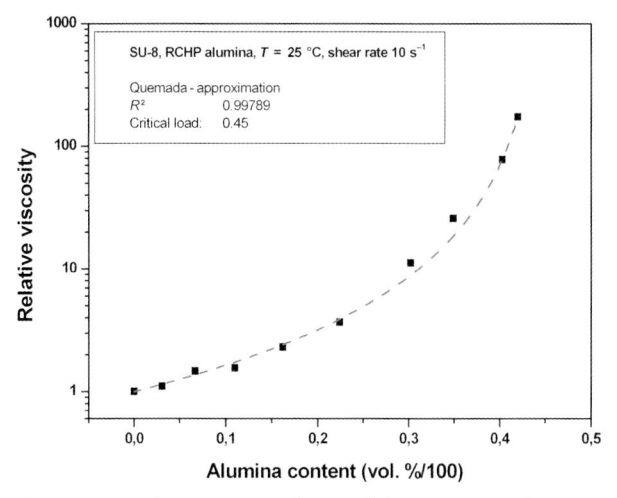

Figure 17.11 Relative viscosity change of the SU-8-RCHP alumina composite with filler load.

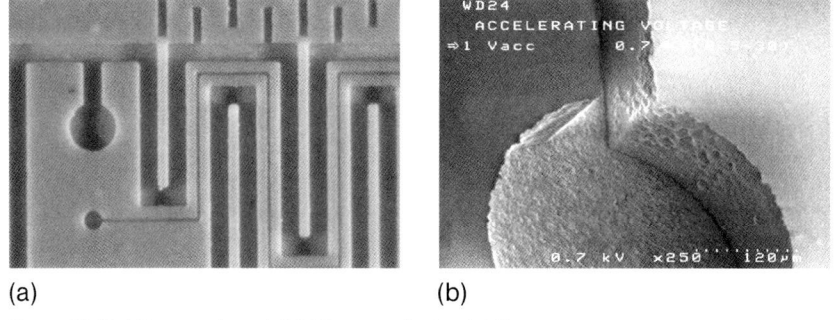

Figure 17.12 Microscopic and SEM images of microfluidic structures after development [31]. Reproduced with permission from Springer.

resist of 15 J cm^{-3}. A dose screening yielded an imperative bottom dose of 400 J cm^{-3} for full crosslinking of the resist. The fluidic microstructures shown in Figure 17.12 are 20 μm wide and about 200 μm deep with an aspect ratio of 10. In Figure 17.12b a circular inlet and the channel opening are magnified, showing the improved smoothness of the layer surface and channel walls. The reduced sidewall surface roughness is caused by the alumina particles at the boundary between exposed and unexposed areas [31].

Thermogravimetric investigations of the pure SU-8 showed the decomposition behavior at elevated temperatures up to 600 °C, which is important for the crack-free realization of the pure alumina microcomponents. After debinding, the final sintering step at 1550 °C followed. Figure 17.13 shows exemplarily free-standing high aspect ratio test structures after debinding. In Figure 17.14, the same micro-

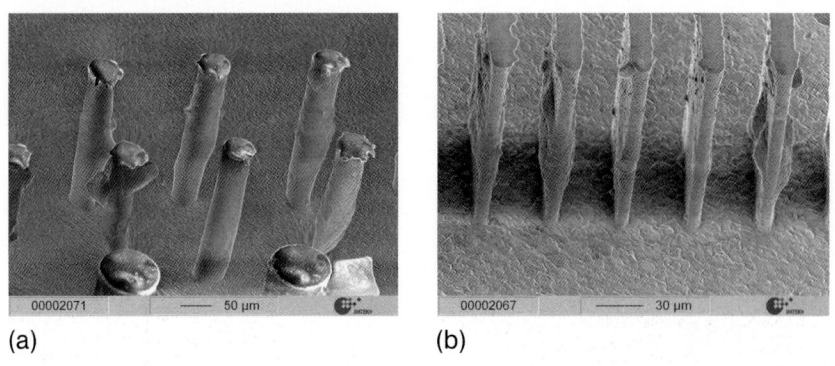

Figure 17.13 Free-standing high aspect ratio posts and walls before thermal treatment [31]. Reproduced with permission from Springer.

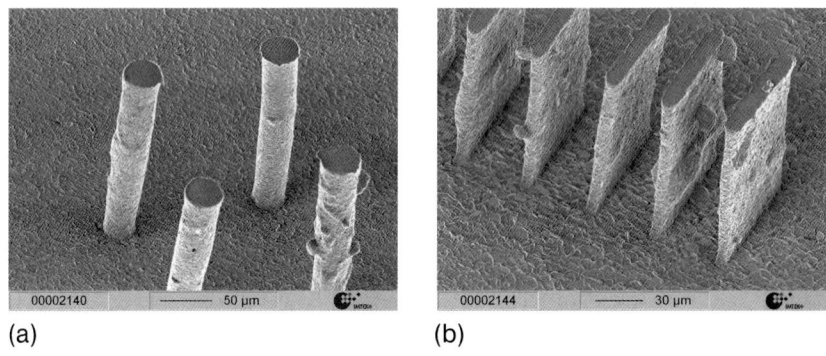

Figure 17.14 Free-standing high aspect ratio posts and walls after thermal treatment [31]. Reproduced with permission from Springer.

structures after sintering are shown. Prior to debinding, the posts had an average height of 360 μm and a diameter of 40 μm. The small walls had a length of 100 μm, a thickness of 10 μm and a height of 160 μm, gaining an aspect ratio of 16. Especially in the sintered structures, agglomerates of the ceramic particles are visible at the side-walls.

17.4.3
Nanoparticle-filled Resists for e-Beam Lithography

Nanocomposite resist systems have been developed recently containing fullerenes (C_{60}) and silica nanoparticles based on established standard e-beam resists (ZEP520) [39]. It was found that the incorporation of small amounts of SiO_2 nanoparticles improved the accessible resolution. Figure 17.15 shows line patterns with the

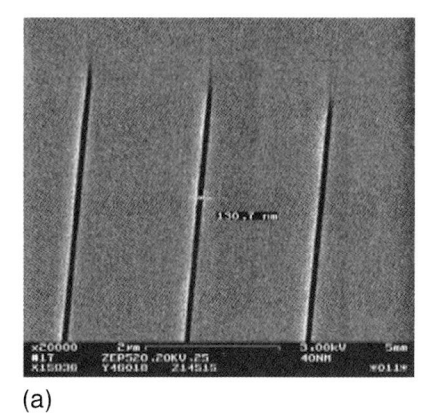

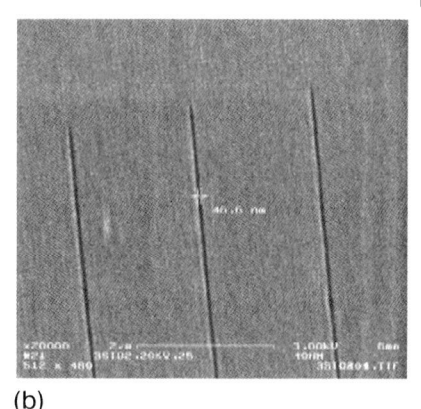

Figure 17.15 SEM images of trenches after writing on pure ZEP520 (a) and ZEP-silica composite films (b) [39]. Reproduced with permission from Elsevier

objective of obtaining 40 nm structures, in the pure ZEP520-resist a linewidth of 131 nm could be achieved (a) and in the composite system 47 nm lines were possible (b). A comprehensive overview of nanolithography using organic–inorganic nanocomposites is given in [40].

Currently, the majority of e-beam resists for the realization of metallic nanostructures consist of ligand-coated metal colloids, molecular organometallic species or purely inorganic materials [39–42]. In the exposed regions, the material undergoes a chemical change that cleaves the stabilizing ligands from the metal cluster or atoms. The unprotected metal atoms tend to undergo nucleation and therefore build the metallic structure; this means that the resists operate in the negative-tone mode. Highly metallized polyferrocene-based resists allow the fabrication of patterned metal-containing ceramics with special magnetic properties [41]. Metal salts that are soluble in organic solvents can serve as a precursor to metal atoms, if a suitable photoreactive dye is added as sensitizer. The ligand-coated metal nanoparticles then act as seeds [42].

Iron nanoparticles can be fabricated by e-beam lithography of a calix(6)arene-based negative resist doped with the iron complex $Fe(acac)_3$ [43]. The exposure to the e-beam and subsequent thermal treatment form iron nanoparticles only at the positions where the original resist pattern had been generated. The metal nanoparticles with an average height of 1.7 nm can be used as a catalyst for the growth of carbon nanotubes via ethanol CVD in an evacuated quartz tube.

17.4.4
Filled Resists for Microstereolithography

Resins or resist systems for microstereolithography must fulfill the following basic requirements [44]:

- low viscosity for a better resist flow and layer thickness control
- large curing speed
- isotropic polymerization
- good structural accuracy
- adjustable and advantageous mechanical properties.

During the last 10 years, a large number of papers dealing with different variants of microstereolithography applied to filled resist systems, in most of the cases alumina or PZT, have been published. Griffith and Halloran extended the established stereolithography process to the fabrication of ceramics by adding silica, alumina or silicon nitride to photocurable aqueous acrylamide and non-aqueous diacrylate resists [25]. As described in Section 17.3.2, the curing depth is a function of, among others, the filler type, the filler's refractive index, the particle size and the load in the resist. For the commercial equipment used (SLA-250, 3D Systems), the composite viscosity must be lower than 3 Pa s, which was achieved at an alumina load of 40 vol.% in the diacrylate system.

In the case of a silica load of 50 vol.% in different aqueous UV-curable systems, curing depths between 250 and 700 µm could be achieved with a dose of $1.5\,J\,cm^{-2}$, which is more than three times larger than that for the pure resists. The curing behavior at larger loads is dominated by scattering effects. With increasing solid load, the curing depth is reduced. The particle size affects the cure depth significantly; in the case of different aluminas, an average particle size reduction from 0.61 to 0.34 µm yielded a cure depth increase from 300 to 450 µm. Using an SLA-250 stereolithography apparatus and composites based on the diacrylate system, simple windowpane test structures or curved line profiles could be solidified to green bodies. This could be transferred into solid, dense alumina by sintering at 1550 °C [25, 45].

The first ceramic microcomponents, namely alumina microgears with diameters of 400 and 1000 µm (thickness 20 µm), were fabricated from aqueous and non-aqueous alumina-filled resists with a solid load of 33 vol.% using a microstereolithography apparatus and subsequent debinding and sintering [46]. The density achieved was only 56% of the theoretical value of $3.9\,g\,cm^{-3}$.

A similar approach using a highly viscous photocurable ceramic paste instead of a low-viscosity resin as in stereolithography was described by Doreau et al. [47]. The modified process allowed resists with a ceramic load up to 65 vol.%. A further process improvement allowed the fabrication of microstructured alumina parts with a density of 97% of the theoretical value (Figure 17.16). The best dimensional resolution is around 230 µm with an applied irradiation dose (laser beam irradiation, $\lambda = 351–364\,nm$) of $0.05\,J\,cm^{-2}$ [48] in an acrylate-based resist. PZT transducer elements have been fabricated by stereolithography as first active microcomponents using an acrylate resist filled with microsized PZT filler [19]. The density of the ceramic was around 94% after firing.

The realization of smaller structural details and a reduced layer thickness requires finer ceramic particles in the sub-micrometer range. As described earlier, the addition of nanofillers instead of microsized fillers results in a significant

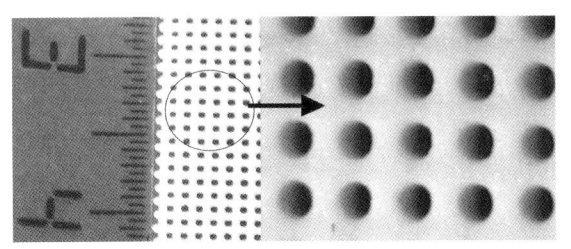

Figure 17.16 Sintered alumina mesh pattern (hole diameter, 600 μm; pitch, 230 μm; layer thickness, 25 μm) [48]). Reproduced with permission from Springer.

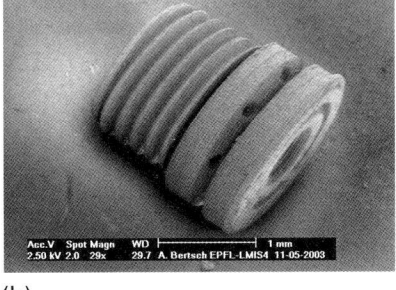

Figure 17.17 (a) Resist–alumina composite teapot; (b) sintered alumina screw [49]. Reproduced with permission from IOP Publisher Limited.

increase in the composite viscosity due to the large specific surface area. The best compromise is a ceramic with a low specific surface area and a sub-micrometer average particle size. As a typical example, Bertsch and coworkers used an alumina filler with an average particle size of 300 nm and a specific surface area around $9 \, m^2 g^{-1}$, dispersed in poly(ethylene glycol) 400 diacrylate [49, 50]. The viscosity of the composite systems used increased significantly with increasing filler content up to values of 50 Pa s at low shear rates. Real 3D microcomponents such as a teapot structure (Figure 17.17a) consisting of 295 layers with an individual thickness of 10 μm or a screw can be fabricated within 90 min. After firing, a final ceramic microcomponent with a shrinkage of around 20% can be obtained (Figure 17.17b).

Similar requirements as in microstereolithography are valid for UV-curable composites used in tape casting. Chartier *et al.* [51] reported the development of composites consisting of either alumina or zirconium dioxide with an average particle size around 0.6 μm, dispersed in hydroxyethyl methacrylate, using Darocur 1173 from Ciba as photoinitiator. The dispersion was cured using a wide band emission UV lamp (200–450 nm). The authors investigated systematically the influence of the incident light energy, the photoinitiator concentration, the exposure conditions, the difference in the resist and filler refractive index and the filler volume fraction on the curing behavior and the resulting penetration depth in the composite film as described in Section 17.3.2 [51]. The following basic correlations were found:

- An increase in the irradiation energy results in an increase in the curing depth.
- For each composite, an optimized photoinitiator concentration exists.
- The curing depth decreases with increasing filler volume fraction due to scattering.

Sheets of alumina (30 vol.%)-filled methacrylate with a thickness up to 700 μm have been photocured after tape casting.

Resist systems filled with metal powders were described for the first time recently [52]. Lee *et al.* used fine copper powder (average particle size around 3 μm), dispersed in a photocurable mixture of 1,6-hexanediol diacrylate (HDDA), trimethylolpropane triacrylate (TMPTA) or *N*-vinylpyrrolidone (NVP) as diluent, dimethoxyphenylacetophenone (DMPA) as photoinitiator and a dispersant for improved laser light penetration, producing a scattering effect [52]. The resulting composite containing 60 wt% metal filler shows a viscosity of 100 mPa s at 26 °C and can be photocured using either a UV lamp (290–450 nm) or an argon ion laser (351.1 nm) integrated in a microstereolithography system. As first simple examples, a microtube (Figure 17.18a) with an inner diameter of 320 μm was generated by 20 curing steps of each 50 μm layer thickness. The helical gear (Figure 17.18b) was fabricated by a rotation of 3° between each layer formation and consists of 30 layers with an individual height of 50 μm. The resulting total height is around 1550 μm. The final copper microcomponents were obtained by debinding and vacuum sintering (Figure 17.19). Unfortunately, the authors gave no information about the resulting sinter density.

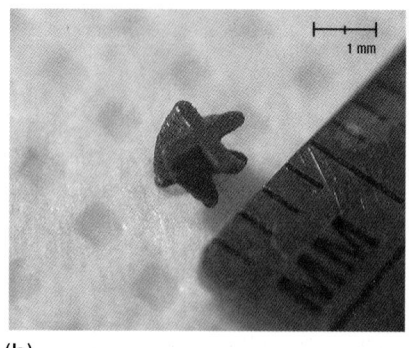

Figure 17.18 Images of small photopolymer–copper components [52]. Reproduced with permission from Elsevier.

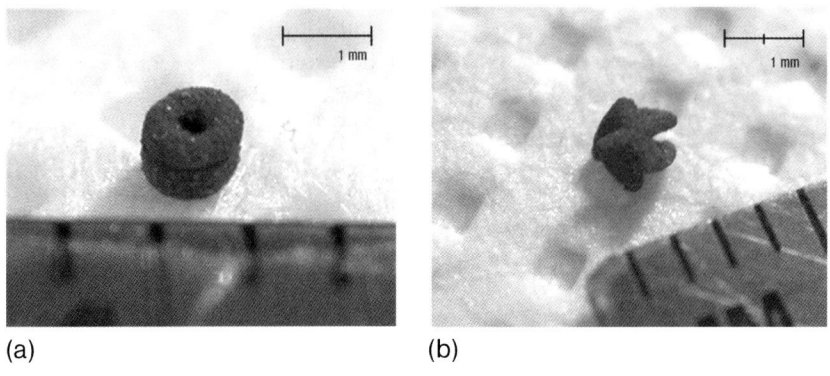

Figure 17.19 Images of small copper components after sintering [52]. Reproduced with permission from Elsevier.

17.5
Miscellaneous

17.5.1
Photosensitive Polyimide–Silica Composites

Photosensitive polyimides are widely used in microelectronics such as in packaging or in printed circuit boards. The large volume shrinkage of up to 50% hinders the use of this polymer class in MEMS or in optoelectronic devices as resist material due to a pronounced distortion of the resulting pattern, such as rounded edges. Wang *et al.* [53] described the synthesis of a novel ionic salt-type photopatternable fluorinated polyimide–silica hybrid material with improved physical properties. Thin films of different polyimide–silica compositions were prepared via spin coating, then cured with UV radiation at 365 nm with a dose of $1.8\,\text{J}\,\text{cm}^{-2}$ using a chromium mask and finally post-exposure baked at 90 °C. After developing in a mixture of γ-butyrolactone, dimethyl sulfoxide and water, the final pattern was imidized at temperatures up to 300 °C. The finally obtained polyimides show glass transition temperatures up to 280 °C, a reduced coefficient of thermal expansion and good optical transmission properties in the near-infrared range (1319 nm). Increasing amounts of silica reduce the refractive index. Figure 17.20 shows examples of waveguides prepared by photolithography. The composites used have potential for use as optical waveguides prepared by UV lithography.

17.5.2
Water-based Filled Photoresists

Water-based photopolymerizable epoxy resist formulations can be filled with organic or inorganic fillers such as poly(vinyl acetate)–poly(vinyl chloride) copolymers, starch, silica, kaolin or titanium dioxide for improvement of the dried film

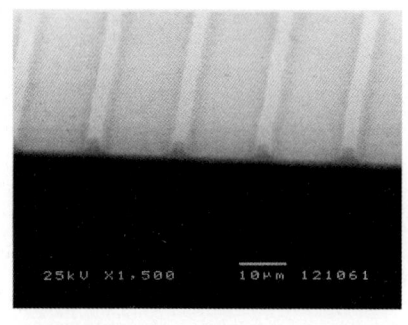

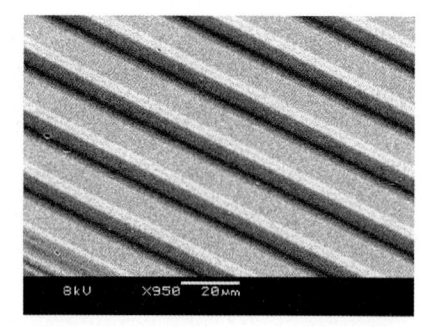

Figure 17.20 SEM images of developed patterns using different polyimide–silica compositions [53]. Reproduced with permission from Elsevier.

quality after coating [54]. A similar approach using an ethylenically unsaturated dextrin oligomer, doped with large surface area inorganics such as silica for improved scratch resistance and wet strength, has been described [55]. During patterning using UV sources, the resist becomes water insoluble. Both systems are designed for the realization of lithographic printing plates.

17.5.3
Three-dimensional Lithography Using Sol–Gel-based Compositions

Organic–inorganic hybrid polymers such as ORMOCERs carrying light-sensitive organic moieties such as acrylate, styryl or epoxy units can be patterned in standard photolithography [56]. Real 3D nanostructures such as photonic crystals can be generated in a two-photon absorption process by using femtosecond laser pulses.

17.6
Conclusion

Filled resist systems containing mainly inorganic materials are widely used in almost all established variants of 2.5D lithography starting with EBL and ending with the generation of real 3D microcomponents via microstereolithography. The addition of fillers is motivated by such different aspects as film stabilization, resolution enhancement, controlled synthesis of carbon nanotubes and the direct realization of ceramic and metal components via lithography avoiding intermediate molding steps. The successful use of filled resists depends strongly on the adaptation of standard resist technology to the physical properties of the composite resist systems, especially the flow behavior with significantly higher viscosity, which affects directly the film formation and the reduced curing depth during irradiation due to scattering and absorption.

Acknowledgments

We thank J. Goettert (LSU-CAMD), M. Boerner (FZKa-ANKA) and G. Motz (University of Bayreuth) for their continous support during our common research in recent years.

References

1 Peiffer, R.W. (1997) Application of photopolymer technology. *ACS Symposium Series*, **673**, 1–14.
2 Ito, H., Reichmanis, E., Nalamasu, O. and Ueno, T. (1999) Micro- and nanopatterning polymers. *ACS Symposium Series*, **706**.
3 Menz, W., Mohr, J. and Paul, O. (2001) *Microsystem Technology*, Wiley-VCH Verlag GmbH, Weinheim.
4 Campbell, S.A. (1996) *The Science and Engineering of Microelectronic Fabrication*, Oxford University Press, New York.
5 Meyer, P., Schulz, J. and Hahn, L. (2003) DoseSim: Microsoft Windows graphical user interface for using synchrotron X-ray exposure and subsequent development in the LIGA process. *Review of Scientific Instruments*, **74**, 1113.
6 Hanemann, T., Bauer, W., Knitter, R. and Woias, P. (2006) Rapid prototyping and rapid tooling techniques for the manufacturing of silicon, polymer, metal and ceramic microdevices, in *MEMS/NEMS Handbook: Techniques and Applications* (ed. C.T. Leondes), Springer, Berlin, pp. 187–255.
7 Takagi, T. and Nakajima, N. (1994) Architecture combination by micro photoforming process, in IEEE Proceedings of Micro Electro Mechanical Systems MEMS '94, 25–28 January 1994, Oiso, Japan, pp. 211–16.
8 Ikuta, K., Hirowatari, K. and Ogata, T. (1994) Three dimensional micro integrated fluid systems (MIFS) fabricated by stereo lithography, in IEEE Proceedings of Micro Electro Mechanical Systems MEMS'94, 25–28 January 1994, Oiso, Japan, pp. 1–6.
9 Hull, C.H. (1984) Apparatus for production of 3D objects by stereolithography, 3D Systems, US Patent 4 575 330.
10 Bertsch, A., Zissi, S., Jezequel, J.Y., Corbel, S. and Andre, J.C. (1997) Microstereolithography using a liquid crystal display as dynamic mask-generator. *Microsystem Technologies*, **3**, 42–7.
11 Bertsch, A., Lorenz, H. and Renaud, P. (1999) 3D-microfabrication by combining microstereolithography and thick resist UV lithography. *Sensors and Actuators*, **73**, 14–23.
12 Straub, M., Nguyen, L.H., Fazlic, A. and Gu, M. (2004) Complex-shaped three-dimensional microstructures and photonic crystals generated in a polysiloxane polymer by two-photon stereolithography. *Optical Materials*, **27**, 359–64.
13 Einstein, A. (1906) Eine neue Bestimmung der Moleküldimension. *Annalen der Physik*, **19**, 289–306.
14 Einstein, A. (1911) Berichtigung zu meiner Arbeit: eine neue Bestimmung der Moleküldimension. *Annalen der Physik*, **34**, 591–2.
15 German, R.G. (1990) *Powder Injection Molding*, Metal Powder Industries Federation, Princeton, NJ.
16 Thomas, D.G. (1965) Transport characteristics of suspension: VIII. A note on the viscosity of Newtonian suspensions of uniform spherical particles. *Journal of Colloid Science*, **20**, 267–77.
17 Krieger, I.M. and Dougherty, T.J. (1959) A mechanism for non-Newtonian flow in suspensions of rigid spheres. *Transactions of the Society of Rheology*, **3**, 137–52.
18 Quemada, D. (1977) Rheology of concentrated disperse systems and minimum energy dissipation principle – I. Viscosity–concentration relationship. *Rheologica Acta*, **16**, 82–94.

19 Dufaud, O., Marchall, P. and Corbel, S. (2002) Rheological properties of PZT suspensions for stereolithography. *Journal of the European Ceramic Society*, **22**, 2081–92.

20 Agote, I., Odriozola, A., Gutierrez, M., Santamaria, A., Quintanilla, J., Coupelle, P. and Soares, J. (2001) Rheological study of waste porcelain feedstocks for injection molding. *Journal of the European Ceramic Society*, **21**, 2843–53.

21 Hanemann, T. (2006) Influence of dispersants on the flow behavior of unsaturated polyester-alumina composites. *Composites A: Applied Science and Manufacturing*, **37**, 735–41.

22 Hanemann, T., Heldele, R. and Hausselt, J. (2006) Particle size dependent viscosity of polymer-silica composites. Proceedings of 4M – 2nd International Conference on Multi-Material-Micro-Manufacture (4M), 20–22 September 2006, Grenoble.

23 Hanemann, T., Honnef, K. and Hausselt, J. (2004) Rapid prototyping of ceramic microcomponents using nanosized alumina, nanofair, *VDI-Berichte*, **1839**, 175–8.

24 Ajayan, P., Schadler, L. and Braun, P. (2003) *Nanocomposite Science and Technology*, Wiley-VCH Verlag GmbH, Weinheim, p. 140.

25 Griffith, M.L. and Halloran, J.W. (1996) Freeform fabrication of ceramics via stereolithography. *Journal of the American Ceramic Society*, **79**, 2601–8.

26 Chartier, T., Hinczewski, C. and Corbel, S. (1999) UV curable systems for tape casting. *Journal of the European Ceramic Society*, **19**, 67–74.

27 Kroke, E., Li, Y.L., Konetschny, C., Lecomte, E., Fasel, C. and Riedel, R. (2000) Silazane derived ceramics and related materials. *Materials Science and Engineering R-Reports*, **26**, 97–199.

28 Schulz, M., Boerner, M., Hausselt, J. and Heldele, R. (2004) Polymer derived ceramic microparts from X-ray lithography: crosslinking behavior and process optimization. *Journal of the European Ceramic Society*, **25**, 199–204.

29 Schulz, M., Haußelt, J. and Heldele, R. (2008) Structuring ceramics using lithography. Mechanical Properties and Performance of Engineering Ceramics and Composites: *Ceramic Engineering and Science Proceedings*, **26**(2), 177–85.

30 Motz, G., Hacker, J. and Ziegler, G. (2000) Special modified silazanes for coatings, fibers and CMCs. *Ceramic Engineering and Science Proceedings*, **4**, 21.

31 Mueller, C., Hanemann, T., Wiche, G., Kumar, C. and Goettert, J. (2005) Fabrication of ceramic microcomponents using deep X-ray lithography. *Microsystem Technologies*, **11**, 271–7.

32 Schulz, M., Börner, M., Göttert, J., Hanemann, T., Hausselt, J. and Motz, G. (2004) Crosslinking behavior of preceramic polymers effected by UV- and synchrotron radiation. *Advanced Engineering Materials*, **6**, 676–80.

33 Hanemann, T., Ade, M., Borner, M., Motz, G., Schulz, M. and Hausselt, J. (2002) Microstructuring of preceramic polymers. *Advanced Engineering Materials*, **4**, 869–73.

34 Jiguet, S., Bertsch, A., Hofmann, H. and Renaud, P. (2004) SU8–silver photosensitive nanocomposite. *Advanced Engineering Materials*, **6**, 719–24.

35 Jiguet, S., Judelewicz, M., Mischler, S., Bertsch, A. and Renaud, P. (2006), Effect of filler behavior on nanocomposite SU8 photoresist for moving micro-parts. *Microelectronic Engineering*, **83**, 1273–6.

36 Damean, N., Parviz, B.A., Lee, J.N., Odom, T. and Whitesides, G.M. (2005) Composite ferromagnetic photoresist for the fabrication of microelectromechanical systems. *Journal of Micromechanics and Microengineering*, **15**, 29–34.

37 Bedenbecker, M. and Gatzen, H.H. (2005) Herstellung polymergebundener Hartmagnete durch in SU-8 eingebrachtes Magnetpulver. Proceedings of MST-Kongress, 10–12 October 2005, Freiburg.

38 Markovich, V.R., Mehta, A.A., Skarvinko, E.R. and Wang, D.W. (1991) Thermally stable photoimaging composition, European Patent EP 480154.

39 Merhari, L., Gonsalves, K.E., Hu, Y., He, W., Huang, W.-S., Angelopoulus, M., Bruenger, W.H., Dzionk, C. and Torkler, M. (2002) Nanocomposite resist systems for next generation lithography. *Microelectronic Engineering*, **63**, 391–403.

40 Gonsalves, K.E., Merhari, L., Wu, H.P. and Hu, Y.Q. (2001) Organic–inorganic

nanocomposites: unique resists for nanolithography. *Advanced Materials*, **13**, 703–14.

41 Clendenning, S.B., Aouba, S., Rayat, M.S., Grozea, D., Sorge, J.B., Brodersen, P.M., Sodhi, R.N.S., Lu, Z.-H., Yip, C.M., Freeman, M.R., Ruda, H.E. and Manners, I. (2004) Direct writing of patterned ceramics using electron-beam lithography and metallopolymer resists. *Advanced Materials*, **16**, 215–9.

42 Stellacci, F., Bauer, C.A., Meyer-Friedrichsen, T., Wenseleers, W., Alain, V., Kuebler, S.M., Pond, S.J.K., Zhang, Y.D., Marder, S.R. and Perry, J.W. (2002) Laser and electron-beam induced growth of nanoparticles for 2D and 3D metal patterning. *Advanced Materials*, **14**, 194–8.

43 Ishida, M., Hongo, H., Nihey, F. and Ochiai, Y. (2004) Diameter-controlled carbon nanotubes grown from lithographically defined nanoparticles. *Japanese Journal of Applied Physics*, **43**, L1356–8.

44 Hagiwara, T. (2001) Recent progress of photo-resin for rapid prototyping, 'resin for stereolithography'. *Macromolecular Symposia*, **175**, 397–402.

45 Brady, G.A. and Halloran, J.W. (1997) Stereolithography of ceramic suspensions. *Rapid Prototyping Journal*, **3**, 61–5.

46 Zhang, X., Jiang, X.N. and Sun, C. (1999) Microstereolithography of polymeric and ceramic microstructures. *Sensors and Actuators A*, **77**, 149–56.

47 Doreau, F., Caput, C. and Chartier, T. (2000) Stereolithography for manufacturing ceramic parts. *Advanced Engineering Materials*, **2**, 493–6.

48 Chartier, T., Caput, C., Doreau, F. and Loiseau, M. (2002) Stereolithography of structural complex ceramic parts. *Journal of Materials Science*, **37**, 3141–7.

49 Bertsch, A., Jiguet, S. and Renaud, P. (2004) Microfabrication of ceramic components by microstereolithography. *Journal of Micromechanics and Microengineering*, **14**, 197–203.

50 Bertsch, A., Jiguet, S., Hofmann, H. and Renaud, P. (2004) Ceramic microcomponents by microstereolithography, in IEEE International Conference on Micro Electro Mechanical Systems, Technical Digest, 17th, Maastricht, TheNetherlands, Institute of Electrical and Electronics Engineers, New York, pp. 725–8.

51 Chartier, T., Hinczewski, C. and Corbel, S. (1999) UV curable systems for tape casting. *Journal of the European Ceramic Society*, **19**, 67–74.

52 Lee, J.W., Lee, I.H. and Cho, D.-W. (2006) Development of micro-stereolithography technology using metal powder. *Microelectronic Engineering*, **83**, 1253–6.

53 Wang, Y.-W., Yen, C.-T. and Chen, W.-C. (2005) Photosensitive polyimide/silica hybrid optical materials: synthesis properties and patterning. *Polymer*, **46**, 6959–67.

54 Dickinson, P. and Ellwood, M. (1983) Water-based photopolymerizable compositions and their use, British Patent 2 137 626.

55 Fohrenkamm, E.A. and Rousseau, A.D. (1985) Ethylenically-unsaturated dextrin composition for preparing a durable hydrophilic photopolymer, US Patent 4 511 646.

56 Houbertz, R. (2005) Laser interaction in sol-gel based materials – 3D lithography for photonic applications. *Applied Surface Science*, **247**, 504–12.

18
Dramatic Downsizing of Soft X-ray Synchrotron Light Source from Compact to Tabletop

Hironari Yamada, Norio Toyosugi, Dorian Minkov, and Yoshiko Okazaki

18.1 Introduction to the Tabletop Synchrotron 443
18.2 EUV and Soft X-ray Generation Process 444
18.3 Transition Radiation Theory 445
18.4 The Tabletop Synchrotron 447
18.5 Generation of Coherent Transition Radiation by our 6 MeV Tabletop Synchrotron 449
18.6 Summary and Trend of Microfabrication 451
References 452

18.1
Introduction to the Tabletop Synchrotron

Demand for X-ray lithography and LIGA processes (microfabrication) has been increasing along with the trend of nanotechnology. These technologies became feasible and useful when synchrotron radiation sources became available, and are now popular in this field. Currently some compact synchrotrons, such as Ritsumeikan University AURORA (http://www.ritsumei.ac.jp/acd/re/src/index.htm), Sumitomo Heavy Industries AURORA-2S (http://www.shi.co.jp/srmicro/en/index.html), NTT Super Alice (https://www.keytech.ntt-at.co.jp/nano/prd_e0006.html), Hyogo University New Subaru (www.lasti.u-hyogo.ac.jp), University of Wisconsin–Madison Aladdin (www.src.wisc.edu), National University of Singapore HELIOS (ssls.nus.edu.sg), Louisiana State University CAMD (www.camd.lsu.edu), ANKA (ankaweb.fzk.de) and the Swiss Light Source (http://sls.web.psi.ch/view.php/about/index.html) support the wide activity of LIGA.

World competition in compact synchrotron development started about 15 years ago, aiming at X-ray lithography for the next generation of VLSI fabrication at semiconductor factories. The superconducting synchrotrons AURORA [1], and HELIOS were a result of this competition. Semiconductor manufacturers did not, however, appreciate this technology until recently, because of the large machine size and the insufficient photon flux. Another difficulty was in fabricating a super-accurate mask for proximity exposure. Because of these problems, the semiconductor community decided to develop extreme ultraviolet (EUV) lithography [2] for the next-generation very large-scale integration (VLSI) fabrication, instead of the soft X-ray approach. EUV enables projection lithography to reduce the line and space with the aid of optical elements.

After finding the above problems of compact synchrotrons, Yamada and coworkers started to consider further downsizing of the X-ray source without using superconducting technology. The superconducting technology was also problematic because of handling and maintaining the He liquefier engine. Yamada proposed the use of a tiny target in the electron orbit of the synchrotron [3] to generate X-rays by low-energy electrons. The physical processes involved are the bremsstrahlung for hard X-ray generation and the transition radiation for soft X-ray and EUV generation. These radiations are of the same kind as the synchrotron radiation, in the sense that radiation is emitted from the incident electron, and is focused in the forward direction due to special relativity. In this chapter, we describe resent achievements with the tabletop synchrotron, which can possibly lead to the generation of more than 10 W EUV, and soft X-rays, from only a 20 MeV synchrotron. We call this tabletop synchrotron the MIRRORCLE type.

18.2
EUV and Soft X-ray Generation Process

Transition radiation (TR) has been known for a long time, as described in the textbook on electrodynamics by Jackson [4]. This radiation mechanism is similar to Cherenkov radiation in gases and liquids. The presence of blue light in a reactor is well known. Blue light in the atmosphere, due to high-energy particles from outer space, is another example. TR occurs when a charged particle passes through the interface between two media. Electron linear accelerators (LINACs) have been used most often [5]. The radiation process at the medium surface is as follows. The medium is polarized by the incident charged particle and generate photons when relaxation occurs. These photons have energies between the plasma energy E_{pl} of the medium and γE_{pl} (γ is the relativistic factor). The number of photons emitted decreases dramatically with increase in the photon energy. In the case of 20 MeV electrons, the maximum photon energy is 1.5 keV. Most photons are emitted along a $1/\gamma$ cone with respect to the charged particle direction. The photon energy is roughly defined by the radiation angle. Lower energy photons are emitted at larger emission angles. The transition radiation is a collective phenomenon, which occurs by the cooperation of all electrons near the charged particle. The electrons in a boundary layer instantly polarize around the incident electron. Consequently, all photons are generated with the same phase, and are emitted coherently. Therefore, the transition radiation is a coherent radiation. When multiple thin foils are aligned periodically, the photons emitted from all foils interact resonantly.

18.3
Transition Radiation Theory

The spectrum of the TR is calculated electrodynamically. When a charged particle at relativistic energy traverses the interface between two media having different dielectric constants, TR occurs. The basic formalism is given in [6]. The radiation generated on both sides of the interface (as shown in Figure 18.1) between the media i (i = 1, 2) within the so-called formation lengths z_i is given by [7]

$$z_i(E,\theta) = \left| \frac{l}{1/\gamma^2 + \theta^2 + 2(\delta_i - i\beta_i)} \right| \qquad (1)$$

where $\dot{\varepsilon}_i(E) = 1 - 2[\delta_i(E) - i\beta_i(E)]$ is the dielectric constant of the media, which is calculated from the known complex scattering factor [8], E (eV) is the energy of the emitted photon, $l = 4\hbar c/E$, and θ is the photon emission angle with respect to the particle direction.

If the charged particle is one relativistic electron, the number of photons emitted per unit photon energy interval, and unit solid angle, from a single interface is given by

$$F_1 = \frac{\alpha \sin^2 \theta}{\pi^2 E l^2} |\dot{z}_1 - \dot{z}_2|^2 \qquad (2)$$

where $\alpha = e^2/\hbar c \cong 1/137$ is the fine-structure constant.

The TR differential cross-section for a periodic structure target consisting of M bi-layers with spacing l_1 and thickness l_2 is given by

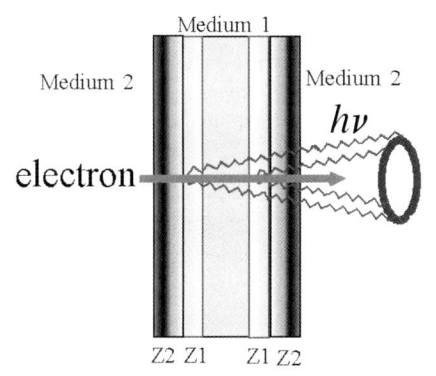

$$\frac{d^2 N(E)}{dE d\Omega} = F_1 F_2 F_3 \qquad (3)$$

Figure 18.1 A single foil target for transition radiation. Formation zones Z1 and Z2 are defined at the interface of two media. Radiation is generated within the formation zone.

where $N(E)$ is the number of photons per unit energy interval and unit solid angle. The second factor F_2 corresponds to the interference effect between the transition radiations emitted from the two surfaces of a single layer. If the attenuation, due to the electron collisions within the layer, and the photon absorption in the layer are negligible, then

$$F_2 = 4\sin^2\left(\operatorname{Re}\frac{l_2}{\dot{z}_2}\right) \qquad (4)$$

The third factor F_3 describes the coherent sum of the contributions from the each layer in the stack:

$$F_3 = \frac{\sin^2 MX}{\sin^2 X} \qquad (5)$$

where $X = \operatorname{Re}(l_1/\dot{z}_1) + (l_2/\dot{z}_2)$.

Taking into account the soft X-ray absorption by a target as in [6], then

$$F_3 = \frac{1 + \exp(-M\sigma) - 2\exp(-M\sigma/2)\cos(2MX)}{1 + \exp(-\sigma) - 2\exp(-\sigma/2)\cos(2X)} \qquad (6)$$

where $\sigma = \mu_1 l_1 + \mu_2 l_2$ and μ_i are the X-ray absorption coefficients. In this case, the resonance conditions for transition radiation are

$$X = r\pi \quad \text{and} \quad \operatorname{Re}\left(\frac{l_2}{\dot{z}_2}\right) = \left(m - \frac{1}{2}\right)\pi \qquad (7)$$

where r and m are positive integers. For the first resonance $r = m = 1$, the transition radiation gives highest intensity.

Figure 18.2 gives the calculated emitted TR energy per 1 eV band for one pass of one electron through one 384 nm thick Al foil target for electron energies of 6 and 20 MeV. Each of the electrons, injected in the synchrotron hits the target many times. The number of such hits represents a product of the lifetime of the circulating beam and the circulation frequency. The circulating electron beam has a lifetime limited by losses in the target. The formalism for calculating lifetime and beam current is given in [3].

A single foil of Al provides a relatively low radiation power. We expect that the use of carbon or beryllium targets will increase the TR power by one order of magnitude, and the multiple layer targets would increase it by one more order of magnitude. If the target is made of a lighter material, the bremsstrahlung yield is relatively smaller than that of TR, and the beam lifetime increases.

Later, we show experimental results for one Al foil, which is well established. At present the stored beam current, which depends strongly on the synchrotron design, is of the order of 1 A, for our tabletop synchrotrons. Our tabletop synchrotron is described in the next section.

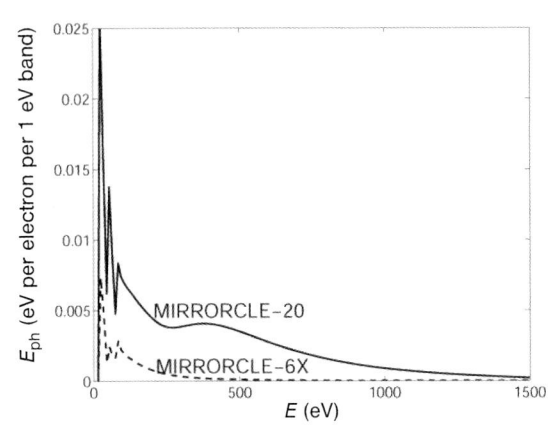

Figure 18.2 Theoretically obtained TR spectra for 384 nm thick Al foil target for the tabletop synchrotrons MIRRORCLE-20 and -6X. These are 20 and 6 MeV machines, respectively.

18.4
The Tabletop Synchrotron

The MIRRORCLE-type tabletop synchrotron is the key issue in this technology [9, 10]. We already have the 6 MeV operating synchrotron MIRRORCLE-6X (see Figure 18.3) [11], and discuss the nature of the synchrotron known from that machine [12]. We measured transition radiation from this synchrotron. The 20 MeV synchrotron, MIRRORCLE-20SX [13], is being commissioned, as shown in Figure 18.4. This machine is fabricated for commercial use for EUV lithography, and micromachining. We are establishing a micromachining center based on this machine.

The technology of the tabletop synchrotron is not same as that of conventional synchrotrons; it is not just a miniature version of a large synchrotron. The MIRRORCLE type has the following features [9–12]:

1. The MIRRORCLE synchrotron has an exactly circular electron orbit with less than 15 cm orbit radius.

2. We inject the beam continuously at an injection rate of more than 400 Hz to compensate for the short beam lifetime. The beam lifetime is of the order of 1 ms.

3. The 1 ms lifetime is very short compared with that of a large synchrotron, because the beam energy is very low and the target is installed. If, however, we place the target in the orbit of a conventional synchrotron, the lifetime will be of the order of a 1 μs. This happens because MIRRORCLE has a very large dynamic aperture, and the lifetime is limited by the physical aperture of 10 cm horizontally and 5 cm vertically. Also, this machine has a very large momentum aperture of 2 MeV out of 6 MeV. This is the reason for the long lifetime.

4. For the reasons described above, MIRRORCLE has a very high injection efficiency, of nearly 100%. We accumulate 3 A in a single-shot injection when

Figure 18.3 The tabletop synchrotron MIRRORCLE-6X. Behind, the 6 MeV injector microtron is seen. The 60 cm outer diameterring is seen in front. One of the X-ray ports is facing outwards.

Figure 18.4 MIRRORCLE-20SX under commissioning. Due to its heavy duty, the shielding concrete is 1 m thick. The synchrotron is placed vertically to provide the X-ray beam downwards, which allows us to set the wafer horizontally.

the injection peak current is 100 mA. This large beam current is a result of the only 1 m short orbit circumference and the 3×10^8 high circulation frequency.

5. The beam size is 20×3 mm. The average beam current hitting the 1 mm wide target is about 100 mA.

The beam injection method is sophisticated and different from that of a conventional synchrotron. The beam injection is based on a switching process. In the conventional synchrotron, the beam approaching the synchrotron orbit from the injector is switched to the central orbit by a switching pulse magnet. This magnetic field must vanish when the beam approaches the switching magnet, otherwise the beam will be deflected out of the central orbit. This is possible when the circumference is large. In the 1 m orbit circumference ring, however, the circulation time is only 3 ns. There is no such fast switching magnet available. Therefore, MIRRORCLE introduced an exotic injection scheme, named a half-integer resonance method. We invoke the half-integer resonance on the horizontal betatron motion for a period of 100 ns. The injected electrons are captured to this resonance state. When the resonance condition is terminated, the electron beam settles in the normal stable orbit. One might say that the MIRRORCLE orbit is a kind of a gigantic atomic orbit. This atomic-like orbit is excited, and decays to the stable orbit in the resonance process. When the injected electron energy satisfies the resonance condition, the electron is captured.

18.5
Generation of Coherent Transition Radiation by our 6 MeV Tabletop Synchrotron

In this section, we describe the results from our TR experiment using MIRRORCLE-6X, shown in Figure 18.3. The size of MIRRORCLE-6X is $1 \times 2 \times 1.5$ m.

Our detector includes a plastic scintillator (PS), an optical bundle and a photomultiplier (PMT). The PS is driven to points x_j, located along a horizontal line (see Figure 18.5) that is perpendicular to the axis of the TR emission cone, and

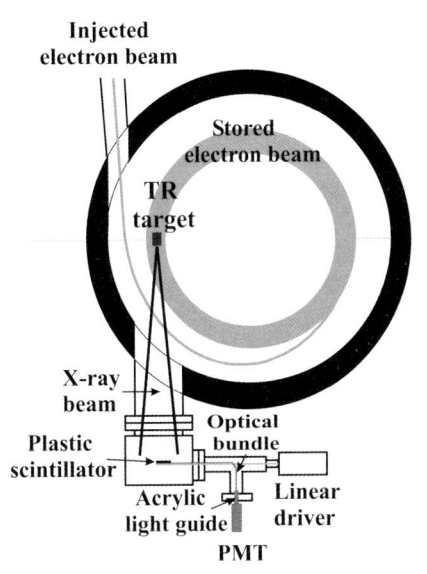

Figure 18.5 Experimental setup for measurement of the TR power distribution.

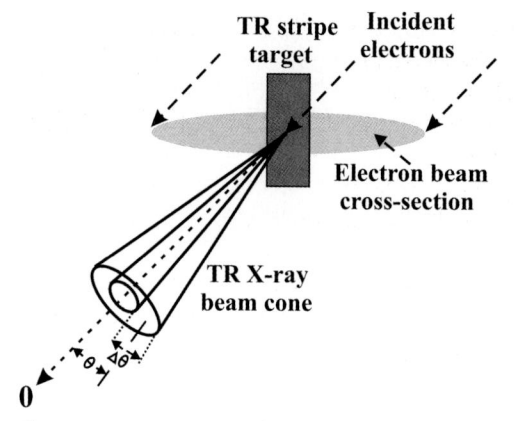

Figure 18.6 Cross-section of the electron beam in the synchrotron, and positioning of the TR target. The emitted TR has a conical spatial distribution with a half-width of $\Delta\theta \cong \theta \sim 1/X$.

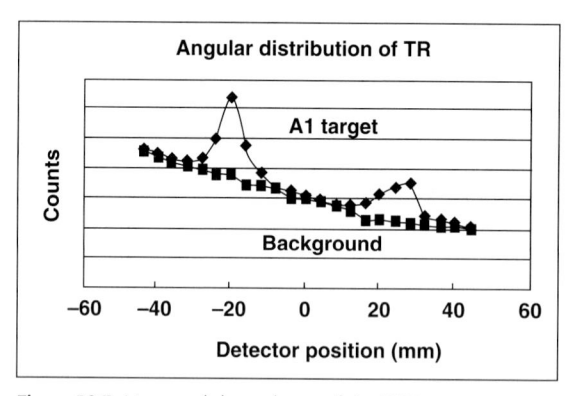

Figure 18.7 Measured dependence of the PMT output current counts Y as a function of the PS location x with respect to the emission axis 0. The injected beam current was $I_{Bi} = 0.8\,\mu A$. The TR target was an Al foil strip, with a thickness $d = 384\,nm$. The PMT was operated at 1 kV.

crosses it at $x = 0$. The current at the PMT output is recorded in counts for each PS location.

The electron beam in the storage ring has an elliptical cross-section (see Figure 18.6) with a length of 20 mm along its horizontal long axis and a height of 3 mm. Our TR target represented a vertical strip of one 2.5 mm wide Al foil. When an electron hits the target, it emits TR with a conical shape.

Figure 18.7 shows the observed horizontal distribution of TR. This represents a typical TR distribution. The two peaks are part of the cone-shaped TR emission.

From the above typical spatial distribution of TR and the calibration of our detector over the entire emitted photon spectrum, we determined the emitted TR power to be $P^{6\text{MeV}}_{384\text{nm\_Al}} = 0.39\,\text{mW}$ in our reference experiment. Based on our analytical expressions for the lifetime of the electrons injected in the storage ring [13], and the TR energy emitted for one pass of one electron through the target for different TR targets, we calculated the ratio between the emitted power for optimum TR emission conditions and the above reference conditions. Knowing that one of the very best TR emitting targets for MIRRORCLE-6X is a 35 nm C foil strip, we calculated that a TR power of $P^{6\text{MeV}}_{35\text{nm\_C}} = 82.8\,\text{mW}$ can be emitted from MIRRORCLE-6X.

At present, Photon Production Laboratory is commissioning the synchrotron MIRRORCLE-20SX as shown in Figure 18.4, which produces 20 MeV electrons, and will be the most powerful tabletop synchrotron. Knowing that one of the very best TR emitting targets for MIRRORCLE-20SX is a 240 nm Be foil stripe, we calculated that a TR power of $P^{20\text{MeV}}_{240\text{nm\_Be}} = 1.2\,\text{W}$ can be emitted from MIRRORCLE-20SX.

18.6
Summary and Trend of Microfabrication

We have demonstrated the potential of the MIRRORCLE-type tabletop synchrotron regarding its power in the EUV and soft X-ray domains. The small 80 cm outer diameter of MIRRORCLE-20SX and the 1.2 W power in the present scheme (100 mA injector peak current and 400 Hz injection repetition) are attractive for commercializing the microfabrication. The total radiation power is dependent on the injection repetition rate, and hence the power supply. The 4 kHz repetition promises a 12 W output. Provided that the heating does not destroy the target, we are able to increase the repetition rate. As the absorption of heat in the target should be less than 0.1%, emission of 100 W TR will be feasible.

The coherence of the MIRRORCLE EUV radiation is fairly high due to its emission mechanism, as explained in Section 18.2. Coherence is an important fact of radiation when we introduce optical elements. The power should be radiated from a few square millimeters wide area. The implies that if we introduce appropriate optics, we should be able to focus the beam down to this size. The special coherence of $1.2\,\text{W}\,\text{mm}^{-2}$ power is attractive.

Moreover, stacked foils with an equal spacing is a possible way to generate quasi-monochromatic radiation. This is similar to undulator radiation. The coherence increases proportionally to the number of foils in the stack [14, 15]. Monochromatic EUV or soft X-rays will ensure a better performance of the optics for produce projection lithography. Precise ordering of the foils at the nano level is the key issue for amplification of the monochromatic radiation power.

MIRRORCLE-type EUV and soft X-ray synchrotron sources will change and extend the means of application. As far as the use of a conventional synchrotron is concerned, occupation of a beamline is not the best approach for mass production processing since the running costs are very high. Therefore, the LIGA process is preferred. The beamline is only used for making the mold, and this mold is used repeatedly for mass production. MIRRORCLE will, however, provide a direct etching and manufacturing of micrometer-sized parts. From one mask we make many final products similarly to lithography. This will be realized at a moderate cost of the photon flux from MIRRORCLE, including in its primary installation, operation and maintenance.

References

1 Yamada, H. (1990) *Journal of Vacuum Science and Technology B*, **8**, 1628–31.
2 Shriever, G. (2004) Gas discharge produced plasma sources for EUV generation. 326th Heraeus Seminar, 7 June 2004, Bad Honnef, Germany.
3 Yamada, H. (1996) *Japanese Journal of Applied Physics*, **35**, L182–5.
4 Jackson, J.D. (1999) *Classical Electrodynamics*, John Willey & Sons, Inc., New York.
5 Piestrup, M.A., Kephart, J.O., Park, H., Klein, R.K., Pantell, R.H., Ebert, P.J., Moran, M.J., Dahling, B.A. and Berman, B.L. (1985) *Physical Review A*, **32**, 917–27.
6 Cherry, M.L., Hartmann, G., Muller, D. and Prince, T.A. (1974) *Physical Review D*, **10**, 3594–607.
7 Minkov, D., Yamada, H., Toyosugi, N., Yamaguchi, T.Y., Kadono, T. and Morita, M. (2006) *Journal of Synchrotron Radiation*, **13**, 336–42.
8 Attwood, D. (1999) *Soft X-Rays and Extreme Ultraviolet Radiation*, Cambridge University Press, Cambridge, pp. 52–61.
9 Yamada, H. (1998) *Journal of Synchrotron Radiation*, **5**, 1326–31.
10 Yamada, H. (2003) *Nuclear Instruments and Methods B*, **199**, 509–16.
11 Hasegawa, D., Yamada, H., Kleev, A.I., Toyosugi, N., Hayashi, T., Yamada, T., Tohyama, I. and Ro, Y.D. (2004) In *Portable Synchrotron Light Sources and Advanced Applications* (eds. H. Yamada, N. Mochizuki-Oda and M. Sasaki), CP716, American Institute of Physics, College Park, MD, pp. 116–9.
12 Yamada, H. (2003) Features of the portable synchrotrons named MIRRORCLE. AIP Conference Proceedings, Vol. 716, pp. 12–17.
13 Toyosugi, N., Yamada, H., Minkov, D., Morita, M., Yamaguchi, T. and Imai, S. (2007) *Journal of Synchrotron Radiation*, **14**, 212–8.
14 Okazaki, Y., Toyosugi, N., Yamada, H., Navazaki, Y., Takashima, T. and Imai, S. (2004) In *Portable Synchrotron Light Sources and Advanced Applications*, CP716, American Institute of Physics, College Park, MD, pp. 124–7.
15 Sugano, K., Sun, W., Tsuchiya, T. and Tabata, O. (2006) Design and analysis of soft X-ray source using resonance transition radiation for tabletop synchrotron. Asia–Pacific Conference of Transducers and Micro-Nano Technology, 25–28 June 2006, Singapole, 95-OMN-A0593.

19
PTFE Photo-fabrication by Synchrotron Radiation

Takanori Katoh and Yanping Zhang

19.1 Introduction *453*
19.2 SR Fabrication of PTFE: TIEGA Process *454*
19.3 Mechanism of SR Photo-fabrication of PTFE *462*
19.4 Application: X-ray Lenses Made of PTFE *463*
19.5 Conclusions *467*
    References *467*

19.1
Introduction

Synchrotron radiation (SR) has been applied to deep X-ray lithography to generate microstructures with a very high aspect ratio. This is generally known as the LIGA-process [1]. In the LIGA process, a very thick layer (more than 50 µm) of resists is necessary. A positive photo-resist of poly(methyl methacrylate) (PMMA) has mainly been used from the beginning [2] and a negative photo-resist of SU-8 was also introduced later [3]. No matter which kind of these resists (some other negative resists such as DuPont Pyralin and TG-P have also been applied to the LIGA process, but are less popular than SU-8) may be used for deep SR lithography, a development process is needed. We have found that Teflon polymers such as polytetrafluoroethylene (PTFE), polytetrafluoroethylene-co-hexafluoropropylene (FEP) and polytetrafluoroethylene-co-perfluoroalkoxy vinyl ether (PFA) can be directly etched by SR, generating structures with a very high aspect ratio [4]. This direct etching process looks like laser ablation and needs no development process.

The reason why we chose Teflon polymers, especially PTFE, are as follows. First, the LIGA process cannot be applied to Teflon, since there is no solvent to dissolve Teflon and heating over its melting point does not result in sufficient fluidity to mold the microparts. Second, laser ablation is the only method for microprocessing Teflon, using either vacuum ultraviolet (VUV) pulsed lasers, with wavelengths of 160 nm or less, or ultrashort pulsed lasers, which can give rise to strong multi-photon absorption. The laser beam had to be strongly focused to achieve the threshold flux so that the ablation depth would be limited by the focus depth. The aspect ratio is too small (<1). Further, it may be one of the most suitable materials for making microparts for bio-science and medical applications, as it is thermo-stable and chemically resistant and also its surface has very low adhesion and very low frictional resistance.

Like the LIGA process, the Teflon-included etched (TIE) process by SR can be further combined with electroforming and molding and is named the TIEGA process. In this chapter, we first introduce the TIEGA process and compare it with the LIGA process to demonstrate the advantages and disadvantages of the TIEGA process in the first part. Next, we discuss the mechanism of the SR etching PTFE in comparison with laser ablation to illustrate why it is different from the latter. Finally, we describe the fabrication of a refractive lens for the generation of X-ray microbeams using the TIEGA process, to demonstrate the application potential of this process.

19.2
SR Fabrication of PTFE: TIEGA Process

A home-made compact warm ring AURORA-2S synchrotron with self-radiation shielding installed at our Tanashi factory (Figure 19.1) was used for SR exposure [5]. The energy is 0.7 GeV and it is operated under a routine beam current between 300 and 500 mA. The SR beam has a continuous spectrum from infrared to X-ray with a critical wavelength of 1.5 nm and bunched beams have a duration of 170 ps with a repetition rate of 191 MHz. In the beamline used for direct SR etching, there is no mirror or other optical system, but pinholes or slits may be used to

Figure 19.1 AURORA-2S compact synchrotron radiation source.

select the beam size or to improve the beam uniformity and variable metallic filters to select the synchrotron X-rays in different wavelength ranges. The SR beam that covers the whole spectrum is called white light for simplicity and its photon flux per unit beam current on the resist surface is on the order of a few 10×10^{14} photons $s^{-1} mA^{-1} cm^{-2}$. The SR beam with a spectrum selected by the metallic filter is called X-rays for simplicity. An aluminum filter (1 μm thick), for example, was used to select the X-rays mainly in the spectral range between 0.4 and 5 keV peaked at about 1 keV, and their photon flux on the sample surface was about one-fifth of that of the white light. The Teflon-included etch process by SR was reproducible on replacing the white light with X-rays and no consistent differences were found except that the etching rate became lower. The etching depth was measured with either an optical microscope or a stylus profilometer. The etching rate was given by the ratio of the depth to the irradiation time, which was usually set to be 1 min to prevent any errors due to the decay of the beam current and due to the temperature rise induced by the SR irradiation. The SR etching process was usually carried out under vacuum (its base pressure was 1×10^{-6} Pa) for the white light, but could also be done under a helium atmosphere, especially for the X-rays.

Commercial sheets (thickness between 0.1 and 2 mm) of the Teflon-like polymers (PTFE, PFA and FEP) were polished and cleaned with organic agents before use. The resist sheet was set on to a substrate holder to face the SR beam and heated by a hot-plate attached to the rear of the holder. Its temperature was measured by a thermocouple. A contact metallic stencil mask made of nickel or copper was used for the white light and a proximity X-ray mask was used in some cases for the X-rays.

The main results are summarized in Table 19.1 and typical scanning electron microscopy (SEM) observations of microstructures made of PTFE are shown in Figure 19.2. In Figure 19.2a, the honeycomb height is 500 μm and its width is 20 μm, that is, the aspect ratio is 25. In Figure 19.2b, the hole height is 12 μm and its diameter is 1 μm, that is, the aspect ratio is 12. Figure 19.3 shows the etching rates (given in units of both μm min^{-1} and Å per pulse) versus the beam current between 18 and 310 mA at a PTFE temperature of room temperature and 100 and 200 °C in the case of PTFE [6]. The etching rate becomes higher as the beam

Table 19.1 Direct synchrotron radiation etching for micromachining.

Sample substrate	PTFE	PFA[a]	FEP[b]
Sample temperature, T_s (°C)	100–200	150	100
SR spectra (keV)	0.4–5.5	0.4–5.5	0.4–5.5
Etching rate (μm min^{-1})	60	36	17
Beam current, I (mA)	250	290	275
Max. processing depth (μm)	1500	250	250
Max. aspect ratio	75	12	12

a Teflon PFA (DuPont).
b Teflon FEP (DuPont).

Figure 19.2 Microstructures made of PTFE by direct SR etching: (a) honeycomb structures with 500 μm height and 20 μm width; (b) hole structures with 12 μm height and 1 μm diameter. Courtesy of NTT-AT Nanofabrication Corp.

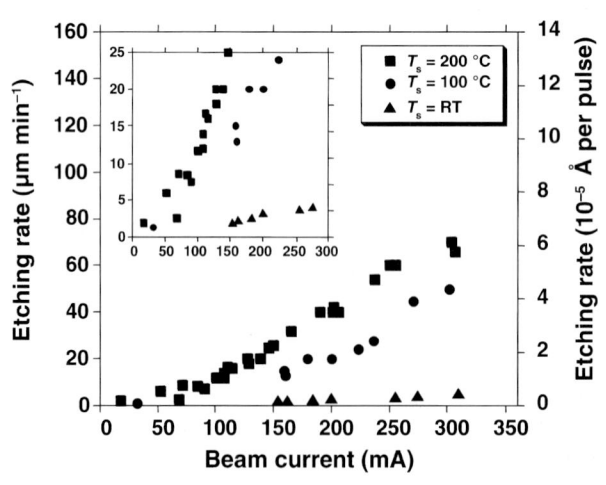

Figure 19.3 Etching rate (shown in two sets of units) versus SR beam current in the case of direct SR etching of PTFE at a sample temperature T_s = room temperature, 100 °C and 200 °C. The inset shows the low-range etching rate details more clearly. Reproduced with permission from *Journal of Synchrotron Radiation*, 1998, **5**, 1153–6. Copyright 1998 International Union of Crystallography.

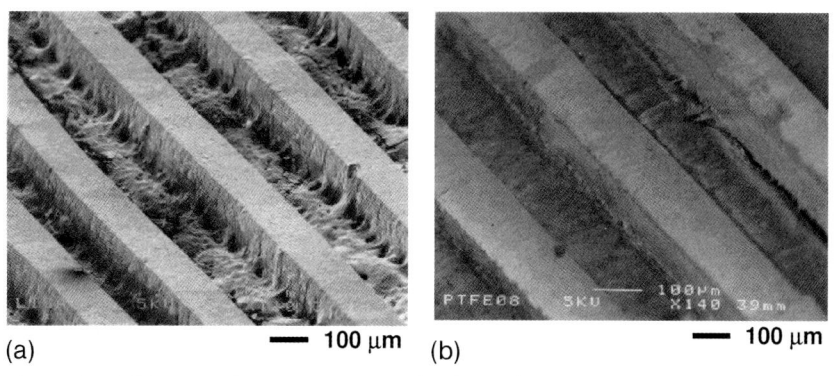

Figure 19.4 SEM image of microstructures made of PTFE by direct SR etching: (a) at room temperature; (b) at 200 °C. Reproduced with permission from *Microsystem Technologies*, 2002, **9**, 1–4. Copyright 2002 Springer-Verlag.

current (photon flux) increases, while the increase in the sample temperature results in significant enhancement of the etching rate and also in an improvement of the surface roughness of the side-wall of the microstructures (Figure 19.4) [7, 8].

In the case where the resist is considered as the material of the final microparts, in other words, no further electroforming and molding are needed, the resist itself will play a very important role. For PMMA, SU-8 and PTFE, since their physical and chemical characteristics are very different, we cannot compare them from the material point of view and the application would decide which kind of resists are needed to generate the microparts. However, we can discuss how long the process may take for these resists, especially when the polymeric layer is very thick. This may be important for industrial fabrication. As seen in Figure 19.3, the etching rate can reach a few tens of $\mu m\,min^{-1}$ so that it may take a few ten minutes for 1000 μm thick PTFE, much faster than the LIGA process with a resist of PMMA, which usually takes a few hours of SR exposure time or otherwise needs an SR source with higher electron energy (e.g. more than 2 GeV) [9]. The SU-8 resist can reduce the SR exposure time by a factor of 1000 compared with the PMMA, but baking before and after the exposure may still take hours. Furthermore, the development time for both PMMA and SU-8 is long for a thick layer, whereas it is not needed at all for PTFE. From the mask point of view, three resists demand different types of X-ray masks. The highly sensitive resist SU-8 demands an X-ray mask with an absorber thicker than that for the PMMA, whereas PTFE etching requires an X-ray mask with a very thin membrane. As a mask with a very thick absorber or with a very thin membrane is very difficult to fabricate, PMMA is preferred rather than SU-8 or PTFE if the application dose not limit the type of polymer. From the photo-energy point of view, three resists prefer different ranges in the SR spectrum, especially for a thick layer (more than 500 μm) of resists. PMMA needs hard X-rays whose energy is more than 4000 eV, whereas PTFE is comfort-

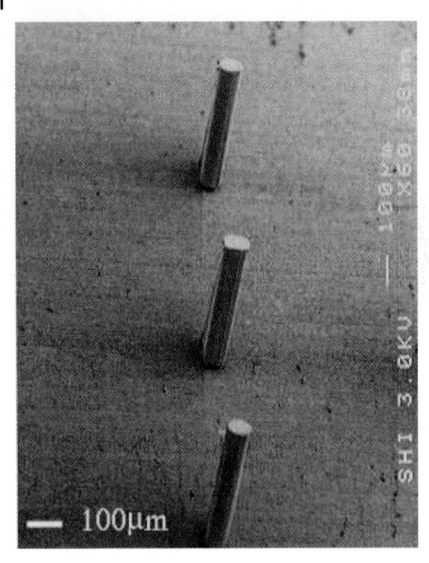

Figure 19.5 Microstructures made of Ni generated from a PTFE template, with a height of 500 μm and a diameter of 80 μm.

able with an energy as low as 7–8 eV provided that a suitable mask can be found. SU-8 can even be fabricated with UV radiation provided that the aspect ratio need not be extremely high.

In the case where the resist is considered as a template for the generation of a metallic mold, the resist has to be bound to a conducting substrate for electroforming and should be removable after the metallic mold has been formed. For this, the SU-8 resist can hardly be used since it is very difficult to remove the thick layer of this resist completely from the mold without causing any damage. PTFE can be used as the template but its sheet has to be coated with a layer of copper, and maybe further with an enhanced metallic layer using electroforming. In contrast to the PMMA template, which can be removed chemically, the PTFE template has to be removed with the SR etching process again. Figure 19.5 shows the Ni patterns generated from a PTFE template.

Like the LIGA process [10, 11], the TIEGA process can also be carried out in a non-planar surface [12]. Figure 19.6 shows a schematic drawing of an SR etching lathe for the generation of a helical structure on PTFE and Figure 19.7 is an SEM image of the structure achieved on PTFE. Here, the wire rotated with an angular speed of $16°\,min^{-1}$ and with a translation speed of $20\,\mu m\,min^{-1}$ under an etching rate of $100\,\mu m\,min^{-1}$.

Although PTFE has many excellent properties, such as high resistivity, high chemical and thermal stability, low surface energy and potential biocompatibility, it has also some poor properties such as low mechanical strength, less transparency to visible light and susceptibility to ionization radiation damage. It has been found that PTFE undergoes crosslinking under electron beam irradiation

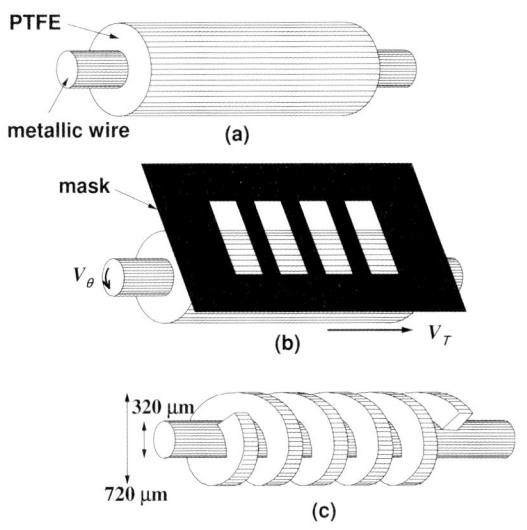

Figure 19.6 Schematic of synchrotron radiation etching lathe for fabrication of three-dimensional structures on PTFE: (a) metallic wire covered with PTFE; (b) etching process; (c) helical structure. Reproduced with permission from *Sensors and Actuators A*, 2001, **89**, 10–15. Copyright 2001 Elsevier Science.

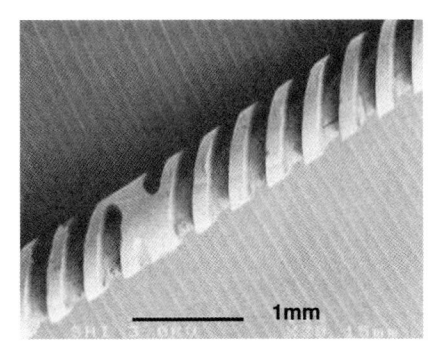

Figure 19.7 SEM image of the helical structure with an SR etching lathe. Reproduced with permission from *Sensors and Actuators A* 2001, **89**, 10–15. Copyright 2001 Elsevier Science BV.

around its melting temperature of 340 °C in an oxygen-free atmosphere [13, 14] and the crosslinked PTFE (RX-PTFE) shows both improved mechanical properties, radiation resistance and optical transparency in the visible to ultraviolet region. We found that the RX-PTFE can also be fabricated by SR with a higher etching rate than PTFE and that crosslinking also occurred under SR irradiation [15]. Figure 19.8 shows an SEM image of the pattern of RX-PTFE etched at 140 °C with a proximity X-ray mask. The exposure time was 4 min under a ring current of 300 mA and the etched depth was 40 µm. The structural quality of the RX-PTFE

Figure 19.8 SEM image of microgears made of radiation crosslinked PTFE (RX-PTFE) created by SR etching at 140 °C with a proximity X-ray mask. Reproduced with permission from *Applied Surface Science*, 2002, **186**, 24–8. Copyright 2002 Elsevier Science.

appears as good as that of PTFE. Our measurement of the etching rate shows that it did not depend directly on the crystallinity and molecular weight of the RX-PTFE. The crosslinking density determined by the abundance of generated radicals and the chain mobility may influence the etching rate of RX-PTFE. A temperature rise can also increase its etching rate. The etching rate of RX-PTFE is 2.1 times higher than that of PTFE at 70 °C and 1.5 times higher at 140 °C. Since the absorbed dose rate in the surface region (depth from surface 1 μm) under SR etching of PTFE is estimated to be 2–3 $MGy\,s^{-1}$, much higher than $1\,kGy\,s^{-1}$ of the electron beam for crosslinking, crosslinking should also occur upon SR etching. Figure 19.9 shows differential scanning calorimetry (DSC) thermograms of PTFE upon SR etching, where the thickness of the resist upon etching is given for each measurement and the PTFE sheet was 500 μm thick before etching. It can be seen that the starting sheet has a single exothermic peak at 315 °C and this peak vanishes and another peak appears at 298 °C. The shift of the exothermic peak to a lower temperature is due to the decrease in the crystallite size upon crosslinking. Similar observations were also made with four stacked pieces of the PTFE sheets with a thickness of 50 μm on exposure to the SR beam, shown in Figure 19.10. The first and second sheets showed the same change in the DSC thermograms, indicating that the depth of this change is on the order of 100 μm for the SR beam used. It is found that modification from the original PTFE also takes place in this case. Two curves of the first and second films have double peaks; the lower temperature peak may be due to modification and the higher temperature peak was attributed to the original PTFE, and the shapes of these curves are broader than

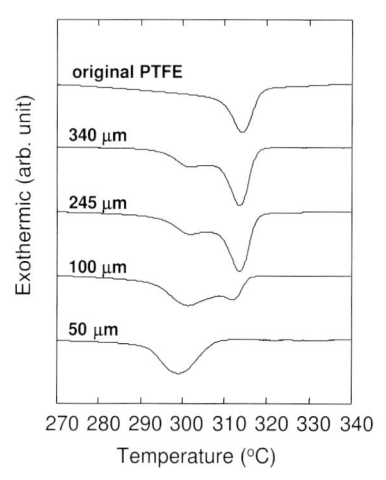

Figure 19.9 DSC thermograms of crystallization for PTFE and modified layers of PTFE irradiated with SR under vacuum at 140°C, Thickness of the PTFE sheet as starting material, 500 μm. The thickness of the PTFE sheet upon etching is given for each measurement. Reproduced with permission from *Applied Surface Science*, 2002, **186**, 24–8. Copyright 2002 Elsevier Science.

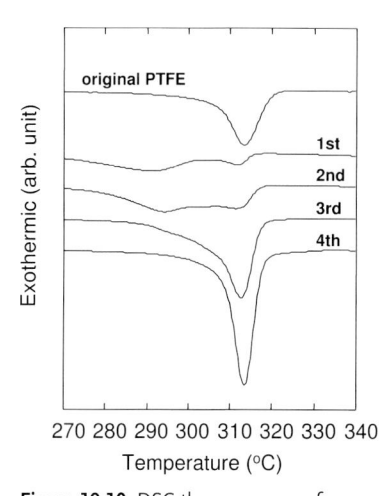

Figure 19.10 DSC thermograms of crystallization for PTFE and modified films of PTFE irradiated with SR under vacuum at 140°C. Thickness of the PTFE film as starting material is 50 μm. Reproduced with permission from *Proceedings of the SPIE*, 2003, **4979**, 493–500. Copyright 2003 The Society of Photo-Optical Instrumentation Engineers.

that of the original. In contrast, the peak temperature of the third and fourth films, which correspond to 150 μm and more in Figure 19.9, was the same as in the original, but the shapes became sharper and the peak area enlarged. Our measurements using Fourier transform infrared (FTIR) spectroscopy and solid-state ^{19}F

NMR spectroscopy further confirmed that crosslinking indeed occurred in those sheets whose DSC spectra showed a shift to the low-temperature side [16, 17]. This means that crosslinking occurred beneath the etched layer with a penetration depth of about 100 μm (mainly depending on the SR wavelength) upon SR etching of PTFE. Since the crosslinking PTFE has a higher etched rate, the SR etching process is an accelerating process. On the other hand, chain scission might be dominantly induced in the deeper layer that is also influenced by SR 'hard' X-rays due to the large penetration depths. As a result, a highly crystallized region of PTFE could be created in the bulk.

19.3
Mechanism of SR Photo-fabrication of PTFE

Here, we compare SR etching with laser ablation. As can be seen in Figure 19.3, several tens of thousands of SR pulses are needed to etch one layer (a few ångstroms) of PTFE, whereas a single laser pulse can ablate more than one layer [18]. In laser ablation [19], there generally exists a fluence threshold for the onset of the photo-etching (also termed 'explosive desorption') [20] which proceeds layer by layer. Such a fluence threshold should not exist for direct SR etching, which does not proceed layer by layer. Our measurement (Figure 19.3, inset) is in agreement with this. In addition, we reduced the photon flux (used in Figure 19.3) by a factor of 30 and indeed found clean pits created by the long X-irradiation with a rate of 0.13 μm min^{-1}. The fact that SR etching proceeds very finely may explain why we did not find any reduction in the rate for larger aspect ratios, namely a transportation effect [21], and why this process can achieve a very high aspect ratio and a nearly zero taper angle for the etched edge. In contrast to SR etching, the pulse etching rate (e.g. 0.1 μm per pulse) of laser ablation is much higher and the ablation plume (a dense cloud of ablated material) is usually generated. There are some effects induced by the ablation plume such as the beam-size effect, that is, the ablation rate for a larger irradiation surface may become smaller due to shielding of the laser pulse by the plume [22]. In contrast, the SR etching rate was found to become larger for a larger irradiation surface. Therefore, the direct SR etching process is essentially different from the ablation induced by laser pulses and therefore we would prefer to call it SR etching rather than SR ablation. Furthermore, polyimide (also called Kapton) can easily be fabricated by laser ablation but not by SR etching. In contrast to PTFE, which is highly susceptible to reactor radiation damage, polyimide shows good radiation resistance since the aromatic imide ring has excellent radiation stability due to p-electron delocalization and aromatic ethers is stable under radiation thanks to lone pair delocalization [23]. This also gives evidence to show the difference between SR fabrication and laser ablation.

To study the decomposition mechanism, mass spectrometric diagnosis was carried out *in situ*, using quadrupole mass spectrometry (QMS) to detect gaseous species both in SR etching and in laser ablation [24]. All QMS patterns were observed to be similar when the substrate temperature was increased or when the SR beam was switched from white light to X-rays. The predominant signal CF_3^+ was detected at 69 amu, due to saturated fluorocarbons $CF_3C_nF_{2n}CF_3$ as the gaseous products on SR etching of PTFE. However, they appeared to be different when the SR beam was replaced with a laser beam. The QMS patterns observed on laser ablation of PTFE look similar to that of the monomer C_2F_4, instead of that of the saturated fluorocarbons. This is further evidence to differentiate SR etching from laser ablation or thermally induced decomposition.

Based on the above study, the photo-reaction under an SR beam with different photon fluxes is understood as follows. Under a low photon flux of less than 10^{15} photons s^{-1} cm^{-2}, absorption of the X-ray photons in the substrate leads to dissociation of chemical bonds to form fragments which may react with neighboring fragments to form a crosslinked network or desorb from the surface if they are small enough [25]. This degradation (also termed 'incubation') process may result in visible discoloration and pitting of the polymer surface. Under the high photon flux of more than 10^{17} photons s^{-1} cm^{-2} and with increasing substrate temperature, although the main dissociation mechanism may not change, the desorption rate becomes considerably larger and the surface is etched into the bulk. This is the so-called photo-etching process, which may result in a very high aspect ratio for the structures formed. Explosive desorption, like laser ablation, however, does not occur under the present SR conditions. The ablation occurs only when a laser pulse reaches a certain fluence threshold, whereas there is no such a threshold for photo-etching. It should be noted that the increase in the substrate temperature may play a considerable role in desorption to enhance the SR etching rate but would not take over the role of the X-ray photons in the dissociation. The Arrhenius-like plot for the etching rate (µm per 100 mA min^{-1}) gave a slope of 0.1 eV in the case of PTFE, much smaller than the activation energy for vacuum pyrolysis of PTFE, 3.6 eV [26]. As mentioned above, mass spectrometric analysis in the case of PTFE has also given evidence that the dissociation of PTFE during the SR process is completely different from thermal dissociation.

19.4
Application: X-ray Lenses Made of PTFE

In this section, as an example of the application of the SR etching, we show how to make a refractive lens of PTFE for the generation of X-ray microbeams. Since the refractive lens has to have a very small radius of curvature for the X-rays and also its aperture has to be large enough for the beam, it becomes very difficult to fabricate with any traditional technique owing to its extremely high aspect ratio (that is, the ratio of the aperture to the radius of curvature) [27]. For X-rays at an energy of 10 keV as an example, the radius of curvature of a refractive lens should be just a few micrometers in order to focus the beam at about 0.5 m. Even if the beam size may be reduced to a few hundred micrometers, the lens has to have an aspect ratio between 40 and 100. Furthermore, the PTFE lens has a shorter focal

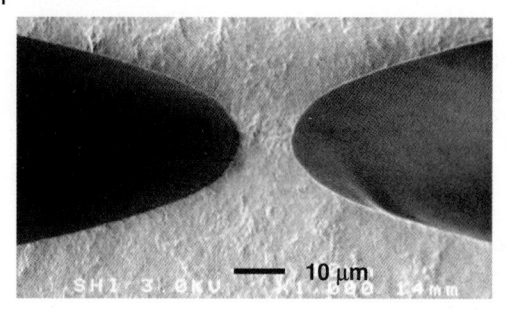

Figure 19.11 SEM image of the one-dimensional single refractive lens (1D lens) made of PTFE. The gap of 10 μm is achieved easily by the single SR exposure. Reproduced with permission from *Sensors and Actuators A*, 2002, **97/98**, 725–8. Copyright 2002 Elsevier Science.

(a) (b)

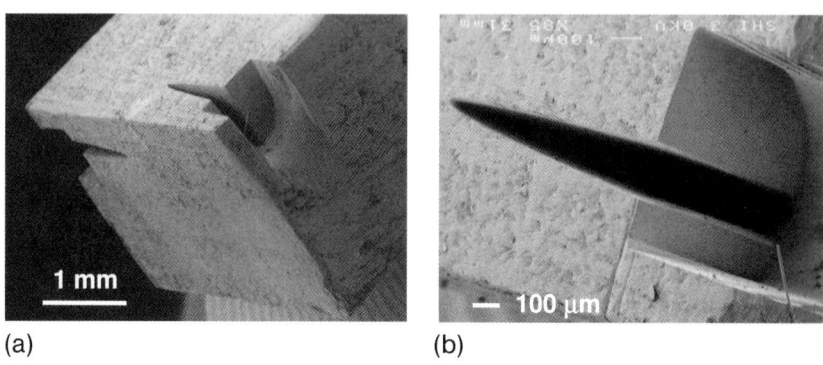

Figure 19.12 SEM image of the two-dimensional single refractive lens (2D lens) made of PTFE (a) and a close-up view (b). Reproduced with permission from *Sensors and Actuators A*, 2002, **97/98**, 725–8. Copyright 2002 Elsevier Science.

length, nearly half that of the PMMA lens, due to the nearly half value of the refractive index decrement compared with that for PTFE. It is just good challenge for the TIEGA process to fabricate the PFTE lens.

X-ray refractive lenses were designed to have two parabolic curvatures with radii of 4 μm and apertures of 179 μm [28]. As shown in Figures 19.11 and 19.12, two curvatures are on the same axis for one-dimensional (1D) focusing, whereas they are vertical and horizontal, respectively, for two-dimensional (2D) focusing. The 1D lens (Figure 19.11) needs only a single exposure with the mask which has two curvature -patterns and the gap between the two curvatures is 10 μm, which has been designed in the mask. To make the 2D lens (Figure 19.12), however, the exposure has to be carried out twice in different directions and therefore the gap is dependent on alignment for the second exposure. For an aperture of 179 μm,

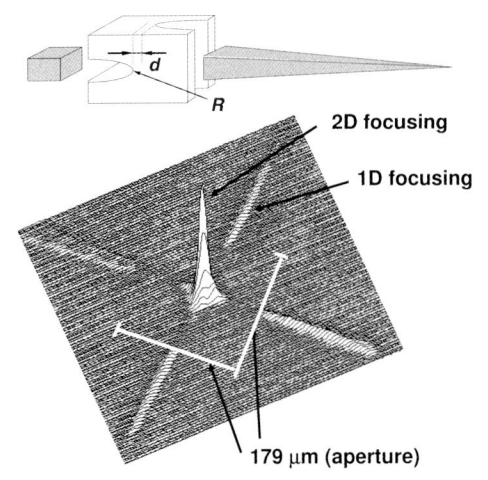

Figure 19.13 Two-dimensional intensity mapping of the beam focused by a 2D lens. Four lines around the focal beam are due to focusing by each of two curvatures and have to be eliminated for direct writing. Reproduced with permission from *Sensors and Actuators A*, 2002, **97/98**, 725–8. Copyright 2002 Elsevier Science.

furthermore, the depth of one exposure should be more than 300 μm and therefore etching up to 1000 μm has been carried out (Figure 19.12).

The performance of these refractive lenses made of PTFE was checked with a beam of hard X-rays (10 keV, $\Delta E/E \approx 2.4 \times 10^{-4}$) from the Hyogo beamline at SPring-8 [29]. For the 10 keV beam, the focal length is 0.41 m for the 1D lens, whereas it is 0.8 m for the 2D lens, in agreement with calculation using the refractive index decrement. Figure 19.13 shows 2D intensity mapping for the beams focused by the 1D and 2D lens, respectively. It can be seen that in the 1D case the beam intensity has been enhanced on the focal line, whereas in the 2D case the intensity has been further enhanced on the focal point by focusing photons around it. The width of the part of the two focal lines that has disappeared corresponds to the lens aperture (~179 μm). The focal width is 3.4–6.7 μm, given by the full width at half-maximum (FWHM), which is obtained from a Gaussian fit to the measured intensity distribution [29]. This shows that our lens can generate the X-ray microbeam simply. The transmissions were 54% for the 1D lens and 89% for 2D lens using a beam with a diameter of 100 μm. The transmissions were obtained by measuring the X-ray intensity with and without the lenses using a PIN photodiode. Our single refractive lenses have transmissions much higher than 10% of the compound refractive lens (CRL). Moreover, we checked for any possible damage caused by long exposure to the hard X-rays. In contrast to the CRL, our lenses did not show any visible damage such as deformation and bubble formation. The reason might be that such damage is mainly due to heat load rather than photo-decomposition, since the CRL has much higher absorption than the single lens.

X-ray images of a copper 2000 mesh were taken using the 1D lens, which has the shortest focal length. In Figure 19.14, a magnification of about 5× in the vertical can be seen by a comparison between the images without and with the lens (inset) and the imaging resolution is evaluated as 0.9 µm with the criterion of 25–75% of the total contrast. This is poorer than the theoretical value of 0.6 µm, eventually limited by the attenuation, no matter how much larger the aperture can be.

Direct writing has been carried out using the X-ray microbeam and Figure 19.15 shows the first demonstration, where the letters 'SHI' have been written with a

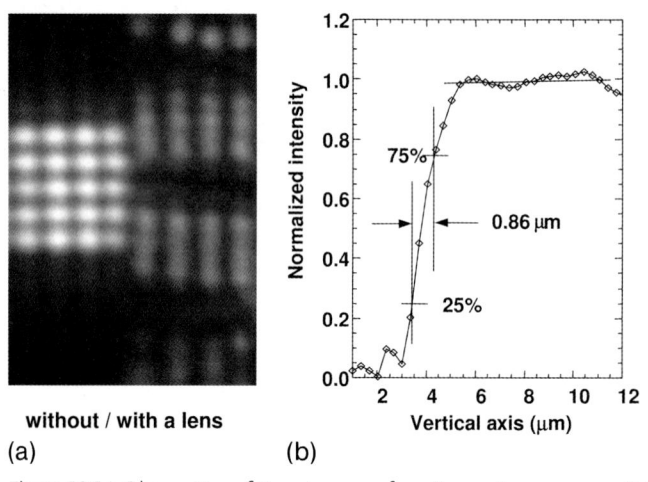

Figure 19.14 Observation of X-ray images of a copper 2000 mesh without and with a lens (a) and evaluation of the imaging resolution after the lens, leading to a resolution of 0.86 µm. Here, the mesh was set to be 0.5 m before the PTFE 1D lens ($F = 0.41$ m) and the X-ray microscope was 2.4 m behind the lens. Reproduced with permission from *Japanese Journal of Applied Physics*, 2001, **40**, L75–7. Copyright 2001 The Japan Society of Applied Physics.

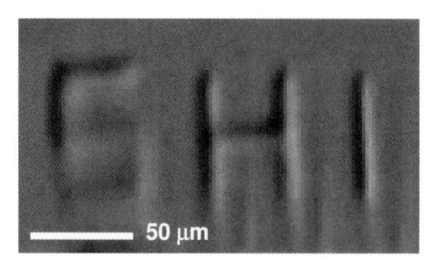

Figure 19.15 Demonstration of the direct writing by etching a fluoropolymer thin film with a microbeam that has been formed by the single refractive lens. Reproduced with permission from *Sensors and Actuators A*, 2002, **97/98**, 725–8. Copyright 2002 Elsevier Science.

linewidth of about 5 μm through etching the 5 μm thin PTFE-like film on the silicon substrate. Before using the microbeam generated with our PTFE lens, we had to eliminate the four lines shown in Figure 19.13, by setting a pinhole of 165 μm diameter in front of the lens.

19.5
Conclusions

We have shown that SR can be applied to etch PTFE polymers directly, generating microstructures with a very high aspect ratio. Like the LIGA process, this TIE can be combined with galvanoforming for fabrication of micromolds and named the TIEGA process. Not only the standard PTFE polymer but also crosslinked PTFE polymer can be applied to the TIEGA process so that various types of fluorocarbon polymers can be useful. Although this etching process appears to be similar to laser ablation, we have shown that the decomposition and the molecular dynamics are essentially different from each other. As an example, an X-ray refractive lens with a radius of curvature of 4 μm and an aperture of 179 μm, that is, an aspect ratio of 45, has be made of PTFE polymer using our technique. A microbeam has been generated with this PTFE lens and its focal length was 0.8 m.

References

1 Ehrfeld, W., Bley, P., Götz, F., Mohr, J., Münchenmeyer, D., Schelb, W., Baving, H.J. and Beets, D. (1988) *Journal of Vacuum Science and Technology B*, **6**, 178–82.

2 Menz, W. and Bley, P. (1993) *Mikrosystemtechnik für Ingenieure*, Wiley-VCH Verlag GmbH, Weinheim.

3 Lorenz, H., Despont, M., Fahrni, N., LaBianca, N., Renaud, P. and Vettiger, P. (1997) *Journal of Micromechanics and Microengineering*, **7**, 121–4.

4 Zhang, Y. and Katoh, T. (1996) *Journal of Applied Physics*, **35**, L186–8.

5 Zhang, Y. and Hori, T. (2000) *Synchrotron Radiation News*, **13**, 32–3.

6 Katoh, T. and Zhang, Y. (1998) *Journal of Synchrotron Radiation*, **5**, 1153–6.

7 Zhang, Y., Katoh, T., Washio, M., Yamada, H. and Hamada, S. (1995) *Applied Physics Letters*, **67**, 872–4.

8 Nishi, N., Katoh, T., Ueno, H. and Sugiyama, S. (2002) *Microsystem Technologies*, **9**, 1–4.

9 Guckel, H. (1996) *Review of Scientific Instruments*, **67**, 1–5.

10 Marques, C., Desta, Y.M., Rogers, J., Murphy, M.C. and Kelly, K. (1997) *Journal of Microelectromechanical Systems*, **6**, 329–35.

11 Feinerman, A.D., Lajos, R.E., White, V. and Denton, D.D. (1996) *Journal of Microelectromechanical Systems*, **5**, 250–5.

12 Katoh, T., Nishi, N., Fukagawa, M., Ueno, H. and Sugiyama, S. (2001) *Sensors and Actuators A*, **89**, 10–15.

13 Sun, J., Zhang, Y., Zhong, X. and Zhu, X. (1994) *Radiation Physics and Chemistry*, **44**, 655–9.

14 Oshima, A., Tabata, Y., Kudoh, H. and Seguchi, T. (1995) *Radiation Physics and Chemistry*, **45**, 269–73.

15 Katoh, T., Sato, Y., Yamaguchi, D., Ikeda, S., Aoki, Y., Washio, M. and Tabata, Y. (2002) *Applied Surface Science*, **186**, 24–8.

16 Katoh, T., Sato, Y., Yamaguchi, D., Ikeda, S., Aoki, Y., Oshima, A., Washio, M. and Tabata, Y. (2003) *Proceedings of the*

SPIE – The International Society for Optical Engineering, **4979**, 493–500.
17 Sato, Y., Yamaguchi, D., Katoh, T., Ikeda, S., Aoki, Y., Oshima, A., Tabata, Y. and Washio, M. (2003) *Nuclear Instruments and Methods in Physics Research B*, **208**, 231–5.
18 Küper, S. (1989) Ablation mit UV-Laserlicht, PhD Dissertation, Göttingen University, pp. 10–66.
19 Srinivasan, R. and Braren, B. (1989) *Chemical Reviews*, **89**, 1303–61.
20 Domen, K. and Chuang, T.J. (1987) *Physical Review Letters*, **59**, 1484–7.
21 El-Kholi, A., Mohr, J. and Stransky, R. (1994) *Microelectronic Engineering*, **23**, 219–22.
22 Eyett, M. and Bäuerle, D. (1987) *Applied Physics Letters*, **51**, 2054–5.
23 Sasuga, T., Hayakawa, N., Yoshida, K. and Hagiwara, M. (1985) *Polymer*, **26**, 1039–45.
24 Zhang, Y., Katoh, T. and Endo, A. (2000) *Journal of Physical Chemistry B*, **104**, 6212–17.
25 Simons, J.K., Frigo, S.P., Taylor, J.W. and Rosenberg, R.A. (1994) *Journal of Vacuum Science and Technology A*, **12**, 681–9.
26 Single, J.G., Muus, L.T., Lin, T.P. and Larsen, H.A. (1964) *Journal of Polymer Science A*, **2**, 391–404.
27 Yang, B.X. (1993) *Nuclear Instruments and Methods in Physics Research Section A – Accelerators Spectrometers Detectors and Associated Equipment*, **328**, 578.
28 Zhang, Y., Katoh, T., Kagoshima, Y., Tsusaka, Y. and Matsui, J. (2001) *Japanese Journal of Applied Physics*, **40**, L75–7.
29 Katoh, T., Zhang, Y., Kagoshima, Y., Tsusaka, Y. and Matsui, J. (2002) *Sensors and Actuators A*, **97/98**, 725–8.

Index

a

ABSE   425
absorbed dose *see* exposure dose
absorber materials   16–19 (*see also* photoresist)
– densities   17
– effective absorption   16–18
– electroplating   18, 19, 40–43
– gold   17–19, 40–43
– material selection   19
– physical vapor deposition   19
absorption contrast   26, 27
acceleration sensor   148–150, 199
achromatic confocal distance sensor   216–218
actuators *see* microactuators
adhesive bonding   174–176
alignment
– design for assembly   170, 172–174
– mask and substrate   23, 24, 31, 33, 35, 38, 173
– optical components   213, 218, 220
alignment markers   23, 24, 31, 33, 35, 38, 173
alumina, thermal expansion coefficient   155
alumina powder fillers   422, 423, 430, 434, 435
AMANDA micropump   148
ambient effects   180
ANKA   108–110, 201
annealing, stress relieving   30, 155
ANSYS   76–78
aspect ratios   82, 86, 168, 206
– filled resist systems   419, 427
– microactuators   299
– piezoelectric composite materials   338, 339
– RF applications   257, 272, 273
– TIEGA process   455
– vertical wall structures   249, 257
assembly
– design rules   170–178
– feeding process   170, 171
– gripping process   171, 172
– joining   174–178
– main steps   169
– positioning   172–174
– rotary bond tool   385–388
AURORA   109–111, 443
AURORA-2S   443, 454
Axsun Technologies   201

b

bandpass filters   209–211
Beer-Lambert law   14
beryllium
– electroplating   22
– mask substrates   21, 22, 44
– – alignment markers   23
– – effective transparency   25
– – fabrication   35–37
– – properties and thickness   26
– – temperature rise   20
– optical transmission characteristics   23
– X-ray attenuation characteristics   15, 16
bidirectional transceiver/receiver   215
biodegradability   152
biodisposables   152
bi-stability principle   149, 150
bi-stable membrane microvalve   149, 150
bonding
– adhesive   174–176
– plastic microfluidic devices   331–333
boron nitride, X-ray attenuation characteristics   15
borosilicate glass, mask substrates   22, 37, 38

Bosch process 192
bulk micromachining 190, 191
– microfabrication costs 196, 197
– throughput rates 197

c

C (alumina powder) 422, 423
cantilever beams 41, 42, 254, 255, 258–260
carbon, X-ray attenuation characteristics 15
cavity resonators 270–277
Center for Advanced Microstructures and Devices (CAMD) 13
ceramic precursors 424, 425
Ceraset 425
clamping elements 177, 178
cleanrooms 200
CMOS thin-film techniques 7, 190
comb drive actuator 225, 226, 306–308
commercial issues 6, 7 (see also costs; productivity)
– commercial centres 193, 194
– foundry operation 199–201
computer software see software
COMS (Commercialization of Micro and Nano Systems Conference) 9
conduction paths, hot embossing 92, 93
coplanar waveguides (CPW) 249–251, 254, 263–270
copper
– densities and absorption characteristics 17, 18
– powder fillers 436
– thermal expansion coefficient 155
costs 6, 7, 44, 192–199
– cost models 194–196
– X-ray sources 193, 194
counter meshing gears discriminator 316, 317
crack avoidance
– deep X-ray lithography (DXRL) 154–160
– molding 168
crosslinking
– polytetrafluoroethylene (PTFE) 458–460
– preceramic polymers 424, 425
CT3000SG 422, 423
curing process
– adhesive bonding 175, 176
– ceramic composites 423–425, 434
cyanide-based electroplating 40
cycle times (see also throughput rates)
– hot embossing 413
– micromolding 288, 327, 330, 411
cylinder lenses 171, 172, 207–209

d

deep reactive ion etching (DRIE) 191, 192
– microfabrication costs 196–199
– throughput rates 197
deep X-ray lithography (DXRL) 5, 6, 51–67
– 3D microstructures 52, 55–67
– costs 192–198
– design for manufacturing 154–167
– – crack avoidance 154–160
– – mechanical stabilization 160–163
– development modelling 118–132
– – photoresist damage theory 119–121
– – photoresist dissolution rate 121, 122
– – topology representation 123–132
– development process 53, 54
– – dissolution rate 55, 60, 61, 121, 122
– exposure dose 53–55
– – two-step dose distribution 60, 61
– exposure modelling 104–118
– – beamline data 109–111
– – beam-line modeling and calculation 105–118
– – object-oriented data structures 113–118
– – process parameters 105, 106
– – software 105, 106, 109, 132–140
– exposure techniques 52, 55–67
– – double exposure 62, 63
– – inclined exposure 57–59
– – low contrast masks 61
– – multi-step dose distribution 63–65
– – non-planar resist substrates 55–57
– – two-step dose distribution 60–63
– masks see X-ray masks
– photoresist see photoresist
– scanning process 418, 419
– substrates see photoresist, substrates
– throughput rates 197
deformation matching 147
demolding
– hot embossing 75–77, 79–82
– injection molding 397
densities, absorber materials 17
design methodology
– design for assembly 168–178
– design for manufacturing 154–168
– – design rules 157–160
– design principles 146–150
– design requirements 179
– design sequence 145, 146
– embodiment design 146, 147
– functional design 178–187
– general microsystems design rules 150–153
'die attach' machines 385

differential expansion 151
diffraction effects 325–327
dimensional deviations 152, 153, 160–163
Direct-LIGA 5, 6, 8
– microgears fabrication 371–373
distance sensors see micro-optical distance sensors
division of tasks 147, 148
dose see exposure dose
DoseSim 132–134
double exposure technique 62, 63
DRIE see deep reactive ion etching (DRIE)
DSAS (Stress Deposit Analyzer System) 42
DXRL see deep X-ray lithography (DXRL)
dynamic thermomechanical analysis (DMA) 77

e

e-beam lithography (EBL) 419
– nanoparticle-filled resists 432, 433
effective absorption 16, 18
effective transparency 15, 16
eikonal partial differential equation 129
electrical contact probe see microcontact probe
electroforming 303
– nickel mold inserts 306, 307, 326, 327, 341
– TIEGA process 458
electromagnetic actuators 299–302, 308
– variable reluctance 310–312
electromagnetic microchopper 227, 228
electron beam lithography 419
– nanoparticle-filled resists 432, 433
electro-optical systems 218–225
electroplating 18, 19
– beryllium 22
– circles 166
– cyanide-based 40
– cylindrical holes 166, 167
– deformations 163, 164
– design rules 162–167
– gold 40–42, 274, 275
– – internal stress 41–43
– graphite 22
– internal stress 41–43
– mechanical stabilization 160–163
– microgear wheels 372
– microstructured frames 163–165
– nickel 70, 89, 256, 258–260, 266, 275
– polishing 268
– subdivision of structures 165, 166
electrostatic microactuators 225, 226, 299, 300, 306

– micropositioning 315
– rotary 308–310
electrostatic motors 226, 227, 310–312
electrostatic switching matrix 226, 227
embodiment design 146, 147
EPON SU-8 see SU-8
EUV see extreme ultraviolet (EUV) lithography
EVG 97, 98
excimer laser ablation 191, 192
– compared with SR etching 462, 463
– microfabrication costs 196–198
– throughput rates 197
exposure dose 24, 53–55
– dissolution rate function 55, 121
– distribution
– – filled resists 425, 426
– – multi-step or arbitrary 63–65
– – simulation 104–106, 118
– – two-step 60–63
– dose ratio 25
– relation to pattern depth 53, 325
exposure techniques 52–67
– 3D resist substrates 55–57
– absorber thickness distribution 64
– inclined exposure 57–59
– multi-step or arbitrary dose distribution 63
– planar movement of mask or resist substrate 64, 65
– two-step dose distribution 60–63
– UV lithography 417, 418
exposure times 193, 196, 457
extreme ultraviolet (EUV) lithography 1, 2, 444

f

fabrication (see also manufacturing)
– design issues 152–154
– process sequence and variations 5
– task subdivision 147, 148
fast marching method 122–124
– topology representation 125, 126, 128–130
fault-free design 150, 184–186
FELIG 8, 201
fiber optics
– intensity coupling element 212, 213
– multifiber connector 213, 214
– switch 308, 316
filler–polymer matrix composites 420–437
– alumina powder fillers 422, 423, 430, 434, 435
– curing depth 423, 424

– filled SU-8   428–432
– metal powder fillers   436
– nanoparticle-filled resists   432, 433
– polyimide–silica composites   437, 438
– preceramic polymers   424–428
– silica fillers   430, 432, 433
finite element modelling (FEM)   76
fluidic devices *see* microfluidic devices
fluorescence radiation   15, 119
fluorinated polyimide–silica composites   437
force transmission   147, 149
Forschungszentrum Karlsruhe (FZK)   2, 8, 96, 97
– ANKA   108–110, 201
foundry operation   199–201
Fourier transform infrared (FTIR) spectrometer   82, 83, 86, 177, 229, 230, 316
friction, design for   151
front normal   128
functional design   178–187
functional division   147, 148

g

gears *see* microgears
G-G developer   54, 55, 121
glass
– lenses   176
– mask substrates   37, 38
– – properties and thickness   26
– optical transmission characteristics   23
– X-ray attenuation characteristics   15, 16
glassy carbon *see* vitreous carbon
gold
– absorber layer fabrication   19
– absorber materials   17–19, 40–43
– densities and absorption characteristics   17, 18
– electroplating   40–42, 274, 275
– – internal stress   41–43
graphite
– electroplating   22
– mask substrates   44
– – fabrication   33–35
– – properties and thickness   26
– optical transmission characteristics   23
– X-ray attenuation characteristics   16
grippers   171, 172, 174

h

Halback arrays   312
hard disk drives   315
hard tool–soft countertool combination   91, 92
harmonic drive gear   352–393
– advantages   369–371
– applications   384–393
– – microassembly   385–388
– – microrobots   390–393
– – nanostage   388–390
– fabrication   371–373
– flexible properties   361, 362
– gear profiles   373–382
– hollow shaft   366, 367, 383, 385
– operating principle   358–361
– properties   366–369
– simulation   363–365
HARMST (High Aspect Ratio Micro-Structure Technology) workshop   9
heap data structure   130
HELIOS   13, 443
heterodyne receiver   215–217
HEX03   96–98
hole making   90–92
hollow-cavity waveguide   288
hook joint   177, 178
hot embossing
– applications
– – high aspect ratio replication   86, 87
– – large area replication   86, 87
– – optical microstructures   83–86, 212
– basic principle   72–74
– compared with injection molding   73
– composite foils   90
– conduction paths   92, 93
– cooling time   95
– cycle times   413
– damage structures   77
– deformations   82
– demolding   75
– – simulation   76, 77, 79–82
– design rules   167, 168
– diced microstructures   90
– facilities   96–98
– heating and cooling blocks   95, 96
– mold filling
– – pressure distribution   79
– – simulation   77, 79
– multilayer replication   91–93
– polymer mold inserts   93, 94
– process optimization   82, 83
– process simulation   76–83
– – in practice   82, 83
– – material properties basis   76, 77
– replication on a substrate   88, 89
– residual layer   88, 90

– through-holes 90–92
– tools 94–96
hot punching 90, 91
hydraulic actuators 314
hydrogen bubbles, photoresist 155, 156, 166, 167
hysteresis motor 313, 314

i

IMTEK, X3D 109
inclined exposure 57–59
inclined sidewalls 57–59, 325
– freely shaped 52, 60–67
industrial production *see* manufacturing
infrared spectrometer *see* Fourier transform infrared (FTIR) spectrometer
injection compression process 413, 414
injection molding 70, 71
– applications
– – microfluidic devices 327–330
– – microspectrometers 287, 288
– – permanent magnets 304, 305
– clamping unit 396–398, 403
– compared with hot embossing 73
– components and design concepts 396–407
– cycle times 288, 327, 330, 411
– design rules 167, 168
– driers 404
– ejection from mold 397
– electric drive 403, 404
– flow control 398, 399, 406, 409, 410
– history 395, 396
– hydraulic drive 403, 404
– injection unit 398, 399, 408–410
– materials
– – lead zirconate titanate (PZT) 341
– – liquid crystal polymer (LCP) 400
– – polyacetal (POM) 400
– – polycarbonate (PC) 400, 404
– – poly(ether ether ketone) (PEEK) 400
– – poly(methyl methacrylate) (PMMA) 70, 71, 329, 330, 400
– – polypropylene (PP) 400–402
– – polystyrene 332, 333
– melt film 401
– melt pressure transducer 405
– melt temperature sensors 405, 406
– microinjection molding machines 287, 288, 407–414
– mold inserts *see* mold inserts
– multi-cavity molds 412
– non-return valves 402, 403
– parallel processing 410, 411

– physical process 412, 413
– piston injection 414
– plastification unit 399–402, 407, 408
– processing temperatures 400
– quality assurance 406, 407, 411
– temperature control 404–406
– variotherm process 405
– venting 404
INNOLIGA 8
inspection devices, piezoelectric composites 337–344
integrated circuit testing, microcontact probe 344–349
intensity coupling element 212, 213

j

Jenoptik Mikrotechnik 96–98
joining (*see also* bonding)
– design for 174–178

k

kinoform lens 238

l

laser micromachining 191, 192
– compared with SR etching 462, 463
– microfabrication costs 196–198
– throughput rates 197
lateral comb structures 253, 254
lead, properties 17, 18
lead zirconate titanate (PZT)
– impedance curve 342
– injection molding 341
– piezoelectric composites 337–340
– resist filler 434
lead–tin alloy, absorber materials 19
lenses *see* microlenses; X-ray lenses
level sets 123–125
– topology representation 126–132
LEX-D 133
liquid crystal polymer (LCP), injection molding 400
Lito 2 108
loads, fabrication 153 (*see also* thermal stress)

m

M2DXL *see* moving mask deep X-ray lithography (M2DXL)
'magic rings' 312
malfunction prevention 148
manufacturing 7, 8
– commercial centres 193, 194
– commercial issues 6, 7 (*see also* costs)

– design methodology 154–168
– – design rules 157–160
– foundry operation 199–201
masks *see* X-ray masks
materials matching 147
mechanical microclamping 176–178
MEMSCOST 194
mesh-free topology representation 124, 125
metal powder fillers 436, 437
μFEMOS 180
micro harmonic drive gears *see* harmonic drive gear
microactuators 297–317 (*see also* microgears; micromotors)
– applications 314–317
– backlash-free 383, 384
– electrostatic 225, 226, 299, 300, 306
– – micropositioning 315
– – rotary 308–310
– materials 303–305
– scaling 298–302
– types 306–314
microassembly, rotary bond tool 385–388
microchopper 227, 228
microclamping 176–178
microcontact probe 344–349
– design of tip shape 346–349
– materials 345, 346
– production process 345
microcracking 154–160
microdrive systems 384–393 (*see also* harmonic drive gear; microactuators)
microfabrication 152
microfluidic devices 323–334
– actuators 317
– bonding 331–333
– injection molding 327–330
– microflow paths 331–333
– nickel tool part 324–327
– polystyrene 332, 333
microgears 6, 90, 151, 434, 460 (*see also* harmonic drive gear)
microlenses 84, 85 (*see also* X-ray lenses)
– cylinder lenses 171, 172, 207–209
micromotors 301, 309, 310, 312–314, 317, 318
– electrostatic 226, 227, 310–312
– pancake 383
micro-optical benches (MOB) 84, 88, 170, 174, 181, 211–218
micro-optical bypass switch 225, 226
micro-optical components 205–230

– electro-optical systems 218–225
– materials 207–209
– optical benches 211–218
– optical components 208–211
– optical MEMS 225–230
micro-optical distance sensors 151, 173, 216–218
– modeling and simulation 180–187
– triangulation 219–222
micro-optical switches 160–162
micropositioning (*see also* microgears)
– electrostatic microactuators 315
– nanostage 388–390
microprisms 208, 209
micropumps 148
microresonators 308
microrobots 390–393
microscopes 316
– X-ray 233, 240, 241
microspectrometers 85, 86, 90, 221–225, 281–296 (*see also* Fourier transform infrared (FTIR) spectrometer)
– assembly 289
– characteristics 285–287
– fabrication 287–289
– marketing and sales 295, 296
– products and applications 290–295
microstereolithography 419, 420
– exposure techniques 55–66
– filled resists 433–436
microvalves 149, 150
mikroFEMOS 96
MIRRORCLE 444–451
MOB *see* micro-optical benches (MOB)
modeling 179–187 (*see also* simulation)
mold inserts (*see also* hot embossing; injection molding)
– electroforming 306, 307, 326, 327, 341
– high aspect ratio 86
– large area replication 87, 88
– loads and stresses 153
– multi-cavity molds 412
– nickel 306, 307, 326, 327, 339–341
– optical component fabrication 83, 84, 211–214
– polymer 93, 94
MOLDFLOW 76, 78
molecular imprint 97
mosaic lens 238
moving mask deep X-ray lithography (M2DXL) 63, 64, 111, 112
multifiber connectors 213, 214
multi-mode fibers 212, 213
multi-step dose distribution 63–65

n

nanoparticle-filled resists 432, 433
nanoscopes 240, 241
nanostage 388–390
Nanotek 422, 423
NCP-200 425
negative resist 119
NEUTRONEX 309 40, 43
nickel
– densities and absorption characteristics 17, 18
– electroplating 70, 89, 256, 258–260, 266, 275
– mold inserts 339, 340
– – electroforming 306, 307, 326, 327, 341
– thermal expansion coefficient 151
– varactors 256–263
– X-ray lenses 237
nickel–iron electrolyte 372
Ni—Mn alloy, microspring 346, 347
non-planar resist substrates 55–57
non-return valves, injection molding 402, 403

o

Obducat 97
object-oriented data structures 113–118
optical components see micro-optical components
optical coupling elements 83
optical fibers
– intensity coupling element 212, 213
– multifiber connector 213, 214
– switch 308, 316
optical lenses see microlenses
optical lithography 2
optical MEMS 225–230
optical switch matrix 160–162
ORMOCERs 438
overload protection 149

p

parallel-plate variable capacitor 248, 249
Parvus 390–393
PEEK 84, 90, 93, 94
– injection molding 400
permanent magnets 312–314
– injection molding 304, 305
photoresist
– bonding 155
– casting 430
– corners 159, 160
– cracking 154–160
– damage theory 119–121
– degradation mechanism 120
– development of irradiated resist 53–55, 120, 121
– – dissolution rate 55, 121–125
– – simulation 133–139
– exposure dose see exposure dose
– exposure techniques see exposure techniques
– exposure times see exposure times
– flow properties 421–423
– hydrogen bubbles 155, 156, 166, 167
– mechanical stabilization 160–162
– optical properties 423, 424
– spin coating 421
– subdivision 157
– substrates
– – 3D 55–57
– – clearance 111, 211
– – planar movement 63–65, 112
– – thermal expansion 155
– swelling 160–163
– tapered structures 159
– tension stress 155
– thermal expansion 155, 160–163
– viscosity 421–423
– water uptake and swelling 160–162
– X-ray absorption 418, 419
photoresist materials
– compared 457, 458
– filler–polymer matrix composites 420–437
– – filled SU-8 428–432
– – metal powder fillers 436, 437
– – nanoparticle-filled resists 432, 433
– – polyimide–silica composites 437, 438
– – preceramic polymers 424–428
– – properties 420–424
– – silica fillers 430, 432, 433
– PMMA see poly(methyl methacrylate) (PMMA)
– polytetrafluoroethylene see polytetrafluoroethylene (PTFE)
– sol–gel-based composites 438
– SU-8 see SU-8
– types 119, 416, 417
– water-based fillers 437, 438
physical vapor deposition (PVD) 19, 22, 288, 289
– titanium nitride 400
piezoelectric composites 337–344
– characteristics 342–344
– impedance curve 342
– production process 339–341
platinum, properties 17, 18

PMMA *see* poly(methyl methacrylate) (PMMA)
polishing, electroplated nickel 268
polyacetal (POM), injection molding 400
polycarbonate (PC), injection molding 400, 404
polydimethylsiloxane (PDMS) devices 323
poly(ether ether ketone) (PEEK) 84, 90, 93, 94
– injection molding 400
polyimide
– flexible mask substrates 38, 39
– optical transmission characteristics 23
– X-ray attenuation characteristics 15
polyimide–silica composites 437
polymer mold inserts
– fabrication loads and stresses 153
– hot embossing 93, 94
polymer-based masks 44
polymers
– optical components 207–209, 211, 212
– plastification 399–402
poly(methyl methacrylate) (PMMA) 5, 53
– characteristics 119
– copolymer 340
– degradation mechanism 120
– development of irradiated resist 55, 120, 121, 326
– – dissolution rate 55, 121–125
– exposure dose 53–55
– exposure times 193, 196
– hot embossing 76, 77
– hydrogen bubbles 155, 156
– injection molding 70, 71, 329, 330, 400
– microfabrication costs 195–197
– swelling 258
– tension stress 155
– thermal expansion 258
– – coefficient 155
– throughput rates 197
– Young's modulus 77, 78
polyoxymethylene (POM) 90
polystyrene, microfluidic devices 332, 333
polytetrafluoroethylene (PTFE)
– crosslinking 458–460
– direct etching 453–467
– exposure times 457
POM (polyacetal), injection molding 400
positioning *see* alignment; micropositioning
positive resist 119
PP (polypropylene), injection molding 400–402
preceramic polymers 424–428

precision (*see also* tolerance)
– dimensional deviations 152, 153, 160–163
– X-ray mask pattern 345
pre-polymerization, substrate 155
press fits 176, 177
pressure sensors 298
process sequence and variations 5
process simulation *see* simulation
production *see* manufacturing
productivity 7
– throughput rates 193, 197
prototyping 7 (*see also* rapid prototyping)
proximity exposure 417
PTFE *see* polytetrafluoroethylene (PTFE)
PVD *see* physical vapor deposition (PVD)
pyrolytic graphite, substrate materials 22

q
quality management 8

r
radio frequency applications *see* RF applications
rapid prototyping 38, 44, 419
RCHP 422, 423
reactive ion etching (RIE) 19 (*see also* deep reactive ion etching (DRIE))
recycling, design for 152
refractive index, X-ray 234
residual stress, photoresist 155
resist *see* photoresist
resonant actuators 308, 316
resonant filters 209–211
resonant gratings 209, 316
resonant microbeams 298
RF applications 243–277
– cavity resonators 270–277
– coplanar waveguide couplers 263–270
– design approach 248–251
– reactive elements 245
– 'software' radios 247
– substrates 248, 251
– switches 245
– variable capacitor (varactor) 252–263
– vertical wall structures 249–251
– wireless transceiver 246, 247
RIE (reactive ion etching) 19 (*see also* deep reactive ion etching (DRIE))
Ritsumeikan University (Japan)
– AURORA 109–111, 443
– AURORA-2S 443, 454
– X3D 109, 133–139
rotary bond tool 385–388

roughness, sidewalls 34, 38, 44, 206
Rowland arrangement 223
RX-PTFE 459–462

s
SCARA robots 390–393
self-compensation 148
self-enforcement 148
self-help principle 148
self-protection 149
sensors 297, 298
– acceleration 148–150, 199
– distance
– – micro-optical 151, 173, 180–187, 216–222
– pressure 298
servo actuators *see* actuators
Shadow (program) 105
shadow exposure 417
sidewalls (*see also* aspect ratios)
– inclined 57–60, 325
– inclined and freely shaped 52, 60–67
– roughness 34, 38, 44, 206
silica composite fillers 430, 432–434, 437, 438
silicon
– mask substrates 39
– microfabrication technologies 248, 249
– thermal expansion coefficient 155
silicon carbide, optical properties 23
silicon carbonitride 424
silicon nitride
– mask substrates 22, 44
– – fabrication 31–33
– – properties and thickness 26
– optical transmission characteristics 23
– X-ray attenuation characteristics 15, 16
simulation 179–187
– development of irradiated resist 133–139
– exposure dose distribution 104–106, 118
– harmonic drive gear 363–365
– hot embossing 76–83
– micro-optical distance sensor 180–187
– software 76, 78, 105, 132–139
SLM465012VP 425
soft X-ray lithography 20, 28
soft X-ray sources 13, 444–451
software
– cost modeling 194
– simulation 105
– – development of irradiated resist 132–139
– – hot embossing 76, 78

spatial structure encoding and tracking 123–132
spectrometers *see* microspectrometers
spin coating 421
sputter deposition, titanium membranes 22
SR *see* synchrotron radiation sources
stability, design for 149
stereolithography 419, 420
Stoney's equation 41
stop faces 173
Stress Deposit Analyzer System (DSAS) 42
stress measurement 41, 42
stresses (*see also* thermal stress)
– demolding 79–82
– fabrication 153
string methods 123
structure inversion 152
SU-8 7, 8, 34, 37, 119
– exposure time 193, 195
– filled resists 428–432
– flexible mask substrates 38
– gold coating 275
– microfabrication costs 195–198
– throughput rates 197
– UV LIGA 192
– X-ray lenses 237
SU-8TM 52
subfunctions 147, 148
substrate materials
– beryllium 35–37, 44
– effective transparency 15, 16
– glass 37, 38
– graphite 33–35, 44
– material selection 21, 22
– polymer-based 44
– properties and thickness 25, 26
– silicon 39
– silicon nitride 44
– thermal expansion coefficients 155
– titanium 29–31, 44, 274
– vitreous carbon 44
– X-ray transmission characteristics 14–16
sulfite-based electroplating 40–42
surface curvature measure 128
surface microengineering 192
– microfabrication costs 196–198
– throughput rates 197
switching matrixes 226, 227
synchrotron radiation sources 4, 12–13
– ANKA 109, 110
– AURORA 109–111
– commercial centres 193, 194
– commercial issues 200

– compact 193, 443–451
– costs 193
– MIRRORCLE-type 444–451
– power spectra 13
– radiation spectrum 15
– spectral distribution 17

t

tantalum, properties 17, 18
tape casting 426, 429
task subdivision 147, 148
TECHNI-GOLD 25 (TG 25) 40
Teflon-included etched (TIE) process 454–467
tension stress, photoresist 155
thermal expansion
– design for 151
– photoresist substrates 155
thermal stress
– analysis 186, 187
– fabrication 153
– photoresist 155, 156
thermo-elastic properties, mask substrates 19–21
thermopneumatic micropump 148
three-dimensional (3D) microfabrication 52, 55–67
three-dimensional (3D) resist substrates 55–57
through-holes 90–92
throughput rates 193, 197 (see also cycle times)
TIEGA 454–467
tin, properties 17, 18
titanium
– mask substrates 22, 44, 256, 257, 274
– – alignment markers 23
– – fabrication 29–31
– – properties and thickness 26
– – temperature rise 20
– optical transmission characteristics 23
– X-ray attenuation characteristics 15, 16
titanium nitride, physical vapor deposition 400
TM DAR 422, 423
tolerance 180, 184–186, 201, 249 (see also precision)
– dimensional deviations 152, 153, 160–163
topology representation 123–132
transition radiation (TR) 444–446, 449–451
transmission contrast 26
Transmit (program) 105
triangulation distance sensor 219–222

tunable capacitors 252, 253, 315
tungsten, properties 17, 18
two-step dose distribution 60–63

u

ultrasonic diagnosis 337–344
uranium separation nozzles 2, 3
UV LIGA 5, 190, 192
– microfabrication costs 196–198
– throughput rates 197
– X-ray sources 444–451
UV lithography 52, 416, 417
– extreme ultraviolet (EUV) 1, 2, 444
UV–VIS microspectrometer 85, 86, 287, 289

v

valves see microvalves
variable capacitors (varactors) 252–263
vertical wall structures
– RF applications 249–251
– vertical wall varactors 254–263
vitreous carbon
– mask substrates 22, 44
– optical transmission characteristics 23
– properties and thickness 26
– X-ray attenuation characteristics 16
VL20 425

w

wafer bonding 192
– microfabrication costs 196–198
– throughput rates 197
water-based fillers 437, 438
waveguides see coplanar waveguides (CPW)
wear, design for 151

x

X3D 109, 133–139
XOP 105
X-Ray (program) 105
X-ray attenuation characteristics
– absorber materials 16, 17
– resist materials 419
– substrate materials 14–16
X-ray lenses 6, 233–241
– characteristics 239, 240
– geometry 235–239
– materials 237
– PTFE 463–467
X-ray masks 11–44
– absorbers
– – materials see absorber materials
– – thickness 26, 27

– absorption contrast 26, 27
– alignment markers 35
– architecture 24
– costs 193
– critical dimension (CD) 27–29
– fabrication 27–42
– – alternative approaches 38, 39
   beryllium masks 35, 37
– – borosilicate glass masks 37, 38
– – commercial issues 200
– – gold electroplating 39–43
– – graphite masks 33–35
– – loads and stresses 153
– – silicon nitride masks 31–33
– – titanium membrane masks 29–31
– intermediate 28–30
– low contrast 61, 62
– metal mesh 38
– metal stencil 38
– multi-step absorber thickness 63
– planar movement technique 63, 64, 112
– resolution 28, 29
– silicon nitride masks, fabrication 31–33

– substrates (*see also* substrate materials)
– – alignment markers 23, 24
– – dose ratio 25
– – effective transparency 25
– – flexible 38
– – membrane expansion 21
– – membrane-type 25
   optical transmission 22, 24
– – surface quality 21, 22
– – therm-elastic properties 19–21
– – thickness 16, 25, 26
– temperature rise 19, 20
– transmission contrast 26
– working masks 28–31
X-ray microscopes 233, 240, 241
X-ray sources 12 (*see also* synchrotron radiation sources)
X-ray transmission characteristics
– absorber materials 16–18
– substrate materials 14–16

z
ZEP520 432, 433